Titles in This Series

20 **R. L. Dobrushin, editor,** Probability contributions to statistical mechanics, 1994
19 **V. A. Marchenko, editor,** Spectral operator theory and related topics, 1994
18 **Oleg Viro, editor,** Topology of manifolds and varieties, 1994
17 **Dmitry Fuchs, editor,** Unconventional Lie algebras, 1993
16 **Sergei Gelfand and Simon Gindikin, editors,** I. M. Gelfand seminar, Parts 1 and 2, 1993
15 **A. T. Fomenko, editor,** Minimal surfaces, 1993
14 **Yu. S. Il′yashenko, editor,** Nonlinear Stokes phenomena, 1992
13 **V. P. Maslov and S. N. Samborskiĭ, editors,** Idempotent analysis, 1992
12 **R. Z. Khasminskiĭ, editor,** Topics in nonparametric estimation, 1992
11 **B. Ya. Levin, editor,** Entire and subharmonic functions, 1992
10 **A. V. Babin and M. I. Vishik, editors,** Properties of global attractors of partial differential equations, 1992
9 **A. M. Vershik, editor,** Representation theory and dynamical systems, 1992
8 **E. B. Vinberg, editor,** Lie groups, their discrete subgroups, and invariant theory, 1992
7 **M. Sh. Birman, editor,** Estimates and asymptotics for discrete spectra of integral and differential equations, 1991
6 **A. T. Fomenko, editor,** Topological classification of integrable systems, 1991
5 **R. A. Minlos, editor,** Many-particle Hamiltonians: spectra and scattering, 1991
4 **A. A. Suslin, editor,** Algebraic *K*-theory, 1991
3 **Ya. G. Sinaĭ, editor,** Dynamical systems and statistical mechanics, 1991
2 **A. A. Kirillov, editor,** Topics in representation theory, 1991
1 **V. I. Arnold, editor,** Theory of singularities and its applications, 1990

Probability Contributions to Statistical Mechanics

Advances in

SOVIET MATHEMATICS

Volume 20

Probability Contributions to Statistical Mechanics

R. L. Dobrushin
Editor

American Mathematical Society
Providence, Rhode Island

Translation edited by A. B. SOSSINSKY

1991 *Mathematics Subject Classification.* Primary 82Bxx.

ABSTRACT. This collection contains papers written by representatives of the Moscow school of mathematical statistical mechanics, and illustrates certain aspects of the developing interaction between statistical mechanics on the one hand and the theories of probability and of dynamical systems on the other. It includes papers on random walks, on phase transition phenomena for Gibbs random fields, on the existence of nonstandard motion integrals in statistical physics models, and on the analysis of the Frenkel-Kontorova model.

Library of Congress Catalog Card Number: 91-640741

ISBN 0-8218-4120-3

ISSN 1051-8037

∞ The paper used in this book is acid-free and falls within the guidelines
established to ensure permanence and durability.
♻ Printed on recycled paper.

This publication was typeset using AMS-TEX,
the American Mathematical Society's TEX macro system.

10 9 8 7 6 5 4 3 2 1 98 97 96 95 94

Contents

Foreword

The title of this volume, *Probability Contributions to Statistical Mechanics,* reflects its intent and contents only in part. It would have been more precise to call it *Contributions of Probability and Dynamical Systems Theories to Statistical Mechanics and Contributions of Statistical Mechanics to Probability and Dynamical Systems Theories*, but this, of course, is too unwieldy for a title.

The relationship between probability theory and the theory of dynamical systems on one hand and statistical mechanics on the other has not always been cloudless. In the period when the modern theories of probability and of dynamical systems were being created (at the turn of the century), many of the main ideas came from statistical mechanics. It suffices to recall that a Wiener process, the etalon process in the theory of random processes, is still often called Brownian motion, despite the fact that the description of actual Brownian motion requires more complicated models, taking into account the inertia of the particle, while the notion of ergodicity arose from efforts to give meaning to the main postulates of statistical mechanics. However, by the middle of the century, probabilists and statistical physicists practically lost their mutual understanding as well as their interest in each other. One should recall with warmth the heroic efforts undertaken by M. Kac and A. Khinchin to reestablish the lost contacts between probability theory and statistical mechanics, but the seeds that they planted were to sprout only when the time for it had come.

At the end of the sixties, a new branch of science, which might well be called "mathematical statistical mechanics", came into being. Its aim was to clarify the ideas of statistical mechanics from the point of view of rigorous mathematical methods and of the latest mathematical achievements. Since then hundreds of articles and several books in this field have been published. To summarize, one may say that now all the main notions of statistical mechanics have been reformulated in a rigorous mathematical language, and statements that were long established on a level of rigor usual in theoretical physics have been reinterpreted as theorems or as mathematical conjectures awaiting their proofs. There are also new results, or, at least, mathematical theorems that allow one to make a unique choice among several points of view appearing in the physics literature, although in fact there aren't too many such cases. The opinions of various representatives of traditional statistical mechanics about the new rigorous approaches to the field are diverse. Some admit that the mathematical results give a deeper understanding of the essense of the problems. The other extreme viewpoint is that they are merely a very complicated way to present simple and well-known facts. Who is right? Only time will tell.

But there are also other aspects. Physics has always been an important source of new mathematical notions and ideas, and in the past decade this stream of ideas and notions from physics to mathematics has increased drastically. If at the beginning, the majority of experts in probability theory were rather skeptical about what they felt were not very understandable excursions into statistical mechanics, now no doubts seem to remain. It is clear that this interaction with statistical mechanics gave rise to a new field of probability theory – the theory of random fields, i.e., random functions of many variables. Such notions as Gibbs random fields, Markov processes with interaction, the hydrodynamical limit approach have found their mathematical formalization and have firmly established themselves in probability theory. Moreover, it turned out that a natural path of physical ideas to other sciences passes through mathematics. While a quarter century ago it seemed that the notion of Gibbs distribution is something intimately connected with the specifics of systems appearing in statistical mechanics, and that this notion is useless outside this specific environment, it is now clear that the Gibbs approach is simply a natural way of describing a large class of random fields, i.e., random functions of many variables. Since it was necessary to give mathematical descriptions of such random functions in many fields of science, the spectrum of developing applications of the notion of Gibbs random field has become immense: from pattern recognition and the theory of communication networks with queues to the description of systems of economic or even sociological nature. The same path from physics through rigorous mathematics into other sciences was taken by the notion of phase transition behaviour. Now one can claim that the possibility of phase transition is a common property of all large multidimensional systems.

The present collection, written by representatives of the Moscow school of mathematical statistical mechanics, illustrates certain aspects of the developing interaction of statistical mechanics on the one hand and the theories of probability and of dynamical systems on the other.

The series of three articles written by R. Minlos and coauthors is devoted to the study of an important and natural probabilistic object – random walks in a random medium, for different definitions of this object. The results are obtained by applying the method of cluster expansions. This deep analytic method has become the foundation for obtaining many rigorous results both in statistical mechanics and in quantum field theory, but it seems that it is still far from being appreciated at its true value and has not been adopted by the experts in probability theory.

The long joint article by S. Shlosman and the editor of this collection deals with the theory of large deviations for Gibbs random fields. Its main goal is to demonstrate the cardinal changes in the familiar probabilistic picture that occur when phase transitions are allowed. It turns out that this probability theory question is closely connected with the natural physical question of the level of concentration at which condensation of droplets of one phase within another begins. Conceivably, there is something new about this even from the physical point of view. The proofs are based on a combination of probabilistic methods from the theory of large deviations and the use of a special version of the method of cluster decompositions, applicable to the contour ensembles that appear in the case of multiple phases. Chapter 3 of the article, devoted to the exposition and development of the method of cluster decompositions, can be read separately.

The large article by B. Gurevich is devoted to the development of an important series of studies of his, related to certain fundamental problems of statistical mechanics. Along the road to establishing its main postulate, claiming that the statistical properties of a system in equilibrium are described by a Gibbs distribution with potential determined by the physical energy of the system, there is a difficult mathematical problem. One must prove that the classical integrals of motion (the number of particles, the energy, and the total momentum) are the only asymptotically additive translation-invariant integrals for classical one-dimensional mechanical systems. However, this is not always so. It has been established that in dimension $d = 1$ for pair potentials of the form

$$U(r) = \frac{a}{r^2}, \qquad U(r) = \frac{a}{(\sinh(br))^2},$$

where $a > 0$ and $b > 0$ are constants, there exists an infinite series of additional first integrals. The study of such integrable one-dimensional systems has become a pretty fashionable affair: here certain deep relationships with modern mathematical theories arise, but from the point of view of statistical mechanics, this is an ugly case. Indeed, in such situations the main postulates of statistical mechanics break down. The only hope is that such ugliness can only occur in exceptional, unstable situations. In the works of Gurevich it is shown that this is indeed the case. When the dimension d is 2 or more, there are no exceptions whatsoever, when $d = 1$, they are limited to the class of potentials described above. The corresponding mathematical study requires certain *a priori* restrictions on the class of interactions under consideration. The main restriction used by Gurevich in his previous work was the assumption that only pairwise interaction takes place. In the present article he rejects this harsh restriction and considers three-particle interactions. This case is already quite difficult and requires the introduction of original analytic techniques.

L. Pustyl′nikov's article deals with the ground states for a well-known one-dimensional model in statistical mechanics, the Frenkel-Kontorova model. Ya. Sinai reduced this problem to the study of a special dynamical system. This is the object considered by the author, who finds new cases when phase transitions for it exist.

R. L. Dobrushin

Translated by A. S. SOSSINSKY

ADVANCES IN SOVIET MATHEMATICS
Volume 20, 1994

An Extension of the Ising Model

J. ABDULLAEV AND R. A. MINLOS

§1. Introduction

The aim of this paper is to investigate a one-parameter family of Gibbs fields on the two-dimensional lattice $\mathbb{Z}^2$ at high values of temperature. These fields (introduced, as far as we know, by F. Dunlop [1]) take three values $\sigma_x = -1, 0, 1$ at each lattice point $x \in \mathbb{Z}^2$ and can be obtained by the Gibbs modification [2] of independent translation-invariant fields on $\mathbb{Z}^2$ with the following probabilities of spin values at each point:

$$\Pr(\sigma_x = 1) = \Pr(\sigma_x = -1) = c, \qquad \Pr(\sigma_x = 0) = 1 - 2c, \tag{1.1}$$

where $0 \leqslant c \leqslant 1/2$. The Gibbs modification is defined by a (formal) Hamiltonian of Ising type

$$H(\sigma) = \beta \sum_{|x-x'|=1} \sigma_x \sigma_x', \tag{1.2}$$

where $\sigma = \{\sigma_x, x \in \mathbb{Z}^2\}$ is a field configuration, and β (the inverse temperature) is a small parameter. Note that for $c = 1/2$ the resulting field is isomorphic to the ordinary Ising field on $\mathbb{Z}^2$.

In the main part of our paper, we investigate leading branches of the spectrum of the transfer matrices of the fields in question, namely "one-particle" and "two-particle" branches, and also branches of "two-particle bound states". These topics are of interest for several reasons. First, the fields under consideration can be regarded as simplified models of lattice quantum fields; thus the description of the spectrum of the transfer matrix of such a field is equivalent to the description of the spectrum of the Hamiltonian of the corresponding field model. With this approach to field models, the most interesting and finest point is the structure of the spectrum of bound states. Secondly, from the standpoint of statistical physics, we can readily use the transfer-matrix spectrum data to find the asymptotics (for $|x| \to \infty$) of correlations of the form

$$\langle \mathcal{F}(\sigma), \mathcal{F}(\sigma + x) \rangle,$$

1991 *Mathematics Subject Classification.* Primary 82B20.

where $\mathscr{F}$ is a local functional on the field, and $\sigma + x$ is the shift of the configuration by a vector $x \in \mathbb{Z}^2$. The type of asymptotics for these correlations varies for different sets of values of the parameter c: although the leading term of the asymptotics is always of exponential form, the degree in the factor preceding the exponential function can vary; this is connected with the fact that for some values of c, the spectrum of bound two-particle states is located to the right of the main two-particle spectrum. Finally, this work is of methodological significance: we adhere to a general procedure of investigating the leading branches of the spectrum of the transfer matrix for the Gibbs field. This procedure (the so-called "Moscow method") was developed over the last ten years by a group of Moscow mathematicians (see [3–7] and also the survey paper [8]). The method in question can be best illustrated by the use of a simple model and a detailed computation. In our proofs and computational work, we shall briefly describe the essential elements of the general procedure, referring the reader for further details to the papers listed above.

§2. Formulation of the results

Let $\sigma = \{\sigma_x, x = (x_0, x_1) \in \mathbb{Z}^2\}$ be the Gibbs field on the lattice $\mathbb{Z}^2$ described above and specified by some parameter value c: $0 < c < 1/2$. We denote by μ_c the probability distribution of field values, defined in the configuration space $\Omega = \{-1, 0, 1\}^{\mathbb{Z}^2}$. Since the field σ is a Markov random field (see [2]), along the axis $x_0 = t$ it can be regarded as a stationary Markov chain

$$\sigma = \{\sigma_t, t \in \mathbb{Z}^1\}, \tag{2.1}$$

with values σ_t in the configuration space $\Sigma = \{-1, 0, 1\}^{\mathbb{Z}^1}$ on the lattice $\mathbb{Z}^1$:

$$\sigma_t = \{\sigma_t(x), x \in \mathbb{Z}^1\}.$$

Denote by ν_c the stationary probability distribution for this chain (in the space Σ) and by $\mathscr{H}$ the Hilbert space

$$\mathscr{H} = L_2(\Sigma, \nu_c) \tag{2.2}$$

of functions in configurations $\sigma \in \Sigma$. The stochastic operator $\mathscr{T}$ for the chain, acting on the functions $f \in \mathscr{H}$ according to the relation

$$(\mathscr{T} f)(\sigma) = \int_\Sigma f(\sigma')\, d\mathscr{P}(\sigma'/\sigma), \tag{2.3}$$

will be called, as usual, the *transfer matrix* of the field. Here $\{\mathscr{P}(\cdot \mid \sigma), \sigma \in \Sigma\}$ is the family of transition probabilities for the chain. The operator $\mathscr{T}$ on $\mathscr{H}$ is a self-adjoint bounded operator with $\mathscr{T} 1 = 1$, where $1(\sigma) = 1$, $\sigma \in \Sigma$.

Consider next the group of unitary operators $\{\mathscr{U}_s, s \in \mathbb{Z}^1\}$ acting on $\mathscr{H}$ and defining the representation in $\mathscr{H}$ of the shift group on $\mathbb{Z}^1$:

$$(\mathscr{U}_s f)(\sigma) = f(\tau_s \sigma), \qquad s \in \mathbb{Z}^1, \ \sigma \in \Sigma, \tag{2.4}$$

where $(\tau_s \sigma)(x) = \sigma(x+s)$, $x \in \mathbb{Z}^1$. It is readily seen that the operators $\mathscr{U}_s$ commute with the operator $\mathscr{T}$. It follows that there exist expansions of the space $\mathscr{H}$ and the operators $\mathscr{T}$ and $\mathscr{U}_s$ into the direct integrals (see [10])

$$\mathscr{H} = \int_{T^1} \mathscr{H}(\lambda)\, d\lambda, \qquad \mathscr{T} = \int_{T^1} \mathscr{T}(\lambda)\, d\lambda, \qquad \mathscr{U}_s = \int_{T^1} \mathscr{U}_s(\lambda)\, d\lambda, \tag{2.5}$$

where T^1 is a unit circle, $\{\mathscr{H}(\lambda), \lambda \in T^1\}$ is a family of Hilbert spaces, and $\{\mathscr{T}(\lambda), \lambda \in T^1\}$ and $\{\mathscr{U}_s(\lambda), \lambda \in T^1\}$ are families of operators acting in these spaces, with

$$\mathscr{U}_s(\lambda) = e^{is\lambda} E_\lambda. \tag{2.6}$$

Here E_λ is the identity operator in $\mathscr{H}(\lambda)$. (Physically, λ can be interpreted as the total quasi-momentum of a system and $\mathscr{H}(\lambda)$ as the function space over a field with a "specified quasi-momentum".) In $\mathscr{H}$ we have a self-adjoint involution I induced by the involution $\sigma \mapsto -\sigma$ in the configuration space. The operator I also commutes with $\mathscr{T}$ and $\mathscr{U}_s$. Denote by $\mathscr{H}_0 = \{c1\}$ the one-dimensional subspace of constants. The theorem below describes the "leading" invariant spaces of the transfer matrix and the way it acts on these subspaces.

Theorem 1. I. *For any c, $0 < c < 1/2$, and all sufficiently small β ($|\beta| < \beta_0(c)$) there exist two subspaces $\mathscr{H}_1$ and $\mathscr{H}_2$ that are orthogonal to each other and to the subspace $\mathscr{H}_0$ and are invariant with respect to the operators $\mathscr{T}$, $\mathscr{U}_s$, and I, the subspace $\mathscr{H}_1$ having odd parity and $\mathscr{H}_2$ even parity with respect to I:*

$$If = (-1)^k f, \qquad f \in \mathscr{H}_k, \quad k = 1, 2. \tag{2.7}$$

II. *Denote by $\mathscr{T}^{(k)} = \mathscr{T}|_{\mathscr{H}_k}$, $k = 1, 2$, the restrictions of the transfer matrix $\mathscr{T}$ to the subspaces $\mathscr{H}_k$, $k = 1, 2$, and let*

$$\mathscr{H}_k = \int_{T^1} \mathscr{H}_k(\lambda)\, d\lambda, \quad \mathscr{T}^{(k)} = \int_{T^1} \mathscr{T}^{(k)}(\lambda)\, d\lambda, \qquad k = 1, 2,$$

be the corresponding expansions of these spaces and the operators $\mathscr{T}^{(k)}$ that are generated by the complete expansions (2.5): $\mathscr{T}^{(k)}(\lambda) = \mathscr{T}(\lambda)|_{\mathscr{H}_k(\lambda)}$, $k = 1, 2$. *Then*:

(i) *For each $\lambda \in T^1$ the space $\mathscr{H}_1(\lambda) = \{ce(\lambda)\}$ is one-dimensional, and the operator $\mathscr{T}^{(1)}(\lambda)$ acts on it as multiplication by a real number*:

$$\mathscr{T}^{(1)}(\lambda) e(\lambda) = \widehat{b}(\lambda) e(\lambda), \qquad \lambda \in T^1, \tag{2.8}$$

where $\widehat{b}(\lambda)$ is an even analytic (periodic) function in λ defined in a complex neighborhood $\{|\operatorname{Im}\lambda| < d\}$ of the real axis; the function in question can be expanded in this neighborhood in the series

$$\widehat{b}(\lambda) = 2\sum_{k=1}^{\infty} b_k \cos k\lambda + b_0, \tag{2.9}$$

where the coefficients $b_k = b_k(\beta, c)$ do not exceed

$$|b_k| < (D\beta)^{k+1}, \tag{2.10}$$

D being an absolute constant. In particular

$$\begin{aligned} b_0 &= -2c\beta - (1/3 - 4c^2)c\beta^3 + O(\beta^5), \\ b_1 &= 8c^2\beta^2 + O(\beta^4), \qquad b_2 = -16c^3\beta^3 + O(\beta^5). \end{aligned} \tag{2.11}$$

(ii) *For each $\lambda \in T^1$, denote by $L_2^\lambda \subset L_2(T^1, d\lambda)$ ($d\lambda$ is the normalized Haar measure on the circle T^1), the subspace of functions $f \in L_2(T^1, d\lambda)$ that are invariant with respect to the involution of the circle*

$$\kappa \to \lambda - \kappa, \qquad \kappa \in T^1$$

(reflection in the point $\lambda/2$). Then for any $\lambda \in T^1$ there exists a unitary mapping $V_\lambda \colon \mathscr{H}_2(\lambda) \to L_2^\lambda$ such that the operator $\mathscr{T}^{(2)}(\lambda)$ is taken to the operator $\widetilde{\mathscr{T}}^{(2)}(\lambda) = V_\lambda \mathscr{T}^{(2)}(\lambda) V_\lambda^{-1}$ acting according to the formula

$$(\widetilde{\mathscr{T}}^{(2)}(\lambda) f)(\kappa) = \omega_\lambda(\kappa) f(\kappa) + \int_{T^1} K_\lambda(\kappa, \mu) f(\mu)\, d\mu, \tag{2.12}$$

where

$$\omega_\lambda(\kappa) = \widehat{b}(\kappa)\widehat{b}(\lambda - \kappa), \tag{2.13}$$

the function $\widehat{b}(\cdot)$ being the same as in (2.8). The kernel $K_\lambda(\kappa, \mu)$ is of the form

$$\begin{aligned} K_\lambda(\kappa, \mu) = {} & -[\omega_\lambda(\kappa) + \omega_\lambda(\mu)] + [\overline{b}_\lambda(\kappa) + b_\lambda(\mu)] \\ & + S_\lambda(\kappa, \mu) + \gamma(\lambda) + \int_{T^1} \omega_\lambda(\nu)\, d\nu. \end{aligned} \tag{2.14}$$

Here

$$\begin{aligned} b_\lambda(\kappa) &= -2e^{i\lambda/2}\beta^3 c\sqrt{2c(1-2c)}\,(1 + 2c + 4ce^{i\lambda})\cos(\kappa - \lambda/2) + O(\beta^4), \\ \gamma(\lambda) &= c(1-2c)\beta^2[1 - 4c\beta^2(1 - 8c) + 2\beta^2(2 + 3c + 2c^2)\cos\lambda] \\ &\quad + O(\beta^5), \\ S_\lambda(\kappa, \mu) &= 16c^2\beta^4(1 - 6c)(1 - c + 2c\cos\lambda)\cos(\kappa - \lambda/2)\cos(\mu - \lambda/2) \\ &\quad + O(\beta^5). \end{aligned} \tag{2.15}$$

III. *In the orthogonal complement*

$$\widetilde{\mathscr{H}}_2 = \mathscr{H} \ominus (\mathscr{H}_0 \oplus \mathscr{H}_1 \oplus \mathscr{H}_2)$$

(which is invariant with respect to $\mathscr{T}$ and $\mathscr{U}_s$) the spectrum of the operator $\mathscr{T}|_{\widetilde{\mathscr{H}}_2}$ is contained inside the interval $[-C\beta^3, C\beta^3]$, where C is an absolute constant.

REMARKS. 1. From Theorem 1 it is clear that the operators $\mathscr{T}^{(1)} = \mathscr{T}|_{\mathscr{H}_1}$ and $\mathscr{U}^{(1)} = \mathscr{U}|_{\mathscr{H}_1}$ are simultaneously unitarily equivalent to the operators of multiplication in $L_2(T^1, d\lambda)$ by the functions $\widehat{b}(\lambda)$ and $\exp\{is\lambda\}$, respectively. Note that the spectrum of $\mathscr{T}^{(1)}$ in $\mathscr{H}_1$ is contained in a β^2-neighborhood of the value $-2c\beta$.

2. It can be shown that the two-particle subspace $\mathscr{H}_2$ can be unitarily mapped onto the subspace $L_2^{\text{sym}}(T^1 \times T^1, d\lambda_1 d\lambda_2)$ of symmetric functions in two variables $\lambda_1, \lambda_2 \in T^1$ in such a way that the operator $\mathscr{T}^{(2)} = \mathscr{T}|_{\mathscr{H}_2}$ is taken to an operator in $L_2^{\text{sym}}(T^1 \times T^1, d\lambda_1 d\lambda_2)$, namely

$$(\widetilde{\mathscr{T}}^{(2)}(\lambda)f)(\lambda_1, \lambda_2) = \widehat{b}(\lambda_1)\widehat{b}(\lambda_2)f(\lambda_1, \lambda_2) + \int_{T^1}\int_{T^1} G(\lambda_1, \lambda_2; \mu_1, \mu_2)\delta(\lambda_1 + \lambda_2 - \mu_1 - \mu_2)(\mu_1, \mu_2)\, d\mu_1\, d\mu_2, \tag{2.16}$$

where the kernel $G(\lambda_1, \lambda_2; \mu_1, \mu_2)$ defined on the manifold

$$\{\lambda_1 + \lambda_2 = \mu_1 + \mu_2\} \subset (T^1)^4$$

is equal to

$$G(\lambda_1, \lambda_2; \mu_1, \mu_2) = K_{\lambda_1+\lambda_2}(\lambda_1, \mu_1),$$

and $K_\lambda(\cdot, \cdot)$ is defined by (2.14). Moreover, the operators $\mathscr{U}_s$ are taken to the operators of multiplication by the functions $\exp\{i(\lambda_1 + \lambda_2)s\}$. After passing to the variables $\lambda = \lambda_1 + \lambda_2$, $\kappa = \lambda_1$, the operator $\mathscr{T}^{(2)}$ can be decomposed into a family of operators (a direct integral) of the form (2.12), and $\mathscr{U}_s$ takes the form (2.6).

3. Formulas (2.12) and (2.13) imply that the spectrum of $\mathscr{T}^{(2)}$ is contained in the β^3-neighborhoods of the points $4c^2\beta^2$ and $c(1-2c)\beta^2$; thus for small β the spectrum in question does not overlap with the spectra of the operators $\mathscr{T}^{(1)}$ and $\mathscr{T}|_{\widetilde{\mathscr{H}}_2}$. It follows from this that the subspaces $\mathscr{H}_1$ and $\mathscr{H}_2$ indicated in Theorem 1 are uniquely defined.

Observe further that the spectrum of an operator of the form (2.12) consists of a continuous (Lebesgue) part constituting an interval of values of the function $\omega_\lambda(\kappa)$, and, probably, of a finite number of eigenvalues

$$\varepsilon_0(\lambda), \ \varepsilon_1(\lambda), \ \ldots, \ \varepsilon_s(\lambda)$$

lying outside the continuous spectrum, with eigenvectors

$$e_0(\lambda), \ e_1(\lambda), \ \ldots, \ e_s(\lambda) \in L_2^\lambda.$$

(For certain exceptional values of λ, eigenvalues lying inside the continuous spectrum can appear.) The eigenvectors $e_i(\lambda)$ are called the *bound states* of the operator $\mathscr{T}^{(2)}(\lambda)$ (with quasi-momentum λ). Each of the eigenvalues $\varepsilon_i(\lambda)$ changes continuously as λ traverses some interval $G_i \subseteq T^1$, thus forming supplementary branches of the continuous spectrum of the operator $\mathscr{T}^{(2)}$ (the branches of bound states). At the end points of the interval G_i, the eigenvalue ε_λ coincides with the ends of the continuous spectrum $\max_\kappa \omega_\lambda(\kappa) = \tau_\lambda^{\max}$ or $\min_\kappa \omega_\lambda(\kappa) = \tau_\lambda^{\min}$.

The next theorem describes the branches of the bound states $\{\varepsilon(\lambda)\}$ of the operator $\mathscr{T}^{(2)}$ of the form (2.16). We make the following preliminary observation: for $c \neq 1/6$, the value of $\gamma(\lambda)$ for all λ is at a distance of order β^2 from the ends of the continuous spectrum $[\tau_\lambda^{\min}, \tau_\lambda^{\max}]$, $\gamma(\lambda)$ being to the right of $\tau_\lambda^{\max}$ for $c < 1/6$ and to the left of $\tau_\lambda^{\min}$ for $c > 1/6$.

THEOREM 2. I. *Let $c \neq 1/6$ and $0 < c < 1/2$. Then for sufficiently small β ($|\beta| < \beta_0(c)$) and for all λ there exists an eigenvalue $\varepsilon_0(\lambda)$ of the form*

$$\varepsilon_0(\lambda) = \gamma(\lambda) + m(\lambda)\beta^4 + O(\beta^6), \tag{2.17}$$

where

$$m(\lambda) = (1-6c)^{-1} 4c^2(1-2c)[1 + 4c + 20c^2 + (8c + 16c^2)\cos\lambda]. \tag{2.18}$$

II. *There exist three intervals of values of the parameter c:*

$$\begin{gathered} I_1 = (0, c_1), \quad I_2 = (c_2, 1/6), \quad I_3 = (1/6, c_3), \\ c_1 = (11 - \sqrt{41})/80 < c_2 < 1/6, \quad c_3 = (11 + \sqrt{41})/80 > 1/6, \end{gathered} \tag{2.19}$$

and for each value of c in these intervals there exist neighborhoods $\Delta = \Delta(c)$ of the value $\lambda = \pi$ such that for any $\lambda \in \Delta$ there is exactly one eigenvalue $\varepsilon_1(\lambda)$ of the operator $\widetilde{\mathscr{T}}^{(2)}(\lambda)$ which is at a distance of order $O(\beta^4)$ from the edge of the continuous spectrum $\tau_\lambda^{\min}$ or $\tau_\lambda^{\max}$. For $c \in I_1 \cup I_2$ we have

$$\varepsilon_1(\lambda) < \tau_\lambda^{\min}, \tag{2.20}$$

and for $c \in I_3$ we have

$$\varepsilon_1(\lambda) > \tau_\lambda^{\max}. \tag{2.21}$$

III. *For $c = 1/6$ and for any $\lambda \in T^1$ there exist two eigenvalues $\varepsilon_\pm(\lambda)$ located on either side of the continuous spectrum: $\varepsilon_+(\lambda) > \tau_\lambda^{\max}$ and $\varepsilon_-(\lambda) > \tau_\lambda^{\min}$. They are*

$$\varepsilon_\pm(\lambda) = \frac{1}{9}\beta^2 \pm \frac{4}{27}\beta^3 u(\lambda) + O(\beta^4), \tag{2.22}$$

where

$$u(\lambda) = \frac{5 + 4\cos\lambda}{\sqrt{18 + 14\cos\lambda}}. \tag{2.23}$$

For all values of $c \neq c_i$, $i = 1, 2, 3$, $c \neq 0, 1/2$, the eigenvalues $\varepsilon_0(\lambda)$, $\varepsilon_1(\lambda)$, and $\varepsilon_\pm(\lambda)$ (described above) constitute the whole spectrum of the bound states of the operator (2.12).

REMARKS. 1. To define the supplementary spectrum $\varepsilon_2(\lambda)$ (for values of λ in the neighborhood of $\lambda = \pi$) corresponding to the parameter values

$c = c_i$, $i = 1, 2, 3$, the operator $\mathcal{T}^{(2)}(\lambda)$ must be evaluated with greater accuracy — up to the terms of order β^5. As noted above, for $c = 1/2$ the model under consideration is isomorphic to the Ising model, for which the spectrum of the transfer matrix was, in effect, evaluated by Onsager and Kaufman (see the survey paper [9]); there are no bound states in this case.

2. On any interval of values of the parameter c, excluding the exceptional values $c = 0, c_1, c_2, 1/6, c_3, 1/2$, the estimates of order $O(\beta^k)$, $k = 3, 4, 5, 6$, in all the formulas are uniform. It would be of interest to investigate in greater detail the behavior of the eigenvalue $\varepsilon_0(\lambda)$ in a small (β-dependent) neighborhood of the value $c = 1/6$ for a small, but fixed, value of β: in this neighborhood $\varepsilon_0(\lambda)$ is "absorbed" by the continuous spectrum on the right and is "emitted" on the left. It seems reasonable to say that the "absorption" takes place for parameter values to the right of the value $c = 1/6$, and the "emission" occurs for those to the left of $c = 1/6$; thus for $c = 1/6$ the two eigenvalues $\varepsilon_\pm(\lambda))$ exist simultaneously.

As mentioned in the Introduction, the results for the spectra $\mathcal{T}^{(1)}$ and $\mathcal{T}^{(2)}$ obtained above can be used to find the asymptotics of correlations of the form

$$\langle \mathcal{F}^{(1)}(\sigma_0), \mathcal{F}^{(2)}(\sigma_t)\rangle_{\mu_c} \equiv \langle \mathcal{F}^{(1)}(\sigma_0)\cdot \mathcal{F}^{(2)}(\sigma_t)\rangle_{\mu_c} - \langle \mathcal{F}^{(1)}(\sigma_0)\rangle_{\mu_c}\cdot \langle \mathcal{F}^{(2)}(\sigma_t)\rangle_{\mu_c}, \tag{2.24}$$

where $\mathcal{F}^{(1)}$ and $\mathcal{F}^{(2)}$ are some local functionals on Σ (σ_0 being the field configuration at time zero, and σ_t the field configuration at time t). We represent the functionals in the form

$$\mathcal{F}^{(i)} = \mathcal{F}_0^{(i)} + \mathcal{F}_1^{(i)} + \mathcal{F}_2^{(i)} + \widetilde{\mathcal{F}}_2^{(i)}, \qquad i = 1, 2, \tag{2.25}$$

where

$$\mathcal{F}_0^{(i)} = \langle \mathcal{F}^{(i)}\rangle_{\mu_c} = \text{const}, \quad \mathcal{F}_1^{(i)} \in \mathcal{H}_1, \quad \mathcal{F}_2^{(i)} \in \mathcal{H}_2, \quad \widetilde{\mathcal{F}}_2^{(i)} \in \widetilde{\mathcal{H}}_2.$$

Further, we denote

$$\alpha_0 = \max_{\lambda\in T^1} \widehat{b}(\lambda), \qquad \alpha_1 = \begin{cases} \max_\lambda \varepsilon_0(\lambda), & \text{for } c < 1/6, \\ \max_\lambda \varepsilon_+(\lambda), & \text{for } c = 1/6. \end{cases} \tag{2.26}$$

THEOREM 3. I. *In the case when two components* $\mathcal{F}_1^{(i)}$, $i = 1, 2$, *are nonzero, the asymptotics (for* $t \to \infty$*) of the correlations* (2.24) *is of the form*

$$\langle \mathcal{F}^{(1)}(\sigma_0), \mathcal{F}^{(2)}(\sigma_t)\rangle = B_1 t^{-1/2}\alpha_0^t(-\operatorname{sgn}\beta)^t(1 + o(1)), \tag{2.27}$$

where B_1 *is a constant independent of* t.

II. *In the case when* $\mathcal{F}^{(1)} = 0$ *or* $\mathcal{F}^{(2)} = 0$, *but* $\mathcal{F}_2^{(i)} \neq 0$, $i = 1, 2$, *the asymptotics for* $t \to \infty$ *is the following*:

$$\langle \mathcal{F}^{(1)}(\sigma_0), \mathcal{F}^{(2)}(\sigma_t)\rangle_{\mu_c} = \begin{cases} B_2 t^{-1/2}\alpha_1^t(1 + o(1)), & \textit{for } c \leqslant 1/6, \\ \widetilde{B}_2 t^{-2}\alpha_0^{2t}(1 + o(1)), & \textit{for } c > 1/6, \end{cases} \tag{2.28}$$

where B_2 *and* $\widetilde{B}_2$ *are constants independent of* t.

REMARK. For functionals $\mathscr{F}^{(1)}$ and $\mathscr{F}^{(2)}$ that are even with respect to the involution I, we have $\mathscr{F}_1^{(1)} = \mathscr{F}_1^{(2)} = 0$; consequently the asymptotics has the form (2.28).

§3. Proofs of the results

3.1. Proof of Theorem 1.

3.1.1. *Multiplicative basis in $\mathscr{H}$ and cluster expansion of the matrix elements.* Let

$$\psi^{(0)}(\sigma) \equiv 1, \quad \psi^{(1)}(\sigma) = \frac{\sigma}{\sqrt{2c}}, \quad \psi^{(2)}(\sigma) = \frac{\sigma^2 - 2c}{\sqrt{2c(1-2c)}}$$

be a basis in the space of functions defined on the set of spin values $\{-1, 0, 1\}$, the basis being orthonormal with respect to the distribution (1.1). For each $s \in \mathbb{Z}^1$, we denote by $\sigma_{<s} = \sigma|_{-\infty}^{s-1}$ the "past" values of the configuration $\sigma \in \Sigma$, and by $\nu_c(\cdot \mid \sigma_{<s})$ the conditional probability distribution (subject to the condition that the values of $\sigma_{<s}$ are kept constant). Further, let $u_s^{(i)}(\sigma) = u_s^{(i)}(\sigma(s), \sigma_{<s})$, $i = 0, 1, 2$, be the functions obtained by orthogonalizing the functions $\psi^{(i)}(\sigma(s))$, $i = 0, 1, 2$, with respect to the distribution $\nu_c(\cdot \mid \sigma_{<s})$:

$$u_s^{(0)} \equiv 0,$$
$$u_s^{(i)} = \frac{\psi^{(i)}(\sigma(s)) - \langle \psi^{(i)}(\sigma(s)) \mid \nu_c(\cdot \mid \sigma_{<s})\rangle}{\{\langle [\psi^{(i)}(\sigma(s))]^2 \mid \nu_c(\cdot \mid \sigma_{<s})\rangle - \langle \psi^{(i)}(\sigma(s)) \mid \nu_c(\cdot \mid \sigma_{<s})\rangle^2\}^{1/2}}, \tag{3.1}$$
$$i = 1, 2,$$

where $\langle \cdot \mid \nu_c(\cdot \mid \sigma_{<s})\rangle$ is the conditional average computed according to the distribution $\nu_c(\cdot \mid \sigma_{<s})$. A rather simple computation shows that

$$u_s^{(1)}(\sigma) = \frac{1}{\sqrt{2c}}\left[\sigma(s)\left(1 - \frac{3}{2}\beta^2 c(1-2c)\right) + 2c\beta\sigma(s-1) - \frac{1}{4}\beta^2(1-6c)\sigma(s)\sigma^2(s-1)\right] + O(\beta^3), \tag{3.2}$$

$$u_s^{(2)}(\sigma) = \frac{1}{\sqrt{2c(1-2c)}}\left[\sigma^2(s)\left(1 - \frac{3}{2}\beta^2 c(1-4c)\right) - (2c + 3c^2\beta^2) - \frac{1}{2}c\beta^2\sigma^2(s-1) - \frac{1}{4}\beta^2(1-4c)\sigma^2(s)\sigma^2(s-1)\right] + O(\beta^3). \tag{3.3}$$

To obtain these results, we have used the following convenient formula for computing the conditional averages $\langle \cdot \mid \sigma_{<s}\rangle$:

$$\langle F(\sigma) \mid \sigma_{<s} \rangle = \sum_{n=0}^{\infty} \frac{(-\beta)^n}{n!} \sum_{(b_1, \dots, b_n)}' \langle F(\sigma), V_{b_1}(\sigma), \dots, V_{b_n}(\sigma) \rangle_0, \tag{3.4}$$

where $\langle \cdot, \cdot, \dots, \cdot \rangle_0$ is the semiinvariant computed according to the distribution of a free (independent) field on the lattice $\mathbb{Z}^2$ with the distribution (1.1) for a single spin; $b = b(x, y)$, where $|x - y| = 1$, is a nonoriented bond of the lattice $\mathbb{Z}^2$ with end points $x, y \in \mathbb{Z}^2$, and $V_b(\sigma) = \sigma(x)\sigma(y)$; the sum $\sum'_{(b_1, \dots, b_n)}$ is taken over all ordered collections of n bonds not belonging to the ray $I_s = \{x = (x_0, x_1) \in \mathbb{Z}^2 : x_0 = 0, x_1 < s\}$ (the field values on I being held constant and equal to $\sigma(0, x_1) = \sigma(x_1)$, $x_1 < s$).

Using the functions $\{u_s^{(i)}, s \in \mathbb{Z}^1, i = 0, 1, 2\}$, we introduce the following basis in $\mathscr{H}$:

$$u_\gamma = \prod_{s \in \mathbb{Z}^1} u_s^{\gamma(s)}(\sigma), \tag{3.5}$$

where $\Gamma = \{\gamma(s), s \in \mathbb{Z}^1\}$ is a multi-index which is a finite function with values $\gamma(s) = 0, 1, 2$. The functions (3.5) form an orthonormal basis in $\mathscr{H}$ (see [10]). In the basis (3.5), the matrix elements of the transfer matrix $\mathscr{T}$ can be represented in the form

$$(\mathscr{T} u_\gamma, u_{\gamma'}) = \sum_{\substack{\{(\Gamma_1, \Gamma_1'), \dots, (\Gamma_k, \Gamma_k')\} \\ k = 1, 2, \dots}} \prod_{i=1}^{k} a_{\Gamma_i, \Gamma_i'}, \tag{3.6}$$

where the sum is taken over all unordered collections of the pairs $\{(\Gamma_i, \Gamma_i'), i = 1, \dots, k\}$, $k = 1, 2, \dots$, which specify a disjoint partitioning of the pair (Γ, Γ') (for more details, see [9]), and the quantities $a_{\Gamma, \Gamma'}$ are, in fact, semiinvariants (see [2])

$$a_{\Gamma_i, \Gamma_i'} = \left\langle \prod_s{}' u_s^{\gamma(s)}(\sigma_0), \prod_{s'}{}' u_{s'}^{\gamma'(s')}(\sigma_1) \right\rangle_{\mu_c} \tag{3.6a}$$

(σ_0 and σ_1 being the field configurations at $t = 0$ and $t = 1$, respectively). Then the following cluster estimate is valid:

$$|a_{\Gamma_i, \Gamma_i'}| < B(C\beta)^{d_{\{\operatorname{supp}\Gamma\} \cup \{\operatorname{supp}\Gamma'\} + e_0}}, \tag{3.7}$$

where B and C are absolute constants, $d_{\mathscr{D}}$, $\mathscr{D} \subset \mathbb{Z}^2$, is the length of the minimal connected graph constructed on the set $\mathscr{D}$, $\operatorname{supp}\Gamma = \{x : \gamma(x) \neq 0\}$, and $e_0 = (1, 0) \in \mathbb{Z}^2$ (for more details, see [4]). It is readily seen that

$$I u_\Gamma = (-1)^{|\Gamma|} u_\Gamma, \tag{3.8}$$

where $|\Gamma| = \sum_s \gamma(s)$.

3.1.2. *Invariant one-particle subspace of the transfer matrix.* We denote by $\overline{\mathscr{H}}_1 \subset \mathscr{H}$ the subspace spanned by the vectors $\{u_s^{(1)}, s \in \mathbb{Z}^1\}$. By the obvious relation

$$u_s^{(1)} = \mathscr{U}_s u_0^{(1)} \tag{3.9}$$

(recall that $\mathscr{U}_s$ is the group of translation operators in $\mathscr{H}$), the space $\overline{\mathscr{H}}_1$ is invariant with respect to u_s. By relation (3.9), the matrix elements are of the form

$$(\mathscr{T} u_s^{(1)}, u_{s'}^{(1)}) = b_{s-s'}^{(0)}, \tag{3.10}$$

where $\{b_s^{(0)}, s \in \mathbb{Z}^1\}$ is an even function on the lattice $\mathbb{Z}^1$, satisfying, by (3.7), the following estimate:

$$|b_s^{(0)}| < B(\beta c)^{|s|+1}. \tag{3.11}$$

In particular,

$$\begin{gathered} b_0^{(0)} = -2\beta c - \beta^3 c(1/3 - 8c + 12c^2) + O(\beta^5), \\ b_1^{(0)} = 4\beta^2 c^2 + O(\beta^4), \qquad b_2^{(0)} = -8\beta^3 c^3 + O(\beta^5). \end{gathered} \tag{3.12}$$

In order to compute the values of $b_s^{(0)}$, $s = 0, 1, 2$, we have used formula (3.6a), the expansion (3.2), and also the following general formula similar to (3.4) for the semiinvariants of a Gibbs field:

$$\langle \cdot, \dots, \cdot \rangle_{\mu_c} = \sum_{n=0}^{\infty} \frac{(-\beta)^n}{n!} \sum_{(b_1, \dots, b_n)} \langle \cdot, \cdot, \dots, \cdot, V_{b_1}, \dots, V_{b_n} \rangle_0, \tag{3.13}$$

where the sum $\sum_{(b_1, \dots, b_n)}$, in contrast with (3.4), is taken over all ordered collections of the bonds $(b_1, \dots, b_n)$. Finally, to compute semiinvariants of the form $\langle P_1(\sigma), \dots, P_n(\sigma) \rangle_0$ ($P_1, \dots, P_n$ being monomials in the field variables), we must use the known formula expressing the semiinvariants of the field in terms of its moments (see [2]).

According to [4] (see also [8]), the one-particle invariant subspace $\mathscr{H}_1$ is sought for as the graph

$$\mathscr{H}_1 = \{v : v = u + S^{(1)} u, \ u \in \overline{\mathscr{H}}_1\} \tag{3.14}$$

of an operator

$$S^{(1)} : \overline{\mathscr{H}}_1 \to \widehat{\mathscr{H}}_1, \tag{3.15}$$

where $\widehat{\mathscr{H}}_1$ is the subspace spanned by the vectors $\{u_\Gamma, |\Gamma| > 1, |\Gamma| \text{ odd}\}$. The subspace $\overline{\mathscr{H}}_1 \oplus \widehat{\mathscr{H}}_1$ consists of all odd vectors with respect to the involution I and is thus invariant with respect to the operators $\mathscr{T}$ and $\mathscr{U}_s$. The operator $\mathscr{T}$ in $\overline{\mathscr{H}}_1 \oplus \widehat{\mathscr{H}}_1$ can be represented as an operator matrix

$$\begin{pmatrix} A_{11}^{(1)} & A_{12}^{(1)} \\ A_{21}^{(1)} & A_{22}^{(1)} \end{pmatrix}, \tag{3.16}$$

where $A_{11}^{(1)} = \mathscr{P}_{\overline{\mathscr{H}}_1} \mathscr{T} \mathscr{P}_{\overline{\mathscr{H}}_1}$, $A_{12}^{(1)} = \mathscr{P}_{\overline{\mathscr{H}}_1} \mathscr{T} \mathscr{P}_{\widehat{\mathscr{H}}_1}$, and so on ($\mathscr{P}_{\overline{\mathscr{H}}_1}$ and $\mathscr{P}_{\widehat{\mathscr{H}}_1}$ are the projectors onto the corresponding subspaces). The invariance condition for the subspace (3.14) with respect to $\mathscr{T}$ yields the following equation for the operator $S^{(1)}$:

$$S^{(1)} = A_{21}^{(1)}(A_{11}^{(1)})^{-1} + A_{22}^{(1)} S^{(1)} (A_{11}^{(1)})^{-1} - S^{(1)} A_{12}^{(1)} S^{(1)} (A_{11}^{(1)})^{-1}. \tag{3.17}$$

Using the general cluster estimates (3.7) and the straightforward computation of matrix elements of the form

$$(\mathscr{T} u_s^{(1)}, u_s^{(2)} \cdot u_{s\pm 1}^{(1)}) = (2c)^{3/2}(1-2c)^{1/2}\beta^2 + O(\beta^4), \tag{3.18}$$

we can easily obtain from (3.17) the formula

$$S^{(1)} = A_{21}^{(1)}(A_{11}^{(1)})^{-1} + O(\beta^2). \tag{3.19}$$

If in the invariant subspace $\mathscr{H}_1$ (3.14) we consider the basis $\{w_s^{(1)}, s \in \mathbb{Z}^1\}$ obtained by orthogonalizing the system of functions $\{u_s^{(1)} + S^{(1)} u_s^{(1)}, s \in \mathbb{Z}^1\}$, then the elements of the operator $\mathscr{T}$ in this basis are

$$\begin{aligned}(\mathscr{T} w_s^{(1)}, w_{s'}^{(1)}) = b_{s-s'} &= b_{s-s'}^{(0)} + (A_{12}^{(1)} S^{(1)})_{s,s'} + O(\beta^4) \\ &= b_{s-s'}^{(0)} + (A_{12}^{(1)} A_{21}^{(1)} (A_{11}^{(1)})^{-1})_{s,s'} + O(\beta^4).\end{aligned} \tag{3.20}$$

A straightforward computation of $(A_{12}^{(1)} A_{21}^{(1)} (A_{11}^{(1)})^{-1})_{s,s'}$ by means of formulas (3.18) and (3.12) gives the following result:

$$\begin{gathered} b_0 = b_0^{(0)} - 8c^2(1-2c)\beta^3 + O(\beta^4) = -2c\beta - c\beta^3(1/3 - 4c^2) + O(\beta^4), \\ b_1 = b_1^{(0)} + O(\beta^4), \quad b_2 = b_2^{(0)} + O(\beta^4), \quad b_k = O(\beta^{k+1}), \quad k > 2. \end{gathered} \tag{3.21}$$

Consider a unitary mapping

$$V^{(1)} \colon \mathscr{H}_1 \to L_2(T^1, d\lambda), \qquad V^{(1)} w_s^{(1)} = e^{i\lambda s}, \quad s \in \mathbb{Z}^1, \ \lambda \in T^1. \tag{3.22}$$

This mapping takes the operators $\mathscr{T}|_{\mathscr{H}_1}$ and $\mathscr{U}|_{\mathscr{H}_1}$ to the operators of multiplication by the function

$$\widehat{b}(\lambda) = \sum_{s \in \mathbb{Z}^1} b_s e^{i\lambda s} = b_0 + 2\sum_{s=1}^{\infty} b_s \cos \lambda s \tag{3.23}$$

and by $\exp\{i\lambda s\}$, respectively. The estimates (3.11) and similar estimates for b_s (for more details, see [4]) imply that the function $\widehat{b}(\lambda)$ is analytic in a small complex neighborhood of the real axis. The fact that the space $\mathscr{H}_1$ has odd parity with respect to I follows from the construction scheme. Hence all the statements of Theorem 1 about the one-particle space $\mathscr{H}_1$ are proved.

3.1.3. *Two-particle invariant space.* Here we follow the same pattern of construction as with the one-particle subspace $\mathscr{H}_1$. For the "initial" space $\overline{\mathscr{H}}_2$ we select the span of the vectors

$$\{u_{s_1}^{(1)} \cdot u_{s_2}^{(2)}, \ s_1 < s_2, \ s_1, s_2 \in \mathbb{Z}^1; \ u_{s_2}^{(2)}, \ s \in \mathbb{Z}^1\}.$$

Then the cluster expansion (3.6) for the matrix elements

$$(\mathscr{T}(u_{s_1}^{(1)} \cdot u_{s_2}^{(1)}), u_{s_1'}^{(1)} \cdot u_{s_2'}^{(1)}) = B_{s_1, s_2; s_1', s_2'}, \qquad s_1 < s_2, \; s_1' < s_2',$$

will be of the form

$$B_{s_1, s_2; s_1', s_2'} = b_{s_1 - s_1'}^{(0)} \cdot b_{s_2 - s_2'}^{(0)} + b_{s_1 - s_2'}^{(0)} \cdot b_{s_2 - s_1'}^{(0)} + a_{\{s_1, s_2; s_1', s_2'\}}^{(1,1),(1,1)}, \qquad (3.24)$$
$$s_1 < s_2, \; s_1' < s_2',$$

where $b_s^{(0)}$ is the function defined in (3.10), and the values of $a_{\{s_1, s_2; s_1', s_2'\}}^{(1,1),(1,1)}$ satisfy the general cluster estimate (3.7). In particular, for any $S \in \mathbb{Z}^1$ we obtain

$$a_{\{s, s+1; s, s+1\}}^{(1,1),(1,1)} = 4c^2 \beta^4 (1 - 6c) + O(\beta^6), \qquad (3.25a)$$

$$a_{\{s, s+1; s-1, s\}}^{(1,1),(1,1)} = a_{\{s, s+1; s+1, s+2\}}^{(1,1),(1,1)} = 8c^3 \beta^4 (1 - 6c) + O(\beta^6), \qquad (3.25b)$$

and the values of $a_{\{s_1, s_2; s_1', s_2'\}}^{(1,1),(1,1)}$ for other collections $\{s_1, s_2; s_1', s_2'\}$ are of order not less than $O(\beta^5)$.

The matrix elements

$$(\mathscr{T}(u_{s_1}^{(1)} \cdot u_{s_2}^{(1)}), u_s^{(2)}) = a_{\{s_1, s_2; s\}}^{(1,1),(2)}, \qquad (\mathscr{T} u_s^{(2)}, u_{s'}^{(2)}) = a_{\{s; s'\}}^{(2),(2)} \qquad (3.26)$$

have the following leading terms:

$$\begin{aligned}
a_{\{s, s+1; s+1\}}^{(1,1),(2)} &= -c\beta^3 (1 + 2c)\sqrt{2c(1 - 2c)} + O(\beta^5), \\
a_{\{s, s+1; s\}}^{(1,1),(2)} &= -4c^2 \beta^3 \sqrt{2c(1 - 2c)} + O(\beta^5), \\
a_{\{s, s+2; s+2\}}^{(1,1),(2)} &= 2c^2 \beta^4 (1 + 6c)\sqrt{2c(1 - 2c)} + O(\beta^6), \qquad (3.27) \\
a_{\{s, s+2; s\}}^{(1,1),(2)} &= 8c^3 \beta^4 \sqrt{2c(1 - 2c)} + O(\beta^6), \\
a_{\{s, s+2; s+1\}}^{(1,1),(2)} &= 16c^3 \beta^4 \sqrt{2c(1 - 2c)} + O(\beta^6),
\end{aligned}$$

and

$$\begin{aligned}
a_{\{s; s\}}^{(2),(2)} &= \beta^2 c(1 - 2c) + 4c^2 \beta^4 (1 - 2c)(8c - 1) + O(\beta^6), \\
a_{\{s; s+1\}}^{(2),(2)} &= a_{\{s; s-1\}}^{(2),(2)} = c(1 - 2c)(2 + 3c + 2c^2)\beta^4 + O(\beta^6).
\end{aligned} \qquad (3.28)$$

All other matrix elements of the form (3.26) are of order not less than $O(\beta^5)$.

Further, we consider the subspace $\widehat{\mathscr{H}}_2$ spanned by the vectors $\{u_\Gamma, |\Gamma| > 2, |\Gamma| \text{ even}\}$. The subspace $\overline{\mathscr{H}}_2 \oplus \widehat{\mathscr{H}}_2$ is invariant with respect to $\mathscr{T}$ and $\mathscr{U}_s$, and in this subspace the operator $\mathscr{T}$ can be represented by a matrix

$$\begin{pmatrix} A_{11}^{(2)} & A_{12}^{(2)} \\ A_{21}^{(2)} & A_{22}^{(2)} \end{pmatrix}, \qquad (3.29)$$

which is similar to the matrix (3.16). The two-particle invariant subspace $\mathscr{H}_2$ is again sought in the form

$$\mathscr{H}_2 = \{v : v = u + S^{(2)} u, \; u \in \overline{\mathscr{H}}_2\}, \qquad (3.29a)$$

where the operator $S^{(2)}$ satisfies an equation similar to (3.17). As above, using a cluster expansion for the matrix elements of $\mathscr{T}$ and the cluster estimates (3.7), we obtain

$$S^{(2)} = A_{21}^{(2)}(A_{11}^{(2)})^{-1} + O(\beta^2). \tag{3.30}$$

Again choosing in $\mathscr{H}_2$ the orthonormal basis

$$\{w_{s_1,s_2}^{(1,1)},\ s_1 < s_2,\ s_1, s_2 \in \mathbb{Z}^1;\ w_s^{(2)},\ s \in \mathbb{Z}^1\}$$

obtained by orthogonalizing the elements

$$u_{s_1}^{(1)} \cdot u_{s_2}^{(1)} + S^{(2)}(u_{s_1}^{(1)} \cdot u_{s_2}^{(1)}) \quad \text{and} \quad u_s^{(2)} + S^{(2)} u_s^{(2)},$$

we find that in this basis the matrix $\tilde{A}_{11}^{(2)}$ of the operator $\mathscr{T}|_{\mathscr{H}_2}$ is of the form

$$\tilde{A}_{11}^{(2)} = A_{11}^{(2)} + A_{12}^{(2)} A_{21}^{(2)} (A_{11}^{(2)})^{-1} + O(\beta^5). \tag{3.31}$$

A nonzero contribution to the second summand in (3.31) (in the term of order β^3) is provided by matrix elements of the form

$$\begin{aligned}(\mathscr{T}(u_s^{(1)} \cdot u_{s'}^{(1)}), u_s^{(1)} \cdot u_{s'\pm 1}^{(1)} \cdot u_{s'}^{(2)}) &= (\mathscr{T}(u_s^{(1)} \cdot u_{s'}^{(1)}), u_{s\pm 1}^{(1)} \cdot u_s^{(2)} \cdot u_{s'}^{(1)}) \\ &= -4c^2\beta^3\sqrt{2c(1-2c)} + O(\beta^5)\end{aligned}$$

for $s' > s+1$, and also by the following matrix elements:

$$\begin{aligned}(\mathscr{T}(u_s^{(1)} \cdot u_{s+1}^{(1)}), u_s^{(1)} \cdot u_{s+1}^{(2)} \cdot u_{s+2}^{(1)}) &= (\mathscr{T}(u_s^{(1)} \cdot u_{s+1}^{(1)}), u_{s-1}^{(1)} \cdot u_s^{(2)} \cdot u_{s+1}^{(1)}) \\ &= -4c^2\beta^3\sqrt{2c(1-2c)} + O(\beta^5)\end{aligned}$$

and

$$(\mathscr{T}(u_s^{(1)} \cdot u_{s+1}^{(1)}), u_s^{(2)} \cdot u_{s+1}^{(2)}) = -4c^3\beta^2\sqrt{2c(1-2c)} + O(\beta^5).$$

Finally, with the above corrections, from (3.31) we obtain

$$(\tilde{A}_{11}^{(2)})_{s_1,s_2;s_1',s_2'} = b_{s_1-s_1'} \cdot b_{s_2-s_2'} + b_{s_1-s_2'} \cdot b_{s_2-s_1'} + \tilde{a}_{\{s_1,s_2;s_1',s_2'\}}^{(1,1),(1,1)}, \tag{3.32}$$

where b_s is the same function as in (3.20), and

$$\tilde{a}_{\{s,s+1;s,s+1\}}^{(1,1),(1,1)} = a_{\{s,s+1;s,s+1\}}^{(1,1),(1,1)} + 4c^2\beta^4(1-2c)(1-6c)\beta^4 + O(\beta^5),$$

but

$$\tilde{a}_{\{s_1,s_2;s_1',s_2'\}}^{(1,1),(1,1)} = a_{\{s_1,s_2;s_1',s_2'\}}^{(1,1),(1,1)} + O(\beta^5)$$

for other collections s_1, s_2, s_1', s_2'. Besides,

$$\tilde{a}_{\{s_1,s_2;s\}}^{(1,1),(2)} = a_{\{s_1,s_2;s\}}^{(1,1),(2)} + O(\beta^5), \qquad \tilde{a}_{\{s;s'\}}^{(2),(2)} = a_{\{s;s'\}}^{(2),(2)} + O(\beta^5), \tag{3.33}$$

Further, we consider a unitary mapping

$$\begin{aligned} V^{(2)}&: \mathscr{H}_2 \to L_2^{\mathrm{sym}}(T^1 \times T^1, d\lambda_1\, d\lambda_1): \\ V^{(2)} w_{s_1,s_2}^{(1),(1)} &= (1/\sqrt{2})(e^{is_1\lambda_1 + is_2\lambda_2} + e^{is_1\lambda_2 + is_2\lambda_1}), \qquad s_1 < s_2, \\ V^{(2)} w_s^{(2)} &= e^{is(\lambda_1+\lambda_2)}. \end{aligned} \tag{3.34}$$

By a rather simple computation we can show that this mapping takes the operator $\mathscr{T}^{(2)} = \mathscr{T}|_{\mathscr{H}_2}$ to an operator of the form (2.16) and the operators $\mathscr{U}_s^{(2)}$ to the operators of multiplication by the functions $\exp\{is(\lambda_1+\lambda_2)\}$. When the complete quasi-momentum $\lambda = \lambda_1 + \lambda_2$ is isolated, the operator (2.16) is decomposed into a direct integral of operators of the form (2.12). The parity of the space $\mathscr{H}_2$ with respect to the involution I follows from the construction scheme.

Theorem 1 is proved.

3.2. **Proof of Theorem 2.** Consider the decomposition of the space L_2^λ into a direct sum

$$L_2^\lambda = \widetilde{L}_2^\lambda \oplus \mathbb{C}^1,$$

where $\widetilde{L}_2^\lambda = \{f \in L_2^\lambda : \int_{T^1} f(x)\,dx\} = 0$. With this in mind, the equation for the eigenvector of the operator (2.12), corresponding to the eigenvalue z, can be written as

$$\begin{gathered}(\omega_\lambda(\kappa) - z) f(\kappa) - \int_{T^1} \omega_\lambda(\mu) f(\mu)\,d\mu + \int_{T^1} S_\lambda(\kappa, \mu) f(\mu)\,d\mu + b_\lambda(\kappa) C = 0, \\ (\gamma(\lambda) - z) C + \int_{T^1} f(\mu) \overline{b}_\lambda(\mu)\,d\mu = 0, \qquad (3.35)\end{gathered}$$

where C is a constant, and $f \in \widetilde{L}_2^\lambda$. On elimination of the constant C, we obtain

$$\begin{aligned}(\omega_\lambda(\kappa) - z) f(\kappa) &- \int_{T^1} \omega_\lambda(\mu) f(\mu)\,d\mu \\ &+ \int_{T^1} \left[S_\lambda(\kappa, \mu) - \frac{b_\lambda(\kappa)\overline{b}_\lambda(\mu)}{\gamma(\lambda) - z} \right] f(\mu)\,d\mu = 0. \qquad (3.36)\end{aligned}$$

Using the results of [9], we can show that the eigenvalues $z \in \Pi_\lambda$ of the operator (2.12), where $\Pi_\lambda = C^1 \backslash [\tau_\lambda^{\min}, \tau_\lambda^{\max}]$, $\tau_\lambda^{\text{ext}} = \text{ext}_k\, \omega_\lambda(\kappa)$, ext = min, max, coincide with the zeros of the following function (modified Fredholm determinant):

$$\begin{aligned}\Delta_\lambda(z) = &\int_{T^1} \frac{d\kappa}{\omega_\lambda(\kappa) - z} \\ &+ \sum_{n=2}^{\infty} \frac{1}{n!} \int_{T^1} \cdots \int_{T^1} \left[P_n^\lambda(\kappa_1, \dots, \kappa_n) - \frac{1}{\gamma(\lambda) - z} Q_n^\lambda(\kappa_1, \dots, \kappa_n) \right] \\ &\qquad \times \frac{1}{(\omega_\lambda(\kappa) - z) \cdots (\omega_\lambda(\kappa) - z)} \prod_{i=1}^{n} d\kappa_i,\end{aligned}$$

where for $n \geqslant 2$ we have

$$P_n^{\lambda}(\kappa_1, \dots \kappa_n) = \sum_{i=1}^{n} \det \begin{pmatrix} S_\lambda(\kappa_1, \kappa_1) & \dots & 1 & \dots & S_\lambda(\kappa_1, \kappa_n) \\ S_\lambda(\kappa_2, \kappa_1) & \dots & 1 & \dots & S_\lambda(\kappa_2, \kappa_n) \\ \vdots & \ddots & \vdots & \ddots & \vdots \\ S_\lambda(\kappa_n, \kappa_1) & \dots & 1 & \dots & S_\lambda(\kappa_n, \kappa_n) \end{pmatrix}, \tag{3.37}$$

$$Q_n^{\lambda}(\kappa_1, \dots \kappa_n) \tag{3.38}$$
$$= \sum_{i \neq j} \overline{b}_\lambda(\kappa_j) \det \begin{pmatrix} S_\lambda(\kappa_1, \kappa_1) & \dots & b_\lambda(\kappa_1) & \dots & 1 & \dots & S_\lambda(\kappa_1, \kappa_n) \\ S_\lambda(\kappa_2, \kappa_1) & \dots & b_\lambda(\kappa_2) & \dots & 1 & \dots & S_\lambda(\kappa_2, \kappa_n) \\ \vdots & \ddots & \vdots & \ddots & \vdots & \ddots & \vdots \\ S_\lambda(\kappa_n, \kappa_1) & \dots & b_\lambda(\kappa_n) & \dots & 1 & \dots & S_\lambda(\kappa_n, \kappa_n) \end{pmatrix}.$$

In the sum (3.37), each column of the matrix is replaced consecutively by a column of 1's, and in the sum (3.38), the remaining columns are, in addition, replaced consecutively by a column of values $\{b_\lambda(\kappa_m),\ m = 1, \dots, n\}$.

We now consider the following integrals for $n \geqslant 2$:

$$I_n^{1,\lambda}(z) = \int \frac{P_n^{\lambda}(\kappa_1, \dots \kappa_n)}{\prod_{i=1}^n (\omega_\lambda(\kappa_i) - z)} \prod_{i=1}^{n} d\kappa_i,$$

$$I_n^{2,\lambda}(z) = \int \frac{Q_n^{\lambda}(\kappa_1, \dots \kappa_n)}{\prod_{i=1}^n (\omega_\lambda(\kappa_i) - z)} \prod_{i=1}^{n} d\kappa_i,$$

which are analytic functions in $z \in \Pi_\lambda$. It follows from [9] and also from further analysis that all these functions have similar singularities in a neighborhood of the cut $[\tau_\lambda^{\min}, \tau_\lambda^{\max}]$. More precisely,

$$|I_n^{i,\lambda}(z) g| < (C n^{1/2})^n \beta^{6(n-2)} |I_2^{1,\lambda}(z)|, \qquad i = 1, 2, \quad n > 2, \tag{3.39}$$

where C is an absolute constant.

Furthermore, a straightforward computation gives the following result:

$$I_2^{1,\lambda}(z) = A'(\lambda) \int_{(T^1)^2} \frac{[\cos(\kappa_1 - \lambda/2) - \cos(\kappa_2 - \lambda/2)]^2}{(\omega_\lambda(\kappa_1) - z)(\omega_\lambda(\kappa_2) - z)} d\kappa_1\, d\kappa_2,$$

$$I_2^{2,\lambda}(z) = B'(\lambda) \int_{(T^1)^2} \frac{[\cos(\kappa_1 - \lambda/2) - \cos(\kappa_2 - \lambda/2)]^2}{(\omega_\lambda(\kappa_1) - z)(\omega_\lambda(\kappa_2) - z)} d\kappa_1\, d\kappa_2,$$

where

$$A'(\lambda) = 16\beta^4 c^2 (1 - 6c)(1 - c + 2c\cos\lambda) + O(\beta^6),$$
$$B'(\lambda) = 8\beta^6 c^3 (1 - 2c)[1 + 4c + 20c^2 + 8c(1 + 2c)\cos\lambda] + O(\beta^8),$$

Thus, with (3.39), the equation for z takes the form

$$\int_{T^1} \frac{d\kappa}{\omega_\lambda(\kappa) - z} + \frac{1}{2}\left(A(\lambda) - \frac{B(\lambda)}{\gamma(\lambda) - z}\right)$$
$$\times \int_{(T^1)^2} \frac{[\cos(\kappa_1 - \lambda/2) - \cos(\kappa_2 - \lambda/2)]^2}{(\omega_\lambda(\kappa_1) - z)(\omega_\lambda(\kappa_2) - z)} d\kappa_1\, d\kappa_2 = 0, \tag{3.40}$$

where $A(\lambda) = A'(\lambda) + O(\beta^6)$ and $B(\lambda) = B'(\lambda) + O(\beta^8)$.

I. Consider first the case $c \neq 1/6$. In this case the value of $\gamma(\lambda)$ is at the distance $c(1-6c)\beta^2 + O(\beta^3)$ from the ends of the continuous spectrum $[\tau_\lambda^{\min}, \tau_\lambda^{\max}]$. It readily follows that for all $\lambda \in T^1$ there exists an eigenvalue $\varepsilon_0(\lambda)$ of the form (2.17) located near $\gamma(\lambda)$. For $c < 1/6$, this eigenvalue is located to the right of $\gamma(\lambda) > \tau_\lambda^{\max}$, and for $c > 1/6$, it is located to the left of $\gamma(\lambda) < \tau_\lambda^{\max}$. Besides, for the range (2.19) of values of c and for λ lying in some neighborhood of $\lambda = \pi$, there exist eigenvalues $\varepsilon_1^{\pm}(\lambda)$ that are located near the ends of the continuous spectrum $\tau_\lambda^{\min}$ or $\tau_\lambda^{\max}$:

$$\varepsilon_1^+(\lambda) = \tau_\lambda^{\max} + 64c^4\beta^4 s^+(\lambda), \quad \varepsilon_1^-(\lambda) = \tau_\lambda^{\min} - 64c^4\beta^4 s^-(\lambda), \qquad s^{\pm}(\lambda) > 0.$$

In particular, for $\lambda = \pi$ the values of $s^{\pm}(\pi) = s^{\pm}$ can be found from

$$\int_{T^1} \frac{\sin^2\kappa}{\cos^2\kappa + s^-}\, d\kappa = -\frac{8c^2(1-6c)}{2(1-6c)^2(1-3c) - (1-2c)^3} \equiv -F(c),$$

$$\int_{T^1} \frac{\sin^2\kappa}{\sin^2\kappa + s^+}\, d\kappa = F(c).$$

The first of these equations has a (unique) solution for all c satisfying $F(c) < 0$, and the second for all c satisfying $0 < F(c) < 1$. It is readily seen that we have $F(c) < 0$ on the interval I_2 (2.19) and $0 < F(c) < 1$ on the intervals I_1 and I_3. Hence the eigenvalue $\varepsilon_1^- < \tau_\pi^{\min}$ exists on the interval I_2, and the eigenvalue $\varepsilon_1^+ > \tau_\pi^{\max}$ exists on the intervals I_1 and I_3. Due to the continuity of the family of operators $\mathscr{T}(\lambda)$ with respect to the parameter λ, these eigenvalues also exist in some neighborhood $\Delta = \Delta(c)$ of the point $\lambda = \pi$. Outside the intervals I_1, I_2, and I_3 there are no eigenvalues of the type described above.

II. Consider the case $c = 1/6$, in which the value of $\gamma(\lambda)$ is either within the continuous spectrum (for all λ sufficiently remote from the value $\lambda = \pi$) or at a distance of β^4 from its left edge (for λ belonging to some neighborhood of the point $\lambda = \pi$). It readily follows that there exist two values $\varepsilon^{\text{right}}(\lambda)$ and $\varepsilon^{\text{left}}(\lambda)$, which are at a distance of $O(\beta^3)$ from the ends of the continuous spectrum:

$$\varepsilon_1^{\text{right}}(\lambda) = \tau_\lambda^{\max} + \frac{4}{27}\beta^3 t_\lambda^{\text{right}}, \qquad \varepsilon_1^{\text{left}}(\lambda) = \tau_\lambda^{\min} - \frac{4}{27}\beta^3 t_\lambda^{\text{left}}.$$

If we substitute these expressions in (3.40) (for $c = 1/6$) and take into account the relations

$$\tau_\lambda^{\max} = \frac{1}{9}\beta^2 + \frac{4}{27}\beta^3\left|\cos\frac{\lambda}{2}\right| + O(\beta^4), \qquad \tau_\lambda^{\min} = \frac{1}{9}\beta^2 - \frac{4}{27}\beta^3\left|\cos\frac{\lambda}{2}\right| + O(\beta^4),$$

and also the relation

$$\gamma(\lambda) = \frac{1}{9}\beta^2 + O(\beta^4),$$

after a comparatively simple computation we obtain (2.22). Theorem 2 is proved.

3.3. Proof of Theorem 3.

3.3.1. In formula (2.25) we let $\mathscr{F}_1^{(i)} \neq 0$, $i = 1, 2$. It is then readily seen that in the case of a local functional $\mathscr{F}$, the identification (3.22) of the space $\mathscr{H}_1$ with $L_2(T^1, d\lambda)$ takes the projection $\mathscr{F}_1$ to a smooth (analytic) function $f(\lambda) \in L_2(T^1)$. It follows from (2.25) that

$$\begin{aligned}\langle \mathscr{F}^{(1)}(\sigma_0) \cdot \mathscr{F}^{(2)}(\sigma_t)\rangle_{\mu_c} &- \langle \mathscr{F}^{(1)}(\sigma_0)\rangle_{\mu_c} \cdot \langle \mathscr{F}^{(2)}(\sigma_t)\rangle_{\mu_c} \\ &= (\mathscr{T}^t \mathscr{F}_1^{(2)}, \mathscr{F}_1^{(1)})_{\mathscr{H}_1} + (\mathscr{T}^t \mathscr{F}_2^{(2)}, \mathscr{F}_2^{(1)})_{\mathscr{H}_2} + (\mathscr{T}^t \widetilde{\mathscr{F}}_2^{(2)}, \widetilde{\mathscr{F}}_2^{(1)})_{\widetilde{\mathscr{H}}_2}. \quad (3.41)\end{aligned}$$

The last two summands satisfy the estimates

$$|(\mathscr{T}^t \mathscr{F}_2^{(2)}, \mathscr{F}_2^{(1)})_{\mathscr{H}_2}| < \text{const} \cdot (C\beta^2)^t, \quad |(\mathscr{T}^t \widetilde{\mathscr{F}}_2^{(2)}, \widetilde{\mathscr{F}}_2^{(1)})_{\widetilde{\mathscr{H}}_2}| < \text{const} \cdot (C\beta^2)^t.$$

The first summand can be represented in the form

$$((\mathscr{T}^{(1)})^t \mathscr{F}_1^{(2)}, \mathscr{F}_1^{(1)})_{\mathscr{H}_1} = \int_{T^1} (\widehat{b}(\lambda))^t f^{(2)}(\lambda) f^{(1)}(\lambda)\, d\lambda.$$

Computing the asymptotics of this integral by the Laplace method, we get (2.27).

3.3.2. Consider the case $\mathscr{F}_1^{(1)} = 0$ (or $\mathscr{F}_1^{(2)} = 0$). Let $\mathscr{F}_2^{(1)} \neq 0$ and $\mathscr{F}_2^{(2)} \neq 0$. Then the leading term of the correlation asymptotics is defined by the summand

$$((\mathscr{T}^{(2)})^t \mathscr{F}_2^{(2)}, \mathscr{F}_2^{(1)}) = \int_{T^1} ((\mathscr{T}^{(2)}(\lambda))^t f_\lambda^{(2)}, f_\lambda^{(1)})_{L_2^\lambda}\, d\lambda,$$

where $f_\lambda^{(1)} = f_\lambda^{(1)}(\kappa)$ and $f_\lambda^{(2)} = f_\lambda^{(2)}(\kappa)$ are the images of the components $\mathscr{F}_2^{(i)}$, $i = 1, 2$, under their mapping into $L_2^{\text{sym}}(T^1 \times T^1)$ (and the subsequent expansion in an integral with respect to λ). We can show once again that $f_\lambda^{(i)}(\kappa)$, $i = 1, 2$, are smooth functions in λ and κ. We shall now use the general formula

$$\mathscr{T}^t = \frac{1}{2\pi i} \int_\gamma z^t R(z)\, dz, \qquad t \in \mathbb{Z}^1,$$

where $R(z) = (\mathscr{T} - zE)^{-1}$ is the resolvent of the operator $\mathscr{T}$, and the integration is along the contour γ (traversed clockwise) containing the spectrum of this operator. Consider the case $c < 1/6$, in which the eigenvalue $\varepsilon_0(\lambda)$ is to the right of the continuous spectrum $\mathscr{T}^{(2)}$. In this case

$$\begin{aligned}((\mathscr{T}^{(2)}(\lambda))^t f_\lambda^{(2)}, f_\lambda^{(1)})_{L_2^\lambda} &= \frac{1}{2\pi i} \int_\gamma z^t (R_\lambda(z) f_\lambda^{(2)}, f_\lambda^{(1)})_{L_2^\lambda}\, dz \\ &= \frac{1}{2\pi i} \int_{\gamma_1} z^t (R_\lambda(z) f_\lambda^{(2)}, f_\lambda^{(1)})\, dz \\ &\quad + \varepsilon_0^t(\lambda)(\overline{f}^{(2)}, e_0(\lambda)) \cdot (f^{(1)}, \overline{e}_0(\lambda)),\end{aligned}$$

where $e_0(\lambda)$ is the normalized eigenvector associated with the eigenvalue $\varepsilon_0(\lambda)$, and γ_1 is a contour containing the spectrum $\mathscr{T}^{(2)}(\lambda)$ without the point $\varepsilon_0(\lambda)$. Observe that for $t \to \infty$,

$$\int_{\gamma_1} z^t (R_\lambda(z) f_\lambda^{(2)}, f_\lambda^{(1)})\, dz = o(\varepsilon_0^t(\lambda))$$

and, consequently, uniformly in λ,

$$((\mathscr{T}^{(2)}(\lambda))^t f_\lambda^{(2)}, f_\lambda^{(1)})_{L_2^\lambda} = \varepsilon_0^t(\lambda)[(\overline{f}_\lambda^{(2)}, e_0(\lambda)) \cdot (f_\lambda^{(1)}, \overline{e}_0(\lambda)) + o(1)].$$

It is easily verified that the maximum point λ_0 of $\varepsilon_0(\lambda)$ is close to $\lambda = 0$ (or coincides with zero), and that the maximum of $\varepsilon_0(\lambda)$ at this point is nondegenerate, i.e.,

$$\varepsilon_0(\lambda) = \varepsilon_0(\lambda_0) - a(\lambda - \lambda_0)^2/2 + \cdots, \qquad a > 0,$$

for $|\lambda - \lambda_0| < \delta$. Hence

$$\begin{aligned}
&\int_{T^1} ((\mathscr{T}^{(2)}(\lambda))^t f_\lambda^{(2)}, f_\lambda^{(1)})_{L_2^\lambda}\, d\lambda \\
&= \text{const} \cdot (\varepsilon_0(\lambda_0))^t \int_{|\lambda-\lambda_0|<\delta} \left(1 - a(\varepsilon_0(\lambda))^{-1} \frac{(\lambda-\lambda_0)^2}{2} + \cdots\right)^t d\lambda + o(\varepsilon_0(\lambda_0))^t \\
&= \frac{\text{const}}{\sqrt{t}} (\varepsilon_0(\lambda_0))^t [1 + o(1)] \quad \text{as } t \to \infty.
\end{aligned}$$

3.3.3. Finally, consider the case $c > 1/6$, in which $\varepsilon_0(\lambda) < \tau_\lambda^{\min}$. Observe that in all cases

$$\max_\lambda \tau_\lambda^{\max} = \max_{\lambda_1, \lambda_2} \widehat{b}(\lambda_1)\widehat{b}(\lambda_2) = \alpha_0^2$$

with the maximum (nondegenerate) $\tau_\lambda^{\max}$ being at the point $\lambda = \lambda_0 \sim \beta$. Therefore we must investigate the matrix element $(R_\lambda(z) f_\lambda^{(2)}, f_\lambda^{(1)})_{L_2^\lambda}$ for the values of λ that are close to zero.

LEMMA 1. I. *The matrix element* $(R_\lambda(z) f_\lambda^{(2)}, f_\lambda^{(1)})_{L_2^\lambda}$ *can be represented in the form*

$$\begin{aligned}
(R_\lambda(z) f_\lambda^{(2)}, f_\lambda^{(1)})_{L_2^\lambda} = &\int_{T^1} \frac{f_\lambda^{(2)}(\kappa) \overline{f}_\lambda^{(1)}(\kappa)\, d\kappa}{\omega_\lambda(\kappa) - z} \\
&- \frac{1}{D_\lambda(z)} \int_{(T^1)^2} \frac{D_\lambda(\kappa, \mu; z) f_\lambda^{(2)}(\mu) \overline{f}_\lambda^{(1)}(\kappa)}{(\omega_\lambda(\kappa) - z)(\omega_\lambda(\mu) - z)}\, d\kappa\, d\mu,
\end{aligned} \tag{3.42}$$

where

$$D_\lambda(z) = 1 + \int_{T^1} \frac{K_\lambda(\kappa, \kappa)}{\omega_\lambda(\kappa) - z}\, d\kappa + \sum_{n=2}^\infty \frac{1}{n!} \int_{(T^1)^n} \frac{\det |K_\lambda(\kappa_i, \kappa_j)|}{\prod_{i=1}^n (\omega_\lambda(\kappa_i) - z)} \prod_{i=1}^n d\kappa_i,$$

$$D_\lambda(\kappa, \mu; z) = K_\lambda(\kappa, \mu) + \sum_{n=1}^\infty \frac{1}{n!} \int_{(T^1)^n} \frac{V_n^\lambda(\kappa, \mu; \kappa_1, \ldots, \kappa_n)}{\prod_{i=1}^n (\omega_\lambda(\kappa_i) - z)} \prod_{i=1}^n d\kappa_i,$$

and

$$V_n^\lambda(\kappa, \mu; \kappa_1, \dots, \kappa_n) = \det \begin{pmatrix} K_\lambda(\kappa, \mu) & K_\lambda(\kappa, \kappa_1) & \dots & K_\lambda(\kappa, \kappa_n) \\ K_\lambda(\kappa_1, \mu) & K_\lambda(\kappa_1, \kappa_1) & \dots & K_\lambda(\kappa_1, \kappa_n) \\ \vdots & \vdots & \ddots & \vdots \\ K_\lambda(\kappa_n, \mu) & K_\lambda(\kappa_n, \kappa_1) & \dots & K_\lambda(\kappa_n, \kappa_n) \end{pmatrix}.$$

II. *Let λ be such that the maximum of the function $\omega_\lambda(\kappa)$ is nondegenerate. Then in a small neighborhood of $\tau_\lambda^{\max}$: $\widetilde{U}_\varepsilon = \{z : |z - \tau_\lambda^{\max}| < \varepsilon,\ z \in \Pi_\lambda\}$, the function $(R_\lambda(z) f_\lambda^{(2)}, f_\lambda^{(1)})_{L_2^\lambda}$ can be represented in the form*

$$(R_\lambda(z) f_\lambda^{(2)}, f_\lambda^{(1)})_{L_2^\lambda} = \varphi_1^\lambda(z) + \varphi_2^\lambda(z)(z - \tau_\lambda^{\max})^{1/2}, \tag{3.42a}$$

where φ_1^λ and φ_2^λ are analytic functions in the circle $U_\varepsilon = \{|z - \tau_\lambda^{\max}| < \varepsilon\}$, and $|z - \tau_\lambda^{\max}|^{1/2}$ is positive for real $z > \tau_\lambda^{\max}$.

PROOF OF THE LEMMA. The representation (3.42) was proved in [8]. Further, by the methods of [8] it is not difficult to show that

$$D_\lambda(z) = \frac{c}{(z - \tau_\lambda^{\max})^{1/2}} \times \left[K_\lambda(\kappa_0, \kappa_0) + \sum_{n=1}^\infty \frac{1}{n!} \int_{(T^1)^n} \frac{W_{n+1}^\lambda(\kappa_0, \dots, \kappa_n)}{\prod_{i=1}^n (\omega_\lambda(\kappa_i) - \tau_\lambda^{\max})} \prod_{i=1}^n d\kappa_i\right] + \widetilde{\varphi}_1^\lambda(z) + (z - \tau_\lambda^{\max})^{1/2} \widetilde{\varphi}_2^\lambda(z), \qquad z \in \widetilde{U}_\varepsilon,$$

where $W_n^\lambda(\kappa_1, \dots, \kappa_n) = \det|K_\lambda(\kappa_i, \kappa_j)|$, $\kappa_0 = \kappa_0(\lambda)$ is the maximum point of $\omega_\lambda(\kappa)$, and $\widetilde{\varphi}_1^\lambda$ and $\widetilde{\varphi}_2^\lambda$ are analytic functions in $z \in U_\varepsilon$; finally,

$$c = c(\lambda) = \pi(-\omega_\lambda''(\kappa_0)/2)^{-1/2}. \tag{3.43}$$

By analogy with the earlier argument, we obtain

$$\int_{(T^1)^2} \frac{D_\lambda(\kappa, \mu; z) f_\lambda^{(2)}(\mu) \cdot \overline{f}_\lambda^{(1)}(\kappa)}{(\omega_\lambda(\kappa) - z)(\omega_\lambda(\mu) - z)} d\kappa\, d\mu = \frac{f_\lambda^{(1)}(\kappa_0) \overline{f}_\lambda^{(2)}(\kappa_0)}{z - \tau_\lambda^{\max}} c^2 \left[K_\lambda(\kappa_0, \kappa_0) + \sum_{n=1}^\infty \frac{1}{n!} \int_{(T^1)^n} \frac{W_{n+1}^\lambda(\kappa_0, \dots, \kappa_n)}{\prod_{i=1}^n (\omega_\lambda(\kappa_i) - \tau_\lambda^{\max})} \prod_{i=1}^n d\kappa_i\right] + \frac{\widetilde{\widetilde{\varphi}}_1^\lambda(z)}{(z - \tau_\lambda^{\max})^{1/2}} + \widetilde{\widetilde{\varphi}}_2^\lambda(z), \qquad z \in \widetilde{U}_\varepsilon, \tag{3.44}$$

where $\widetilde{\widetilde{\varphi}}_1^\lambda$ and $\widetilde{\widetilde{\varphi}}_2^\lambda$ are analytic functions in U_ε. Finally,

$$\int_{T^1} \frac{f_\lambda^{(2)}(\kappa) \overline{f}_\lambda^{(1)}(\kappa)}{\omega_\lambda(\kappa) - z} d\kappa = \frac{c f_\lambda^{(2)}(\kappa_0) f_\lambda^{(1)}(\kappa_0)}{(z - \tau_\lambda^{\max})^{1/2}} + \widetilde{\widetilde{\varphi}}_1^\lambda + \widetilde{\widetilde{\varphi}}_2^\lambda (z - \tau_\lambda^{\max})^{1/2}, \tag{3.45}$$

where $\tilde{\tilde{\varphi}}_i^\lambda$, $i = 1, 2$, are analytic functions in U_ε. If we substitute (3.43), (3.44), and (3.45) in formula (3.42) and take into account the inequality $K_\lambda(\kappa_0, \kappa_0) > \tau_\lambda^{\max} > C\beta^2$, where $C > 0$ is a constant, we obtain (3.42a). The lemma is proved.

Further, the integration along the contour γ can be reduced to the integration over the "cut" $(\tau_\lambda^{\max} - \varepsilon, \tau_\lambda^{\max})$ and over the circle $|z| = \tau_\lambda^{\max} - \varepsilon$. The integration over the cut can be reduced to the following integral:

$$\text{const} \cdot \int_{\tau_\lambda^{\max} - \varepsilon}^{\tau_\lambda^{\max}} z^t (\tau_\lambda^{\max} - z)^{1/2}\, dz = \frac{(\tau_\lambda^{\max})^{1/2}}{t^{3/2}} (\text{const} + o(1)).$$

Since $\tau_\lambda^{\max}$ has a nondegenerate maximum at the point λ_0, the additional integration with respect to λ leads to (2.28). The proof of Theorem 3 is complete.

References

1. F. Dunlop, Personal communication.
2. V. A. Malyshev and R. A. Minlos, *Gibbs random fields*, Kluwer Academic Publishers, Dordrecht, 1991.
3. R. A. Minlos and Ya. G. Sinai, *Investigations of the spectra of stochastic operators arising in lattice models of a gas*, Teoret. Mat. Fiz. **2** (1970), no. 2, 230–243; English transl. in Theoret. and Math. Phys. **2** (1970).
4. V. A. Malyshev and R. A. Minlos, *Invariant subspaces of clustering operators*, I, J. Stat. Phys. **21** (1979), 231–242; II, Commun. Math. Phys. **82** (1981), 211–226.
5. ——, *Clustering operators*, Trudy Sem. Petrovsk. **9** (1983), 63–80; English transl. in J. Soviet Math. **33** (1986), no. 5.
6. R. A. Minlos, *Spectral expansion of the transfer matrices of Gibbs fields*, Soviet Sci. Rev. Sect. C, Math. Phys. Rev. **7** (1988), 235–288.
7. ——, *Invariant k-particle subspaces of the Hamiltonian of chiral lattice fields*, Acta Appl. Math. **22** (1991), 55–75.
8. Sh. Mamatov and R. A. Minlos, *Bound states of a two-particle clustering operator*, Teoret. Mat. Fiz. **79** (1989), no. 2, 163–179; English transl. in Theoret. and Math. Phys **79** (1989).
9. Yu. B. Rumer, *The thermodynamics of a plane dipole lattice*, Uspekhi Fiz. Nauk **53** (1954), no. 2, 245–284. (Russian)
10. V. A. Malyshev, *Cluster expansions in lattice models of statistical physics and the quantum theory of fields*, Uspekhi Mat. Nauk **35** (1980), no. 2, 3–53; English transl. in Russian Math. Surveys **35** (1980), no. 2, 7–62.

Moscow State University, Mechanics and Mathematics Department, Moscow 119899, Russia

Translated by N. K. KULMAN

ADVANCES IN SOVIET MATHEMATICS
Volume 20, 1994

Central Limit Theorem for the Random Walk of One and Two Particles in a Random Environment, with Mutual Interaction

C. BOLDRIGHINI, R. A. MINLOS, AND A. PELLEGRINOTTI

§1. Introduction

In this paper we study some models of particles performing a discrete time random walk on the lattice $\mathbb{Z}^\nu$, $\nu = 1, 2, \dots$, and interacting with a random environment that is represented by a random field in evolution $\{\xi_t(x) : t \in \mathbb{Z}_+, x \in \mathbb{Z}^\nu\}$. The one-particle model that we consider here is practically the same as in the paper [1], where the local limit theorem was proved for the displacement of the particle from its initial position. However, the methods used there differ from those in the present paper, which seem to be more convenient for studying random walks of two or more particles and the behavior of time and space correlations of the field.

In this paper we consider, in addition to the one-particle case, a similar model for the random walk of two particles in a random environment. We assume interaction with the environment, and, in the case of two particles, interaction between the particles as well. All interactions are supposed to be localized, and sufficiently small, in some precise sense.

In the one-particle case we prove the local central limit theorem for the displacement of the particle for any dimension of the lattice ν. In the two-dimensional case we prove the central limit theorem (in integral form) for the joint distribution of the displacements of the two particles from their initial positions, for $\nu \geqslant 3$. One should note that in the one-dimensional case the central limit theorem simply does not hold in general, and in the two-dimensional case it holds, but has nonstandard "tails" and the proof is much more complicated. The reader can compare with the results of the paper [9].

We plan to study other problems concerning the behavior of the correlations of the field, or convergence to an equilibrium measure in the region around the particle (for one-particle models) in future papers.

1991 *Mathematics Subject Classification.* Primary 60J15, 60K35, 82B41.

In order to avoid confusion, it is worth noting that the expression "random environment" in our context differs from the usual one, which corresponds to taking a random field of transition probabilities for the random walk of particles (independent of the particles), and studying the asymptotic behavior of the random walk for typical configurations of the field. In the present paper, as in the paper [1], we consider instead a random environment evolving in time as a Markov random field, and particles interacting with each other and with the field.

The methods applied here are based on a detailed investigation of the maximal branch of the spectrum of the transfer matrix (stochastic operator) of the system. We introduce constructions that are independent of previous work, but reflect the experience obtained in studying the spectrum of the transfer matrix for Gibbs random fields and for lattice models of quantum field theory (we refer to the review paper of R. A. Minlos [2]).

We are grateful to Professor E. Presutti for some useful remarks. In particular, he observed that our results are presumably also true if the influence of the particle on the evolution of the environment is not small (i.e., without assuming that the parameter ε_1 defined below is small). We plan to return to this point in a later paper.

The paper consists of five sections and an Appendix. The first section is this introduction, and in the second section we describe the model and state the results. The third section is devoted to the study of the one-particle case, for which we reproduce the result of the paper [1]. In the fourth and fifth sections we study the random walk of two particles. The Appendix contains the proof of some technical lemmas.

Note that the situation we consider in this paper includes the case of the random walk of two identical particles which do not interact with any environment and are subject to a local interaction (i.e., only when their distance is less than some interaction range) with each other. This particular situation, in a nonperturbative case, is studied in a yet unpublished paper by E. A. Zhizhina and R. A. Minlos ([9], see also [10]), which is in some sense a natural continuation of the present paper, and makes use of the same analytical techniques (see §5).

§2. Description of the model and statement of the main results

A. Random walk of one particle

Let $\xi \equiv \{\xi_t(x),\ (x, t) \in \mathbb{Z}^\nu \times \mathbb{Z}^1\}$ be a random field (environment) taking values in some finite set S, and $X_t \in \mathbb{Z}^\nu$, $t \in \mathbb{Z}_+$, the position of a particle performing a random walk. We assume that the whole system $\{X_t, \xi_t\}$ behaves as a Markov chain with conditionally independent transition probabilities:

$$\begin{aligned}&\mathbb{P}(X_t = z\,,\ \xi_t \in A \mid X_{t-1} = x\,,\ \xi_{t-1} = \eta)\\&\quad = \mathbb{P}(X_t = z \mid X_{t-1} = x\,,\ \xi_{t-1} = \eta)\,\mathbb{P}(\xi_t \in A \mid X_{t-1} = x\,,\ \xi_{t-1} = \eta). \qquad (2.1)\end{aligned}$$

Here $\eta \equiv \{\eta(x)\}_{x\in\mathbb{Z}^\nu} \in \Omega \equiv S^{\mathbb{Z}^\nu}$ is a fixed configuration of the field and $A \subset \Omega$ is an arbitrary measurable set of configurations of the environment. For the factors on the right-hand side of equation (2.1) we assume the following hypotheses.

I. The random walk transition probability is

$$\mathbb{P}(X_t = z \mid X_{t-1} = x\,, \xi_{t-1} = \eta) = P_0(z-x) + c(z-x\,,\ \eta(x)). \qquad (2.2)$$

Here P_0 is a probability distribution on $\mathbb{Z}^\nu$ (jump distribution of the "unperturbed" random walk), and $c(\cdot\,,\cdot)$ is a function on $\mathbb{Z}^\nu \times S$ such that

$$1)\quad \sum_{u\in\mathbb{Z}^\nu} c(u\,,\ s) = 0 \qquad \text{for all } s \in S; \qquad (2.3a)$$

$$2)\quad P_0(u) + c(u\,,\ s) \geqslant 0 \quad \text{for all } u \in \mathbb{Z}^\nu\,,\ \ s \in S. \qquad (2.3b)$$

In what follows we assume that

$$\max_{u\,,s} |c(u\,,\ s)| \equiv \varepsilon_0 \ll 1\,. \qquad (2.3c)$$

We assume also that both P_0 and c have finite range, namely, there is $D > 0$ such that

$$P_0(u) = c(u\,,\ s) = 0 \quad \text{for } |u| > D \text{ and all } s \in S\,, \qquad (2.3d)$$

$$|u| \equiv \sum_{i=1}^{\nu} |u^{(i)}|\,, \qquad u = (u^{(1)}\,, \dots\,, u^{(\nu)})\,.$$

Moreover, we assume that the characteristic function

$$\widetilde{p}_0(\lambda) = \sum_{u\in\mathbb{Z}^\nu} P_0(u) e^{i(\lambda\,,u)}\,, \qquad \lambda = (\lambda_1\,, \dots\,, \lambda_\nu) \in T^\nu\,, \qquad (2.3e)$$

satisfies

$$|\widetilde{p}_0(\lambda)| < 1 \qquad (2.3f)$$

for all $\lambda \neq 0$, the Taylor expansion of the logarithm for λ in a neighborhood of 0 is

$$\log(\widetilde{p}_0(\lambda)) = i\sum b_k^0 \lambda_k - \frac{1}{2}\sum a_{i\,,j}^0 \lambda_i \lambda_j + \dots\,, \qquad (2.3g)$$

and the quadratic form $\sum a_{i\,,j}^0 \lambda_i\lambda_j$ is strictly positive for $\lambda \neq 0$.

II. The distribution $\mathbb{P}(\xi_t \in \cdot \mid X_{t-1} = z\,,\ \xi_{t-1} = \eta)$ is a product measure for the independent variables $\{\xi_t(x)\,,\ x \in \mathbb{Z}^\nu\}$, each of which is distributed according to the law

$$\mathbb{P}(\xi_t(z) = s \mid X_{t-1} = x\,,\ \xi_{t-1} = \eta) = \begin{cases} q_0(\eta(z)\,,\ s) & \text{if } z \neq x\,,\\ q_1(\eta(z)\,,\ s) & \text{if } z = x\,.\end{cases} \qquad (2.4)$$

Here $q_0(s, s')$, $q_1(s, s')$, $s, s' \in S$, are two stochastic matrices. We set

$$\Delta(s, s') = q_1(s, s') - q_0(s, s') \tag{2.5a}$$

and assume that

$$\max_{s, s'} |\Delta(s, s')| \equiv \varepsilon_1 \ll 1 . \tag{2.5b}$$

We assume that the Markov chain with space state S and transition probabilities

$$\mathbb{P}(s' \to s) = q_0(s', s) .$$

is ergodic and denote by π_0 the (unique) stationary measure associated to it. We assume furthermore that the stochastic operator Q_0,

$$(Q_0 f)(s') = \sum_s q_0(s', s) f(s), \tag{2.5c}$$

can be diagonalized, and denote its eigenvectors (whose normalization will be fixed below) by $\{e_i(s), s \in S\}_{i=0}^{|S|-1}$, where $|S|$ is the cardinality of the set S. (The general case, when Q_0 has Jordan cells, can also be treated, but the formulas are more complicated.)

The eigenvalue corresponding to e_j is denoted by μ_j, and we label the eigenvectors in such a way that

$$1 = \mu_0 > |\mu_1| \geqslant \cdots \geqslant |\mu_{|S|-1}| .$$

Clearly $e_0(s) = 1$, $s \in S$. Note that, since the eigenvectors e_i of the matrix Q_0 are orthogonal to the eigenvector π_0 of the transposed matrix, we have

$$\sum_s e_i(s) \pi_0(s) = 0, \quad \text{for } i > 0. \tag{2.6}$$

Consider the Hilbert space $l_2(S, \pi_0)$. The operator Q_0^* adjoint to the operator Q_0 given by (2.5c) acts on $l_2(S, \pi_0)$ by the formula

$$(Q_0^* f)(s) = \sum_{s'} q_0(s', s) \frac{\pi_0(s')}{\pi_0(s)} f(s'), \qquad s \in S. \tag{2.6a}$$

The eigenvalues of this operator are

$$\mu_0^* = 1, \ \mu_1^* = \overline{\mu}_1, \ \ldots, \ \mu_{|S|-1}^* = \overline{\mu}_{|S|-1}.$$

The corresponding eigenvectors are denoted by $e_i^*(s)$. We have $e_0^*(s) \equiv 1$, and clearly the eigenvectors e_i^* are biorthogonal to the eigenvectors e_i:

$$(e_i, e_j^*) = 0 \quad \text{if } i \neq j .$$

We normalize the eigenvectors in such a way that

$$(e_i, e_j^*) = \delta_{ij}, \quad \|e_i\| = \|e_i^*\|, \qquad i = 0, \ldots, |S| - 1 . \tag{2.6b}$$

Without loss of generality we can assume that

$$\sum_s c(x, s) \pi_0(s) = 0 . \tag{2.6c}$$

If (2.6c) does not hold, we can replace $P_0(x)$ by $P_0 + \sum_s c(x, s)\pi_0(s)$.

Observe that for $\varepsilon_0 = \varepsilon_1 = 0$ (i.e., $c = 0$, $q_0 = q_1$) the particle moves independently of the environment. We call this the "unperturbed random walk". Conditions (2.3c) and (2.5a) mean that the perturbation, i.e., the interaction of the particle and the environment, is small.

The initial position of the particle at time $t = 0$ is fixed, $X_0 = u$, $u \in \mathbb{Z}^\nu$, and the initial distribution of the random field ξ_0 is denoted by Π. We denote by $\widehat{\Omega} \equiv \Omega \times \mathbb{Z}^\nu$ the phase space of the states of the system. The conditional average (according to the distribution (2.1)) of a bounded function $f(\xi, x)$, $(\xi, x) \in \widehat{\Omega}$ (measurable with respect to the variable ξ), is denoted by

$$(\mathcal{T} f)(\eta, z) \equiv \langle f \mid \eta, z\rangle. \tag{2.7}$$

$\mathcal{T}$ is "the stochastic operator" or "the transfer matrix" of the system.

A convenient expression of the probability that the particle is at the site x at time t is given by the following lemma.

Lemma 1. *Set $\Phi_x^0(\xi, y) = \delta_{x,y}$, where $\delta_{x,y}$ is the Kronecker symbol. Then*

$$\mathbb{P}(X_t = x \mid X_0 = u) = \int_\Omega (\mathcal{T}^t \Phi_x^0)(\xi, u)\, d\Pi(\xi). \tag{2.8}$$

Proof. It is easy to check that the formula holds for $t = 1$. For $t > 1$ this follows by the matrix multiplication rule. □

In the first part of the paper we establish the asymptotic behavior of the probabilities $\mathbb{P}(X_t = x \mid X_0 = z)$ as $t \to \infty$.

Theorem 1 (Local limit theorem). *For fixed P_0, q_0, and any initial distribution Π, the following asymptotics holds for ε_0, ε_1 sufficiently small*

$$\mathbb{P}(X_t = x \mid X_0 = u) = \frac{C}{(\sqrt{2\pi t})^\nu} \exp\left\{ -\frac{1}{2t}(A(x-u-bt), x-u-bt)\right\}(1+o(1)), \tag{2.9}$$

where $b \in \mathbb{R}^\nu$ (drift, or average shift of the particle in a single jump) is a vector, $(\cdot,\cdot)$ is the scalar product in $\mathbb{R}^\nu$,

$$(Av, v) = \sum a_{ij} v^i v^j, \qquad v = (v^1, \dots, v^\nu),$$

is a positive definite quadratic form with matrix $A = \{a_{ij}\}$, and $C = \sqrt{\det A}$. The asymptotics (2.9) is uniform with respect to $x \in \mathbb{Z}^\nu$ in the region

$$|x - u - bt| < t^{1/2+\varepsilon}, \qquad \varepsilon < 1/6.$$

B. Random walk of two particles

In this case the phase space $\widehat{\Omega} \equiv \Omega \times \mathbb{Z}^\nu \times \mathbb{Z}^\nu$ consists of triples $(\xi, x^{(1)}, x^{(2)})$ where $\xi = \{\xi(x), x \in \mathbb{Z}^\nu\}$ is, as above, the configuration of the field on $\mathbb{Z}^\nu$ and $x^{(i)}$, $i = 1, 2$, are the particle positions. As before, we assume that the transition probabilities are conditionally independent. The conditional transition probability for the jump of the ith particle is

$$\begin{aligned}\mathbb{P}(X_t^{(i)} = x^{(i)} \mid X_{t-1}^{(1)} = z^{(1)}, X_{t-1}^{(2)} = z^{(2)}, \xi_{t-1} = \eta) \\ = P_0(x^{(i)} - z^{(i)}) + \widehat{c}(x^{(i)} - z^{(i)}, z^{(1)} - z^{(2)}, \eta(z^{(i)})).\end{aligned} \tag{2.10a}$$

Here P_0 is the same function as in (2.2), and $\widehat{c}$ is given by

$$\widehat{c}(x, z, s) = c(x, s) + c_1(x, z, s), \tag{2.10b}$$

where $c(\cdot, \cdot)$ is the same as in (2.2), and c_1 has finite support in x, z and is even in z: $c_1(x, z, s) = c_1(x, -z, s)$. The random field transition probability is again a countable product of transition probabilities at each site $x \in \mathbb{Z}^\nu$:

$$\begin{aligned}&\mathbb{P}(\xi_t(x) = s \mid X_{t-1}^{(1)} = z^{(1)}, X_{t-1}^{(2)} = z^{(2)}, \xi_{t-1} = \eta) \\ &= \begin{cases} q_0(\eta(x), s) & \text{if } x \neq z^{(1)}, \ x \neq z^{(2)}, \\ \widehat{q}_1(\eta(x), s) & \text{if } x = z^{(i)}, \ i = 1 \text{ or } 2, \ z^{(1)} \neq z^{(2)}, \\ \widehat{q}_2(\eta(x), s) & \text{if } x = z^{(1)} = z^{(2)}, \end{cases}\end{aligned}$$

where q_0, $\widehat{q}_1$, $\widehat{q}_2$ are stochastic matrices close to each other (in the sense of (2.11) below). As above, we write this as

$$\begin{aligned}&\mathbb{P}(\xi_t(x) = s \mid X_{t-1}^{(1)} = z^{(1)}, X_{t-1}^{(2)} = z^{(2)}, \xi_{t-1} = \eta) \\ &= q_0(\eta(x), s) + \Delta(\eta(x), s, x - z^{(1)}, x - z^{(2)}),\end{aligned}$$

where

$$\max_{s_1, s_2, x_1, x_2} |\Delta(s_1, s_2, x_1, x_2)| = \varepsilon_1 \ll 1. \tag{2.11}$$

For $\widehat{c} \equiv 0$ and $\Delta(s_1, s_2, z_1, z_2) = 0$ both particles move independently of each other and independently of the environment (independent random walk of the particles). The transfer matrix $\mathscr{T}^{(2)}$ for the present case is introduced just as in the previous case:

$$(\mathscr{T}^{(2)} f)(\eta, z^{(1)}, z^{(2)}) = \langle f \mid \eta, z^{(1)}, z^{(2)} \rangle.$$

As before, the following formula holds:

$$\begin{aligned}&\mathbb{P}(X_t^{(1)} = x^{(1)}, X_t^{(2)} = x^{(2)} \mid X_0^{(1)} = u^{(1)}, X_0^{(2)} = u^{(2)}) \\ &= \int_\Omega ((\mathscr{T}^{(2)})^t \Phi^0_{x^{(1)}, x^{(2)}})(\xi, u^{(1)}, u^{(2)})\, d\Pi(\xi),\end{aligned} \tag{2.12}$$

where Π is the initial distribution of the field and

$$\Phi^0_{x^{(1)}, x^{(2)}}(\xi, y^{(1)}, y^{(2)}) = \delta_{x^{(1)}, y^{(1)}} \delta_{x^{(2)}, y^{(2)}}.$$

We shall prove the central limit theorem under some additional technical assumptions.

I. We assume that

$$|\widetilde{p}_0(\lambda)|^2 > |\mu_1|, \qquad \lambda \in T^\nu, \tag{2.13}$$

where $\widetilde{p}_0$ is the characteristic function given by (2.3e).

II. We assume that the one-particle random walk, which is obtained by setting $c_1(x, z, s) \equiv 0$ in (2.10b), satisfies the following symmetry condition:

$$P_0(u) = P_0(-u), \quad c(u, s) = c(-u, s), \qquad s \in S, \ u \in \mathbb{Z}^\nu.$$

Moreover, as we mentioned in the Introduction, we assume that $\nu \geqslant 3$.

In the case of the two-particle random walk, the central local limit theorem for the displacements with respect to the initial positions $x^{(1)} - u^{(1)}$, $x^{(2)} - u^{(2)}$ holds, apparently, only under the condition that the distance between the particles is large enough. The central limit theorem in integral form is always true, however.

The main result of the present paper is the following theorem.

THEOREM 2. *Under the above assumptions, the following asymptotics takes place for* $t \to \infty$

$$\mathbb{P}((X_t^{(1)} - u^{(1)}, X_t^{(2)} - u^{(2)}) \in \sqrt{t}G \mid X_0^{(1)} = u^{(1)}, X_0^{(2)} = u^{(2)}) \tag{2.14}$$
$$= \frac{C^2}{(2\pi)^\nu} \iint_G \exp\left\{ -\frac{1}{2}(Av^{(1)}, v^{(1)}) - \frac{1}{2}(Av^{(2)}, v^{(2)}) \right\} dv^{(1)} dv^{(2)} + o(1),$$

where $G \subset \mathbb{R}^\nu \times \mathbb{R}^\nu$ *is any bounded region, the matrix* A *is the same as in Theorem* 1, *and* $C^2 = \det A$. (*The lattice* $\mathbb{Z}^\nu$, *as usual, is considered to be a subset of* $\mathbb{R}^\nu$.)

REMARK. The above theorem shows that asymptotically in time the two particles perform independent random walks, and each random walk is the same as if the other particle were absent. That is, the short range interaction between particles and the dependence arising from the interaction with the environment do not show up in the leading term of the asymptotics (2.14).

§3. The one-particle case

We introduce the following spaces of functions on the phase space $\widehat{\Omega} = \Omega \times \mathbb{Z}^\nu$.

I. The space $C_\infty(\widehat{\Omega})$ of bounded functions $f(\xi, x)$ continuous in ξ, with respect to the product topology, with norm

$$\|f\|_\infty = \sup_{\xi, x} |f(\xi, x)|. \tag{3.1}$$

II. The Hilbert space $\mathscr{H} = L_2(\Omega, \Pi_0) \otimes l_2(\mathbb{Z}^\nu)$ of functions $f(\xi, x)$ with norm

$$\|f\|_{\mathscr{H}}^2 = \sum_{x \in \mathbb{Z}^\nu} \int_\Omega |f(\xi, x)|^2 d\Pi_0(\xi), \tag{3.2}$$

where the distribution $\Pi_0 = \pi_0^{\mathbb{Z}^\nu}$ is a countable product of the probability measures π_0.

III. The space $C_{\text{dec}}(\widehat{\Omega}) = C(\Omega) \otimes S(\mathbb{Z}^\nu)$ of functions $f(\xi, x)$ continuous in ξ and decreasing in x faster than any inverse power $|x|^{-n}$, $n > 0$. The topology on $C_{\text{dec}}(\widehat{\Omega})$ is given by the family of seminorms $\sup_{\xi,x} |f(\xi, x)|\,|x|^n$, $n = 0, 1, \ldots$.

LEMMA 3.1. *The transfer matrix $\mathcal{T}$ of our system transforms each of the spaces introduced above into itself, and defines a bounded operator in each of them. Moreover, the norm of $\mathcal{T}$ in the space $C_\infty(\widehat{\Omega})$ is* 1.

PROOF. Let $L_\infty(\widehat{\Omega})$ be the Banach space of functions $f(\xi, x)$ bounded in the supremum norm (3.1). It is easy to see that $\mathcal{T}$ maps $L_\infty(\widehat{\Omega})$ in itself and $\|\mathcal{T} f\| \leqslant \|f\|$ for all $f \in L_\infty(\widehat{\Omega})$, which shows that $\mathcal{T}$ is a continuous linear operator of $L_\infty(\widehat{\Omega})$ into itself. The space $C_\infty(\widehat{\Omega})$ is a closed subspace of $L_\infty(\widehat{\Omega})$. Actually, it consists of the functions $f(\xi, x) \in L_\infty(\widehat{\Omega})$ that are limits (in the supremum norm) of cylinder functions, i.e., of functions that depend only on the values of the field ξ in a bounded region. It follows from the explicit expression (2.1) that $\mathcal{T}\varphi$ is a cylindric function if φ is cylindric and $\|\mathcal{T}\|_{L_\infty(\widehat{\Omega})} \leqslant 1$. Hence $\mathcal{T}$ maps $C_\infty(\widehat{\Omega})$ into itself, and $\|\mathcal{T}\|_{C_\infty(\widehat{\Omega})} \leqslant 1$. It is easy to see that in fact $\|\mathcal{T}\| = 1$: one just has to consider its action on the functions that are constant on $\widehat{\Omega}$.

For $\mathcal{H}$ and $C_{\text{dec}}(\widehat{\Omega})$ the proof is not as simple. For $\mathcal{H}$ the assertion is proved in the Appendix. As to the proof for $C_{\text{dec}}(\widehat{\Omega})$, it is similar to that of Lemma 4.2 below. We leave out the details. □

Let us investigate the action of $\mathcal{T}$ on $\mathcal{H}$. As a base we introduce in $\mathcal{H}$ the functions

$$\Psi_{\Gamma, z}(\xi, x) = \prod_{y \in \operatorname{supp} \Gamma} e_{\gamma(y)}(\xi(y))\,\delta_{x, z} \tag{3.3}$$

where $\Gamma \equiv \{\gamma(y) : y \in \mathbb{Z}^\nu\}$ is a multi-index, with the single indices $\gamma(x)$ taking values in the set $\{0, \ldots, |S| - 1\}$ and such that

$$\operatorname{supp} \Gamma \equiv \{x : \gamma(x) > 0\}$$

is finite, and $e_i(s)$, $i = 0, \ldots, |S| - 1$, are the eigenvectors of the operator Q_0. In what follows we shall denote by $\Gamma + z$, $z \in \mathbb{Z}$, the shift of the multi-index Γ, i.e., $\widetilde{\Gamma} = \Gamma + z$ is defined by the relation $\widetilde{\gamma}(x) = \gamma(x - z)$.

LEMMA 3.2. *The vectors $\{\Psi_{\Gamma, z}\}$ form a basis in $\mathcal{H}$, i.e., any vector $f \in \mathcal{H}$ can be represented as*

$$f = \sum_{\Gamma, z} \hat{f}_{\Gamma, z} \Psi_{\Gamma, z}, \tag{3.3a}$$

and the representation is unique.

PROOF. It is enough to prove that the functions

$$\Psi_\Gamma(\xi) = \prod_{x\in\operatorname{supp}\Gamma} e_{\gamma(x)}(\xi(x)), \qquad \Gamma = \{\gamma(x),\ x\in\mathbb{Z}^\nu\}, \tag{3.3b}$$

form a basis in $L_2(\Omega, \Pi_0)$. Note that they are not orthogonal, their scalar products being

$$a_{\Gamma,\Gamma'} = (\Psi_\Gamma, \Psi_{\Gamma'}) = \prod_{x\in\operatorname{supp}\Gamma} (e_{\gamma(x)}, e_{\gamma'(x)})\,\delta_{\operatorname{supp}\Gamma,\operatorname{supp}\Gamma'}, \tag{3.4}$$

where the scalar products $(e_n, e_{n'})$ are the elements of the Gram matrix for the vectors $\{e_j(s),\ j=1,\dots,|S|-1\}$:

$$(e_n, e_{n'}) = \sum_{s\in S} e_n(s)\bar{e}_{n'}(s)\pi_0(s), \qquad n, n' = 1, \dots, |S|-1.$$

For any finite set $Q\subset\mathbb{Z}^\nu$ we denote by $\mathscr{H}_Q \subset L_2(\Omega,\Pi_0)$ the finite-dimensional subspace of functions $f\in L_2(\Omega,\Pi_0)$ depending only on the values $\{\xi(x),\ x\in Q\}$ of the field in the set Q. Clearly any such function can be represented as a linear combination

$$\sum_{\Gamma:\,\operatorname{supp}\Gamma\subseteq Q} \hat{f}_\Gamma\Psi_\Gamma.$$

Since the linear span of the vectors of the subspaces $\mathscr{H}_Q$, for all finite subsets $Q\subset\mathbb{Z}^\nu$, is dense in $L_2(\Omega,\Pi_0)$, this is true also for the system $\{\Psi_\Gamma\}$. We denote by $\widehat{\mathscr{H}_Q}\subset\mathscr{H}_Q$ the space of the functions $f(\xi)$ which can be represented as

$$f(\xi) = \sum_{\operatorname{supp}\Gamma=Q} \hat{f}_\Gamma\Psi_\Gamma(\xi). \tag{3.5}$$

From (3.4) it follows easily that $\widehat{\mathscr{H}_{Q_1}}$ is orthogonal to $\widehat{\mathscr{H}_{Q_2}}$ for $Q_1\neq Q_2$, and therefore $L_2(\Omega,\Pi_0)$ can be decomposed as a direct sum of spaces

$$L_2(\Omega,\Pi_0) = \bigoplus_{Q\subset\mathbb{Z}^\nu} \widehat{\mathscr{H}_Q}.$$

Hence, taking into account the expansion (3.5) (which is unique), we get the assertion of the lemma. □

For $f\in\mathscr{H}\equiv L_2(\Omega,\Pi_0)$ we have

$$\|f\|^2 = \sum_{\Gamma,\Gamma'} \hat{f}_\Gamma\overline{\hat{f}_{\Gamma'}}a_{\Gamma,\Gamma'}. \tag{3.6a}$$

Therefore the representation of any vector $f\in\mathscr{H}$ in the basis $\{\Psi_{\Gamma,z}\}$

$$f = \sum_{\Gamma,z} \hat{f}_{\Gamma,z}\Psi_{\Gamma,z}$$

generates a canonical isomorphism $\mathscr{H} \to \hat{l}_2(\mathfrak{M} \times \mathbb{Z}^\nu)$ of the space $\mathscr{H}$ to the Hilbert space of sequences $\hat{f} = \{\hat{f}_{\Gamma, z}\}$ defined on $\mathfrak{M} \times \mathbb{Z}^\nu$, where $\mathfrak{M}$ is the collection of multi-indices Γ with norm

$$\|\hat{f}\|^2 = \sum_{Q, z} \sum_{\substack{\Gamma, \Gamma' : \\ \operatorname{supp}\Gamma = \operatorname{supp}\Gamma' = Q}} \hat{f}_{\Gamma, z} \overline{\hat{f}_{\Gamma', z}} a_{\Gamma, \Gamma'}. \tag{3.6b}$$

The operator $\mathscr{T}$ in the basis $\Psi_{\Gamma, z}$ is represented as

$$\mathscr{T}\Psi_{\Gamma, z} = \sum_{\Gamma', z'} R_{\Gamma, z; \Gamma', z'} \Psi_{\Gamma', z'}.$$

A tedious but straightforward computation gives the following result:

$$R_{\Gamma, z; \Gamma', z'} = \mu_{\Gamma_{z'}} \delta_{\Gamma_{z'}, \Gamma'_{z'}} L(\gamma'(z), \gamma(z), z' - z), \tag{3.7a}$$

where $\mu_\Gamma = \prod_{x \in \mathbb{Z}^\nu} \mu_{\gamma(x)}$, the multi-index $\Gamma_z \equiv \{\gamma_z(x),\ x \in \mathbb{Z}^\nu\}$ is given by

$$\Gamma_z(x) = \begin{cases} \gamma(x), & x \neq z, \\ 0, & x = z, \end{cases}$$

and the quantity $L(j, j'; u)$ is given by

$$\begin{aligned} L(j, j'; u) &= \mu_j \delta_{jj'} P_0(u) + \alpha_j^{j'} P_0(u) \\ &\quad + \mu_{j'} \sum_{m=0}^{|S|-1} c_m(u) b_{j'm}^j + \sum_{m, m'=0}^{|S|-1} c_m(u) \alpha_j^{j'} b_{m, m'}^j, \end{aligned} \tag{3.7b}$$

where the coefficients α_j^i, $b_{i,k}^j$, $c_j(u)$ are given by the relations

$$\sum_s \Delta(s', s') e_j(s) = \sum_{j'=0, \dots, |S|-1} \alpha_{j'}^j e_{j'}(s'), \qquad j = 1, \dots, |S| - 1,$$

$$e_m(s) e_{m'}(s) = \sum_j b_{m, m'}^j e_j(s), \qquad m, m' = 0, \dots, |S| - 1,$$

$$c(u; s) = \sum_{m=1}^{|S|-1} c_m(u) e_m(s), \qquad u \in \mathbb{Z}^\nu.$$

Note that $b_{m, m'}^j = b_{m', m}^j$, $b_{m,0}^j = \delta_{jm}$, $\alpha_j^0 = 0$, and $c_0(u) = 0$ (by condition (2.6c)). The operator induced by $\mathscr{T}$ on $\hat{l}_2(\mathfrak{M} \times \mathbb{Z}^\nu)$ is denoted, by abuse of notation, again by $\mathscr{T}$, and is given by the formula

$$(\mathscr{T}\hat{f})_{\Gamma, z} = \sum_{\Gamma', z'} R^T_{\Gamma, z; \Gamma', z'} \hat{f}_{\Gamma', z'}, \tag{3.8a}$$

where R^T is the transposed of the matrix R, and by formula (3.7) is given by

$$R^T_{\Gamma, z; \Gamma', z'} = R_{\Gamma', z'; \Gamma, z} = \mu_{\Gamma_z} \delta_{\Gamma_z, \Gamma'_z} L(\gamma(z), \gamma'(z), z - z'). \tag{3.8b}$$

In the space $\mathscr{H}$ the unitary group of operators $\{\mathscr{U}_v, v \in \mathbb{Z}^\nu\}$ generated by the translations in $\mathbb{Z}^\nu$ acts in the following way:

$$(\mathscr{U}_v f)(\xi, x) = f(\xi + v, x + v),$$

where $\xi + v$ is the shift of the configuration ξ by v: $(\xi + v)(x) = \xi(x - v)$, $x \in \mathbb{Z}^\nu$. It is easy to see that the isomorphism between $\mathscr{H}$ and $\hat{l}_2(\mathfrak{M} \times \mathbb{Z}^\nu)$ transforms the operator $\mathscr{U}_v$ into the operator

$$(U_v \hat{f})(\Gamma, z) = \hat{f}_{\Gamma+v, z+v}. \tag{3.9}$$

As it is easily seen from (3.6b), the space $\hat{l}_2(\mathfrak{M} \times \mathbb{Z}^\nu)$ can be represented as (i.e., is isomorphic to) the tensor product $\hat{l}_2(\mathfrak{M}) \otimes l_2(\mathbb{Z}^\nu)$, where $\hat{l}_2(\mathfrak{M})$ is the Hilbert space of the functions $\hat{f} = \{\hat{f}_\Gamma, \Gamma \in \mathfrak{M}\}$ with scalar product

$$(\hat{f}, \hat{g})_{\hat{l}_2(\mathfrak{M})} = \sum_{Q \subset \mathbb{Z}^\nu} \sum_{\substack{\Gamma, \Gamma' : \\ \operatorname{supp}\Gamma = \operatorname{supp}\Gamma' = Q}} \hat{f}_\Gamma \overline{\hat{g}_{\Gamma'}} a_{\Gamma, \Gamma'}. \tag{3.10}$$

Consider the Hilbert space $\hat{l}_2(\mathfrak{M}) \otimes L_2(T^\nu, d\lambda)$ of functions $\varphi_\Gamma(\lambda)$ with scalar product

$$(\varphi_1, \varphi_2) = \int_{T^\nu} \left(\sum_Q \sum_{\substack{\Gamma, \Gamma' : \\ \operatorname{supp}\Gamma = \operatorname{supp}\Gamma' = Q}} (\varphi_1)_\Gamma(\lambda) \overline{(\varphi_2)_{\Gamma'}(\lambda)} a_{\Gamma, \Gamma'} \right) d\lambda, \tag{3.11}$$

where T^ν is the ν-dimensional torus and $d\lambda$ is the normalized Haar measure on it. The application $\hat{f} \mapsto \varphi$ of $\hat{l}_2(\mathfrak{M} \times \mathbb{Z}^\nu)$ in $\hat{l}_2(\mathfrak{M}) \otimes L_2(T^\nu, d\lambda)$ given by the formula

$$\varphi_\Gamma(\lambda) = \sum_z \hat{f}_{\Gamma+z, z} e^{i(\lambda, z)} \tag{3.12a}$$

is unitary and its inverse is

$$\hat{f}_{\Gamma, z} = \int_{T^\nu} \varphi_{\Gamma - z}(\lambda) e^{-i(\lambda, z)} d\lambda. \tag{3.12b}$$

By the commutation rule $\mathscr{T} \mathscr{U}_v = \mathscr{U}_v \mathscr{T}$ we have

$$R^T_{\Gamma, z; \Gamma', z'} = R^T_{\Gamma - z, 0; \Gamma' - z, z' - z}. \tag{3.12c}$$

LEMMA 3.3. *The map* (3.12a) *transforms the operators* U_v *and* $\mathscr{T}$ *respectively into the operators* $\widetilde{U}_v$ *and* $\widetilde{\mathscr{T}}$ *acting in* $\hat{l}_2(\mathfrak{M}) \otimes L_2(T^\nu, d\lambda)$ *as*

$$(\widetilde{U}_v \varphi)_\Gamma(\lambda) = e^{-i(\lambda, v)} \varphi_\Gamma(\lambda), \qquad (\widetilde{\mathscr{T}} \varphi)_\Gamma(\lambda) = (\mathscr{T}_\lambda \varphi(\lambda))_\Gamma,$$

where $\varphi(\lambda) = \{\varphi_\Gamma(\lambda), \Gamma \in \mathfrak{M}\}$, *and* $\{\mathscr{T}_\lambda, \lambda \in T^\nu\}$ *is a family of operators acting in the space* $\hat{l}_2(\mathfrak{M})$ *according to the formula*

$$(\mathscr{T}_\lambda f)_\Gamma = \sum_{\Gamma'} \widehat{R}_{\Gamma, \Gamma'}(\lambda) \hat{f}_{\Gamma'}, \qquad f \in l_2(\mathfrak{M}), \tag{3.13a}$$

where the matrix elements $\widehat{R}_{\Gamma,\Gamma'}(\lambda)$ *are given, according to equations* (3.12c) *and* (3.8b), *by*

$$\widehat{R}_{\Gamma,\Gamma'}(\lambda) = \sum_{u\in\mathbb{Z}^\nu} R_{\Gamma',0;\Gamma+u,u} e^{i(\lambda,u)}. \tag{3.13b}$$

PROOF. The proof is a simple verification. □

REMARK. In the unperturbed case, i.e., for $\varepsilon \equiv \max\{\varepsilon_0, \varepsilon_1\} = 0$, it follows from (3.7a,b) that

$$R^T_{\Gamma,z;\Gamma',z'}\big|_{\varepsilon=0} = \delta_{\Gamma\Gamma'}\mu_\Gamma P_0(z'-z). \tag{3.14a}$$

Hence the operator $\mathcal{T}_\lambda^0 \equiv \mathcal{T}_\lambda|_{\varepsilon=0}$ acts as

$$(\mathcal{T}_\lambda^0 \hat{f})_\Gamma = \mu_\Gamma \sum_{u\in\mathbb{Z}^\nu} \hat{f}_{\Gamma-u} P_0(u) e^{i(\lambda,u)}. \tag{3.14b}$$

Denoting by $\bar{0} \in \mathfrak{M}$ the multi-index with support $\operatorname{supp}\bar{0} = \varnothing$ (i.e., $\bar{0} = \{\gamma(x) = 0,\ x \in \mathbb{Z}^\nu\}$), it follows from (3.13b) that

$$\widehat{R}_{\bar{0},\bar{0}}(\lambda)|_{\varepsilon=0} = \sum_{x\in\mathbb{Z}^\nu} P_0(x) e^{i(\lambda,x)} = \tilde{p}_0(\lambda). \tag{3.14c}$$

From (3.14b) it is easy to see that the function $\chi^0 = \delta_{\Gamma,\bar{0}}$ is an eigenfunction of $\mathcal{T}_\lambda^0$ with eigenvalue $\tilde{p}_0(\lambda)$, hence the subspace $\hat{l}_2^0(\mathfrak{M}) = \{f_\Gamma : f_{\bar{0}} = 0\}$ is invariant with respect to $\mathcal{T}_\lambda^0$. One can then write

$$\hat{l}_2(\mathfrak{M}) = \{c\chi^0\} + \hat{l}_2^0(\mathfrak{M}), \tag{3.15}$$

where $\{c\chi^0\}$ is the linear span of χ^0. Equation (3.15) induces a decomposition of the operator $\mathcal{T}_\lambda$ as

$$\mathcal{T}_\lambda = \begin{pmatrix} \mathcal{T}_\lambda^{0,0} & \mathcal{T}_\lambda^{0,1} \\ \mathcal{T}_\lambda^{1,0} & \mathcal{T}_\lambda^{1,1} \end{pmatrix}, \tag{3.16}$$

and, representing $\hat{f} \in \hat{l}_2(\mathfrak{M})$ as $\hat{f} = c\chi^0 + f^*$ with $f^*_{\bar{0}} = 0$, we have

$$\mathcal{T}_\lambda^{0,0} c = \widehat{R}_{\bar{0},\bar{0}}(\lambda)c, \quad (\mathcal{T}_\lambda^{0,1} f^*)_{\bar{0}} = \sum_{\Gamma\neq\widehat{0}} \widehat{R}_{\bar{0},\Gamma}(\lambda)\hat{f}_\Gamma, \tag{3.16a}$$

$$(\mathcal{T}_\lambda^{1,0} c)_\Gamma = \widehat{R}_{\Gamma,\bar{0}}(\lambda)c, \quad (\mathcal{T}_\lambda^{1,1}\hat{f})_\Gamma = \sum_{\Gamma'\neq\bar{0}} \widehat{R}_{\Gamma,\Gamma'}(\lambda)\hat{f}_{\Gamma'}, \qquad \Gamma \neq \bar{0}. \tag{3.16b}$$

Observe that from the explicit formula (3.7b) it follows that

$$\tilde{p}_0(\lambda) = \widehat{R}_{\bar{0},\bar{0}}(\lambda). \tag{3.16c}$$

Formula (3.14b) above implies that the spectrum of the operator $\mathcal{T}_\lambda^0|_{\hat{l}_2^0(\mathfrak{M})}$ is contained in the disk $\{z : |z| \leqslant |\mu_1|\}$ of the complex plane. Since $\tilde{p}_0(0) = 1$, we can find a neighborhood G of 0 in T^ν such that $|\tilde{p}_0(\lambda)| > 1 - \delta$ for $\lambda \in G$, where $\delta = (1 - |\mu_1|)/2$.

Consider the space $l_M(\mathfrak{M})$ of functions $\varphi = \{\varphi_\Gamma, \Gamma \in \mathfrak{M}\}$ on $\mathfrak{M}$, such that

$$\|\varphi\|_{l_M} \equiv \sum_\Gamma |\varphi_\Gamma| M^{|\operatorname{supp}\Gamma|} < \infty, \tag{3.16d}$$

where $M = \max\{\max_{i,s} |e_i(s)|, 2\}$. It is easy to see that

$$\|\varphi\|_{\hat{l}_2(\mathfrak{M})} \leqslant \|\varphi\|_{l_M(\mathfrak{M})}, \tag{3.16e}$$

i.e., $l_M(\mathfrak{M}) \subset \hat{l}_2(\mathfrak{M})$.

LEMMA 3.4. *For any $\lambda \in T^\nu$ the operator $\mathcal{T}_\lambda$ is a bounded linear operator mapping $l_M(\mathfrak{M})$ in itself. Namely there is a constant C independent of λ such that*

$$\|\mathcal{T}_\lambda\|_{l_M(\mathfrak{M})} < C.$$

PROOF. The proof is given in the Appendix. □

REMARK. The assertions of Lemma 3.4 and Lemma 3.5 hold for a sufficiently large $M > 1$, not larger, however, than $c\varepsilon^{-1}$, where $c > 0$ is some absolute constant.

LEMMA 3.5. *For ε_0, ε_1 sufficiently small and any $\lambda \in G$ the operator $\mathcal{T}_\lambda$ has a unique eigenvector $\chi(\lambda) = \{\chi_\Gamma(\lambda), \Gamma \in \mathfrak{M}\} \in l_M(\mathfrak{M})$ with eigenvalue $\tilde{p}(\lambda)$ such that*

$$|\tilde{p}(\lambda)| > 1 - 3\delta/2,$$

and such that $\chi_{\bar{0}}(\lambda) \neq 0$. The vector $\chi(\lambda)$ depends analytically on $\lambda \in G$, and if $\chi(\lambda)$ is normalized by setting $\chi_{\bar{0}}(\lambda) = 1$, then it satisfies the condition

$$\sum_{\Gamma \neq \bar{0}} |\chi_\Gamma(\lambda)| M^{|\operatorname{supp}\Gamma|} < \kappa, \tag{3.17}$$

where $\kappa = \kappa(\varepsilon_0, \varepsilon_1)$ is a small constant independent of λ.

In addition the operator $\mathcal{T}_\lambda$ has an invariant subspace $h_\lambda \subset \hat{l}_2(\mathfrak{M})$, and a decomposition analogous to the decomposition (3.15) for the unperturbed operator $\mathcal{T}_\lambda^0$ takes place:

$$\hat{l}_2(\mathfrak{M}) = \{c\chi(\lambda)\} + h_\lambda. \tag{3.18a}$$

The norm of the operator $\mathcal{T}_\lambda|_{h_\lambda \cap l_M(\mathfrak{M})}$ in the invariant space $h_\lambda \cap l_M(\mathfrak{M})$ satisfies the estimate

$$\|\mathcal{T}_\lambda|_{h_\lambda \cap l_M(\mathfrak{M})}\|_{l_M(\mathfrak{M})} < |\mu_1| + \delta/2. \tag{3.18b}$$

The eigenvalue $\tilde{p}(\lambda)$, $\lambda \in G$, has the following properties:

1) $\tilde{p}(0) = 1$;
2) *$\tilde{p}(\lambda)$ is analytic in the neighborhood G;*
3) *in the Taylor series at $\lambda = 0$*

$$\log \tilde{p}(\lambda) = i \sum_k b_k \lambda_k - \frac{1}{2} \sum_{k,j} a_{k,j} \lambda_k \lambda_j + \dots, \qquad \lambda \in G, \tag{3.19}$$

the real part of the quadratic form $\sum a_{i,j}\lambda_i\lambda_j$ *is positive definite;*
4) $|\widetilde{p}(\lambda)| < 1$ *for* $\lambda \neq 0$.

PROOF. Given formulas (3.13a), (3.16a,b) and the fact that $\chi_{\bar{0}}(\lambda) = 1$, the eigenvalue $\widetilde{p}(\lambda)$ and the eigenfunction $\chi(\lambda)$ satisfy the following relations:

$$\widetilde{p}(\lambda) = \widehat{R}_{\bar{0},\bar{0}}(\lambda) + \sum_{\Gamma\neq\bar{0}} \widehat{R}_{\bar{0},\Gamma}\chi_\Gamma(\lambda), \tag{3.20a}$$

$$\begin{aligned}\widehat{R}_{\Gamma,\bar{0}}(\lambda) + \sum_{\Gamma'\neq\bar{0}} \widehat{R}_{\Gamma,\Gamma'}(\lambda)\chi_{\Gamma'}(\lambda) &= \widetilde{p}(\lambda)\,\chi_\Gamma(\lambda)\\ &= \Big(\widehat{R}_{\bar{0},\bar{0}}(\lambda) + \sum_{\Gamma'\neq\bar{0}} \widehat{R}_{\bar{0},\Gamma'}(\lambda)\chi_{\Gamma'}(\lambda)\Big)\chi_\Gamma(\lambda), \qquad \Gamma\neq\bar{0}.\end{aligned} \tag{3.20b}$$

Hence

$$\begin{aligned}\chi_\Gamma(\lambda) = {}& \frac{1}{\widehat{R}_{\bar{0},\bar{0}}(\lambda)} \sum_{\Gamma'\neq\bar{0}} \widehat{R}_{\Gamma,\Gamma'}(\lambda)\,\chi_{\Gamma'}(\lambda)\\ & - \frac{\chi_\Gamma(\lambda)}{\widehat{R}_{\bar{0},\bar{0}}(\lambda)} \sum_{\Gamma'\neq\bar{0}} \widehat{R}_{\bar{0},\Gamma'}\chi_{\Gamma'}(\lambda) + \frac{\widehat{R}_{\Gamma,\bar{0}}(\lambda)}{\widehat{R}_{\bar{0},\bar{0}}(\lambda)}, \qquad \Gamma\neq\bar{0}.\end{aligned} \tag{3.21}$$

Let us denote by $\mathfrak{T}_\lambda : l_M(\mathfrak{M}) \to l_M(\mathfrak{M})$ the operator given by the right-hand side of equation (3.21). Then for all $\lambda \in G$ there is a positive constant $\kappa(\varepsilon_0, \varepsilon_1)$ such that the sphere $S_\kappa = \{\varphi \in l_M(\mathfrak{M}) \cap \hat{l}_2^0(\mathfrak{M}) : \|\varphi\|_M < \kappa\}$ is mapped into itself by $\mathfrak{T}_\lambda$, and $\mathfrak{T}_\lambda$ is a contraction in S_κ. The proof goes as follows. For any $\chi \in l_M(\mathfrak{M}) \cap \hat{l}_2^0(\mathfrak{M})$ we have

$$\|\mathfrak{T}_\lambda\chi\| \leqslant \frac{1}{|\widehat{R}_{\bar{0},\bar{0}}(\lambda)|}\left(\|\widehat{R}_{\Gamma,\bar{0}}(\lambda)\| + \|R(\lambda)\|\,\|\chi\| + \|\chi\|^2 \max_\Gamma \frac{|\widehat{R}_{\bar{0},\Gamma}|}{M^{|\operatorname{supp}\Gamma|}}\right).$$

Here all norms refer to the space $l_M(\mathfrak{M}) \cap \hat{l}_2^0(\mathfrak{M})$. For instance $R(\lambda)$ denotes the operator in $l_M(\mathfrak{M}) \cap \hat{l}_2^0(\mathfrak{M})$

$$(R(\lambda)\varphi)(\Gamma) = \sum_{\Gamma'\neq\bar{0}} \widehat{R}_{\Gamma,\Gamma'}(\lambda)\,\varphi(\Gamma'), \qquad \Gamma\neq\bar{0},$$

and $\|R(\lambda)\|$ its norm. The following assertions hold for $\varepsilon \equiv \max\{\varepsilon_0, \varepsilon_1\}$ small enough and all $\lambda \in T^\nu$:

$$\|\widehat{R}_{\cdot,\bar{0}}(\lambda)\| < c_0\varepsilon, \tag{3.22a}$$

$$\max_{\Gamma\neq\bar{0}} \frac{|\widehat{R}_{\bar{0},\Gamma}(\lambda)|}{M^{|\operatorname{supp}\Gamma|}} < c_1\varepsilon, \tag{3.22b}$$

$$\|R(\lambda)\| < |\mu_1| + \delta/6. \tag{3.22c}$$

where c_0, c_1 are constants independent of ε. These estimates follow easily from the explicit expressions (3.13b) for the matrix elements of the operator

$R(\lambda)$. Moreover, by the definition of the interval G for $\lambda \in G$ and by (3.16c) we have $|\widehat{R}_{\bar{0},\bar{0}}(\lambda)| > 1-\delta$.

Hence we see that, for ε small enough, there is some $\kappa > 0$ so small that the sphere S_κ is transformed by $\mathfrak{T}_\lambda$ into itself. In fact, from 1)–4) it follows that

$$\|\mathfrak{T}_\lambda \chi\| \leqslant \frac{1}{1-\delta}\left(c_0\varepsilon + \left(|\mu_1| + \frac{\delta}{6}\right)\|\chi\| + c_1\varepsilon\|\chi\|^2\right).$$

Since

$$\frac{|\mu_1| + \delta/6}{1-\delta} \equiv g < 1$$

for a sufficiently small ε, one can find κ such that $(c_0\varepsilon + g\kappa + c_1\varepsilon\kappa^2) < \kappa$, and, consequently, $\mathfrak{T}_\lambda S_\kappa \subset S_\kappa$. Similar considerations show that $\mathfrak{T}_\lambda$ is contractive in S_κ. Hence, the equation (3.21) has a unique solution in S_κ. Since the right-hand side of (3.21) depends analytically on the parameter $\lambda \in G$, the solution $\chi_\Gamma(\lambda)$ and the eigenvalue

$$\widetilde{p}(\lambda) = \widehat{R}_{\bar{0},\bar{0}}(\lambda) + \sum \widehat{R}_{\bar{0},\Gamma}\chi_\Gamma(\lambda)$$

have the same property. The estimate from below for $\widetilde{p}(\lambda)$ also follows.

Observe that the representation (3.12b) implies that the space $\hat{l}_2(\mathfrak{M}\times\mathbb{Z}^\nu)$, as well as the space $\mathscr{H}$, which is isomorphic to it, can be decomposed as direct integrals

$$\hat{l}_2(\mathfrak{M}\times\mathbb{Z}^\nu) = \int_{T^\nu} \hat{l}_2^\lambda\, d\lambda, \qquad \mathscr{H} = \int_{T^\nu} \mathscr{H}_\lambda\, d\lambda, \tag{3.23}$$

so that the operators $\mathscr{T}$ and $\mathscr{U}_v$ in $\mathscr{H}$ have the form

$$\mathscr{T} = \int_{T^\nu} \mathscr{T}_\lambda\, d\lambda, \qquad \mathscr{U}_v = \int_{T^\nu} e^{i(v,\lambda)} E_\lambda\, d\lambda$$

(E_λ the identity operator in $\mathscr{H}_\lambda$). The space $\hat{l}_2^\lambda$ consists of functions of the type

$$\{\hat{f}_{\Gamma-z} e^{-i(\lambda,z)}\}, \qquad \hat{f} \in \hat{l}_2(\mathfrak{M}), \tag{3.24}$$

and correspondingly $\mathscr{H}_\lambda$ consists of functionals of the type

$$\Phi(\xi, x) = \sum_{\Gamma, z} \hat{f}_{\Gamma-z} e^{-i(\lambda,z)} \Psi_{\Gamma,z} = e^{-i(\lambda,x)} \sum_{\Gamma} \hat{f}_{\Gamma-x}\Psi_\Gamma(\xi), \tag{3.25}$$

where the Ψ_Γ's are defined by (3.3b). For $\hat{f}_\Gamma = \chi_\Gamma(\lambda)$ we set

$$\psi^\lambda(\xi, x) = e^{-i(\lambda,x)} \sum_\Gamma \chi_{\Gamma-x}(\lambda)\Psi_\Gamma(\xi). \tag{3.26a}$$

The estimate (3.17) shows that $\psi^\lambda \in C_\infty(\widehat{\Omega})$. Now for any $\Phi \in \mathscr{H}_\lambda$ given by (3.25) we have

$$\mathscr{T}\Phi = \sum_{\Gamma, z} e^{-i(\lambda,z)} (\mathscr{T}_\lambda \hat{f})_{\Gamma-z} \Psi_{\Gamma,z}. \tag{3.26b}$$

The function $\Phi^*(\xi, x) = \Psi_{\bar{0}} \equiv 1$ is in $\mathscr{H}_0$ and is represented in the form (3.25) for $\hat{f} = f^*$: $f^*_{\bar{0}} = 1$, $f^*_\Gamma = 0$ for $\Gamma \neq \bar{0}$. Since $\Psi_{\bar{0}}$ is an eigenvector of the operator $\mathscr{T}$ with eigenvalue 1, we have

$$1 = \mathscr{T}\Phi^*(\xi, x) = \sum_\Gamma (\mathscr{T}_0 f^*)_{\Gamma - x}\Psi_\Gamma(\xi) = \sum_\Gamma f^*_{\Gamma - x}\Psi_\Gamma(\xi).$$

Since the vectors Ψ_Γ form a basis, we see that $f^* \in \hat{l}_2(\mathfrak{M})$ is an eigenvector of $\mathscr{T}_0$ with eigenvalue 1. But we have shown that the eigenvector of $\mathscr{T}_\lambda$ with $\bar{0}$-component equal to 1 is unique and corresponds to the eigenvalue $\widetilde{p}(\lambda)$. Hence $\widetilde{p}(0) = 1$ and $\chi_\Gamma(0) = 0$ if $\Gamma \neq \bar{0}$.

We prove further that

$$|\widetilde{p}(\lambda)| \leqslant 1. \tag{3.27}$$

Indeed if $|\widetilde{p}(\lambda_0)| > 1$ for some $\lambda_0 \in T^\nu$, then, since ψ^λ is an eigenvector of $\mathscr{T}$ with eigenvalue $\widetilde{p}(\lambda)$, as follows from (3.26b), we would have

$$\|\mathscr{T}\psi^{\lambda_0}\|_{C_\infty} = |\widetilde{p}(\lambda_0)|\,\|\psi^{\lambda_0}\|_{C_\infty} > \|\psi^{\lambda_0}\|_{C_\infty},$$

which contradicts the assertion of Lemma 3.1.

From inequality (3.27) it follows that in the expansion (3.19) the coefficients b_k, $k = 1, \dots, \nu$, are real, and the real part of the quadratic form is nonnegative, $\operatorname{Re}\sum a_{ij}\lambda_i\lambda_j \geqslant 0$.

Furthermore, by conditions (2.3c), (2.5b) this quadratic form is a small perturbation of the quadratic form (2.3g), which is positive definite. Hence it is also positive definite. It follows that in a sufficiently small neighborhood of the point $\lambda = 0$ we have $|\widetilde{p}(\lambda)| < 1$ for $\lambda \neq 0$. Outside this interval condition 4) for $\widetilde{p}(\lambda)$ follows from (2.3f).

All the required properties of the function $\widetilde{p}(\lambda)$, $\lambda \in G$, have been proved.

We now prove the existence of the invariant subspace h_λ. Let us consider the operator $\mathscr{T}^*$ in $\mathscr{H}$. It is convenient to consider its matrix elements in the basis $\Psi^*_{\Gamma, z}$, biorthogonal to the basis $\Psi_{\Gamma, z}$, defined by (3.3):

$$\Psi^*_{\Gamma, z}(\xi, x) = \prod_{y \in \mathbb{Z}^\nu} e^*_{\gamma(y)}(\xi(y))\,\delta_{z, x}, \tag{3.28}$$

where e^*_i are the eigenvectors of the operator (2.6a) in $l_2(S, \pi_0)$. As before, the matrix elements $R^*_{\Gamma, z; \Gamma', z'}$ of the operator $\mathscr{T}^*$ in the basis $\Psi^*_{\Gamma, z}$ have the form shown in (3.7a,b).

We return to the space $\mathscr{H}$ and to the isomorphic space $\hat{l}^*_2 = \hat{l}^*_2(\mathfrak{M}) \otimes l_2(\mathbb{Z}^\nu)$, where $\hat{l}^*_2(\mathfrak{M})$ is the space of the functions on $\mathfrak{M}$ with scalar product of the form (3.6b), in which the matrix $(e_n, e_{n'})$ is replaced by the matrix $(e^*_n, e^*_{n'}) = \sum_s e^*_n(s)e^*_{n'}(s)\pi_0(s)$.

The space $\hat{l}^*_2(\mathfrak{M} \times \mathbb{Z}^\nu)$, as well as the space $\hat{l}_2(\mathfrak{M} \times \mathbb{Z}^\nu)$, can be represented as a direct integral of the spaces $\hat{l}^{*,\lambda}_2$ consisting of functions of the type (3.26a) (with only a change of the scalar product). The operator $\mathscr{T}^*$ is represented as

$$\mathscr{T}^* = \int_{T^\nu} \mathscr{T}_\lambda^* \, d\lambda .$$

Under the canonical identification of the spaces $\hat{l}_2^{*,\lambda}$ with the spaces $\mathscr{H}_\lambda$ (see above) the operators $\{\mathscr{T}_\lambda^*\}$ are transformed into the operators $\{(\mathscr{T}_\lambda)^*\}$ adjoint to the operators $\mathscr{T}_\lambda$. On the other hand, by applying all the considerations above to the operator $(\mathscr{T}_\lambda)^*$ for $\lambda \in G$, we see that it has a (unique) eigenvector $\chi^*(\lambda)$ (with eigenvalue $\overline{\widetilde{p}(\lambda)}$). We denote by h_λ the orthogonal complement of $\chi^*(\lambda)$. Clearly h_λ is an invariant subspace with respect to $\mathscr{T}_\lambda$.

We now prove assertion (3.18a). Since $\chi_{\bar{0}}(\lambda) \equiv 1$, any vector $w \in h_\lambda$ has the form

$$w = u + c(u)\chi(\lambda), \qquad u \in \hat{l}_2^0(\mathfrak{M}),\ c(u) = -\frac{(u\,,\, \chi^*(\lambda))}{(\chi(\lambda)\,,\, \chi^*(\lambda))} \tag{3.29}$$

(inner products are taken in the space $\hat{l}_2(\mathfrak{M})$). Moreover, if $w \in h_\lambda \cap l_M(\mathfrak{M})$, then $u \in l_2^0(\mathfrak{M}) \cap l_M(\mathfrak{M})$. Furthermore $\mathscr{T}_\lambda w = \mathscr{T}_\lambda u + c(u)\widetilde{p}(\lambda)\chi(\lambda)$, so that

$$\|\mathscr{T}_\lambda w\|_{l_M} < \|\mathscr{T}_\lambda u\|_{l_M} + |c(u)|\,|\widetilde{p}(\lambda)|\,\|\chi(\lambda)\|_{l_M}.$$

Moreover, by the estimates 1)–4) for the coefficients $\widehat{R}_{\Gamma,\Gamma'}(\lambda)$ and estimate (3.17), we have that $|c(u)| < \widetilde{\delta}(\varepsilon)\|u\|_{l_M(\mathfrak{M})}$ ($\widetilde{\delta}(\varepsilon)$ is small for small ε) and

$$\|\mathscr{T}_\lambda w\|_{l_M(\mathfrak{M})} < (|\mu_1| + \delta/3)\|u\|_{l_M(\mathfrak{M})}.$$

From (3.29) we find that $\|u\| < \|w\|(1 - \widetilde{\delta}(\varepsilon)\|\chi(\lambda)\|_{l_M})^{-1}$, whence it is easy to see that the inequality (3.18b) follows.

Finally, for any vector $q \in \mathscr{H}_\lambda$ we have

$$q = C\chi(\lambda) + w, \tag{3.30a}$$

where

$$C = \frac{(q\,,\, \chi^*(\lambda))_{l_2(\mathfrak{M})}}{(\chi(\lambda)\,,\, \chi^*(\lambda))_{l_2(\mathfrak{M})}} \tag{3.30b}$$

and $w \in h_\lambda$. Here $l_2(\mathfrak{M})$ is the usual l_2 space on $\mathfrak{M}$, and we mean that the components of q, $\chi(\lambda)$ are taken with respect to the basis $\{\Psi_\Gamma\}$, whereas the components of $\chi^*(\lambda)$ are taken with respect to the basis $\{\Psi_\Gamma^*\}$. This is the meaning of (3.18a). The lemma is proved. □

LEMMA 3.6. *The norm* $\|\mathscr{T}_\lambda\|_{l_M(\mathfrak{M})}$ *for all* $\lambda \in T^\nu \setminus G$ *satisfies the estimate*

$$\|\mathscr{T}_\lambda\|_{l_M(\mathfrak{M})} < (1 - \delta/2). \tag{3.31}$$

PROOF. From (3.22b), (3.22c), and formula (3.16c) we see that for $\varphi \in l_M(\mathfrak{M})$ we have

$$\begin{aligned}\|\mathcal{T}_\lambda \varphi\|_{l_M(\mathfrak{M})} &< |\widehat{R}_{\bar{0},\bar{0}}(\lambda)||\varphi_{\bar{0}}| + \sum_\Gamma |\widehat{R}_{\bar{0},\Gamma}||\varphi_\Gamma| + \sum_{\Gamma\neq\bar{0}} \left|\sum_{\Gamma'} \widehat{R}_{\Gamma,\Gamma'}\varphi_{\Gamma'}\right| M^{|\operatorname{supp}\Gamma|} \\ &\leqslant \max\left\{\max_{\lambda\notin G} |\widetilde{p}_0(\lambda)|,\ c_1\varepsilon_1 + \mu_1 + \delta/6\right\}\|\varphi\|_{l_M(\mathfrak{M})} \\ &< (1-\delta/2)\|\varphi\|_{l_M(\mathfrak{M})}. \quad \square \qquad (3.32)\end{aligned}$$

COROLLARY. *Let the functional* $\varphi(\xi, x) \in \mathcal{H}$ *be such that under the canonical isomorphism* $\mathcal{H} \to \hat{l}_2(\mathfrak{M}) \otimes l_2(\mathbb{Z}^\nu) \to l_2(\mathfrak{M}) \otimes L_2(T^\nu, d\lambda)$ *it goes into a function* $\varphi(\lambda)$ *such that for all* λ

$$\varphi(\lambda) \in l_M(\mathfrak{M}), \quad \text{and} \quad \varphi(\lambda) \in h_\lambda \cap l_M(\mathfrak{M}) \quad \text{for } \lambda \in G,$$

and

$$\int_{T^\nu} \|\varphi(\lambda)\|_{l_M(\mathfrak{M})}\, d\lambda < \infty.$$

Then

$$\sup_{\xi, x} |(\mathcal{T}^t\varphi)(\xi, x)| < (1-\delta/2)^t \int_{T^\nu} \|\varphi(\lambda)\|_{l_M(\mathfrak{M})}\, d\lambda. \qquad (3.33)$$

PROOF. From formulas (3.3a), (3.12b) we get

$$\varphi(\xi, x) = \sum_{\Gamma, z} \Psi_{\Gamma, z} \int \varphi_{\Gamma - z}(\lambda) e^{-i(\lambda, z)} d\lambda = \sum_\Gamma \Psi_\Gamma \int \varphi_{\Gamma - x}(\lambda) e^{-i(\lambda, x)} d\lambda.$$

Hence

$$(\mathcal{T}^t\varphi)(\xi, x) = \sum_\Gamma \Psi_\Gamma(\xi) \int_{T^\nu} (\mathcal{T}_\lambda^t \varphi(\lambda))_{\Gamma - x} e^{-i(\lambda, x)} d\lambda.$$

Since δ is smaller than $|\mu_1| + \delta/2 < 1 - \delta/2$, inequality (3.18b) and Lemma 3.6 imply that

$$\begin{aligned}|(\mathcal{T}^t\varphi)(\xi, x)| &< \int_{T^\nu} \sum_\Gamma M^{\operatorname{supp}\Gamma} |(\mathcal{T}_\lambda^t \varphi(\lambda))_{\Gamma - x}|\, d\lambda \\ &= \int \|\mathcal{T}_\lambda^t \varphi(\lambda)\|_{l_M(\mathfrak{M})} d\lambda < (1-\delta/2)^t \int_{T^\nu} \|\varphi(\lambda)\|_{l_M(\mathfrak{M})} d\lambda. \quad \square \qquad (3.34)\end{aligned}$$

Now we compute the asymptotic behavior of the probability $\mathbb{P}(X_t = x \mid x_0 = z)$.

Observe that $\Phi_y^0(\xi, x) = \delta_{x,y} = \Psi_{\bar{0},y}(\xi, x)$ and therefore its components in the basis $\Psi_{\Gamma,z}$ are given by $\hat{f}^{(y)} \in \hat{l}_2(\mathfrak{M} \times \mathbb{Z}^\nu)$:

$$\hat{f}^{(y)}_{\Gamma, z} = \delta_{\Gamma, \bar{0}} \delta_{x, y}.$$

Its Fourier transform (given by (3.12a)) is equal to

$$\varphi_\Gamma^{(y)}(\lambda) = \delta_{\Gamma, \bar{0}} e^{i(\lambda, y)}. \qquad (3.35)$$

Consider the partition of unity in T^ν

$$S_1(\lambda) + S_2(\lambda) = 1,$$

where S_1, S_2 are smooth functions such that $\operatorname{supp} S_1 \subset G$, and $\operatorname{supp} S_2 \subset T^\nu \setminus G^\delta$, where G^δ is obtained from G by throwing away a "small piece" near the boundary. By the representation (3.30a), we have, for any $\lambda \in G$,

$$\delta_{\Gamma,\bar{0}} = C(\lambda)\chi_\Gamma(\lambda) + \delta_\Gamma(\lambda), \tag{3.36}$$

where $\delta(\lambda) \in h_\lambda \cap l_M(\mathfrak{M})$ (because both $\delta_{\Gamma,\bar{0}}$ and $\chi(\lambda)$ are in $l_M(\mathfrak{M})$), and $\|\delta(\lambda)\|_{l_M(\mathfrak{M})}$ is uniformly bounded. As one can see from its expression (3.30b), the function $C(\lambda)$ is smooth for all $\lambda \in G$. The function $\varphi^{(y)}(\lambda)$ can be decomposed as $\varphi^{(y)}(\lambda) = \varphi^{(y),1}(\lambda) + \varphi^{(y),2}(\lambda)$ with

$$\begin{aligned} \varphi^{(y),1}(\lambda) &= S_1(\lambda)\, C(\lambda)\, \chi(\lambda)\, e^{i(\lambda, y)}, \\ \varphi^{(y),2}(\lambda) &= (S_2(\lambda)\delta_{\bar{0},\Gamma} + S_1(\lambda)\delta(\lambda))\, e^{i(\lambda, y)}, \end{aligned} \tag{3.37a}$$

and correspondingly $\Phi^0_y(\xi, x)$ is written as

$$\Phi^0_y(\xi, x) = \Phi^0_{y,1}(\xi, x) + \Phi^0_{y,2}(\xi, x). \tag{3.37b}$$

As a consequence of the corollary to Lemma 3.6 we have

$$|\mathscr{T}^t\Phi^0_{y,2}(\xi, x)| < \operatorname{const}(1 - \delta/2)^t, \tag{3.38}$$

and the constant does not depend on y, x and t. Furthermore

$$\begin{aligned} (\mathscr{T}^t\Phi^0_{y,1})(\xi, x) &= \sum_\Gamma \Psi_\Gamma(\xi) \int_{T^\nu} (\tilde{p}(\lambda))^t C(\lambda) S_1(\lambda) \chi_{\Gamma-x}(\lambda) e^{i(\lambda, y-x)} d\lambda \\ &= \int_{T^\nu} (\tilde{p}(\lambda))^t C(\lambda) S_1(\lambda) e^{i(\lambda, y-x)} g(\lambda, x, \xi)\, d\lambda, \end{aligned} \tag{3.39a}$$

where

$$g(\lambda, x, \xi) = \sum_\Gamma \chi_{\Gamma-x}(\lambda)\Psi_\Gamma(\xi), \tag{3.39b}$$

and the series converges by the estimate (3.17). Taking the average on ξ with respect to the initial distribution Π, we find that the leading term of the asymptotics is given by the integral

$$\int_G (\tilde{p}(\lambda))^t C(\lambda) S_1(\lambda) e^{i(\lambda, y-x)} \hat{g}(\lambda, x)\, d\lambda, \tag{3.40}$$

$$\hat{g}(\lambda, x) = \int g(\lambda, x, \xi)\, d\Pi(\xi),$$

and the function $\hat{g}(\lambda, x)$ is bounded and analytic in λ. Now by applying to the integral (3.40) the Laplace method (assuming that $|y - u - bt| < t^{1/2+1/6}$, see [3] or [4] for details) we find the asymptotics

$$\left[\frac{D(\operatorname{Det} A)^{-1/2}}{\sqrt{2\pi t}} \exp\left\{-\frac{1}{2t}(A^{-1}(y - u - bt), y - u - bt)\right\}\right](1 + o(1)), \tag{3.41}$$

where $D = C(0)\hat{g}(0, x)$. But, since $\chi_\Gamma(0) = \delta_{\Gamma,\bar{0}}$, we have $C(0) = 1$ and $\hat{g}(0, x) = 1$, i.e., $D = 1$. Using the fact that the expression (3.41) is the asymptotics of a probability, it is not hard to see that A is a real matrix, and therefore by Lemma 3.5, assertion 3) on the expansion of $\widetilde{p}(\lambda)$, $(A^{-1}\eta, \eta)$ is a positive definite quadratic form.

The theorem is proved.

We give a further result, useful for what follows, that concerns the case of a symmetric random walk. We say that the random walk introduced in §2 is *symmetric*, if the following condition is satisfied:

$$\begin{aligned}\mathbb{P}(X_t = z\,,\ \xi_t \in A \mid X_{t-1} = x\,,\ \xi_{t-1} = \eta) \\ = \mathbb{P}(X_t = -z\,,\ \xi_t = \mathscr{V}A \mid X_{t-1} = -x\,,\ \xi_{t-1} = \mathscr{V}\eta),\end{aligned} \tag{3.42a}$$

where $\mathscr{V}$ is the spatial reflection of the configurations

$$(\mathscr{V}\xi)(x) = \xi(-x). \tag{3.42b}$$

The fact that condition (3.42a,b) is satisfied for all ε in some interval is equivalent to the condition

$$P_0(u) = P_0(-u), \quad c(u, s) = c(-u, s), \qquad s \in S,\ u \in \mathbb{Z}^\nu.$$

LEMMA 3.7. *For a symmetric random walk, the function $\widetilde{p}(\lambda)$ is an even real function of $\lambda \in T^\nu$.*

PROOF. Consider the following antilinear transformation (reflection) $\mathscr{P}$ on the space of functions on $\widehat{\Omega}$:

$$(\mathscr{P}f)(\xi, x) = \overline{f(\mathscr{V}\xi, x)}.$$

The symmetry condition (3.42a,b) implies that $\mathscr{T}$ commutes with $\mathscr{P}$, and moreover

$$\mathscr{P}\mathscr{U}_v = \mathscr{U}_{-v}\mathscr{P}, \qquad v \in \mathbb{Z}^\nu,$$

where $\mathscr{U}_v$ is, as above, the translation operator. It follows that $\mathscr{H}_\lambda$, which is an eigenspace of the group $\{\mathscr{U}_v : v \in \mathbb{Z}^\nu\}$, is transformed by $\mathscr{P}$ into itself. Moreover, since $\mathscr{P}$ and $\mathscr{T}$ commute, the vector $\mathscr{P}\chi(\lambda) \in \mathscr{H}_\lambda$ is an eigenvector of $\mathscr{T}_\lambda$ with the eigenvalue $\widetilde{p}(\lambda)$. Since, by Lemma 3.5, the eigenvector of $\mathscr{T}_\lambda$ with eigenvalue of maximum absolute value is unique, it follows that $\mathscr{P}\chi(\lambda) = \alpha\chi(\lambda)$, for some α with $|\alpha| = 1$, and $\widetilde{p}(\lambda) = \overline{\widetilde{p}(\lambda)}$. If we now consider the linear transformation $\widetilde{\mathscr{P}}$:

$$(\widetilde{\mathscr{P}}f)(\xi, x) = f(\mathscr{V}, -x),$$

then, by repeating the above arguments, we find that

$$\widetilde{\mathscr{P}}\mathscr{H}_\lambda = \mathscr{H}_{-\lambda}, \qquad \widetilde{\mathscr{P}}\chi(\lambda) = \pm\chi(-\lambda),$$

which implies, as it is easy to see, $\widetilde{p}(\lambda) = \widetilde{p}(-\lambda)$. The lemma is proved.

§4. Random walk of two particles (proof of Theorem 2). Construction of the invariant two-particle subspace

The outline of the proof of the central limit theorem for the two-particle random walk repeats in part some steps in the proof of Theorem 1, but is somewhat more complicated.

As above, we introduce three function spaces: 1) $\mathscr{C}_\infty(\widehat{\Omega})$, the space of bounded functions $\varphi(\xi, x_1, x_2)$ continuous in $\xi \in \Omega = S^{\mathbb{Z}^\nu}$ for any $x_1, x_2 \in \mathbb{Z}^\nu$; 2) the space $\mathscr{H}^2 \equiv L_2(\Omega, \Pi_0) \otimes l_2(\mathbb{Z}^\nu \otimes \mathbb{Z}^\nu)$; and finally 3) the space $\mathscr{C}_2(\widehat{\Omega}) = \mathscr{C}(\Omega) \otimes \mathscr{S}(\mathbb{Z}^\nu \otimes \mathbb{Z}^\nu)$ of functions $\varphi(\xi, x_1, x_2)$ continuous in ξ and decreasing at infinity faster than any power of $(|x_1| + |x_2|)^{-1}$. All three spaces are invariant with respect to the operator $\mathscr{T}^{(2)}$ and with respect to the group of operators $\{\mathscr{U}_v, v \in \mathbb{Z}^\nu\}$ generated by the shifts

$$(\xi, x_1, x_2) \to (\xi + v, x_1 + v, x_2 + v),$$

where $\xi + v$ is the configuration shifted by v. On each of the three spaces $\mathscr{T}^{(2)}$ is continuous and moreover $\|\mathscr{T}^{(2)}\|_{\mathscr{C}_\infty(\widehat{\Omega})} = 1$ (see Lemma 3.1).

The operators $\{\mathscr{U}_v, v \in \mathbb{Z}^\nu\}$ act unitarily on $\mathscr{H}^2$.

Furthermore, in the space $\mathscr{H}^2$ we introduce the basis

$$\Psi_{\Gamma, z_1, z_2}(\xi, x_1, x_2) = \Psi_\Gamma(\xi)\delta_{x_1, z_1}\delta_{x_2, z_2}, \tag{4.1}$$

where $\Psi_\Gamma(\xi)$ is defined by (3.3b). Any function $\varphi \in \mathscr{H}^2$ has a unique expansion

$$\varphi = \sum_{\Gamma, z_1, z_2} f_{\Gamma, z_1, z_2} \Psi_{\Gamma, z_1, z_2} \tag{4.2}$$

and expansion (4.2) generates, as before, an isomorphism $\mathscr{H}^2 \mapsto \hat{l}_2^{(2)}$, the Hilbert space $\hat{l}_2^{(2)} \equiv \hat{l}_2(\mathfrak{M} \times \mathbb{Z}^\nu \times \mathbb{Z}^\nu)$ being the space of functions $f = \{f_{\Gamma, z_1, z_2}\}$, $\Gamma \in \mathfrak{M}$, $z_1, z_2 \in \mathbb{Z}^\nu$, with the norm

$$\|f\|^2_{\hat{l}_2^{(2)}} = \sum_{\Gamma, \Gamma', z_1, z_2} a_{\Gamma, \Gamma'} f_{\Gamma, z_1, z_2} \bar{f}_{\Gamma', z_1, z_2}, \tag{4.3}$$

and $a_{\Gamma, \Gamma'}$ are defined by the formula (3.4).

The operator $\mathscr{T}^{(2)}$ acts on $\hat{l}_2^{(2)}$ as follows:

$$\mathscr{T}^{(2)} f_{\Gamma, z_1, z_2} = \sum_{\Gamma', z_1', z_2'} R^{(2)}_{\Gamma, z_1, z_2; \Gamma', z_1', z_2'} f_{\Gamma', z_1', z_2'}, \tag{4.4a}$$

and the matrix elements $R^{(2)}_{\Gamma, z_1, z_2; \Gamma', z_1', z_2'}$ are given by the following formulas. For $z_1 \neq z_2$

$$R^{(2)}_{\Gamma, z_1, z_2; \Gamma', z_1', z_2'} = \mu_{\Gamma_{z_1, z_2}} \delta_{\Gamma_{z_1, z_2}, \Gamma'_{z_1, z_2}} \times L_2(\gamma(z_1), \gamma'(z_1), \gamma(z_2), \gamma'(z_2), z_1 - z_1', z_2 - z_2', z_1 - z_2), \tag{4.4b}$$

where Γ_{z_1, z_2} is a multi-index which coincides with Γ everywhere except perhaps at the points z_1 and z_2, where it takes the value 0, while L_2 is given by

$$L_2(j_1, j_1', j_2, j_2', u_1, u_2, v) = L_1(j_1, j_1', u_1; v) L_1(j_2, j_2', u_2; v), \tag{4.4c}$$

L_1 being given by formula (3.7b) with the coefficients $c_j(u)$ replaced by the coefficients $\widehat{c}_j(u, v) = c_j(u) + c_j'(u, v)$, obtained by the expansion $\widehat{c}(u, v, s) = \sum_j \widehat{c}_j(u, v) e_j(s)$. Hence we see that

$$L_1(j, j', u; v) = L(j, j', u) + L^*(j, j', u; v), \tag{4.4d}$$

where L is the function in formula (3.7b) and L^* is the function L computed by replacing $P_0(u) + c(u, s)$ with $c_1(u, v, s)$ (hence it vanishes for $|v| > R$). Bringing together formulas (4.4c–d), we find that for $v \neq 0$

$$\begin{aligned} L_2(j_1, j_1', j_2, j_2', u_1, u_2, v) \\ = L(j_1, j_1', u_1) L(j_2, j_2', u_2) + \mathscr{L}(j_1, j_1', j_2, j_2', u_1, u_2, v), \end{aligned} \tag{4.5}$$

where

$$\mathscr{L}(j_1, j_1', j_2, j_2', u_1, u_2, v) = 0 \quad \text{for } \max\{|u_1|, |u_2|, |v|\} > D,$$

D being the diameter of the support of the functions P_0, c, and $\widehat{c}$ (see (2.3d)). For $z_1 = z_2 = z$ we have

$$R^{(2)}_{\Gamma, z, z; \Gamma', z_1', z_2'} = \mu_{\Gamma_z} \delta_{\Gamma_z, \Gamma_z'} \widehat{L}_2(\gamma(z), \gamma'(z); z - z_1', z - z_2'), \tag{4.5$'$}$$

where $\widehat{L}_2(j, j'; u_1, u_2)$ can be written explicitly as an expression analogous to the one giving $L(j, j'; u)$ in (3.7b).

Note that in the unperturbed case (independent random walk of two particles) the matrix elements of the transfer matrix $\mathscr{T}^{(2)}$ are

$$R^{(2), 0}_{\Gamma, z_1, z_2; \Gamma', z_1', z_2'} = \mu_\Gamma \delta_{\Gamma, \Gamma'} P_0(z_1 - z_1') P_0(z_2 - z_2'). \tag{4.6}$$

As follows from (4.4b), the action of the operator $\mathscr{T}^{(2)}$ on the function $f = \{f_{\Gamma, z_1, z_2}\}$ can be written as

$$\begin{aligned} (\mathscr{T}^{(2)} f)_{\Gamma, z_1, z_2} \\ = \mu_{\Gamma_{z_1, z_2}} \sum_{\substack{j_1'(z_1), j_2'(z_2) \\ z_1', z_2'}} L_2(\gamma(z_1), j_1'(z_1), \gamma(z_2), j_2'(z_2), z_1 - z_1', z_2 - z_2', z_1 - z_2) \\ \times f_{\Gamma_{z_1, z_2} \cup j'(z_1) \cup j'(z_2), z_1', z_2'} \end{aligned} \tag{4.7a}$$

for $z_1 \neq z_2$. For $z_1 = z_2 = z$

$$(\mathscr{T}^{(2)} f)_{\Gamma, z, z} = \mu_{\Gamma_z} \sum_{j'(z), z_1', z_2'} \widehat{L}_2(\gamma(z), j'(z), z - z_1', z - z_2') f_{\Gamma_z \cup j'(z), z_1', z_2'}. \tag{4.7b}$$

Here $\Gamma_{z_1,z_2} \cup j'(z_1) \cup j'(z_2)$ ($\equiv \Gamma'$) denotes the multi-index that coincides with Γ at all points, except, possibly, z_1, z_2, and at these points it takes the values $j_1'(z_1)$, $j_2'(z_2)$, respectively. The multi-index $\Gamma_z \cup j'(z)$ has a similar meaning.

For what follows we need some additional considerations of the one-particle random walk studied in the preceding section. To avoid confusion, we denote the transfer matrix for the one-particle case by $\mathscr{T}^{(1)}$. Note that by condition (2.13), i.e., $|\widetilde{p}_0(\lambda)|^2 > |\mu_1|$, the region $G \subset T^\nu$ introduced in the preceding section coincides with the whole torus T^ν, i.e., the eigenvectors of the operator $\mathscr{T}_\lambda$ $(= \mathscr{T}_\lambda^{(1)})$ exist for all $\lambda \in T^\nu$ and the function $\widetilde{p}(\lambda)$ is defined on the whole torus T^ν. As we saw above, the vector $\chi(\lambda)$ and the function $\widetilde{p}(\lambda)$ can be extended analytically in the complex region W_d of the torus T^ν:

$$W_d = \{\lambda \equiv (\lambda_1, \dots, \lambda_\nu) : -\pi \leqslant \operatorname{Re}\lambda_k \leqslant \pi,\ |\operatorname{Im}\lambda_k| < d,\ k = 1, \dots, \nu\}. \tag{4.8a}$$

The subspace $\mathscr{H}_1 \subset \hat{l}_2(\mathfrak{M} \times \mathbb{Z}^\nu)$ of elements of type

$$f_{\Gamma,z} = f^{\varphi}_{\Gamma,z} = \int_{T^\nu} \varphi(\lambda)\chi_{\Gamma-z}(\lambda)e^{-i(\lambda,z)}\,d\lambda, \tag{4.8b}$$

where $\varphi(\lambda) \in L_2(T^\nu, d\lambda)$, is clearly an invariant subspace for the operator $\mathscr{T}^{(1)}$, and for the group of operators $\{U_v, v \in \mathbb{Z}^\nu\}$. The action of the operators $\mathscr{T}^{(1)}$ and $\mathscr{U}_v$, $v \in \mathbb{Z}^\nu$, on f^φ reduces to the multiplication of the function φ by $\widetilde{p}(\lambda)$ and $e^{-i(\lambda,v)}$ respectively.

We introduce in $\mathscr{H}_1$ a basis $\{h^u, u \in \mathbb{Z}^\nu\}$ by taking the function φ in (4.8b) equal to

$$\varphi(\lambda) = \varphi^{(u)}(\lambda) = e^{i(\lambda,u)}, \tag{4.9a}$$

i.e.,

$$h^u_{\Gamma,z} = \int_{T^\nu} e^{i(\lambda,u-z)}\chi_{\Gamma-z}(\lambda)\,d\lambda. \tag{4.9b}$$

These functions form a complete system in $\mathscr{H}_1$, and any element $f \in \mathscr{H}_1$ has a unique expansion $f = \sum_u c_u h^{(u)}$. Moreover

$$\mathscr{U}_v h^u = h^{u+v}, \quad \mathscr{T} h^u = \sum_v r(u-v)h^v, \quad r(x) = \int_{T^\nu} \widetilde{p}(\lambda)e^{i(\lambda,x)}\,d\lambda. \tag{4.10a}$$

Since $\widetilde{p}(\lambda)$ is analytic in the region W_d, the function $r(x)$ decays exponentially for $x \to \infty$:

$$|r(x)| < Ce^{-d|x|}, \tag{4.10b}$$

where $C > 0$ is a constant.

The invariant subspace $\mathscr{H}_1$ is usually called the *one-particle* (*one-cluster*) *subspace of the transfer matrix* (see [2]). In this section we shall construct an analogous two-particle invariant subspace for the operator $\mathscr{T}^{(2)}$ and the group $\{\mathscr{U}_v, v \in \mathbb{Z}^\nu\}$. Namely, we shall prove the following theorem.

THEOREM 4.1. *If condition* (2.13) *is satisfied and* ε_0, ε_1 *are small enough, the space* $\mathscr{H}^2$ *can be represented as the direct sum*

$$\mathscr{H}^2 = \mathscr{H}_2 \oplus \overline{\mathscr{H}}_2$$

of two invariant subspaces with respect to $\mathscr{T}^{(2)}$ *and to the group* $\{\mathscr{U}_v, v \in \mathbb{Z}^\nu\}$. *In the space* $\mathscr{H}_2$, *one can construct a basis* $\{\psi_{u_1,u_2}, u_1, u_2 \in \mathbb{Z}^\nu\}$ *such that*

$$\mathscr{U}_v \psi_{u_1,u_2} = \psi_{u_1+v,u_2+v},$$

$$\mathscr{T}^{(2)} \psi_{u_1,u_2} = \sum_{v_1,v_2} [r(u_1 - v_1) r(u_2 - v_2) + D(v_1, v_2, u_1, u_2)] \psi_{v_1,v_2}, \tag{4.10c}$$

where $r(\cdot)$ *is the same function as in* (4.10a,b) *and* $D(v_1, v_2, u_1, u_2)$ *is the matrix satisfying the condition*

$$D(v_1 + v, v_2 + v, u_1 + v, u_2 + v) = D(v_1, v_2, u_1, u_2), \tag{4.11a}$$

the cluster condition

$$|D(v_1, v_2, u_1, u_2)| < Cq^{d_{\{v_1\}\cup\{v_2\}\cup\{u_1\}\cup\{u_2\}}}, \tag{4.11b}$$

and the condition

$$\sum_{u_1,u_2} D(v_1, v_2, u_1, u_2) = 0 \quad \textit{for all } v_1, v_2. \tag{4.11c}$$

Here d_B, $B \subset \mathbb{Z}^\nu$, *is the length of the smallest graph connecting all points of the set* B, $q \in (0, 1)$ *is some constant, and* $C = C(\varepsilon_0, \varepsilon_1) \ll 1$ *for* ε_0, ε_1 *small enough.*

Moreover, both terms in the decomposition

$$\Psi_{\bar{0},x_1,x_2} = \Psi^{(2)}_{\bar{0},x_1,x_2} + \overline{\Psi}^{(2)}_{\bar{0},x_1,x_2}, \quad \Psi^{(2)}_{\bar{0},x_1,x_2} \in \mathscr{H}_2, \ \overline{\Psi}^{(2)}_{\bar{0},x_1,x_2} \in \overline{\mathscr{H}}_2, \tag{4.12}$$

belong to the space $\mathscr{C}_2(\widehat{\Omega})$, *and for any* $t > 0$ *the following estimate holds:*

$$\sup_{\xi\in\Omega} |((\mathscr{T}^{(2)})^t \overline{\Psi}^{(2)}_{\bar{0},x_1,x_2})(\xi, z_1, z_2)| < C\theta^t, \tag{4.13}$$

where $C > 0$ *and* $\theta \in (0, 1)$ *are constants independent of* t *and of* $z_1, z_2 \in \mathbb{Z}^\nu$.

The remaining part of the present section is devoted to the proof of this theorem. As a consequence of the theorem we have, putting together equations (4.12) and (4.13),

$$\begin{aligned} &\mathbb{P}(X_t^{(1)} = x_1, X_t^{(2)} = x_2 \mid X_0^{(1)} = z_1, X_0^{(2)} = z_2) \\ &= \int_\Omega ((\mathscr{T}^{(2)})^t \Psi^{(2)}_{\bar{0},x_1,x_2})(\xi, z_1, z_2)\, d\Pi(\xi) + \mathscr{O}(\theta^t). \end{aligned} \tag{4.14}$$

Hence the asymptotics of this probability is determined by the asymptotic behavior of $(\mathscr{T}^{(2)})^t \Psi^{(2)}_{\bar{0},x_1,x_2}$, which will be studied in the following section.

We first establish the important cluster property for the basis $\{h^u\}$.

LEMMA 4.1. *The following relations hold*:

$$h^u_{\bar{0},z} = \delta_{u,z}, \tag{4.15}$$

$$|h^u_{\Gamma,z}| < C_0 \overline{M}^{-|\operatorname{supp}\Gamma|} \tilde{q}^{d_{\operatorname{supp}\Gamma\cup\{z\}\cup\{u\}}}, \qquad \Gamma \neq \bar{0}, \tag{4.16}$$

where $\tilde{q} \in (0, 1)$, *and* $C_0 = C_0(\varepsilon_0, \varepsilon_1)$ *is small for* $\varepsilon \equiv \max\{\varepsilon_1, \varepsilon_2\}$ *small. The constant* $\overline{M} > 1$ *satisfies* $c_1 < \overline{M} < c_2/\varepsilon$, *where* $c_1, c_2 > 0$ *are absolute constants.*

PROOF. The relation (4.15) follows from the condition $\chi_{\bar{0}}(\lambda) = 1$, and the estimate (4.16) follows from the following lemma, which refines Lemma 3.5 in the previous section. □

LEMMA 4.2. *There is a complex neighborhood* W_d *of the type* (4.8a) *such that the family of vectors* $\{\chi_\Gamma(\lambda)\}$ *in* $\hat{l}_2(\mathfrak{M})$ *is analytic in* $\lambda \in W_d$ *and satisfies the estimate*

$$|\chi_\Gamma(\lambda)| < C\overline{M}^{-|\operatorname{supp}\Gamma|} \tilde{q}^{d_{\operatorname{supp}\Gamma\cup\{0\}}}, \qquad \Gamma \in \mathfrak{M},\ \lambda \in W_d,\ \Gamma \neq \bar{0}, \tag{4.17}$$

where $\tilde{q} \in (0, 1)$, *the constant* $C = C(\varepsilon_0, \varepsilon_1)$ *is small for* ε_0, ε_1 *small, and* $\overline{M}$ *is the constant introduced in the preceding lemma. Moreover, the function* $\tilde{p}(\lambda)$ *is also analytic in* W_d *and satisfies the estimate*

$$|\tilde{p}(\lambda)| > |\mu_1|^{1/2} e^{\max_k |\operatorname{Im}\lambda_k| D}. \tag{4.18}$$

PROOF. Due to estimate (2.13), there is a complex neighborhood of the torus $W_{\bar{d}}$ in which the analytic function $\tilde{p}_0(\lambda)$ satisfies the estimate

$$|\tilde{p}_0(\lambda)| > (|\mu_1|^{1/2} + \bar{\delta}) e^{\max_k |\operatorname{Im}\lambda_k| D}, \tag{4.19}$$

where $\bar{\delta}$ is a positive constant. Consider now the Banach space $\mathscr{L}_{W_d, q^*}$ of vectors $\{\chi_\Gamma(\lambda), \Gamma \in \mathfrak{M}, \Gamma \neq \bar{0}\}$ that are analytic in the neighborhood $W_{\bar{d}}$ and satisfy the estimate (4.17), with the norm

$$\|\chi\|_{\mathscr{L}_{W_{\bar{d}}, q^*}} \equiv \|\chi\| = \sup_{\Gamma, \lambda \in W_{\bar{d}}} \{|\chi_\Gamma(\lambda)| (q^*)^{-d_{\operatorname{supp}\Gamma\cup\{0\}}} \overline{M}^{|\operatorname{supp}\Gamma|}\}.$$

We shall show that for an appropriate q^* the map

$$B\colon \mathscr{L}_{W_{\bar{d}}, q^*} \mapsto \mathscr{L}_{W_{\bar{d}}, q^*} \tag{4.20}$$

defined by the right-hand side of (3.21) for the function χ maps some sphere $\{\|\chi\| < \kappa\}$ in $\mathscr{L}_{W_{\bar{d}}, q^*}$ into itself. The radius of the sphere $\kappa = \kappa(\varepsilon_0, \varepsilon_1)$ is small for ε_0, ε_1 small, and the map is contracting on the sphere.

We set

$$\frac{1}{\tilde{p}_0(\lambda)} \sum_{\Gamma' \neq \bar{0}} \widehat{R}_{\Gamma,\Gamma'}(\lambda) \chi_{\Gamma'}(\lambda) = \frac{1}{\tilde{p}_0(\lambda)} \left[\sum_{\Gamma' \neq \bar{0}} \widehat{R}^0_{\Gamma,\Gamma'}(\lambda) \chi_{\Gamma'}(\lambda) + \sum_{\Gamma' \neq \bar{0}} \Delta\widehat{R}_{\Gamma,\Gamma'}(\lambda) \chi_{\Gamma'}(\lambda) \right], \tag{4.21}$$

and consider the first "unperturbed" term. Using the estimate

$$|\chi_{\Gamma-z}(\lambda)| \leqslant \|\chi\|(q^*)^{d_{\operatorname{supp}\{\Gamma-z\}\cup\{0\}}}\overline{M}^{-|\operatorname{supp}\Gamma|} = \|\chi\|(q^*)^{d_{\operatorname{supp}\Gamma\cup\{z\}}}\overline{M}^{-|\operatorname{supp}\Gamma|},$$

and equation (3.16c), we get

$$\begin{aligned}&\left|\frac{1}{\widetilde{p}_0(\lambda)}\sum_{\Gamma'}\widehat{R}^0_{\Gamma,\Gamma'}\chi_{\Gamma'}(\lambda)\right| \\ &\quad = \left|\frac{\mu_\Gamma}{\widetilde{p}_0(\lambda)}\sum_u \chi_{\Gamma-u}(\lambda)P_0(u)e^{i(\lambda,u)}\right| \\ &\quad < \frac{\|\chi\|\,|\mu_\Gamma|}{\overline{M}^{|\operatorname{supp}\Gamma|}|\widetilde{p}_0(\lambda)|}\sum P_0(u)(q^*)^{d_{\operatorname{supp}\Gamma\cup\{u\}}}e^{-\max_k|\operatorname{Im}\lambda_k|D} \\ &\quad < \frac{\|\chi\|}{\overline{M}^{|\operatorname{supp}\Gamma|}}\frac{|\mu_1|}{|\mu_1|^{1/2}+\delta}(q^*)^{d_{\operatorname{supp}\Gamma\cup\{0\}}}(q^*)^{-D}.\end{aligned}$$

Here we used the inequality $d_{\operatorname{supp}\Gamma\cup\{0\}} \leqslant d_{\operatorname{supp}\Gamma\cup\{u\}} + |u| \leqslant d_{\operatorname{supp}\Gamma\cup\{u\}} + D$. Setting $q^* = |\mu_1|^{1/3D}$ we see that the norm $\|(\mathscr{T}_\lambda^0)^{(1,1)}/\widetilde{p}_0(\lambda)\|_{\mathscr{L}_{W_{\bar{d}},q^*}}$ does not exceed $|\mu_1|^{1/6}$. Similarly we can estimate the norm $\|\Delta\mathscr{T}_\lambda^{(1,1)}/\widetilde{p}_0(\lambda)\|_{\mathscr{L}_{W_{\bar{d}},q^*}}$ (where $\Delta\mathscr{T}_\lambda^{(1,1)}$ is the operator corresponding to the second term on the right-hand side of (4.21)). Hence we have obtained an estimate of the first term on the right-hand side in (3.21). The norm of the last term can be estimated as before. Since the norm of the "free" term $\|\widehat{R}_{\bar{0},\bar{0}}(\lambda)/\widehat{R}_{\bar{0},\Gamma}(\lambda)\|_{\mathscr{L}_{W_{\bar{d}},q^*}}$ in (3.21) is small for small ε, we have shown that the above assertion about the map (4.20), and hence estimate (4.17), hold. The estimate (4.18), for small ε, follows from estimates (4.19) and (4.17) and formula (3.20a) for the function $\widetilde{p}(\lambda)$ (see the preceding section). Lemma 4.2 is proved. □

From this lemma the estimate (4.16) follows. Indeed, by formula (4.9b), and estimate (4.17) for $\chi(\lambda)$, in the neighborhood $W_{\bar{d}}$ we have

$$\begin{aligned}|h^u_{\Gamma,z}| &< \|\chi\|e^{-\bar{d}|u-z|}(q^*)^{d_{\operatorname{supp}\Gamma\cup\{z\}}}\overline{M}^{-|\operatorname{supp}\Gamma|} \\ &\leqslant \overline{M}^{-|\operatorname{supp}\Gamma|}\|\chi\|\widetilde{q}^{\operatorname{supp}\Gamma\cup\{z\}\cup\{u\}},\end{aligned} \tag{4.21'}$$

where $\widetilde{q} = \max\{e^{-\bar{d}}, q^*\}$. Estimate (4.16) is then proved, with $C_0 = \|\chi\|$, which is small for ε small.

Before going to the construction of the two-particle invariant subspace for the operator $\mathscr{T}^{(2)}$, we construct an approximation subspace $l^{(0)} \subset \hat{l}_2^{(2)}$, whose perturbation will give the real invariant subspace of $\mathscr{T}^{(2)}$. For any pair of functions f and g on the set $\mathfrak{M}\times\mathbb{Z}^\nu$ we define the function $f\otimes g$ on $\mathfrak{M}\times\mathbb{Z}^\nu\times\mathbb{Z}^\nu$ according to the formula

$$(f\otimes g)_{\Gamma,z_1,z_2} = \sum_{\Gamma_1,\Gamma_2} f_{\Gamma_1,z_1}g_{\Gamma_2z_2}, \tag{4.22}$$

where the sum is over all ordered pairs of disjoint multi-indices (Γ_1, Γ_2) giving a decomposition of the multi-index Γ. This means that

$$\operatorname{supp}\Gamma_1 \cap \operatorname{supp}\Gamma_2 = \varnothing, \quad \operatorname{supp}\Gamma_1 \cup \operatorname{supp}\Gamma_2 = \operatorname{supp}\Gamma,$$
$$\Gamma|_{\operatorname{supp}\Gamma_i} = \Gamma_i|_{\operatorname{supp}\Gamma_i}, \qquad i = 1, 2.$$

In such a case we shall simply write $\Gamma = \Gamma_1 \cup \Gamma_2$.

Now we consider the system of functions

$$h^{u_1, u_2} = h^{u_1} \otimes h^{u_2}, \tag{4.23}$$

where h^{u_1}, h^{u_2} are defined by (4.9b). The following lemma holds.

LEMMA 4.3. *For any u_1, u_2 the function h^{u_1, u_2} belongs to $\hat{l}_2^{(2)}$.*

PROOF. From the estimate (4.16) and formula (4.22) it is easy to see that

$$\sum_{\Gamma} |h^{u_1, u_2}_{\Gamma, z_1, z_2}| M^{|\operatorname{supp}\Gamma|} < C\overline{q}^{|u_1 - z_1| + |u_2 - z_2|}, \tag{4.24}$$

where $C > 0$ is a constant, $\overline{q} \in (0, 1)$, and M is the constant appearing in the definition of the space $l_M(\mathfrak{M})$ (see (3.16c)). From the estimate (4.24) it is easy to conclude by inequality (3.16d) that $h^{u_1, u_2} \in \hat{l}_2^2$. □

We take the space $l^{(0)} \subset \hat{l}_2^{(2)}(\mathfrak{M} \times \mathbb{Z}^\nu \times \mathbb{Z}^\nu)$ generated by the vectors $\{h^{u_1, u_2}\}$ as an initial approximation for the true two-particle invariant subspace. This choice is justified by the following lemma.

LEMMA 4.4. *For any $u_1, u_2 \in \mathbb{Z}^\nu$*

$$\mathscr{T}^{(2)}(h^{u_1} \otimes h^{u_2}) = (\mathscr{T}^{(1)} h^{u_1}) \otimes (\mathscr{T}^{(1)} h^{u_2}) + g^{u_1, u_2}, \tag{4.25}$$

where the function g^{u_1, u_2} satisfies the estimate

$$|g^{u_1, u_2}_{\Gamma, z_1, z_2}| < C' \widetilde{q}^{d_{\operatorname{supp}\Gamma \cup \{z_1\} \cup \{z_2\} \cup \{u_1\} \cup \{u_2\}}} 2^{|\operatorname{supp}\Gamma|} / \overline{M}^{|\operatorname{supp}\Gamma|}. \tag{4.26}$$

The constant $\overline{M}$ can be made as large as we want if ε_0, ε_1 are sufficiently small, and the constant C' is small for ε_0, ε_1 small.

REMARK. By (4.10a),

$$(\mathscr{T}^{(1)} h^{u_1}) \otimes (\mathscr{T}^{(1)} h^{u_2}) = \sum_{u_1', u_2'} r(u_1 - u_1') r(u_2 - u_2') h^{u_1', u_2'}. \tag{4.27}$$

Therefore, the above lemma implies that the vector $\mathscr{T}^{(2)} h^{u_1, u_2}$ differs from the vectors of $l^{(0)}$ by a small "connected" vector (in the sense of the estimate (4.26)). As we shall see below, it is this circumstance that allows us to construct the invariant subspace $\ell \subset \hat{l}_2^{(2)}$ and a basis $\{b^{u_1, u_2}\}$ in it for which the operator $\mathscr{T}^{(2)}$ has the form (4.10c). They are close respectively to the subspace $l^{(0)}$ and to the basis $\{h^{u_1, u_2}\}$.

Proof of the lemma. In what follows, if $z \notin \operatorname{supp}\Gamma$, we denote by $\Gamma \cup j(z)$ the multi-index obtained by adding to Γ the value $j(z) \in (1, \dots, |S|-1)$ at z. From formulas (4.4a,b), (4.5) and (4.7a) we find that for $z_1 \neq z_2$

$$
\begin{aligned}
&\mathscr{T}^{(2)}(h^{u_1} \otimes h^{u_2})_{\Gamma, z_1, z_2} \\
&\quad = \mu_{\Gamma_{z_1, z_2}} \sum_{z_1', z_2', \Gamma_1, \Gamma_2 : \Gamma_1 \cup \Gamma_2 = \Gamma_{z_1, z_2}} \\
&\qquad \times \Bigg[\sum_{j_1'(z_1), j_2'(z_2) = 0, \dots, |S|-1} L(\gamma(z_1), j_1'(z_1), z_1' - z_1) \\
&\qquad\qquad \times L(\gamma(z_2), j_2'(z_2), z_2' - z_2) h^{u_1}_{\Gamma_1 \cup j_1'(z_1), z_1'} h^{u_2}_{\Gamma_2 \cup j_2'(z_2), z_2'} \\
&\qquad + \sum_{j_1'(z_1) \neq 0, j_2'(z_2) \neq 0} L(\gamma(z_1), j_1'(z_1), z_1' - z_1) \\
&\qquad\qquad \times L(\gamma(z_2), j_2'(z_2), z_2' - z_2) h^{u_1}_{\Gamma_1 \cup j_2'(z_2), z_1'} h^{u_2}_{\Gamma_2 \cup j_1'(z_1), z_2'} \\
&\qquad + \sum_{\substack{j_1'(z_1) = 0, \dots, |S|-1, \\ j_2'(z_2) = 1, \dots, |S|-1}} L(\gamma(z_1), j_1'(z_1), z_1' - z_1) \\
&\qquad\qquad \times L(\gamma(z_2), j_2'(z_2), z_2' - z_2) h^{u_1}_{\Gamma_1 \cup j_1'(z_1) \cup j_2'(z_2), z_1'} h^{u_2}_{\Gamma_2, z_2'} \\
&\qquad + \sum_{\substack{j_1'(z_1) = 1, \dots, |S|-1, \\ j_2'(z_2) = 0, \dots, |S|-1}} L(\gamma(z_1), j_1'(z_1), z_1' - z_1) \\
&\qquad\qquad \times L(\gamma(z_2), j_2'(z_2), z_2' - z_2) h^{u_1}_{\Gamma_1, z_1'} h^{u_2}_{\Gamma_2 \cup j_1'(z_1) \cup j_2'(z_2), z_2'} \Bigg] \\
&\quad + \sum_{j_1'(z_1), j_2'(z_2), z_1', z_2'} \mathscr{L}(\gamma(z_1), j_1'(z_1), \gamma(z_2), j_2'(z_2), \\
&\qquad\qquad z_1' - z_1, z_2' - z_2, z_1 - z_2) \\
&\qquad\qquad \times (h^{u_1} \otimes h^{u_2})_{\Gamma_{z_1, z_2} \cup j_1'(z_1) \cup j_2'(z_2), z_1', z_2'} .
\end{aligned}
\tag{4.28a}
$$

Clearly the first sum in (4.28a) is equal to

$$
\begin{aligned}
&\mu_{\Gamma_{z_1, z_2}} \sum_{\substack{z_1', z_2', \Gamma_1, \Gamma_2 : \\ \Gamma_1 \cup \Gamma_2 = \Gamma_{z_1, z_2}}} \sum_{j_1'(z_1), j_2'(z_2)} L(\gamma(z_1), j_1'(z_1), z_1' - z_1) \\
&\qquad \times L(\gamma(z_2), j_2'(z_2), z_2' - z_2) h^{u_1}_{\Gamma_1 \cup j_1'(z_1), z_1'} h^{u_2}_{\Gamma_2 \cup j_2'(z_2), z_2'} \\
&\quad = \sum_{\Gamma_1 \cup \Gamma_2} \sum_{j_1'(z_1)} \mu_{(\Gamma_1)_{z_1}} L(\gamma(z_1), j_1'(z_1), z_1' - z_1) h^{u_1}_{(\Gamma_1)_{z_1} \cup j_1'(z_1), z_1'} \\
&\qquad \times \sum_{j_2'(z_2)} \mu_{(\Gamma_2)_{z_2}} L(\gamma(z_2), j_2', z_2' - z_1) h^{u_2}_{(\Gamma_2)_{z_2} \cup j_2'(z_2), z_2'}
\end{aligned}
$$

$$= \sum_{\Gamma_1 \cup \Gamma_2 = \Gamma} (\mathcal{T}^{(1)} h^{u_1})_{\Gamma_1, z_1} (\mathcal{T}^{(1)} h^{u_2})_{\Gamma_2, z_2}$$

$$= ((\mathcal{T}^{(1)} h^{u_1}) \otimes (\mathcal{T}^{(1)} h^{u_2}))_{\Gamma, z_1, z_2}.$$

In this way the remaining terms in the sum (4.28a) give the value of the function $g^{u_1, u_2}_{\Gamma, z_1, z_2}$. For $z_1 = z_2 = z$ we set

$$g^{u_1, u_2}_{\Gamma, z, z} = (\mathcal{T}^{(2)}(h^{u_1} \otimes h^{u_2}))_{\Gamma, z, z} - ((\mathcal{T}^{(1)} h^{u_1}) \otimes (\mathcal{T}^{(1)} h^{u_2}))_{\Gamma, z, z}. \quad (4.28b)$$

We now prove the estimate (4.26). Let $z_1 \neq z_2$, and consider, for example, the second sum in (4.28a). By inequality (4.21), it is estimated by the number

$$\sum_{\substack{j_1'(z_1) \neq 0, \\ j_2'(z_2) \neq 0}} |L(\gamma(z_1), j_1'(z_1), z_1' - z_1)| |L(\gamma(z_2), j_2'(z_2), z_2' - z_2)| \, \|\chi\|^2 \overline{M}^{-|\operatorname{supp}\Gamma|}$$

$$\times \sum_{\Gamma_1 \cup \Gamma_2 = \Gamma_{z_1, z_2}} \widetilde{q}^{d_{\operatorname{supp}\Gamma_1 \cup \{z_2\} \cup \{z_1'\} \cup \{u_1\}}} \widetilde{q}^{d_{\operatorname{supp}\Gamma_2 \cup \{z_1\} \cup \{z_2'\} \cup \{u_2\}}}. \quad (4.29)$$

Since $L(\gamma(z_1), j_1'(z_1), z_1' - z_1) = 0$ for $|z_1 - z_1'| > D$, and the same holds for $L(\gamma(z_2), j_2'(z_2), z_2' - z_2)$ when $|z_2 - z_2'| > D$, formula (4.29) yields an estimate of the type (4.26). Estimates for other sums in (4.28a) are obtained in the same way. The case $z_1 = z_2$ is treated similarly, since each term on the right-hand side of (4.28b) satisfies an estimate of the type (4.26). Lemma 4.4 is proved. □

Next we change the natural basis $\{\overline{h}^{Q, u_1, u_2}, \ Q \in \mathfrak{M}, \ u_1, u_2 \in \mathbb{Z}^\nu\}$, where $\overline{h}^{Q, u_1, u_2}_{\Gamma, z_1, z_2} = \delta_{Q\Gamma} \delta_{z_1 u_1} \delta_{z_2 u_2}$, to another basis $\{h^{Q, u_1, u_2}, \ Q \in \mathfrak{M}, \ u_1, u_2 \in \mathbb{Z}^\nu\}$ by setting

$$h^{\overline{0}, u_1, u_2} = h^{u_1, u_2}, \quad h^{Q, u_1, u_2} = \overline{h}^{Q, u_1, u_2}, \qquad Q \neq \overline{0}. \quad (4.30a)$$

Clearly

$$h^{\overline{0}, u_1, u_2} = \overline{h}^{\overline{0}, u_1, u_2} + \sum_{u_1', u_2', Q' \neq \overline{0}} B_{u_1, u_2, Q', u_1', u_2'} \overline{h}^{Q', u_1' u_2'}, \quad (4.30b)$$

where for $Q' \neq \overline{0}$

$$B_{u_1, u_2, Q', u_1', u_2'} = (h^{u_1} \otimes h^{u_2})_{Q', u_1', u_2'}. \quad (4.31)$$

Hence

$$\overline{h}^{\overline{0}, u_1, u_2} = h^{\overline{0}, u_1, u_2} - \sum_{u_1', u_2', Q' \neq \overline{0}} B_{u_1, u_2, Q', u_1', u_2'} h^{Q', u_1', u_2'}. \quad (4.32)$$

The expansion of the vector $f \in \hat{l}_2^{(2)}$ in the series

$$f = \sum_{Q, u_1, u_2} f(Q, u_1, u_2) \overline{h}^{Q, u_1, u_2}$$

after the substitution (4.30a), (4.32) becomes

$$f = \sum_{Q, u_1, u_2} \widetilde{f}(Q, u_1, u_2) h^{Q, u_1, u_2},$$

where

$$\begin{aligned} \widetilde{f}(\overline{0}, u_1, u_2) &= f(\overline{0}, u_1, u_2), \\ \widetilde{f}(Q, u_1, u_2) &= f(Q, u_1, u_2) \\ &\quad - \sum_{u_1', u_2'} f(\overline{0}, u_1', u_2') B_{u_1', u_2', Q, u_1, u_2}, \qquad Q \neq \overline{0}. \end{aligned} \tag{4.33}$$

Note that the introduction of the new coordinates $\widetilde{f} = \widetilde{f}(Q, u_1, u_2)$ is naturally associated to a bilinear form which gives a scalar product in $\hat{l}_2^{(2)}$:

$$(f, g) = \sum_{\substack{Q, u_1, u_2, \\ Q', u_1', u_2'}} \widetilde{a}_{Q, u_1, u_2, Q', u_1', u_2'} \widetilde{f}(Q, u_1, u_2) \overline{\widetilde{g}(Q', u_1', u_2')}, \tag{4.34}$$

where the coefficients can easily be written out. It is convenient to assume that we go from the space $\hat{l}_2^{(2)}$ of functions on $\mathfrak{M} \times \mathbb{Z}^\nu \times \mathbb{Z}^\nu$ with the scalar product (4.3) to the isomorphic space $\hat{\hat{l}}_2^{(2)}$ of functions on $\mathfrak{M} \times \mathbb{Z}^\nu \times \mathbb{Z}^\nu$ with the scalar product given by (4.34). We find now the explicit form of matrix elements $\widetilde{R}_{Q, u_1, u_2, Q', u_1', u_2'}$ of the operator $\mathscr{T}^{(2)}$ when it acts on functions $f \in \hat{\hat{l}}_2^{(2)}$:

$$(\widetilde{\mathscr{T}^{(2)} f})_{Q, u_1, u_2} = \sum_{Q', u_1', u_2'} \widetilde{R}_{Q, u_1, u_2, Q', u_1', u_2'} \widetilde{f}(Q', u_1', u_2').$$

To do this we find how $\mathscr{T}^{(2)}$ acts on elements of the basis $\{h^{Q, u_1, u_2}\}$. From formulas (4.30a), (4.25), (4.27) we find that

$$\begin{aligned} \mathscr{T}^{(2)} h^{\overline{0}, u_1, u_2} &= \sum_{u_1', u_2'} r(u_1 - u_1') r(u_2 - u_2') h^{\overline{0}, u_1', u_2'} \\ &\quad + \sum_{Q', u_1', u_2'} g^{u_1, u_2}_{Q', u_1', u_2'} \overline{h}^{Q', u_1', u_2'} \\ &= \sum_{u_1', u_2'} [r(u_1 - u_1') r(u_2 - u_2') + \widehat{D}(u_1, u_2, u_1', u_2')] h^{\overline{0}, u_1', u_2'} \\ &\quad + \sum_{Q', u_1', u_2', Q' \neq \overline{0}} M_{u_1, u_2, Q', u_1', u_2'} h^{Q', u_1', u_2'}, \end{aligned} \tag{4.35a}$$

where $\widehat{D}(u_1, u_2, u_1', u_2') = g^{u_1, u_2}_{\overline{0}, u_1', u_2'}$ (see 4.25) and for $Q' \neq \overline{0}$,

$$M_{u_1, u_2, Q', u_1', u_2'} = g^{u_1, u_2}_{Q', u_1', u_2'} - \sum_{v_1, v_2} g^{u_1, u_2}_{\overline{0}, v_1, v_2} B_{v_1, v_2, Q', u_1', u_2'}. \tag{4.35b}$$

Furthermore, denoting by $R^{(2)}_{\dots}$ the matrix elements of $\mathcal{T}^{(2)}$ with respect to the basis $\{\overline{h}^{\cdots}\}$, for $Q \neq \overline{0}$ we find that

$$\begin{aligned}
&\mathcal{T}^{(2)} h^{Q, u_1, u_2} \\
&= \mathcal{T}^{(2)} \overline{h}^{Q, u_1, u_2} \\
&= \sum_{Q' \neq \overline{0}, u_1', u_2'} R^{(2)}_{Q', u_1', u_2', Q, u_1, u_2} h^{Q', u_1', u_2'} \\
&\quad + \sum_{u_1', u_2'} R^{(2)}_{\overline{0}, u_1', u_2', Q, u_1, u_2} \\
&\qquad \times \left(h^{\overline{0}, u_1', u_2'} - \sum_{Q' \neq \overline{0}, v_1, v_2} B_{u_1', u_2', Q', v_1, v_2} h^{Q', v_1, v_2} \right) \\
&= \sum_{Q' \neq \overline{0}, u_1', u_2'} \Bigg[R^{(2)}_{Q', u_1', u_2', Q, u_1, u_2} \\
&\qquad - \sum_{v_1, v_2} R^{(2)}_{\overline{0}, v_1, v_2, Q, u_1, u_2} B_{v_1, v_2, Q', u_1', u_2'} \Bigg] h^{Q', u_1', u_2'} \\
&\quad + \sum_{u_1', u_2'} R^{(2)}_{\overline{0}, u_1', u_2', Q, u_1, u_2} h^{\overline{0}, u_1', u_2'}.
\end{aligned} \tag{4.35c}$$

Finally we obtain

$$\widetilde{R}_{\overline{0}, u_1, u_2, \overline{0}, u_1', u_2'} = r(u_1' - u_1) r(u_2' - u_2) + \widehat{D}(u_1', u_2', u_1, u_2), \tag{4.36i}$$

$$\widetilde{R}_{Q, u_1, u_2, \overline{0}, u_1', u_2'} = M_{u_1', u_2', Q, u_1, u_2}, \qquad Q \neq \overline{0}, \tag{4.36ii}$$

$$\widetilde{R}_{\overline{0}, u_1, u_2, Q', u_1', u_2'} = R^{(2)}_{\overline{0}, u_1, u_2, Q', u_1', u_2'}, \qquad Q' \neq \overline{0}, \tag{4.36iii}$$

$$\begin{aligned}
\widetilde{R}_{Q, u_1, u_2, Q', u_1', u_2'} &= R^{(2)}_{Q, u_1, u_2, Q', u_1', u_2'} \\
&\quad - \sum_{v_1, v_2} R^{(2)}_{\overline{0}, v_1, v_2, Q', u_1', u_2'} B_{v_1, v_2, Q, u_1, u_2}, \\
&\qquad Q \neq \overline{0}, \ Q' \neq \overline{0}.
\end{aligned} \tag{4.36iv}$$

The matrix elements that we have written out are the matrix elements of the operators $\mathcal{T}^{(2)}_{00}$, $\mathcal{T}^{(2)}_{10}$, $\mathcal{T}^{(2)}_{01}$, $\mathcal{T}^{(2)}_{11}$ obtained by decomposing the space $\hat{\hat{l}}^{(2)}_2$ into the direct sum of spaces $\hat{\hat{l}}^{(2)}_2 = l^{(0)} \oplus l^{(1)}$, where

$$l^{(0)} = \{f : \widetilde{f}_{Q, u_1, u_2} = 0 \text{ for } Q \neq \overline{0}\}, \qquad l^{(1)} = \{f : \widetilde{f}_{\overline{0}, u_1, u_2} = 0\}.$$

Similarly to the previous section, the invariant two-dimensional subspace will be the graph $\ell = \{f + Sf \mid f \in l^{(0)}\}$ of an operator $S : l^{(0)} \mapsto l^{(1)}$ acting according to the formula

$$(\widetilde{Sf})(Q, u_1, u_2) = \sum_{u_1', u_2'} S_{Q, u_1, u_2, \overline{0}, u_1', u_2'} \widetilde{f}(\overline{0}, u_1', u_2'), \qquad Q \neq \overline{0}.$$

The invariance of the subspace ℓ with respect to the operator $\mathcal{T}^{(2)}$ leads to the following equation for the operator S:

$$S = \mathcal{T}_{10}^{(2)}(\mathcal{T}_{00}^{(2)})^{-1} + \mathcal{T}_{11}^{(2)} S (\mathcal{T}_{00}^{(2)})^{-1} - S\mathcal{T}_{01}^{(2)} S (\mathcal{T}_{00}^{(2)})^{-1}. \tag{4.37a}$$

This equation, which is similar to (3.21), can be rewritten in terms of matrix elements

$$S(Q, u_1, u_2, u_1', u_2') \equiv S_{Q, u_1, u_2, \bar{0}, u_1', u_2'}$$

as follows:

$$\begin{aligned} S(Q, u_1, u_2, u_1', u_2') &= \sum_{v_1, v_2} \widetilde{R}_{Q, u_1, u_2, \bar{0}, v_1, v_2} (\mathcal{T}_{00}^{(2)})^{-1}_{v_1, v_2, u_1', u_2'} \\ &+ \sum_{Q' \neq \bar{0}, v_1, v_2, v_1', v_2'} \widetilde{R}_{Q, u_1, u_2, Q', v_1, v_2} \\ &\qquad \times S(Q', v_1, v_2, v_1', v_2')(\mathcal{T}_{00}^{(2)})^{-1}_{v_1', v_2', u_1', u_2'} \\ &- \sum_{Q' \neq \bar{0}, v_1, v_2, v_1', v_2', v_1'', v_2''} S(Q, u_1, u_2, v_1, v_2) \widetilde{R}_{\bar{0}, v_1, v_2, Q', v_1', v_2'} \\ &\qquad \times S(Q', v_1', v_2', v_1'', v_2'')(\mathcal{T}_{00}^{(2)})^{-1}_{v_1'', v_2'', u_1', u_2'}. \end{aligned} \tag{4.37b}$$

Before studying this equation we must study the operator $(\mathcal{T}_{00}^{(2)})^{-1}$.

LEMMA 4.5. *The matrix* $(\mathcal{T}_{00}^{(2)})^{-1}_{u_1, u_2, u_1', u_2'}$ *is equal to*

$$(\mathcal{T}_{00}^{(2)})^{-1}_{u_1, u_2, u_1', u_2'} = r^{-1}(u_1' - u_1) r^{-1}(u_2' - u_2) + \widetilde{D}(u_1, u_2, u_1', u_2'), \tag{4.38}$$

where

$$r^{-1}(x) = \int_{T^\nu} \frac{e^{i(\lambda, x)}}{\widetilde{p}(\lambda)} d\lambda, \tag{4.39a}$$

and the function $\widetilde{D}(u_1, u_2, u_1', u_2')$ *satisfies the estimate*

$$|\widetilde{D}(u_1, u_2, u_1', u_2')| < \bar{c}\widetilde{q}^{d_{\{u_1\}\cup\{u_2\}\cup\{u_1'\}\cup\{u_2'\}}/2}, \tag{4.39b}$$

where $\widetilde{q} \in (0, 1)$ *is the same constant as in formula* (4.16), *and the constant* $\bar{c} = \bar{c}(\varepsilon_0, \varepsilon_1)$ *is small for* ε_0, ε_1 *small. Moreover*

$$|r^{-1}(x)| < C_1 e^{-d|x|} \tag{4.39c}$$

for some constants C_1, $d > 0$.

PROOF. The estimate (4.39c) follows from formula (4.39a) and Lemma 4.2. Furthermore, by formula (4.36i) we write $\mathcal{T}_{00}^{(2)} = R_0 + \widehat{D}$, where R_0 is the convolution operator with the function $r(u_1)r(u_2)$:

$$(R_0 f)(u_1, u_2) = \sum_{v_1, v_2} r(u_1 - v_1) r(u_2 - v_2) f(v_1, v_2),$$

and $\widehat{D}$ is the operator with matrix coefficients $\widehat{D}(v_1, v_2, u_1, u_2)$. Then

$$(\mathcal{T}_{00}^{(2)})^{-1} = R_0^{-1} + \sum_{k=1}^{\infty}(-1)^k R_0^{-1}\widehat{D}R_0^{-1}\cdots\widehat{D}R_0^{-1}, \tag{4.40}$$

where in the kth term of the sum the product $\widehat{D}R_0^{-1}$ is repeated k times. Clearly

$$\begin{aligned}(R_0^{-1}\widehat{D}R_0^{-1}\cdots\widehat{D}R_0^{-1})_{u_1,u_2,v_1,v_2} &= \sum_{(x_1^{(0)},x_2^{(0)}),\dots,(x_1^{(k)},x_2^{(k)})} r^{-1}(x_1^{(0)}-u_1)r^{-1}(x_2^{(0)}-u_2) \\ &\times \widehat{D}(x_1^{(1)}, x_2^{(1)}, x_1^{(0)}, x_2^{(0)})r^{-1}(x_1^{(2)}-x_1^{(1)})r^{-1}(x_2^{(2)}-x_2^{(1)})\cdots \\ &\times \widehat{D}(x_1^{(k)}, x_2^{(k)}, x_1^{(k-1)}, x_2^{(k-1)})r^{-1}(v_1-x_1^{(k)})r^{-1}(v_2-x_2^{(k)}).\end{aligned} \tag{4.41}$$

Furthermore, using the estimate (4.39c) and the estimate

$$|\widehat{D}(x_1, x_2, y_1, y_2)| < C'\widetilde{q}^{\,d_{\{x_1\}\cup\{x_2\}\cup\{y_1\}\cup\{y_2\}}}, \qquad C' \ll 1,$$

which follows from (4.26) and the definition of $\widehat{D}$ (see (4.35a)), we see that the sum (4.41) can be estimated by the quantity

$$\bar{c}(\sqrt{\widetilde{q}})^{d_{\{x_1\}\cup\{x_2\}\cup\{y_1\}\cup\{y_2\}}}, \tag{4.42}$$

where $\bar{c}$ is small for ε_0, ε_1 small.

The expansion (4.40) and the estimate (4.42) give the representation (4.38) and the estimate (4.39b). Lemma 4.5 is proved. □

LEMMA 4.6. *Let ε_0, ε_1 be sufficiently small. Then there is a unique solution $S(Q, u_1, u_2, u_1', u_2')$ of equation* (4.37a) *satisfying the estimate*

$$|S(Q, u_1, u_2, u_1', u_2')| < c\overline{M}^{-|\operatorname{supp} Q|}\widehat{q}^{\,d_{\operatorname{supp} Q\cup\{u_1\}\cup\{u_2\}\cup\{u_1'\}\cup\{u_2'\}}}, \tag{4.43}$$

where the constant $c = c(\varepsilon_0, \varepsilon_1)$ is small for ε_0, ε_1 small, and the constant $\widehat{q}$ is chosen below.

PROOF. As in the one-particle case, it is convenient for us to introduce the Fourier transform of the function $\widetilde{f}(Q, u_1, u_2)$. That is, setting

$$\widetilde{f}(Q, u_1, u_2) = \int_{T^\nu\times T^\nu}\varphi(Q-u_1, \Lambda, \lambda)e^{-i(\Lambda, u_2)}e^{-i(\lambda, u_1-u_2)}\,d\Lambda\,d\lambda, \tag{4.44}$$

we have

$$\varphi(Q, \Lambda, \lambda) = \sum \widetilde{f}(Q+u_1, u_1, u_2)e^{i(\Lambda, u_2)}e^{i(\lambda, u_1-u_2)}.$$

In passing from $\widetilde{f}(Q, u_1, u_2)$ to $\varphi(Q, \Lambda, \lambda)$, any operator A acting on $\hat{l}_2^{(2)}$ which commutes with the shift operator U_v is transformed into an operator $\widehat{A}$ acting on φ according to the formula

$$(\widehat{A}\varphi)(Q, \Lambda, \lambda) = \sum_{Q'}\int_{T^\nu}\widehat{A}_{Q,Q'}(\Lambda, \lambda, \lambda')\varphi(Q', \Lambda, \lambda')\,d\lambda', \tag{4.45a}$$

with the kernel $\widehat{A}_{Q,Q'}(\Lambda, \lambda, \lambda')$ given by

$$\widehat{A}_{Q,Q'}(\Lambda, \lambda, \lambda') = \sum_{z, u_1, u_2} A_{Q,0,z,Q'+u_1,u_1,u_2} e^{-i(\Lambda, u_2-z)} e^{-i(\lambda, z)} e^{-i(\lambda', u_1-u_2)}, \quad (4.45b)$$

where $A_{Q,u_1,u_2,Q',u'_1,u'_2}$ are the matrix elements of the operator A. Hence, equation (4.37b) for the kernel $S(Q, u_1, u_2, u'_1, u'_2) = S_{Q,u_1,u_2,\bar{0},u'_1 u'_2}$ of the operator S, leads, after applying the Fourier transform, to the following equation for the kernel $\widehat{S}_Q(\Lambda, \lambda, \lambda') = \widehat{S}_{Q,\bar{0}}(\Lambda, \lambda, \lambda')$:

$$\begin{aligned}\widehat{S}_Q(\Lambda, \lambda, \lambda') &= \int_{T^\nu} \widehat{\widehat{R}}_{Q,\bar{0}}(\Lambda, \lambda, \zeta) (\widehat{\mathscr{T}_{00}^{(2)}})^{-1}(\Lambda, \zeta, \lambda')\, d\zeta \\ &+ \sum_{Q'} \iint \widehat{\widehat{R}}_{Q,Q'}(\Lambda, \lambda, \zeta_1) \\ &\qquad \times \widehat{S}_{Q'}(\Lambda, \zeta_1, \zeta_2) (\widehat{\mathscr{T}_{00}^{(2)}})^{-1}(\Lambda, \zeta_2, \lambda')\, d\zeta_1\, d\zeta_2 \\ &- \sum_{Q'} \iiint \widehat{S}_Q(\Lambda, \lambda, \zeta_1) \widehat{\widehat{R}}_{\bar{0},Q'}(\Lambda, \zeta_1, \zeta_2) \\ &\qquad \times \widehat{S}_{Q'}(\Lambda, \zeta_2, \zeta_3) (\widehat{\mathscr{T}_{00}^{(2)}})^{-1}(\Lambda, \zeta_3, \lambda')\, d\zeta_1\, d\zeta_2\, d\zeta_3. \quad (4.46)\end{aligned}$$

Observe that as a result of (4.38), the kernel $(\widehat{\mathscr{T}_{00}^{(2)}})^{-1}(\Lambda, \lambda, \lambda')$ is written as

$$(\widehat{\mathscr{T}_{00}^{(2)}})^{-1}(\Lambda, \lambda, \lambda') = \frac{\delta(\lambda - \lambda')}{\tilde{p}(\lambda)\tilde{p}(\Lambda - \lambda)} + \mathscr{D}(\Lambda, \lambda, \lambda'), \quad (4.47)$$

where the function $\mathscr{D}(\Lambda, \lambda, \lambda')$ is analytic in the complex neighborhood $W_d^{(3)}$ of the product of tori $(T^\nu)^3$:

$$W_d^{(3)} = \{|\operatorname{Im}\Lambda| < d,\ |\operatorname{Im}\lambda| < d,\ |\operatorname{Im}\lambda'| < d\}. \quad (4.48)$$

Moreover, from formulas (4.35b), (4.31) and the estimate (4.26) it follows that for some $\widehat{q} \in (0, 1)$

$$|M_{u'_1,u'_2,Q,u_1,u_2}| < c\overline{M}^{-|\operatorname{supp} Q|} \widehat{q}^{\,d_{\operatorname{supp} Q \cup \{u_1\} \cup \{u_2\} \cup \{u'_1\} \cup \{u'_2\}}}.$$

Hence it is easy to see that the kernel $\widehat{\widehat{R}}_{Q,\bar{0}}(\Lambda, \lambda, \lambda')$ is analytic in the variables $\Lambda, \lambda, \lambda'$ in the neighborhood $W_d^{(3)}$ and in this neighborhood satisfies the inequality

$$|\widehat{\widehat{R}}_{Q,\bar{0}}(\Lambda, \lambda, \lambda')| < C\widehat{q}^{\,d_{\operatorname{supp} Q \cup \{0\}}} \overline{M}^{-|\operatorname{supp} Q|}, \quad (4.49)$$

where c is small for $\varepsilon \equiv \max\{\varepsilon_0, \varepsilon_1\}$ small. We can now state a lemma similar to Lemma 3.5 in the previous section.

LEMMA 4.7. *There is a unique solution* $\widehat{S}_Q(\Lambda, \lambda, \lambda')$ *of equation* (4.46) *which is analytic in the variables* $\Lambda, \lambda, \lambda'$ *in the neighborhood given by* (4.48) *for some* $d > 0$ *and satisfies* (*in this neighborhood*) *the inequality*

$$|\widehat{S}_Q(\Lambda, \lambda, \lambda')| < C\overline{M}^{-|\operatorname{supp} Q|}\widehat{q}^{\,d_{\operatorname{supp} Q \cup \{0\}}}, \tag{4.50}$$

where C *is small for* ε_0, ε_1 *small, and* $\widehat{q} \in (0, 1)$.

PROOF. Here we use the technique that we have already used many times. That is, we consider the Banach space $\mathscr{L}_{W_d^3, \widehat{q}}$ of the functions $\widehat{S}_Q(\Lambda, \lambda, \lambda')$ analytic in the neighborhood (4.48) and satisfying the estimate (4.50). The norm in $\mathscr{L}_{W_d^3, \widehat{q}}$ is given by the following formula:

$$\|\widehat{S}_Q(\Lambda, \lambda, \lambda')\| = \sup_{Q, \Lambda, \lambda, \lambda'} |\widehat{S}_Q(\Lambda, \lambda, \lambda')|\overline{M}^{|\operatorname{supp} Q|}\widehat{q}^{\,-d_{\operatorname{supp} Q \cup \{0\}}}.$$

Furthermore, we consider the right-hand side of (4.46) as a map

$$\mathscr{B}: \mathscr{L}_{W_d^3, \widehat{q}} \mapsto \mathscr{L}_{W_d^3, \widehat{q}} \tag{4.51}$$

and show that this map transforms into itself any sufficiently small sphere $\|\widehat{S}\| < \kappa$, $\kappa = \kappa(\varepsilon_0, \varepsilon_1)$, is small for ε_0, ε_1 small, and acts as a contraction on this sphere. The first important step is to prove that the second term on the right-hand side of (4.46), denoted $\mathscr{T}_{11}^{(2)} S (\mathscr{T}_{00}^{(2)})^{-1}$, which is linear in S, being regarded as an operator in $\mathscr{L}_{W_d^{(3)}, \widehat{q}}$, is smaller than 1 for some choice of d and $\widehat{q}$. As before, one must consider first the contribution to the operator (4.51) $\mathscr{T}_{11}^{(2),0} S (\mathscr{T}_{00}^{(2)})^{-1}_{\text{main}}$ where $\mathscr{T}_{11}^{(2),0}$ is the unperturbed operator and $(\mathscr{T}_{00}^{(2)})^{-1}_{\text{main}}$ corresponds to the operator of multiplication by $1/(\widetilde{p}_0(\lambda)\widetilde{p}_0(\Lambda - \lambda))$. This contribution, as follows from (4.45a) and (4.6), is equal to

$$\mu_Q \sum_x p_0(x) \widehat{S}_{Q-x}(\Lambda, \lambda, \lambda') e^{i(\lambda, x)} \frac{\widetilde{p}_0(\Lambda - \lambda)}{\widetilde{p}_0(\lambda')\widetilde{p}_0(\Lambda - \lambda')}. \tag{4.52}$$

Using the same arguments as in the proof of Lemma 4.2, we find that in the case when $\widehat{q} < 1$ but $\widehat{q}$ is close enough to 1, and d is small enough, the expression (4.52) does not exceed

$$\theta \|\widehat{S}\| M^{-|\operatorname{supp} Q|}\widehat{q}^{\,d_{\operatorname{supp} Q \cup \{0\}}},$$

where $\theta \in (0, 1)$, and is an analytic function of $\Lambda, \lambda, \lambda'$ in the region $W_d^{(3)}$. The estimate of the contribution of the remaining terms of $\mathscr{T}_{11}^{(2)} S (\mathscr{T}_{00}^{(2)})^{-1}$ is carried out similarly and the result follows from the fact that each of the terms is small for small ε_0, ε_1. In the same way one estimates the contribution of the third term $S\mathscr{T}_{01}^{(2)} S (\mathscr{T}_{11}^{(2)})^{-1}$ of equation (4.46).

The second important step in the proof is to note that the first "free" term in (4.46) also belongs to $\mathscr{L}_{W_d^3, \widehat{q}}$ with small norm (see the estimate (4.49)). Hence the assertion of Lemma 4.7 follows.

Since by (4.45b) we have

$$S(Q, u_1, u_2, u_1', u_2') = \int d\Lambda\, d\lambda\, d\lambda' \widehat{S}_{Q-u_1}(\Lambda, \lambda, \lambda') e^{i(\Lambda, u_2'-u_2)} e^{-i(\lambda, u_1-u_2)} e^{i(\lambda', u_1'-u_2')},$$

inequality (4.50) gives the estimate (4.43). Lemma 4.6 is proved. □

Hence the invariant space ℓ has been constructed.

REMARK. Note that since $\chi_\Gamma(0) = 0$ for $\Gamma \neq \overline{0}$, it is easy to see, by the definition of g^{u_1, u_2} given by equations (4.28a,b), that for all Γ, z_1, z_2

$$\sum_{u_1, u_2} g^{u_1, u_2}_{\Gamma, z_1, z_2} = 0. \tag{4.52a}$$

Noting that $(\mathcal{T}^{(2)}_{00})^{-1} 1 = 1$, it is easy to see from (4.36ii), (4.35b) that for

$$\mathbf{S}(Q, u_1, u_2) \equiv \sum_{v_1, v_2} S(Q, u_1, u_2, v_1, v_2)$$

equation (4.37b) yields

$$\begin{aligned}\mathbf{S}(Q, u_1, u_2) = & \sum_{Q', v_1, v_2} \widetilde{R}_{Q, u_1, u_2, Q', v_1, v_2} \mathbf{S}(Q', v_1, v_2) \\ & - \sum_{Q', v_1, v_2, v_1', v_2'} S(Q, u_1, u_2, v_1, v_2) \\ & \times \widetilde{R}_{\overline{0}, v_1, v_2, Q', v_1', v_2'} \mathbf{S}(Q', v_1', v_2').\end{aligned} \tag{4.52b}$$

This equation can be analyzed just as (4.37b) ($S(Q, u_1, u_2, v_1, v_2)$ is assumed to be known) to deduce that it admits a unique solution $\mathbf{S}(Q, u_1, u_2) \equiv 0$.

In the space ℓ we take the basis

$$\begin{aligned} b^{u_1, u_2} &= h^{u_1, u_2} + S h^{u_1, u_2} \\ &= h^{u_1, u_2} + \sum_{Q \neq \overline{0}, v_1, v_2} S(Q, v_1, v_2, u_1, u_2) h^{Q, v_1, v_2}. \end{aligned} \tag{4.53}$$

Any vector $\varphi \in \ell$ can be represented as a series

$$\varphi = \sum f_{u_1, u_2} b^{u_1, u_2},$$

and the action of the operator $\mathcal{T}^{(2)}|_\ell \equiv \mathcal{T}_2$ is given by the formula

$$(\mathcal{T}_2 f)_{u_1, u_2} = \sum_{u_1', u_2'} B_{u_1, u_2, u_1', u_2'} f_{u_1', u_2'}. \tag{4.54a}$$

Here $f \equiv \{f_{u_1, u_2}\}$ and the matrix elements $B_{u_1, u_2, u_1', u_2'}$ can be shown to be

$$\begin{aligned} B_{u_1, u_2, v_1, v_2} &= (\mathcal{T}^{(2)}_{00})_{u_1, u_2, v_1, v_2} \\ &\quad + \sum_{Q \neq \overline{0}, u_1', u_2'} \widetilde{R}_{\overline{0}, u_1, u_2, Q, u_1', u_2'} S(Q, u_1', u_2', v_1, v_2) \\ &= r(v_1 - u_1) r(v_2 - u_2) + D(v_1, v_2, u_1, u_2), \end{aligned} \tag{4.54b}$$

where (by (4.36i))

$$D(v_1, v_2, u_1, u_2) = \widehat{D}(v_1, v_2, u_1, u_2) + \sum_{Q\neq\overline{0}, u_1', u_2'} \widetilde{R}_{\overline{0}, u_1, u_2, Q, u_1', u_2'} S(Q, u_1', u_2', v_1, v_2). \tag{4.54c}$$

Using formulas (4.36iii) and (4.4b), one proves that the series (4.54b) converges for $\overline{M}$ large enough, and that D satisfies the estimate

$$|D(u_1, u_2, u_1', u_2')| < c\overline{q}^{\,d_{\{u_1\}\cup\{u_2\}\cup\{u_1'\}\cup\{u_2'\}}}, \tag{4.54d}$$

where $\overline{q} \in (0, 1)$, and $c \ll 1$ for ε small. The proof follows easily from inequality (4.26) and from (4.35a), if one observes that $\operatorname{supp} Q \subset \{u_1, u_2\}$.

Note that the basis $\{b^{u_1, u_2}\}$ can be written as

$$b^{u_1, u_2} = \overline{h}^{\overline{0}, u_1, u_2} + \sum_{Q'\neq\overline{0}, u_1', u_2'} \mathscr{S}_{u_1, u_2, Q', u_1', u_2'} \overline{h}^{Q', u_1', u_2'}, \tag{4.55a}$$

where, as follows from (4.53) and (4.32),

$$\mathscr{S}_{u_1, u_2, Q', u_1', u_2'} = S(Q', u_1', u_2', u_1, u_2) + B_{u_1, u_2, Q', u_1', u_2'}. \tag{4.55b}$$

This means that going over to the space $\mathscr{H}^2$, and to the invariant subspace $\mathscr{H}_2 \subset \mathscr{H}^2$ corresponding to ℓ, we find a basis $\{\psi_{u_1, u_2}\}$ in $\mathscr{H}_2$ of the form

$$\psi_{u_1, u_2} = \Psi_{\overline{0}, u_1, u_2} + \sum_{Q'\neq\overline{0}, u_1', u_2'} \mathscr{S}_{u_1, u_2, Q', u_1', u_2'} \Psi_{Q', u_1', u_2'}, \tag{4.55c}$$

and from estimate (4.43) it follows that

$$|\psi_{u_1, u_2}(\xi, x_1, x_2)| < \text{const}\, \widetilde{q}^{\,|u_1-x_1|+|u_2-x_2|}, \qquad \widetilde{q} \in (0, 1). \tag{4.55d}$$

The proof is obtained by noting that $|\Psi_{\Gamma, u_1, u_2}| \leqslant M^{|\operatorname{supp}\Gamma|}$ and by applying Proposition A.2 of the Appendix.

From (4.54a,b,c) and translation invariance we get (4.10c) and (4.11a), and inequality (4.54d) is inequality (4.11b). As for relation (4.11c), it can be proved as follows. From (4.31) and the remark following the proof of Lemma 4.6 (in particular from (4.52a)) it follows that for all Q, v_1, v_2

$$\sum_{u_1, u_2} \mathscr{S}_{u_1, u_2, Q, v_1, v_2} = 0. \tag{4.56a}$$

Hence by (4.55c) we get

$$\sum_{u_1, u_2} \psi_{u_1, u_2} = \sum_{u_1, u_2} \Psi_{\overline{0}, u_1, u_2} = 1. \tag{4.56b}$$

Since $\mathscr{T}^{(2)} 1 = 1$, and, by (4.10a), $\sum_u r(u) = 1$, relation (4.11c) follows.

In order to construct an invariant subspace $\overline{\mathscr{H}}_2 \subset \mathscr{H}^2$ which is the complement of $\mathscr{H}_2$, we proceed in the same way as in §3: we consider the operator adjoint to $\mathscr{T}^{(2)}$, $(\mathscr{T}^{(2)})^*$, which it is convenient to consider in the basis

$$\Psi^*_{\Gamma, z_1, z_2}(\xi, x_1, x_2) = \Psi^*_\Gamma(\xi)\delta_{x_1 z_1}\delta_{x_2 z_2}$$

biorthogonal to the basis Ψ_{Γ, z_1, z_2} (the functional $\Psi^*_\Gamma(\xi)$ was defined in the preceding section). Since $(\mathscr{T}^{(2)})^*$ has the same properties as $\mathscr{T}^{(2)}$, we repeat the construction above and define an invariant two-particle subspace $\mathscr{H}_2^* \subset \mathscr{H}_2$ and a basis $\{\psi^*_{u_1, u_2}\}$ in it of the form

$$\psi^*_{u_1, u_2} = \Psi^*_{\bar{0}, u_1, u_2} + \sum_{Q' \neq \bar{0}, u_1', u_2'} \mathscr{S}^*_{u_1, u_2, Q', u_1', u_2'} \Psi^*_{Q', u_1', u_2'},$$

where the coefficients $\mathscr{S}^*_{u_1, u_2, Q', u_1', u_2'}$ are defined in the same way as the coefficients $\mathscr{S}_{u_1, u_2, Q', u_1', u_2'}$, with the only difference that they are written in terms of the operator $(\mathscr{T}^{(2)})^*$ and of its representation in the basis $\{\Psi^*_{\Gamma, z_1, z_2}\}$. It is now easy to see that the vectors

$$\psi_{Q, z_1, z_2} = \Psi_{Q, z_1, z_2} - \sum_{v_1, v_2} \overline{\mathscr{S}^*_{v_1, v_2, Q, z_1, z_2}} \Psi_{\bar{0}, v_1, v_2}, \qquad Q \neq \bar{0}, \tag{4.57a}$$

are orthogonal to all vectors of the type (4.55c), and form a basis in the space $\overline{\mathscr{H}}_2$ orthogonal to the space $\mathscr{H}_2^*$. From the estimates for $S^*_{v_1, v_2, Q, z_1, z_2}$ analogous to the estimates for $S_{v_1, v_2, Q, z_1, z_2}$ we get

$$|\psi_{Q, z_1, z_2}(\xi, x_1, x_2)| < \widetilde{M}^{|\operatorname{supp} Q|} \tilde{q}^{|z_1 - x_1| + |z_2 - x_2|}, \tag{4.57b}$$

where $\widetilde{M} > 1$ is some constant and $\tilde{q} \in (0, 1)$. Clearly the space $\overline{\mathscr{H}}_2$ is invariant with respect to the operator $\mathscr{T}^{(2)}$ and the group $\{U_v, v \in \mathbb{Z}^\nu\}$. For any vector $\varphi \in \mathscr{H}$ we have the decomposition

$$\varphi = \varphi^{(2)} + \overline{\varphi}^{(2)}, \qquad \varphi^{(2)} \in \mathscr{H}_2, \ \overline{\varphi}^{(2)} \in \overline{\mathscr{H}}_2, \tag{4.58}$$

and its uniqueness follows easily from the fact that the spaces $\mathscr{H}_2$ and $\mathscr{H}_2^0 = \mathscr{H}_2|_{\varepsilon=0}$ (the unperturbed invariant two-particle subspace for $\varepsilon = 0$) are close to each other.

Therefore it remains to prove that $\varphi^{(2)}, \overline{\varphi}^{(2)} \in \mathscr{C}_2(\widehat{\Omega})$, and that the last assertion (4.13) of Theorem 1 holds. We write (4.55c) in the form

$$\begin{aligned} \psi_{u_1, u_2} &= \Psi_{\bar{0}, u_1, u_2} \\ &\quad + \sum \mathscr{S}_{u_1, u_2, Q', u_1', u_2'} \left(\Psi_{Q', u_1', u_2'} - \sum \overline{\mathscr{S}^*_{v_1, v_2, Q', u_1', u_2'}} \Psi_{\bar{0}, v_1, v_2} \right) \\ &\quad + \sum_{v_1, v_2} K_{u_1, u_2, v_1, v_2} \Psi_{\bar{0}, v_1, v_2} \\ &= \Psi_{\bar{0}, u_1, u_2} + \sum \mathscr{S}_{u_1, u_2, Q', u_1', u_2'} \psi_{Q', u_1', u_2'} \\ &\quad + \sum K_{u_1, u_2, v_1, v_2} \Psi_{\bar{0}, v_1, v_2}, \end{aligned} \tag{4.59}$$

where

$$K_{u_1,u_2,v_1,v_2} = \sum_{Q'\neq\bar{0},u'_1,u'_2} \mathscr{S}_{u_1,u_2,Q',u'_1,u'_2}\overline{\mathscr{S}^*_{v_1,v_2,Q',u'_1,u'_2}} \tag{4.60a}$$

and in the second equality we use (4.57a). From the estimate (4.43) and (4.16) and formula (4.31) it is easy to deduce that

$$|K_{u_1,u_2,v_1,v_2}| < \bar{c}q^{|u_1-v_1|+|u_2-v_2|}, \tag{4.60b}$$

where $\bar{c} \ll 1$ for ε_0, ε_1 small, and $q \in (0, 1)$. Hence it follows that the matrix $\{\delta_{u_1v_1}\delta_{u_2v_2} + K_{u_1,u_2,v_1,v_2}\}$ is invertible and its inverse has the form

$$\delta_{u_1v_1}\delta_{u_2v_2} + \overline{K}_{u_1,u_2,v_1,v_2},$$

where $\overline{K}_{u_1,u_2,v_1,v_2}$ satisfies an estimate analogous to (4.60b). In this way it follows from (4.59) that

$$\Psi_{\bar{0},u_1,u_2} = \psi_{u_1,u_2} + \sum_{v_1,v_2} \overline{K}_{u_1,u_2,v_1,v_2}\psi_{v_1,v_2} + \sum S'_{u_1,u_2,Q',u'_1,u'_2}\psi_{Q',u'_1,u'_2}, \tag{4.61}$$

where

$$S'_{u_1,u_2,Q',u'_1,u'_2} = -\mathscr{S}_{u_1,u_2,Q',u'_1,u'_2} - \sum_{v_1,v_2} \overline{K}_{u_1,u_2,v_1,v_2}\mathscr{S}_{v_1,v_2,Q',u'_1,u'_2}.$$

The sum of the first two terms on the right in (4.61) is the component $\Psi^{(2)}_{\bar{0},u_1,u_2} \in \mathscr{H}_2$, and the third term is the component $\overline{\Psi}^{(2)}_{\bar{0},u_1,u_2} \in \overline{\mathscr{H}}_2$ in the representation (4.58) for the vector $\Psi_{\bar{0},u_1,u_2}$. From the estimate (4.55d) and an estimate for $\overline{K}_{u_1,u_2,v_1,v_2}$ of the type (4.60b) we get the estimate for $\Psi^{(2)}_{\bar{0},u_1,u_2}(\xi, x_1, x_2)$ similar to estimate (4.55d). Hence, observing that ψ_{u_1,u_2} depends continuously on ξ, we get $\Psi^{(2)}_{\bar{0},u_1,u_2} \in \mathscr{C}_2(\widehat{\Omega})$.

We now prove a similar assertion for $\overline{\Psi}^{(2)}_{\bar{0},u_1,u_2}$. From inequalities (4.43) and (4.24), formulas (4.31), and the definition of S' it follows that the coefficients $f^{u_1,u_2}_{Q,v_1,v_2} = S'_{u_1,u_2,Q,v_1,v_2}$ of the expansion of $\overline{\Psi}^{(2)}_{\bar{0},u_1,u_2}$ in the basis $\{\psi_{Q,v_1,v_2}\}$ satisfy the relation

$$\sum_{Q\neq\bar{0}} |f^{u_1,u_2}_{Q,v_1,v_2}|\widetilde{M}^{|\operatorname{supp} Q|} < c\tilde{q}^{|u_1-v_1|+|u_2-v_2|}, \tag{4.62}$$

where c is small for small ε_0, ε_1, $\tilde{q} \in (0, 1)$, and $\widetilde{M} > 1$ is the same constant as in inequality (4.57b). From this and from inequality (4.57b) it follows that

$$|\overline{\Psi}^{(2)}_{\bar{0},u_1,u_2}(\xi, x_1, x_2)| < c'q_1^{|u_1-x_1|+|u_2-x_2|}, \tag{4.63}$$

where c' is small for ε_0, ε_1 small and $q_1 \in (0, 1)$. Hence it follows that $\overline{\Psi}^{(2)}_{\bar{0},u_1,u_2} \in \mathscr{C}_2(\widehat{\Omega})$.

Note also that matrix elements B_{Q,u_1,u_2,Q',v_1,v_2} of the operator $\mathscr{T}^{(2)}|_{\overline{\mathscr{H}}_2} \equiv \widetilde{\mathscr{T}}$ in the basis ψ_{Q,u_1,u_2} are equal to

$$B_{Q,u_1,u_2,Q',v_1,v_2} = R^{(2)}_{Q,u_1,u_2;Q',v_1,v_2} - \sum_{v_1',v_2'} R^{(2)}_{Q,u_1,u_2,\overline{0},v_1',v_2'} \overline{\mathscr{S}_{v_1',v_2',Q',v_1,v_2}}. \tag{4.64}$$

Now we consider the space $\widetilde{l}_{\widetilde{M}}(\mathfrak{M} \times \mathbb{Z}^\nu \times \mathbb{Z}^\nu)$ of functions $f = \{f_{\Gamma,z_1,z_2}, \Gamma \in \mathfrak{M}, \Gamma \neq \overline{0}, z_1, z_2 \in \mathbb{Z}^\nu\}$ such that

$$\sup_{z_1,z_2} \sum_{\Gamma\neq\overline{0}} |f_{\Gamma,z_1,z_2}| \widetilde{M}^{|\operatorname{supp}\Gamma|} \equiv \|f\|_{\widetilde{l}_{\widetilde{M}}} < \infty. \tag{4.65}$$

LEMMA 4.8. *The operator* $\widetilde{\mathscr{T}}$ *with matrix elements given by* (4.64) *is bounded on* $\widetilde{l}_{\widetilde{M}}(\mathfrak{M} \times \mathbb{Z}^\nu \times \mathbb{Z}^\nu)$ *and its norm satisfies the inequality* $\|\widetilde{\mathscr{T}}\|_{\widetilde{l}_{\widetilde{M}}} < \theta < 1$.

PROOF. The proof is given in the Appendix. □

The proof of inequality (4.13) now goes as follows. Any function $\varphi \in \overline{\mathscr{H}}_2$ can be expanded in the basis $\{\psi_{\Gamma,z_1,z_2}, \Gamma \neq \overline{0}\}$:

$$\varphi = \sum_{\Gamma,z_1,z_2} f_{\Gamma,z_1,z_2} \psi_{\Gamma,z_1,z_2}, \qquad \Gamma \neq \overline{0}.$$

If the coefficients f_{Γ,z_1,z_2} belong to $\widetilde{l}_{\widetilde{M}}$, then $\varphi \in \mathscr{C}_2(\widehat{\Omega})$ (see the considerations above) and hence $\varphi \in \mathscr{C}_\infty(\widehat{\Omega})$. Moreover

$$\|\varphi\|_{\mathscr{C}_\infty(\widehat{\Omega})} \equiv \sup_{\xi,x_1,x_2} |\varphi(\xi, x_1, x_2)| \leqslant C\|f\|_{\widetilde{l}(\widetilde{M})}, \tag{4.66}$$

where C is an absolute constant. From the above lemma it follows that

$$(\mathscr{T}^{(2)})^t \varphi \in \mathscr{C}_\infty(\widehat{\Omega}),$$
$$\sup_{\xi,x_1,x_2} |((\mathscr{T}^{(2)})^t\varphi)(\xi, x_1, x_2)| < C\|(\mathscr{T}^{(2)})^t f\|_{\widetilde{l}(\widetilde{M})} < C\theta^t \|f\|_{\widetilde{l}(\widetilde{M})}. \tag{4.67}$$

Inequality (4.13) follows from this and from inequality (4.62).

As we noted above, Theorem 4.1 implies that the asymptotic behavior of the probability (4.14) is determined by the expression

$$\int ((\mathscr{T}^{(2)})^t \Psi^{(2)}_{\overline{0},x_1,x_2})(\xi, z_1, z_2)\, d\Pi(\xi),$$

where $\Pi(\xi)$ is the initial distribution of the field. The asymptotics will be computed in the following section.

§5. Random walk of two particles (proof of Theorem 2). Conclusion of the proof: Study of the resolvent of $\mathcal{T}^{(2)}$ and computation of the asymptotics

In the preceding section we singled out the leading invariant subspace $\mathcal{H}_2 \subset \mathcal{H}$ of the transfer matrix $\mathcal{T}$ and showed that for an appropriate basis $\{\psi_{u_1,u_2}(\xi, z_1, z_2)\}$ in $\mathcal{H}_2$, the transfer matrix can be written in the form (4.10c). In this section we first express the asymptotics of the probability (2.14) in terms of the resolvent of the operator (4.10c), and then compute the asymptotics itself.

The component $\Psi^{(2)}_{\bar{0},u_1,u_2}$ of the function $\Psi_{\bar{0},u_1,u_2}(\xi, x_1, x_2)=\delta_{u_1,x_1}\delta_{u_2,x_2}$ in the space $\mathcal{H}_2$ is given in terms of the basis $\{\psi_{u_1,u_2}\}$ by (4.61):

$$\Psi^{(2)}_{\bar{0},u_1,u_2} = \psi_{u_1,u_2} + \sum_{v_1,v_2} \overline{K}_{u_1,u_2,v_1,v_2}\psi_{v_1,v_2}, \tag{5.1}$$

where the coefficients $\overline{K}_{\dots}$ satisfy the same inequality (4.60b) as the coefficients $K_{\dots}$.

Therefore

$$\begin{aligned}((\mathcal{T}^{(2)})^t\Psi^{(2)}_{\bar{0},u_1,u_2}) &= \sum_{v_1,v_2,v_1',v_2'} (\delta_{u_1v_1}\delta_{u_2v_2} + \overline{K}_{u_1,u_2,v_1,v_2}) S^t_{v_1,v_2,v_1',v_2'}\psi_{v_1',v_2'} \\ &= \sum_{v_1,v_2,v_1',v_2'} (\delta_{u_1v_1}\delta_{u_2v_2} + \overline{K}_{u_1,u_2,v_1,v_2}) \\ &\quad\times \int_{(T^\nu)^4} e^{i(v_1,\lambda_1)+i(v_2,\lambda_2)-i(v_1',\mu_1)-i(v_2',\mu_2)} \\ &\quad\quad\times \widetilde{S}^t(\lambda_1,\lambda_2,\mu_1,\mu_2)\,d\lambda_1\,d\lambda_2\,d\mu_1\,d\mu_2\,\psi_{v_1',v_2'}.\end{aligned} \tag{5.2}$$

Here $S^t_{v_1,v_2,v_1',v_2'}$ are the matrix elements of the operator $(\mathcal{T}^{(2)})^t$ in the basis ψ_{u_1,u_2} and $\widetilde{S}^t(\lambda_1,\lambda_2,\mu_1,\mu_2)$ is its (generalized) Fourier transform. Note that $S^1_{u_1,u_2,v_1,v_2} = B_{u_1,u_2,v_1,v_2}$ for $t = 1$, where the matrix B is given by (4.54a). Hence, using translation invariance (4.11a), we see that $\widetilde{S}^1(\lambda_1,\lambda_2,\mu_1,\mu_2)$ is given by

$$\begin{aligned}\widetilde{S}^1(\lambda_1,\lambda_2,\mu_1,\mu_2) &= \tilde{p}(\lambda_1)\tilde{p}(\lambda_2)\delta(\lambda_1-\mu_1)\delta(\lambda_2-\mu_2) \\ &\quad + \widetilde{D}(\lambda_1,\lambda_2,\mu_1,\mu_2)\delta(\lambda_1+\lambda_2-\mu_1-\mu_2),\end{aligned} \tag{5.3}$$

where $\delta(\cdot)$ is the δ-function on the ν-dimensional torus T^ν, and where $\widetilde{D}(\lambda_1,\lambda_2,\mu_1,\mu_2)$ is an analytic function defined on the complex manifold

$$\{\lambda_1+\lambda_2=\mu_1+\mu_2 : |\operatorname{Im}\lambda_1|,\ |\operatorname{Im}\lambda_2|,\ |\operatorname{Im}\mu_1|,\ |\operatorname{Im}\mu_2| < d\}, \tag{5.4}$$

and small in absolute value for small ε. Moreover, due to (4.11c),

$$\widetilde{D}(0,0,\mu_1,\mu_2) = 0 \tag{5.5}$$

for all $\mu_1, \mu_2 : \mu_1 + \mu_2 = 0$. Introducing the new variables $\Lambda = \lambda_1 + \lambda_2$, $\lambda = \lambda_1$, $\Lambda' = \mu_1 + \mu_2$, $\mu = \mu_1$ and setting

$$D_\Lambda(\lambda, \mu) = \widetilde{D}(\lambda, \Lambda - \lambda, \mu, \Lambda - \mu), \qquad p_\Lambda(\lambda) = \widetilde{p}(\lambda)\widetilde{p}(\Lambda - \lambda), \tag{5.6}$$

we see that the kernel of the operator $\mathscr{T}^{(2)} - \zeta 1$, in the Fourier representation, is written as

$$(p_\Lambda(\lambda) - \zeta)\left(\delta(\lambda - \mu) + \frac{D_\Lambda(\lambda, \mu)}{p_\Lambda(\lambda) - \zeta}\right)\delta(\Lambda - \Lambda').$$

Using Fredholm theory, the kernel (in the Fourier representation) of the resolvent $R(\zeta) \equiv (\mathscr{T}^{(2)} - \zeta 1)^{-1}$ is written as

$$R_{\Lambda, \Lambda'}(\lambda, \mu; \zeta) = \left(\frac{\delta(\lambda - \mu)}{p_\Lambda(\lambda) - \zeta} + \frac{T_\Lambda(\lambda, \mu; \zeta)}{(p_\Lambda(\lambda) - \zeta)(p_\Lambda(\mu) - \zeta)}\right)\delta(\Lambda - \Lambda'), \tag{5.7}$$

where the Fredholm determinants Δ_Λ and T_Λ are given by

$$\begin{aligned}
\Delta_\Lambda(\zeta) &= 1 + \int_{T^\nu} \frac{D_\Lambda(\nu, \nu)}{(p_\Lambda(\nu) - \zeta)}\, d\nu + \dots \\
&\quad + \frac{1}{n!}\int_{(T^\nu)^n} \frac{\det\{D_\Lambda(\nu_i, \nu_j)\}_{i,j=1,\dots,n}}{\prod_{i=1}^n (p_\Lambda(\nu_i) - \zeta)} + \dots, \\
T_\Lambda(\lambda, \mu; \zeta) &= -\frac{1}{\Delta_\Lambda(\zeta)}\Bigg[D_\Lambda(\lambda, \mu) \\
&\quad + \sum_{n=1}^\infty \frac{1}{n!}\int_{(T^\nu)^n} \frac{K_n^\Lambda(\lambda, \mu, \nu_1, \dots, \nu_n)}{\prod_{i=1}^n (p_\Lambda(\nu_i) - \zeta)}\, d\nu_1 \cdots d\nu_n\Bigg],
\end{aligned} \tag{5.8}$$

$$K_n^\Lambda = \begin{vmatrix} D_\Lambda(\lambda, \mu) & D_\Lambda(\lambda, \nu_1) & \dots & D_\Lambda(\lambda, \nu_n) \\ D_\Lambda(\nu_1, \mu) & D_\Lambda(\nu_1, \nu_1) & \dots & D_\Lambda(\nu_1, \nu_n) \\ \vdots & \vdots & \ddots & \vdots \\ D_\Lambda(\nu_n, \mu) & D_\Lambda(\nu_n, \nu_1) & \dots & D_\Lambda(\nu_n, \nu_n) \end{vmatrix}.$$

Going back to the original variables we find the following formula for the kernel $R(\lambda_1, \lambda_2, \mu_1, \mu_2; \zeta)$ of the resolvent:

$$\begin{aligned}
R(\lambda_1, \lambda_2, \mu_1, \mu_2; \zeta) &= \frac{\delta(\lambda_1 - \mu_1)\delta(\lambda_2 - \mu_2)}{\widetilde{p}(\lambda_1)\widetilde{p}(\lambda_2) - \zeta} \\
&\quad + \frac{T(\lambda_1, \lambda_2, \mu_1, \mu_2; \zeta)\delta(\lambda_1 + \lambda_2 - \mu_1 - \mu_2)}{(\widetilde{p}(\lambda_1)\widetilde{p}(\lambda_2) - \zeta)(\widetilde{p}(\mu_1)\widetilde{p}(\mu_2) - \zeta)},
\end{aligned} \tag{5.9}$$

with

$$T(\lambda_1, \lambda_2, \mu_1, \mu_2; \zeta) = T_{\lambda_1 + \lambda_2}(\lambda_1, \mu_1; \zeta).$$

From (5.5) and (5.8) one can see that for $\Lambda = 0$ and all μ

$$T_0(0, \mu; \zeta) = 0. \tag{5.10}$$

Moreover, the function $\Delta_\Lambda(\zeta)$ is defined for ζ in the set

$$\Pi_\Lambda = \mathbb{C} \setminus \operatorname{Range} p_\Lambda(\lambda),$$

and is analytic there. Finally we use the general formula

$$(\mathscr{T}^{(2)})^t = \frac{1}{2\pi i}\int_\gamma \zeta^t R(\zeta)\, d\zeta, \tag{5.11}$$

where the integral is taken over a contour γ of the complex plane containing the spectrum of $\mathscr{T}^{(2)}$. In this way the kernel of $\widetilde{S}^t(\lambda_1, \lambda_2, \mu_1, \mu_2)$ has the form

$$\begin{aligned}
&\widetilde{S}^t(\lambda_1, \lambda_2, \mu_1, \mu_2)\\
&\quad = \frac{1}{2\pi i}\int_\gamma \frac{\zeta^t \delta(\lambda_1-\mu_1)\delta(\lambda_2-\mu_2)}{\widetilde{p}(\lambda_1)\widetilde{p}(\lambda_2)-\zeta}\, d\zeta\\
&\qquad + \frac{1}{2\pi i}\int_\gamma \zeta^t \frac{T(\lambda_1, \lambda_2, \mu_1, \mu_2; \zeta)\delta(\lambda_1+\lambda_2-\mu_1-\mu_2)}{(\widetilde{p}(\lambda_1)\widetilde{p}(\lambda_2)-\zeta)(\widetilde{p}(\mu_1)\widetilde{p}(\mu_2)-\zeta)}\, d\zeta\\
&\quad = (\widetilde{p}(\lambda_1)\widetilde{p}(\lambda_2))^t \delta(\lambda_1-\mu_1)\delta(\lambda_2-\mu_2)\\
&\qquad + \frac{1}{2\pi i}\int_\gamma \frac{\zeta^t T_\Lambda(\lambda, \mu; \zeta)}{(p_\Lambda(\lambda)-\zeta)(p_\Lambda(\mu)-\zeta)}\, d\zeta\, \delta(\lambda_1+\lambda_2-\mu_1-\mu_2).
\end{aligned} \tag{5.12}$$

Substituting this expression into (5.2), we find that the integral

$$\int_\Omega (\mathscr{T}^{(2)})^t \Psi^{(2)}_{\bar{0}, x_1, x_2}(\xi, z_1, z_2)\, d\Pi(\xi)$$

is represented as

$$\begin{aligned}
&\sum_{u_1, u_2, v_1, v_2} \int d\lambda_1\, d\lambda_2\, d\mu_1\, d\mu_2\, (\delta_{u_1 x_1}\delta_{u_2 x_2} + \overline{K}_{x_1 x_2, u_1, u_2})\\
&\qquad\qquad \times e^{i(u_1, \lambda_1)+i(u_2, \lambda_2)-i(v_1, \mu_1)-i(v_2, \mu_2)}\\
&\quad \times \Big[(\widetilde{p}(\lambda_1)\widetilde{p}(\lambda_2))^t \delta(\lambda_1-\mu_1)\delta(\lambda_2-\mu_2)\\
&\qquad + \frac{\delta(\lambda_1+\lambda_2-\mu_1-\mu_2)}{2\pi i}\int_\gamma \frac{\zeta^t T_\Lambda(\lambda, \mu; \zeta)}{(p_\Lambda(\lambda)-\zeta)(\widetilde{p}_\Lambda(\mu)-\zeta)}\, d\zeta\Big] g_{v_1, v_2}(z_1, z_2),
\end{aligned} \tag{5.13}$$

where

$$g_{v_1, v_2}(z_1, z_2) = \int \psi_{v_1, v_2}(\xi, z_1, z_2)\, d\Pi(\xi).$$

Opening the brackets in the first and second factors of expression (5.13), we get four terms. The first one is

$$\int e^{i(x_1-z_1, \lambda_1)+i(x_2-z_2, \lambda_2)} (\widetilde{p}(\lambda_1)\widetilde{p}(\lambda_2))^t \varphi(\lambda_1, \lambda_2, z_1, z_2)\, d\lambda_1\, d\lambda_2, \tag{5.14}$$

where

$$\varphi(\lambda_1, \lambda_2; z_1, z_2) = \sum_{v_1, v_2} e^{-i(v_1, \lambda_1)-i(v_2, \lambda_2)} g_{v_1+z_1, v_2+z_2}(z_1, z_2).$$

The convergence of the series and the analyticity of $\varphi(\lambda_1, \lambda_2, z_1, z_2)$ in λ_1, λ_2 follow from the estimate (4.55d) for ψ_{u_1, u_2}. By a standard method of computation of such integrals (see [3]), under the condition $|x_1 - z_1 - bt| \approx$

$t^{1/2}$ and $|x_2 - z_2 - bt| \approx t^{1/2}$ the asymptotics of the integral (5.14) is given by

$$\frac{\text{const}}{(2\pi t)^{\nu}} \exp\left\{ -\frac{1}{2t}[(A(x_1 - z_1 - bt), x_1 - z_1 - bt) + (A(x_2 - z_2 - bt), A(x_2 - z_2 - bt))]\right\}, \quad (5.15a)$$

where

$$\text{const} = \varphi(0, 0; z_1, z_2) \det A, \quad (5.15b)$$

and the matrix A and the constant b are the same as in §3. It is easy to see that

$$\varphi(0, 0; z_1, z_2) = 1. \quad (5.15c)$$

(This actually follows from the normalization condition for the asymptotics of the probability and other estimates.) Hence, adding together the asymptotic expression (5.15a) over all x_1, x_2 in the set $\sqrt{t}G + (z_1 + bt, z_2 + bt)$, where G is any bounded region of $\mathbb{R}^{\nu} \times \mathbb{R}^{\nu}$, we get the main term of the central limit theorem.

By Lemma 3.8, $\widetilde{p}(\lambda)$ for $\lambda \in T^{\nu}$ is a real even function. By the assumption (2.13), $\widetilde{p}(\lambda)$ is always positive, $\lambda = 0$ is an absolute (and nondegenerate) maximum for $\widetilde{p}(\lambda)$, and $b = 0$.

We now estimate the remaining terms. We have

$$\begin{aligned} I_t(x_1, x_2) &= \sum_{u_1 u_2} \int d\lambda_1\, d\lambda_2\, \overline{K}_{x_1, x_2, u_1, u_2} e^{i(u_1 - z_1, \lambda_1) + i(u_2 - z_2, \lambda_2)} \\ &\qquad \times (p(\lambda_1) p(\lambda_2))^t \varphi(\lambda_1, \lambda_2; z_1, z_2) \\ &= \int d\lambda_1\, d\lambda_2\, h_{x_1, x_2}(\lambda_1, \lambda_2; z_1, z_2)(\widetilde{p}(\lambda_1)\widetilde{p}(\lambda_2))^t \varphi(\lambda_1, \lambda_2; z_1, z_2), \end{aligned} \quad (5.15'a)$$

where

$$h_{x_1, x_2}(\lambda_1, \lambda_2; z_1, z_2) = \sum_{u_1 u_2} \overline{K}_{x_1, x_2, u_1, u_2} e^{i(u_1 - z_1, \lambda_1) + i(u_2 - z_2, \lambda_2)}. \quad (5.15'b)$$

From the estimate (4.60b), which is valid for the matrix $\overline{K}$ as well, it follows that the series converges and $h_{x_1, x_2}(\lambda_1, \lambda_2; z_1, z_2)$ is an analytic function of λ_1, λ_2 in some strip. Hence we can apply the above method from [3] once more to the integral (5.15′a). The coefficient of the leading term contains, by (5.15c), (5.15′b), the factor

$$\begin{aligned} \varphi(0, 0; z_1, z_2) h_{x_1, x_2}(0, 0; z_1, z_2) &= h_{x_1, x_2}(0, 0; z_1, z_2) \\ &= \sum_{u_1, u_2} \overline{K}_{x_1, x_2, u_1, u_2}. \end{aligned}$$

Now, by (4.60a) one can see that $\sum_{v_1, v_2} K_{u_1, u_2, v_1, v_2} = 0$, since the matrix $\mathcal{S}^*$ has the same property (4.56a) as the matrix $\mathcal{S}$. It is not hard to see that this implies the same property $\sum_{v_1, v_2} \overline{K}_{u_1, u_2, v_1, v_2} = 0$ for the matrix $\overline{K}$. Hence for large t we find that

$$|I_t(x_1, x_2)| < \overline{C} t^{-\nu-1},$$

where $\overline{C}$ is an absolute constant. This implies

$$\sum_{(x_1, x_2) \in \sqrt{t} G + (z_1, z_2)} I_t(x_1, x_2) = O(t^{-1}).$$

We now estimate the third term

$$\begin{aligned} J_t(z_1, z_2) \equiv & \sum_{(x_1, x_2) \in \sqrt{t} G + (z_1, z_2)} \frac{1}{2\pi i} \int_\gamma d\zeta \int d\lambda_1 \, d\lambda_2 \, d\mu_1 \, d\mu_2 \\ & \times \delta(\lambda_1 + \lambda_2 - \mu_1 - \mu_2) e^{i(x_1, \lambda_1) + i(x_2, \lambda_2)} \\ & \times \frac{\zeta^t T(\lambda_1, \lambda_2, \mu_1, \mu_2; \zeta) \varphi^{(1)}(\mu_1, \mu_2; z_1; z_2)}{(\widetilde{p}(\lambda_1)\widetilde{p}(\lambda_2) - \zeta)(\widetilde{p}(\mu_1)\widetilde{p}(\mu_2) - \zeta)} \\ = & \sum_{(x_1, x_2) \in \sqrt{t} G + (z_1, z_2)} \frac{1}{2\pi i} \int_\gamma d\zeta \int d\Lambda \, e^{i(x_2, \Lambda)} \int d\lambda \, d\mu \\ & \times \frac{\zeta^t T_\Lambda(\lambda, \mu; \zeta) e^{i(x_1 - x_2, \lambda)} \varphi_\Lambda^{(1)}(\mu; z_1, z_2)}{(p_\Lambda(\lambda) - \zeta)(p_\Lambda(\mu) - \zeta)}, \end{aligned} \tag{5.16}$$

where

$$\varphi^{(1)}(\mu_1, \mu_2; z_1, z_2) = \sum_{v_1, v_2} e^{-i(v_1, \mu_1) - i(v_2, \mu_2)} g_{v_1, v_2}(z_1, z_2),$$

$$\varphi_\Lambda^{(1)}(\mu; z_1, z_2) = \varphi^{(1)}(\mu, \Lambda - \mu; z_1, z_2).$$

Introduce new variables (w_1, w_2) and functions $f_G^t(\Lambda, \lambda)$:

$$w_1 = x_2, \qquad w_2 = x_1 - x_2, \tag{5.17a}$$

$$\begin{aligned} f_G^t(\Lambda, \lambda) &= \sum_{(w_1, w_2) \in \sqrt{t} \widehat{G} + (z_2, z_1 - z_2)} e^{i(w_1, \Lambda) + i(w_2, \lambda)} \\ &= \sum_{(\alpha, \beta) \in \widehat{G} + (z_2/\sqrt{t}, (z_1 - z_2)/\sqrt{t})} e^{i\sqrt{t}(\alpha, \Lambda) + i\sqrt{t}(\beta, \lambda)}, \end{aligned} \tag{5.17b}$$

where $\widehat{G} \subset \mathbb{R}^\nu \times \mathbb{R}^\nu$ is the region obtained from G by the change of variables (5.17a), and α and β belong to the lattice with spacing $1/\sqrt{t}$. We introduce the function

$$\varphi_G(\Lambda, \lambda) = \int_{\widehat{G} + (z_2/\sqrt{t}, (z_1 - z_2)/\sqrt{t})} e^{i(\alpha, \Lambda) + i(\beta, \lambda)} d\alpha \, d\beta,$$

which is defined for all $(\Lambda, \lambda) \in \mathbb{R}^\nu \times \mathbb{R}^\nu$, and is analytic. Then

$$f_G^t(\Lambda, \lambda) = t^\nu [\varphi_G(\Lambda\sqrt{t}, \lambda\sqrt{t}) + \overline{\varphi}_G^t(\Lambda, \lambda)],$$

with φ_G bounded and $\overline{\varphi}_G^t(\Lambda, \lambda) = o(1)$ as $t \to \infty$. We have

$$J_t(z_1, z_2) = \frac{1}{2\pi i} \int_\gamma d\zeta\, \zeta^t \int_{T^\nu} d\Lambda\, t^\nu F_\Lambda(\zeta; t) \tag{5.18}$$

with

$$\begin{aligned} F_\Lambda(\zeta; t) &= \frac{1}{t^\nu} \int \frac{f_G^t(\Lambda, \lambda)\, T_\Lambda(\lambda, \mu; \zeta)\, \varphi_\Lambda^{(1)}(\mu; z_1, z_2)}{(p_\Lambda(\lambda) - \zeta)(p_\Lambda(\mu) - \zeta)}\, d\lambda\, d\mu \\ &= [F'_\Lambda(\zeta) + F''_\Lambda(\zeta)], \end{aligned} \tag{5.19}$$

where

$$F'_\Lambda(\zeta) = \int \frac{\varphi_G(\Lambda\sqrt{t}, \lambda\sqrt{t})\, T_\Lambda(\lambda, \mu; \zeta)\, \varphi_\Lambda^{(1)}(\mu; z_1, z_2)}{(p_\Lambda(\lambda) - \zeta)(p_\Lambda(\mu) - \zeta)}\, d\lambda\, d\mu, \tag{5.20a}$$

$$F''_\Lambda(\zeta) = \int \frac{\overline{\varphi}_G^t(\Lambda\sqrt{t}, \lambda\sqrt{t})\, T_\Lambda(\lambda, \mu; \zeta)\, \varphi_\Lambda^{(1)}(\mu; z_1, z_2)}{(p_\Lambda(\lambda) - \zeta)(p_\Lambda(\mu) - \zeta)}\, d\lambda\, d\mu. \tag{5.20b}$$

We shall only estimate the contribution

$$J'_t(z_1, z_2) = \frac{1}{2\pi i} \int_\gamma d\zeta\, \zeta^t \int_{T^\nu} d\Lambda\, t^\nu F'_\Lambda(\zeta; t). \tag{5.21}$$

The estimate of the other term follows the same lines.

Since, as we noted before, $\Delta_\Lambda(\zeta)$ is analytic in $\Pi_\Lambda = \mathbb{C} \setminus \operatorname{Range} p_\Lambda$, F'_Λ is also analytic in Π_Λ. We have $\operatorname{Range} p_\Lambda = [s(\Lambda), S(\Lambda)]$, where $s(\Lambda) = \min_\lambda p_\Lambda(\lambda)$, $S(\Lambda) = \max_\lambda p_\Lambda(\lambda)$, and by the assumptions on $\widetilde{p}$, $0 < s(\Lambda) < S(\Lambda) \leqslant 1$. Moreover, it is not hard to see that if $|\Lambda| < \delta$ for some $\delta > 0$ small enough, then the function $p_\Lambda(\lambda)$ takes the maximal value at $\lambda = \Lambda/2$, so that $S(\Lambda) = (\widetilde{p}(\Lambda/2))^2$, and the maximum is not degenerate. It follows that for small $|\Lambda|$, $S(\Lambda)$ is strictly increasing as $\Lambda \to 0$ and $S(\Lambda) \to 1$ as $\Lambda \to 0$. We consider first the contribution to the integral (5.21) of the region $|\Lambda| > \delta'$, $0 < \delta' < \delta$. For such Λ we have $S_\Lambda < 1 - \delta_1$ for some $\delta_1 > 0$, and we can choose a contour γ which goes around the segment $I \equiv [0, 1 - \delta_1]$ on the real axis, and such that $\operatorname{dist}(\zeta, I) = \delta_1/2$ for all $\zeta \in \gamma$. Since φ_G is a bounded function, if δ_1 is small enough, we find that for some $c_1 > 0$ and all t large enough

$$\left| \int_\gamma d\zeta \int_{|\Lambda| > \delta'} d\Lambda\, t^\nu \zeta^t F'_\Lambda(\zeta; t) \right| < e^{-c_1 t}.$$

For $|\Lambda| < \delta'$, using the fact that the function $\widetilde{D}$ in (5.3) is small and the dimension is $\nu \geqslant 3$, it is not hard to prove (see, e.g., [6]) that $\Delta_\Lambda(s) \neq 0$ for all $s \in \operatorname{Range} p_\Lambda$ sufficiently close to $S(\Lambda)$, and the same holds for the limit values $\Delta_\Lambda^\pm(s) = \lim_{\varepsilon \to 0} \Delta_\Lambda(s \pm i\varepsilon)$. Moreover, we shall prove below that for such values of s the limits $\lim_{\varepsilon \to 0} F'(s \pm i\varepsilon; t)$ and the jump

$$\delta^\Lambda(s; t) = \lim_{\varepsilon \to 0} [F'(s + i\varepsilon; t) - F'(s - i\varepsilon; t)]$$

also exist and are bounded in absolute value by const $\sqrt{t}$. Hence for $|\Lambda| < \delta'$ we can write

$$\int_\gamma d\zeta\, \zeta^t F'_\Lambda(\zeta;t) = \int_{\gamma'} d\zeta\, \zeta^t F'_\Lambda(\zeta;t) + \int_{S(\Lambda)-\bar{\delta}}^{S(\Lambda)} s^t \delta^\Lambda(s;t)\,ds, \tag{5.21'}$$

where γ' is a contour that can be chosen as consisting of two vertical segments

$$\{\zeta = 0 + iy,\ |y| \leqslant \bar{\delta}/2\}, \qquad \{\zeta = S(\Lambda) - \bar{\delta} + iy,\ |y| \leqslant \bar{\delta}/2\}$$

and two horizontal segments

$$\{\zeta = s \pm i\bar{\delta}/2,\ s \in [0, S(\Lambda) - \bar{\delta}]\}.$$

Since $|\zeta| \leqslant \beta < 1$ for all $\zeta \in \gamma'$, we see that the contribution of the first term on the right-hand side of (5.21′) is exponentially small as $t \to \infty$, while for the integral on the cut we find for any $\bar{\varepsilon} > 0$

$$\int_{S(\Lambda)-\bar{\delta}}^{S(\Lambda)} s^t \delta^\Lambda(s;t)\,ds = \int_{S(\Lambda)(1-t^{-1+\bar{\varepsilon}})}^{S(\Lambda)} s^t \delta^\Lambda(s;t)\,ds + \mathscr{O}(\exp\{-ct^{\bar{\varepsilon}}\}), \tag{5.22}$$

i.e., the main contribution to the final asymptotics is given by a small neighborhood of the point $S(\Lambda) \equiv \max p_\Lambda(\lambda)$.

We make use of the representation of $T_\Lambda(\lambda, \mu; \zeta)$ as a series (given by (5.8)) and study the jump $\delta_n^\Lambda(s)$ of each term (we consider only the case $n > 0$)

$$\Phi_n^\Lambda(\zeta) = \frac{1}{\Delta_\Lambda(\zeta)} I_n^\Lambda(\zeta), \tag{5.23}$$

$$I_n^\Lambda(\zeta) = \int d\lambda\, d\mu \prod_{i=1}^n d\nu_i \frac{\varphi_G(\sqrt{t}\Lambda, \sqrt{t}\lambda)\varphi_\Lambda^{(1)}(\mu; z_1, z_2) K_n^\Lambda(\lambda, \mu, \nu_1, \dots, \nu_n)}{(p_\Lambda(\lambda) - \zeta)(p_\Lambda(\mu) - \zeta)\prod_{i=1}^n (p_\Lambda(\nu_i) - \zeta)}.$$

Observe that

$$\delta_n^\Lambda(s) = \delta'^\Lambda(s) I_n^\Lambda(s + i0) + \delta''^\Lambda_n \frac{1}{\Delta_\Lambda(s - i0)},$$

$$\delta'^\Lambda(s) = \frac{1}{\Delta_\Lambda(s + i0)} - \frac{1}{\Delta_\Lambda(s - i0)}, \qquad \delta''^\Lambda_n(s) = I_n^\Lambda(s + i0) - I_n^\Lambda(s - i0).$$

Consider the function

$$\begin{aligned} &u_n^\Lambda(t_1, t_2, t_3, \dots, t_{n+2}) \\ &\quad = \int_{p_\Lambda(\lambda)=t_1} d\lambda \int_{p_\Lambda(\mu)=t_2} d\mu \int_{p_\Lambda(\nu_1)=t_3} d\nu_1 \cdots \int_{p_\Lambda(\nu_n)=t_{n+2}} d\nu_n \\ &\quad \times \varphi_G(\sqrt{t}\Lambda, \sqrt{t}\lambda)\varphi_\Lambda^{(1)}(\mu; z_1, z_2) K_n^\Lambda(\lambda, \mu, \nu_1, \dots, \nu_n). \end{aligned} \tag{5.24}$$

Then the integral I_n^Λ is written as

$$I_n^\Lambda(\zeta) = \int \frac{u_n^\Lambda(t_1, t_2, \dots, t_{n+2})}{(t_1 - \zeta)(t_2 - \zeta)\cdots(t_{n+2} - \zeta)} dt_1\, dt_2 \cdots dt_{n+2}. \tag{5.25}$$

By iterating the formula of Sokhotskiĭ (see [7]), we find that the limiting values of $I_n^\Lambda(s \pm i0)$ on the cut $\operatorname{Range} p_\Lambda(\lambda)$ are equal to

$$\sum_{L\subset\{1,\dots,n+2\}} (\pm i\pi)^{|L|} \mathscr{P}\int dt_{i_1}\cdots dt_{i_k} \frac{u_n^\Lambda(t_j = s,\ j\in L,\ t_{i_1},\dots,t_{i_k})}{\prod_{p=1}^k (t_{i_p} - s)}, \tag{5.26}$$

where $\mathscr{P}\int\dots$ denotes the principal part, the summation is over all subsets $L \subset \{1, \dots, n+2\}$, the variables t_j with $j \in L$ are set to be equal to s, and the remaining variables are integrated (in the sense of the principal part) over the set $\operatorname{Range} p_\Lambda(\lambda)$. Therefore the jump δ''^Λ_n of the function I_n^Λ on the cut is equal to twice the sum (5.26) taken only over the sets L with an odd number of elements. We divide the terms of this sum into two classes, with the first class consisting of terms with $1 \in L$, and the second class consisting of terms with $1 \notin L$. We first estimate the contribution of the terms from the first class. For $|\Lambda| < \delta'$ and s close to $S(\Lambda)$ we can perform a smooth change of variables, taking $S(\Lambda) - p_\Lambda(\lambda)$ to be one of the new variables, and we find in particular that

$$\int_{p_\Lambda(\lambda)=s} d\lambda \asymp \operatorname{const}(S(\Lambda) - s)^{\nu/2-1}. \tag{5.27}$$

Suppose now that $\nu \geqslant 4$. Then $(S(\Lambda) - s)^{\nu/2-1}$ is a C^1-function in s, and using the standard estimate of the (multiple) principal part of the type

$$\int dt_1\cdots dt_n \frac{f(t_1,\dots,t_n)}{\prod_{j=1}^n (t_j - s)}$$

by the supremum of the mixed derivative $\|\partial^k/\partial t_1 \cdots \partial t_n\|_\infty$ and estimating the determinants using the Hadamard inequality (see [8]) and the relation (5.5), we find that the sum of the terms of the first class is less than

$$B|\Lambda| C^n n^{n/2} (S(\Lambda) - s)^{\nu/2-1}(1 + \mathscr{O}(S(\Lambda) - s)), \tag{5.28}$$

where C and B are absolute constants, which are small if $\varepsilon \equiv \max\{\varepsilon_1, \varepsilon_2\}$ is small (due to the fact that $\widetilde{D}$ defined in (5.3) is small). The contribution of the terms in the second class has a similar estimate, with the factor $B|\Lambda|$ in (5.28) replaced by the factor $(B|\Lambda| + D/\sqrt{t})$ (D is an absolute constant). This estimate follows from the following properties of the function $\varphi_G(\sqrt{t}\Lambda, \sqrt{t}\lambda)$:

$$\begin{gathered} \int_{|\lambda|<\delta} \varphi_G(\sqrt{t}\Lambda, \sqrt{t}\lambda)\, d\lambda < K_1, \\ \int_{|\lambda|<\delta} (b, \lambda)\varphi_G(\sqrt{t}\Lambda, \sqrt{t}\lambda)\, d\lambda < K_2/\sqrt{t}, \end{gathered} \tag{5.28$'$}$$

where the constants $K_1 = K_1(G)$ and $K_2 = K_2(G, b)$ do not depend on t and Λ ($b \in \mathbb{R}$ is any vector). Hence we find that

$$|\delta''^\Lambda_n| \leqslant (\overline{B}|\Lambda| + D/\sqrt{t})\, C^n n^{n/2} (S(\Lambda) - s)^{\nu/2-1}. \tag{5.29}$$

A similar estimate holds for the terms δ'^{Λ}_{n} associated to the jump δ'^{Λ} according to the first line of formula (5.8):

$$|\delta'^{\Lambda}_{n}| \leqslant B|\Lambda|C^n n^{n/2}(S(\Lambda)-s)^{\nu/2-1}. \tag{5.30a}$$

Moreover, it is not difficult to see that

$$|I^{\Lambda}_{n}(s\pm i0)| \leqslant c^n n^{n/2}, \tag{5.30b}$$

where c is also small for small ε.

Hence we find finally that

$$|\delta^{\Lambda}_{n}| \leqslant (\widehat{B}|\Lambda| + \widehat{D}/\sqrt{t})\overline{C}n^{n/2}(S(\Lambda)-s)^{\nu/2-1}, \tag{5.31}$$

where $\overline{C}$ is small for small ε, and $\widehat{B}$ and $\widehat{D}$ are absolute constants. For $\nu = 3$ the estimate (5.31) is obtained in a somewhat more complicated way, by taking the Hölder norm of $(S(\Lambda)-s)^{\nu/2-1}$ instead of the derivative in the estimate of the principal part.

We finally get for large t

$$\begin{aligned}
&\int_{S(\Lambda)(1-t^{-1+\bar{\varepsilon}})}^{S(\Lambda)} s^t\delta_{\Lambda}(s\,;t)\,ds \\
&\quad < C[\widehat{B}|\Lambda| + \widehat{D}/\sqrt{t}](S(\Lambda))^{t+\nu/2}\int_{1-t^{-1+\bar{\varepsilon}}}^{1} u^t(1-u)^{\nu/2-1}du \\
&\quad < \widetilde{C}[\widehat{B}|\Lambda| + \widehat{D}/\sqrt{t}](S(\Lambda))^t t^{-\nu/2},
\end{aligned} \tag{5.32}$$

where $\widetilde{C} > 0$ is an absolute constant. Using the explicit expression of $S(\Lambda)$ for small Λ, we find that

$$|J'_t(z_1, z_2)| \leqslant \frac{t^{\nu}}{2\pi}\left[O(\exp\{-ct^{\bar{\varepsilon}}\}) + \frac{\widetilde{C}}{t^{\nu/2}}\int_{|\Lambda|<\delta}(\widehat{B}|\Lambda| + \widehat{D}/\sqrt{t})S^t(\Lambda)\,d\Lambda\right] \leqslant \frac{M}{\sqrt{t}},$$

where $M > 0$ is an absolute constant. The contribution of $\int d\Lambda F''_{\Lambda}(\zeta)$ is estimated in a similar way. We get finally that expression (5.16) is bounded by a quantity which is $o(1)$ as $t\to\infty$. The contribution of the fourth term in (5.13) is treated similarly. Theorem 2 is proved.

REMARK. A more detailed proof of some estimates that are used in this section (e.g., of estimates (5.28), (5.29), etc.) can be found in [9].

Appendix

PROPOSITION A.1. *$\mathscr{T}$ is a bounded operator in $\mathscr{H}$.*

PROOF. We prove the equivalent assertion that $\mathscr{T}$ is a bounded operator in $\hat{l}_2(\mathfrak{M}\times\mathbb{Z}^{\nu})$. This amounts to showing that for some positive constant $K<\infty$ and any $\hat{f}$, $\hat{g}\in\hat{l}_2(\mathfrak{M}\times\mathbb{Z}^{\nu})$ we have

$$|(\mathscr{T}\hat{f}, \hat{g})| \leqslant K\|\hat{f}\|\,\|\hat{g}\|. \tag{A.1}$$

Consider the subspace $\hat{l}_0 \subset \hat{l}_2(\mathfrak{M} \times \mathbb{Z}^\nu)$ consisting of all functions with compact support, i.e., such that $\hat{f}_{\Gamma, z} = 0$ if $\max\{|z|, \max\{|z'| : z' \in \operatorname{supp}\Gamma\}\} > R$, for some $R < \infty$ depending on $\hat{f}$. We shall prove inequality (A.1) for $\hat{f}, \hat{g} \in \hat{l}_0$.

Let $\Lambda \subset \mathbb{Z}^\nu$ be a finite set, so large that $\hat{f}_{\Gamma, z} = \hat{g}_{\Gamma, z} = 0$ for all $z \in \mathbb{Z}^\nu$ if $\operatorname{supp}\Gamma \cap \Lambda^c \neq \varnothing$, where Λ^c is the complement of Λ. We have

$$
\begin{aligned}
(\mathcal{T}\hat{f}, \hat{g}) &= \sum_{z, \Gamma, \Gamma'} (\mathcal{T}\hat{f})_{\Gamma, z} \overline{\hat{g}}_{\Gamma', z} a_{\Gamma\Gamma'} \\
&= \sum_{z, z'} \sum_{\Gamma, \Gamma', \Gamma''} \delta_{\Gamma_z, \Gamma''_z} \mu_{\Gamma_z} a_{\Gamma\Gamma'} L(\gamma(z), \gamma''(z), z' - z) \hat{f}_{\Gamma'', z'} \overline{\hat{g}}_{\Gamma', z}. \qquad \text{(A.2)}
\end{aligned}
$$

Let A be the $|S| \times |S|$ Gram matrix with elements (e_i, e_j), $i, j = 0, \ldots, |S| - 1$. The matrix A is Hermitian, positive definite, and hence has positive eigenvalues $0 < \lambda_0 \leqslant \cdots \leqslant \lambda_{|S|-1}$. By (3.4)

$$
a_{\Gamma, \Gamma'} = \prod_{x \in \Lambda} A_{\gamma(x), \gamma'(x)}.
$$

Let V be the matrix whose columns are the components of the (normalized) eigenvectors of A; this matrix diagonalizes A. Changing variables, we set

$$
\begin{aligned}
\widehat{h}_{\Gamma, z} \equiv \mu_{\Gamma_z} \hat{f}_{\Gamma, z} &= \sum_J \prod_{x \in \Lambda} V_{\gamma(x), j(x)} \widehat{\varphi}_{J, z}, \\
\hat{g}_{\Gamma, z} &= \sum_J \prod_{x \in \Lambda} V_{\gamma(x), j(x)} \widehat{\psi}_{J, z},
\end{aligned} \qquad \text{(A.3)}
$$

where $J \equiv \{j(x), x \in \Lambda\}$ is a multi-index defined on Λ. Substituting (A.3) into (A.2) we have

$$
\begin{aligned}
(\mathcal{T}\hat{f}, \hat{g}) &= \sum_{z, z'} \sum_{J, J'} \sum_{\Gamma_z, \Gamma'_z} \prod_{x \in \Lambda \setminus \{z\}} V_{\gamma(x), j(x)} \overline{V}_{\gamma'(x), j'(x)} A_{\gamma(x), \gamma'(x)} \\
&\quad \times \sum_{\gamma(z), \gamma'(z), \gamma''(z)} A_{\gamma(z), \gamma'(z)} V_{\gamma''(z), j(z)} \\
&\qquad \times \overline{V}_{\gamma'(z), j'(z)} L(\gamma(z), \gamma''(z), z' - z) \widehat{\varphi}_{J, z'} \overline{\widehat{\psi}}_{J', z} \\
&= \sum_{z, z'} \sum_{J, J'} \prod_{x \in \Lambda \setminus \{z\}} \lambda_{j(x)} \delta_{j(x), j'(x)} \\
&\quad \times \sum_{k, l, m} A_{k, l} V_{m, j(z)} \overline{V}_{l, j'(z)} L(k, m, z' - z) \widehat{\varphi}_{J, z'} \overline{\widehat{\psi}}_{J', z}. \qquad \text{(A.4)}
\end{aligned}
$$

Let M_z be the smallest constant such that for any $(x_0, \ldots, x_{|S|-1}) \in \mathbb{C}^{|S|}$,

$$
\left| \sum_{r, s=0}^{|S|-1} B_{r, s}(z) x_r \overline{x}_s \right| \leqslant M_z \sum_{s=0}^{|S|-1} |x_s|^2,
$$

where

$$B_{r,s}(z) = \sum_{k,l,m} V_{m,r}\overline{V}_{l,s}L(k, m; z)A_{k,l}.$$

Clearly $M_z = 0$ for $|z| > D$ and

$$\begin{aligned}|(\mathcal{T}\hat{f}, \hat{g})| &\leqslant \sum_{z,z'} M_{z'-z} \sum_J \prod_{x\in\Lambda\setminus\{z\}} \lambda_{j(x)}|\widehat{\varphi}_{J,z'}||\widehat{\psi}_{J,z}| \\ &\leqslant \sum_{z,z'} \frac{M_{z'-z}}{2\lambda_0} \sum_J \prod_{x\in\Lambda\setminus\{z\}} \lambda_{j(x)}(|\widehat{\varphi}_{J,z'}|^2 + |\widehat{\psi}_{J,z}|^2) \\ &\leqslant \mathrm{const}(\|\widehat{h}\|^2_{\hat{l}_2(\mathfrak{M}\times\mathbb{Z}^\nu)} + \|\hat{g}\|^2_{\hat{l}_2(\mathfrak{M}\times\mathbb{Z}^\nu)}),\end{aligned} \tag{A.5}$$

where $\mathrm{const} = \sum_z M_z/(2\lambda_0)$. Now note that $\|\widehat{h}\|^2_{\hat{l}_2(\mathfrak{M}\times\mathbb{Z}^\nu)} \leqslant \|\hat{f}\|^2_{\hat{l}_2(\mathfrak{M}\times\mathbb{Z}^\nu)}$. In fact we have

$$\|\widehat{h}\|^2_{\hat{l}_2(\mathfrak{M}\times\mathbb{Z}^\nu)} = \sum_z \sum_{\Gamma,\Gamma'} a_{\Gamma\Gamma'}\mu_{\Gamma_z}\overline{f}_{\Gamma,z}\overline{\mu}_{\Gamma'_z}\overline{f}_{\Gamma,z} = \sum_z \sum_{\Gamma,\Gamma'} b^{(z)}_{\Gamma,\Gamma'} f_{\Gamma,z}\overline{f}_{\Gamma',z},$$

where $b^{(z)}_{\Gamma,\Gamma'}$ is again a product of Gram matrices:

$$b^{(z)}_{\Gamma,\Gamma'} = A_{\gamma(z),\gamma'(z)} \prod_{x\in\Lambda\setminus\{z\}} (Q_0 e_{\gamma(x)}, Q_0 e_{\gamma'(x)}).$$

When the support of $\hat{f}_{\Gamma,z}$ reduces to one point $\{x_0\}$, the inequality $\|\widehat{h}\|^2 \leqslant \|\hat{f}\|^2$ follows from the inequality $\|Q_0 v\|^2 \leqslant \|v\|^2$, valid for any vectors $v = \sum_{j=0}^{|S|-1} v_j e_j$. The inequality can then be generalized to any support $\operatorname{supp}\Gamma$.

Therefore, inequality (A.5) implies that

$$\sup_{\|\hat{f}\|=\|\hat{g}\|=1,\,\hat{f},\hat{g}\in\hat{l}_0} |(\mathcal{T}\hat{f}, \hat{g})| = K < \infty, \tag{A.6}$$

where K does not depend on $\hat{f}, \hat{g} \in \hat{l}_0$. Proposition A.1 follows from inequality (A.6).

PROOF OF LEMMA 3.4. For $\hat{f} \in l_M(\mathfrak{M})$ we have, by (3.13a),

$$\|\mathcal{T}_\lambda \hat{f}\|_{l_M(\mathfrak{M})} \leqslant \sum_{\Gamma,\Gamma'} M^{|\operatorname{supp}\Gamma|}|R_{\Gamma,\Gamma'}(\lambda)||\hat{f}_{\Gamma'}|. \tag{A.7}$$

On the other hand, by definition (3.13b) we have

$$|R_{\Gamma,\Gamma'}(\lambda)| \leqslant \sum_{u\in\mathbb{Z}^\nu} |\mu_{\Gamma'_u}|\delta_{\Gamma'_u,(\Gamma+u)_u}|L(\gamma(u), \gamma'(u); u)|,$$

which implies

$$\sum_\Gamma M^{|\operatorname{supp}\Gamma|}|R_{\Gamma,\Gamma'}(\lambda)| < \mathrm{const}\, M^{|\operatorname{supp}\Gamma'|+1}.$$

Substituting into (A.7), we get

$$\|\mathcal{T}_\lambda \hat{f}\|_{l_M(\mathfrak{M})} \leqslant \mathrm{const}\, M \|\hat{f}\|_{l_M(\mathfrak{M})}.$$

Lemma 3.4 is proved.

PROOF OF LEMMA 4.8. By (4.64), (4.65) we need to evaluate, for $f \in \tilde{l}_{\widetilde{M}}$,

$$\sup_{z_1, z_2} \sum_{\Gamma \neq \bar{0}} |(\widetilde{\mathscr{T}} f)_{\Gamma, z_1, z_2}| \widetilde{M}^{|\operatorname{supp}\Gamma|} \tag{A.8}$$

with

$$(\widetilde{\mathscr{T}} f)_{\Gamma, z_1, z_2} = \sum_{\Gamma' \neq \bar{0}, z_1', z_2'} \Big[R^{(2)}_{\Gamma, z_1, z_2; \Gamma', z_1', z_2'} - \sum_{v_1, v_2} R^{(2)}_{\Gamma, z_1, z_2; \bar{0}, v_1, v_2} \overline{S_{v_1, v_2, \Gamma', z_1', z_2'}} \Big] f_{\Gamma', z_1', z_2'}. \tag{A.9}$$

The contribution of the first term on the right-hand side of (A.9) is bounded in absolute value by

$$\sum_{\Gamma \neq \bar{0}} \sum_{\Gamma' \neq \bar{0}; z_1', z_2'} |R^{(2)}_{\Gamma, z_1, z_2; \Gamma', z_1', z_2'}| |f_{\Gamma', z_1', z_2'}| \widetilde{M}^{|\operatorname{supp}\Gamma|}. \tag{A.10}$$

Now we can write

$$R^{(2)}_{\Gamma, z_1, z_2; \Gamma', z_1', z_2'} = \mu_\Gamma \delta_{\Gamma, \Gamma'} P_0(z_1 - z_1') P_0(z_2 - z_2') + \delta_{\Gamma_{z_1, z_2}, \Gamma'_{z_1, z_2}} G(\Gamma, \Gamma', z_1 - z_1', z_2 - z_2', z_1 - z_2), \tag{A.11}$$

where G has the following properties: its supremum over all values of the arguments is small for small ε, and it is zero for $|z_1 - z_1'|$, $|z_2 - z_2'|$ large enough. The contribution of the first term of (A.11) to (A.10) is bounded by

$$\sum_{\Gamma \neq \bar{0}} \sum_{\Gamma' \neq \bar{0}, z_1', z_2'} |\mu_\Gamma| \delta_{\Gamma, \Gamma'} P_0(z_1 - z_1') P_0(z_2 - z_2') |f_{\Gamma', z_1', z_2'}| \widetilde{M}^{|\operatorname{supp}\Gamma|} \leqslant |\mu_1| \|f\|_{\tilde{l}_{\widetilde{M}}}. \tag{A.12}$$

The contribution of the second term is bounded by

$$\sum_{\Gamma \neq \bar{0}} \sum_{\Gamma' \neq \bar{0}, z_1', z_2'} \delta_{\Gamma_{z_1, z_2}, \Gamma'_{z_1, z_2}} |G(\Gamma, \Gamma', z_1 - z_1', z_2 - z_2', z_1 - z_2)| \times |f_{\Gamma', z_1', z_2'}| \widetilde{M}^{|\operatorname{supp}\Gamma'|+2}. \tag{A.13}$$

The sum over Γ reduces to a finite sum over $\gamma(z_1)$, $\gamma(z_2)$, so that by the properties of G expression (A.13) is bounded by

$$C_\varepsilon \|f\|_{\tilde{l}_{\widetilde{M}}},$$

where $C_\varepsilon > 0$ is a constant which can be made as small as we want for ε small. Summing up, we find that the contribution of the first term of (A.11) to the quantity (A.10) is bounded by

$$(|\mu_1| + C_\varepsilon) \|f\|_{\tilde{l}_{\widetilde{M}}}. \tag{A.14}$$

As for the contribution of the second term of (A.11), by definition (4.54c) it splits into two parts. We write the first part as

$$\sum_{\Gamma'\neq\bar{0},z_1',z_2'}\sum_{v_1,v_2}|S(\Gamma',z_1',z_2',v_1,v_2)||f_{\Gamma',z_1',z_2'}| \times\sum_{\Gamma\neq\bar{0}}|R^{(2)}_{\Gamma,z_1,z_2;\bar{0},z_1',z_2'}|\widetilde{M}^{|\operatorname{supp}\Gamma|}. \quad (A.15)$$

By formulas (4.4), (4.5′), and (4.6) it can be seen that only the multi-indices Γ with $|\operatorname{supp}\Gamma|\leqslant 2$ give a contribution, and the sum

$$G_\varepsilon^*(z_1-v_1,z_2-v_2,z_1-z_2)\equiv\sum_{\Gamma\neq\bar{0}}|R^{(2)}_{\Gamma,z_1,z_2;\bar{0},z_1',z_2'}|\widetilde{M}^{|\operatorname{supp}\Gamma|}$$

is small for small ε and zero if $|z_1-v_1|$, $|z_2-v_2|$ are large enough. Hence, using estimate (4.43), we see that expression (A.15) is bounded by

$$\sum_{\Gamma'\neq\bar{0},z_1',z_2'}\sum_{v_1,v_2}G_\varepsilon^*(z_1-v_1,z_2-v_2,z_1-z_2)C\overline{M}^{-|\operatorname{supp}\Gamma'|} \times\widehat{q}^{\,d_{\operatorname{supp}\Gamma'\cup\{z_1'\}\cup\{z_2'\}\cup\{v_1\}\cup\{v_2\}}}|f_{\Gamma',z_1',z_2'}|. \quad (A.16)$$

Using the estimate $d_{\operatorname{supp}\Gamma'\cup\{z_1'\}\cup\{z_2'\}\cup\{v_1\}\cup\{v_2\}}\geqslant[|z_1'-v_1|+|z_2'-v_2|]/2$, and multiplying the expression under the summation sign by $\widetilde{M}^{|\operatorname{supp}\Gamma'|}$, for expression (A.15) we get the bound

$$C_\varepsilon'(\widehat{q})\|f\|_{\tilde{l}_{\widetilde{M}}}, \quad (A.17)$$

where the constant C_ε' is small for small ε. For the remaining term one can repeat similar considerations using the estimate (4.24) for the matrix coefficients $B_{u_1,u_2,\Gamma,v_1,v_2}$. One finds an estimate analogous to (A.17) with a constant C_ε'' which is also small for small ε. Summing up, we find that the norm of the operator $\widetilde{\mathscr{T}}$ in the space $\tilde{l}_{\widetilde{M}}$ is bounded by

$$\|\widetilde{\mathscr{T}}\|\leqslant|\mu_1|+\overline{C}_\varepsilon$$

where $\overline{C}_\varepsilon$ is small for small ε. Since $|\mu_1|<1$, we have $\theta\equiv|\mu_1|+\overline{C}_\varepsilon<1$ if ε is small enough. Lemma 4.8 is proved.

PROPOSITION A.2. *For any $q\in(0,1)$ the following inequality holds:*

$$\sum_{v_1,v_2,Q:|\operatorname{supp}Q|=n} q^{d_{\operatorname{supp}Q\cup\{v_1\}\cup\{v_2\}\cup\{u_1'\}\cup\{u_2'\}}+|u_1-v_1|+|u_2-v_2|} \leqslant q^{(|u_1-u_1'|+|u_2-u_2'|)/2}\overline{C}^n, \quad (A.18)$$

where $\overline{C}$ is an absolute constant depending only on q.

The first factor on the right-hand side of (A.18) comes from the estimate

$$d_{\operatorname{supp}Q\cup\{v_1\}\cup\{v_2\}\cup\{u_1'\}\cup\{u_2'\}}+|u_1-v_1|+|u_2-v_2| > [d_{\operatorname{supp}Q\cup\{v_1\}\cup\{v_2\}\cup\{u_1'\}\cup\{u_2'\}}+|u_1-u_1'|+|u_2-u_2'|]/2 \quad (A.19)$$

and the second factor comes from the estimate

$$\sum_{Q:\,|\operatorname{supp} Q|=n,\,v_1,\,v_2} q^{d_{\operatorname{supp} Q\cup\{v_1\}\cup\{v_2\}\cup\{u_1'\}\cup\{u_2'\}}} \leqslant \sum_{Q:\,|\operatorname{supp} Q|=n,\,v_1,\,v_2} \sum_{\mathcal{T}} \prod_{(x,y)\in\mathcal{T}} q^{|x-y|}, \qquad \text{(A.20)}$$

where $\sum_{\mathcal{T}}$ denotes the sum over all "trees" (tree-graphs) with vertices at the points of the set $\operatorname{supp} Q\cup\{v_1\}\cup\{v_2\}\cup\{u_1'\}\cup\{u_2'\}$. If some of the points v_1, v_2, u_1', u_2' coincide, or coincide with points of $\operatorname{supp} Q$, it is convenient to consider them as distinct, so that the tree $\mathcal{T}$ has always $n+4$ vertices (the distance between some of them may be zero). Interchanging the summation order in (A.20) and passing from summation over $Q\cup\{v_1\}\cup\{v_2\}$ to summation over the ordered set of points $\{x_1, \dots, x_n, x_{n+1}, x_{n+2}\}$ which consists of the points of $\operatorname{supp} Q$, plus v_1 and v_2, we find, setting $x_{n+3}=u_1'$, $x_{n+4}=u_2'$, that the right-hand side of (A.20) is bounded by

$$\sum_{\widetilde{\mathcal{T}}_{n+4}} \frac{1}{(n+2)!} \sum_{\{x_1,\dots,x_{n+2}\}} \prod_{(i,i')\in\widetilde{\mathcal{T}}_{n+4}} q^{|x_i-x_{i'}|}. \qquad \text{(A.21)}$$

Here the sum is over all abstract trees $\widetilde{\mathcal{T}}_{n+4}$ with $n+4$ vertices labeled from 1 to $n+4$. Now the sum over $\{x_1, \dots, x_{n+2}\}$, for $\widetilde{\mathcal{T}}_{n+4}$ and $x_{n+3}=u_1'$, $x_{n+4}=u_2'$ fixed, is bounded by

$$\sum_{\{x_1,\dots,x_{n+2}\}} \prod_{(i,i')\in\widetilde{\mathcal{T}}_{n+4}} q^{|x_i-x_{i'}|} \leqslant \overline{C}^{\,n+2}, \qquad \text{(A.22)}$$

where

$$\overline{C} = \sum_{x\in\mathbb{Z}^\nu} q^{|x|}.$$

Estimate (A.22) is obtained by summing over an external vertex of the tree $\widetilde{\mathcal{T}}_{n+4}$ (except u_1', u_2'), and then applying the same procedure to the tree obtained by canceling the vertex which we have already summed over. If the only external vertices are the points u_1', u_2' (i.e., the tree is a "chain" with endpoints u_1', u_2'), then we can "cancel" one of them and carry out the summation as in the previous case (for details see the book [11]). Taking into account the fact that the number of trees with $n+4$ vertices is $(n+4)^{n+2}$, we get the estimate (A.18). Choosing $\overline{M} > \overline{C}$, we get the estimate (4.59).

References

1. C. Boldrighini, I. A. Ignatyuk, V. A. Malyshev, and A. Pellegrinotti, *Random walk in dynamic environment with mutual influence*, Stochastic Processes Appl. **41** (1992), 157–177.

2. R. A. Minlos, *Spectral expansion of the transfer matrices of Gibbs fields*, Soviet Sci. Rev. Sect. C: Math. Phys. Rev., vol. 7, Harwood Academic Publ., Chur, 1987, pp. 236–277.

3. I. I. Gikhman and A. V. Skorokhod, *Theory of random processes.* Vol 1, "Nauka", Moscow, 1971; English transl., Springer-Verlag, Berlin and New York, 1974.
4. M. V. Fedoryuk, *Asymptotics, integrals and series*, "Nauka", Moscow, 1987. (Russian)
5. V. I. Arnol′d, A. N. Varchenko, and S. M. Gusein-zade, *Singularities of differentiable maps.* Vol. II, "Nauka", Moscow, 1987; English transl., Birkhäuser, Basel, 1988.
6. S. Mamatov and R. A. Minlos, *Bound states of the two-particle cluster operator*, Teoret. Mat. Fiz. **79** (1989), no. 2, 163–179; English transl. in Theoret. and Math. Phys. **79** (1989).
7. M. A. Lavrent′ev and B. V. Shabat, *Methods in the theory of functions of a complex variable*, Fizmatgiz, Moscow, 1958, German transl. of 3rd ed., VEB Deutscher Verlag Wiss., Berlin, 1967.
8. W. V. Lovitt, *Linear integral equations*, McGraw-Hill, New York, 1924; reprint, Dover, New York, 1950.
9. E. A. Zhizhina and R. A. Minlos, *Local limit theorem for the random walk of two identical particles interacting at finite distance* (to appear).
10. ———, *Local limit theorem for a nonhomogeneous symmetric random walk on the lattice* (to appear).
11. V. A. Malyshev and R. A. Minlos, *Gibbs random fields*, Kluwer Academic Publishers, Dordrecht, 1991.

Dipartimento di Matematica e Fisica, Università di Camerino, Via Madonna delle Carceri, 62032 Camerino, Italy

Moscow State University, Mechanics and Mathematics Department, Moscow 119899, Russia

Dipartimento di Matematica, Università di Roma La Sapienza, P. A. Moro 2, 00187 Roma, Italy

ADVANCES IN SOVIET MATHEMATICS
Volume 20, 1994

Random Walk of a Particle Interacting with a Random Field

R. A. MINLOS

Introduction

In the existing literature the concept of a "random walk in a random medium" has several meanings.

First, it means that a random field of transition probabilities $\Pr(x-y)$, $x, y \in \mathbb{Z}^\nu$, on the ν-dimensional lattice is considered, and for a fixed "typical" configuration of the field, the random walk on the lattice of a particle governed by this configuration is investigated (see, for example, [1, 2]). A "quantum version" of the above approach was developed in [3, 4]. This means that a random potential $\{V(x, t)\}$ is introduced, and the probability distribution dW^T for the path $\{x(t), 0 \leqslant t \leqslant T\}$ of the particle performing a random walk is described by the Feynman–Kac relation (see [5, 6])

$$dW^T(x(\cdot)) = \frac{1}{Z_T} \exp\left\{ -\sum_{t=0}^{T} V(x(t), t)\right\} dW_0^T(x(\cdot)), \tag{0.1}$$

where dW_0^T is the probability distribution for the path of the ordinary walk with constant (nonrandom) transition probabilities, and Z_T is the normalizing multiplier. ($V(x, t)$ in (0.1) is a sample of the "typical" configuration of a random potential V.)

Finally, in [7–9] another approach was developed, in which the particle and the random medium (described by a Markov random field) are regarded as a single Markov system with interaction; such a system is characterized by transition probabilities that depend on the interaction of the field and the particle (see the article [8], in this volume). In this paper, we investigate a "quantum" version of this model in which the interaction of the field and the particle is introduced again by the Feynman–Kac relation (0.1). More precisely, let W_0^{T, x_0} be the probability distribution for a system without interaction in which the particle's walk and the random field $\xi = \{\xi_x(t), x \in \mathbb{Z}^\nu, t \in Z_+^1\}$ are independent of each other (x_0 being the initial position of

1991 *Mathematics Subject Classification.* Primary 60J15, 60K35, 82B41.

the particle, $\mathbb{Z}^\nu$ the "space", and $\mathbb{Z}^1_+$ the "time"), and let $\mathscr{L}_T(x(\cdot), \xi)$ be the "Lagrangian" defining the interaction of the particle and the field, taken in the form

$$\mathscr{L}(x(\cdot), \xi) = \beta \sum_{t=0}^{T} f(\xi_{x(t)}(t)), \tag{0.2}$$

where f is a real-valued function defined on the set of field values ξ. Then the probability distribution for the system with interaction is specified again by the Feynman–Kac relation:

$$dW^{T, x_0}(x(\cdot), \xi) = Z_T^{-1} \exp\{-\mathscr{L}(x(\cdot), \xi)\} \, dW_0^{T, x_0} \tag{0.3}$$

(Z_{T, x_0} being the normalizing multiplier).

In this paper, we prove that there exists a limiting distribution

$$W^{x_0} = \lim_{T\to\infty} W^{T, x_0}, \tag{0.4}$$

and compute the stochastic operator (transfer matrix) of the limit system using the appropriate cluster expansion of the partition function Z_{T, x_0}. It turns out that the transfer matrix in question is of the form investigated earlier in the articles referred to (in particular, [8, 9]). Specifically, the structure of the leading branch of the transfer matrix spectrum can be computed, and consequently the asymptotics as $t \to \infty$ of the particle's position at time t can be obtained as well as the asymptotics of the decreasing correlations of functions in field values

$$\langle F_1(\xi_0(x_1)), F_2(\xi_t(x_2))\rangle, \tag{0.5}$$

where x_1 and x_2 are two fixed points, and F_1 and F_2 are two functions of the field values at these points. We restrict ourselves to deducing the required form of the transfer matrix of the system, since the subsequent steps, namely the investigation of the system's spectrum and the deduction of the desired asymptotics, are treated in the articles referred to above (see [8, 9]). As in the articles [7–9], we consider the simplest case in which the initial (unperturbed) field ξ is a collection of identical independent Markov chains assigned to the points of the lattice $\mathbb{Z}^\nu$ and taking values in a finite set S. However, the results of this article may be generalized to the case of Markov fields of a more general type (such as Gibbs fields at high temperatures or Markov fields with weak local interaction).

§1. Description of the model and statement of the results

Let $\xi = \{\xi_x(t), x, t \in \mathbb{Z}^\nu \times \mathbb{Z}^1_+\}$ be a random field on the semilattice $\mathbb{Z}^\nu \times \mathbb{Z}^1_+$ ("space-time") taking on values in a finite set S. Suppose the field ξ is a collection of identical and independent ergodic Markov chains $\xi = \{\xi_x(t), x \in \mathbb{Z}^\nu\}$, $\xi_x = \{\xi_x(t), t \in \mathbb{Z}^1_+\}$, that are assigned to the points of the lattice $\mathbb{Z}^\nu$, with one-step transition probabilities

$$\Pr(\xi_x(t) = s \to \xi_x(t+1) = s') = q(s, s'), \qquad s, s' \in S, \tag{1.1}$$

where $\{q(s, s')\} \equiv Q$ is a stochastic matrix. We denote by π_0 the stationary measure for each chain ξ_x. The configuration space Ω of the field ξ is of the form $\Omega = \Sigma^{\mathbb{Z}^1_+}$, where Σ is the set of field configurations on the lattice $\mathbb{Z}^\nu$: $\Sigma = S^{\mathbb{Z}^\nu}$. Denote by $\mathscr{P}_{\text{field}}$ the probability distribution of the field defined on the space Ω, and assume that the stochastic operator

$$(Qf)(s) = \sum_{s' \in S} q(s, s') f(s') \tag{1.2}$$

(f being a function defined on S) can be diagonalized, i.e., there exist $|s|$ eigenvectors $e_0 \equiv 1, e_1, \dots, e_{|s|-1}$ of the operator Q with eigenvalues μ_i, $i = 0, \dots, |s|-1$, such that

$$\mu_0 = 1 > |\mu_1| \geqslant |\mu_2| \geqslant \cdots \geqslant |\mu_{|s|-1}|. \tag{1.3}$$

Consider the space $l_2(S, \pi_0)$ and the operator Q^* adjoint to Q,

$$(Q^* f)(s) = \frac{1}{\pi_0(s)} \sum_{s' \in S} f(s') \pi_0(s') q(s', s). \tag{1.4}$$

This operator also has $|s|$ eigenvectors

$$e_0^* \equiv 1, \; e_1^*, \; \dots, \; e_{|s|-1}^*$$

with eigenvalues $\mu_i^* = \bar{\mu}_i$. The basis $\{e_i\}$ is orthogonal to the basis $\{e_i^*\}$:

$$(e_i, e_j^*)_{l_2(S, \pi_0)} = 0, \qquad i \neq j. \tag{1.5}$$

We shall normalize the vectors $\{e_i\}$ and $\{e_i^*\}$ so that

$$(e_i, e_i^*) = 1 \quad \text{and} \quad \|e_i\| = \|e_i^*\|, \qquad i = 0, \dots, |s|-1. \tag{1.6}$$

Suppose that the particle performs a random walk on $\mathbb{Z}^\nu$ characterized by the following one-step transition probabilities:

$$\Pr(x - y) = p_0(x - y), \tag{1.7}$$

where $p_0(u)$, $u \in \mathbb{Z}^\nu$, is a finite function on the lattice $\mathbb{Z}^\nu$. Further, we assume that the characteristic function $\widetilde{p}_0(\lambda) = \sum_u p_0(u) e^{i(u, \lambda)}$ of the distribution (1.7) is such that

$$|\widetilde{p}_0(\lambda)| < 1, \qquad \lambda \in T^\nu \quad (\nu\text{-dimensional torus}) \tag{1.8}$$

for all $\lambda \neq 0$. Besides, in the Taylor expansion

$$\log \widetilde{p}_0(\lambda) = i \sum_{k=1}^{\nu} b_k^{(0)} \lambda_k - \frac{1}{2} \sum_{k, k'} a_{k,k'}^{(0)} \lambda_k \lambda_{k'} + \cdots, \qquad \lambda = (\lambda_1, \dots, \lambda_\nu) \in T^\nu,$$

the quadratic form

$$\sum_{k, k'} a_{k,k'}^{(0)} \lambda_k \lambda_{k'} > 0 \tag{1.9}$$

is strictly positive for $\lambda \neq 0$. Geometrically, this means that the random walk occurs over the whole lattice $\mathbb{Z}^\nu$ and is not contained within some sublattice. We denote by y_0 the initial position of the particle and by $\mathfrak{M}_{y_0} \subset (\mathbb{Z}^\nu)^{\mathbb{Z}^1_+}$ the set of all the trajectories of the particle starting from y_0. The probability distribution on the space of paths $\mathfrak{M}_{y_0}$ is denoted by $\mathscr{P}_{\text{part}}$. When the walk of the particle and the field are independent of each other, the system under consideration is said to be *unperturbed* or "*free*". For such a system, the probability distribution (defined on the space $\Omega \times \mathfrak{M}_{y_0}$) is equal to

$$\mathscr{P}^0_{y_0} = \mathscr{P}_{\text{field}} \times \mathscr{P}^{y_0}_{\text{part}}. \tag{1.10}$$

Consider a "perturbed" probability distribution on $\Omega \times \mathfrak{M}_{y_0}$ specified by the interaction of the field and the particle of the form

$$\mathscr{L}_T(\xi, Y) = \beta \sum_{t=0}^{T} f(\xi_{y_t}(t)) \tag{1.11}$$

(cf. (0.2)). In this formula, $\xi = \{\xi_x(t)\}$ is the configuration of the field, $Y = \{y_t, t \in \mathbb{Z}^1_+\}$ is the path of the particle, f is a real-valued function defined on the set S, $T > 0$ is an integer, and β is a real number. Further, define the distribution $\mathscr{P}^T_{y_0}$ on $\Omega \times \mathfrak{M}_{y_0}$ as the Gibbsian modification of the measure $\mathscr{P}^0_{y_0}$ due to the interaction (1.11). Namely

$$\frac{d\mathscr{P}^T_{y_0}}{d\mathscr{P}^0_{y_0}}(\xi, Y) = \frac{1}{Z_{T, y_0}} \exp\{-\mathscr{L}_T(\xi, Y)\}, \tag{1.12}$$

where Z_{T, y_0} is the normalizing multiplier (partition function):

$$Z_{T, y_0} = \int \exp\{-\mathscr{L}_T(\xi, Y)\}\, d\mathscr{P}^0_{y_0}. \tag{1.13}$$

THEOREM 1. *Let β be sufficiently small ($|\beta| < \beta_0$). Then*
I. *For all T and $y_0 \in \mathbb{Z}^\nu$ we have*

$$0 < Z_{T, y_0} < \infty. \tag{1.14}$$

II. *There exists a weak limit*

$$d\mathscr{P}_{y_0} = \lim_{T \to \infty} d\mathscr{P}^T_{y_0}. \tag{1.15}$$

III. *The distribution $\mathscr{P}^T_{y_0}$ and also the limiting distribution $\mathscr{P}_{y_0}$ define Markov chains with state space $\Sigma \times \mathbb{Z}^\nu$.*

A Markov chain with distribution $\mathscr{P}_{y_0}$ is said to describe a *perturbed system*. Now let us consider the probability distribution of the particle's position at time t computed according to the distribution $\mathscr{P}_{y_0}$:

$$\Pr(y_t = y) \equiv p_t^{y_0}(y). \tag{1.16}$$

THEOREM 2. *The distribution* $p_t^{y_0}(y)$ *is asymptotically Gaussian*:

$$p_t^{y_0}(y) = \frac{C}{t^{\nu/2}} \exp\left\{ -\frac{1}{2t} \sum_{i,j} a_{i,j}(y_i - y_i^0 - b_i t)(y_j - y_j^0 - b_j t)\right\}(1 + o(1)) \quad (t \to \infty), \tag{1.17}$$

where $y = (y_1, \dots, y_\nu)$, $y_0 = (y_1^0, \dots, y_\nu^0)$, $\{a_{ij}\} = A$ *is a positive definite matrix of rank* ν, $b = (b_1, \dots, b_\nu) \in R^\nu$, *and*

$$C = (2\pi)^{-\nu/2}(\det A)^{1/2}.$$

Consider a correlation of the form (0.5):

$$\langle F_1(\xi_{x_1}(x_1)), F_2(\xi_{x_2}(x_2))\rangle_{P_{y_0}}, \tag{1.18}$$

where F_1 and F_2 are two functions on the space S, and x_1 and x_2 are two fixed points of the lattice $\mathbb{Z}^\nu$. In addition, suppose that the following condition is satisfied: $\min_{\lambda \in t^\nu} |\widetilde{p}(\lambda)| > 0$.

THEOREM 3. I. *When the "drift"* b *in formula* (1.17) *is nonzero, the following estimate is valid*:

$$|\langle F_1(\xi_{x_1}(x_1)), F_2(\xi_{x_2}(x_2))\rangle_{P_{y_0}}| < C\theta^t, \tag{1.19}$$

where $C > 0$ *and* $\theta > 0$ *are constants, and* $\theta < 1$.

II. *If the function* $p_0(u)$ *specifying the jump distribution for the particle's walk is even* (*i.e.*, $b = 0$), *the correlation asymptotics* (1.18) *is of the form*

$$Ct^{-\nu/2-1}(1 + o(1)), \tag{1.20}$$

where $C = C(f_1, f_2, x_1, x_2, y_0)$ *is a constant.*

§2. Proof of Theorem 1

The Markov property of the finite perturbed chain $\mathscr{P}_{y_0}^T$ is obvious. We shall try to find the transition probability $Q^T(\cdot \mid \xi(0), y_0)$ for this chain (at step one), i.e., the conditional distribution for the state $(\xi(1), y_1)$ of the chain at step one given the state $(\xi(0), y_0)$ at step zero. Suppose that $\mathscr{P}^0(\cdot \mid \xi(0), y_0)$ is the conditional probability distribution for the unperturbed chain under the same assumption, and

$$Z_T(\xi(0), y_0) = \int \exp\{-\mathscr{L}_T'(\xi, Y)\}\, d\mathscr{P}^0(\xi, Y \mid \xi(0), y_0). \tag{2.1}$$

is the conditional partition function. In this formula

$$\mathscr{L}_T'(\xi, Y) = \beta \sum_{t=1}^{T} f(\xi_{y_t}(t)).$$

It is readily seen that

$$dQ^T\{(\xi(1), y_1) \mid (\xi(0), y_0)\} = e^{-\beta f(\xi_{y_1}(1))} \frac{Z_{T-1}(\xi(1), y_1)}{Z_T(\xi(0), y_0)} dQ^0\{(\xi(1), y_1) \mid (\xi(0), y_0)\}, \quad (2.2)$$

where $Q^0\{(\xi(1), y_1) \mid (\xi(0), y_0)\}$ is the transition probability for the unperturbed system.

LEMMA 1. *Under the assumptions of Theorem* 1, *the following limit exists:*

$$\lim_{T\to\infty} \frac{Z_{T-1}(\xi(1), y_1)}{Z_T(\xi(0), y_0)} = \widehat{\gamma} \frac{K(\xi(1), y_1)}{K(\xi(0), y_0)}, \quad (2.3)$$

where $\widehat{\gamma} > 0$ *is a constant, and* $K(\xi, y)$ *is a functional on the space* $\Sigma \times \mathbb{Z}^\nu$ (*this functional will be described below in greater detail*).

This lemma and formula (2.2) imply the existence of the limiting transition probability $Q\{\cdot \mid (\xi(0), y_0)\}$:

$$dQ\{(\xi(1), y_1) \mid (\xi(0), y_0)\} = e^{-\beta f(\xi_{y_1}(1))} \widehat{\gamma} \frac{K(\xi(1), y_1)}{K(\xi(0), y_0)} dQ^0\{(\xi(1), y_1) \mid (\xi(0), y_0)\}.$$

It follows from this that the limit (1.15) exists.

PROOF OF LEMMA 1 (cluster expansion of the conditional partition function). We have

$$Z_T(\xi(0), y_0) = \sum_{Y=\{y_t,\, 0\leqslant t\leqslant T\}} \prod_{t=1}^{T} p_0(y_t - y_{t-1}) \sum_{\{\xi_x(t),\, 1\leqslant t\leqslant T,\, x\in\Pi(Y)\}} \prod_{t=1}^{T} e^{-\beta f(\xi_{y_t}(t))} \times \prod_{\substack{t=1\\ x\in\Pi(Y)}}^{T} q(\xi_x(t-1), \xi_x(t)), \quad (2.4)$$

where $Y = \{y_t,\ 0 \leqslant t \leqslant T\}$ is the particle's path, and

$$\Pi(Y) = \{x = y_t,\ x \in \mathbb{Z}^\nu,\ 0 \leqslant t \leqslant T\}$$

is the set of lattice points the particle passes through in the time interval $[1, T]$. For an arbitrary function $\psi(s)$ on S we introduce the expansion

$$\psi = p_0\psi + p\psi, \quad (2.5)$$

where

$$p_0\psi = \langle\psi\rangle_{\pi_0}, \qquad p\psi = \psi - \langle\psi\rangle_{\pi_0} \in \tilde{l}_2(S, \pi_0).$$

In this relation, $\tilde{l}_2(S, \pi_0) \subset l_2(S, \pi_0)$ is the subspace of functions ψ defined on S such that $\langle\psi\rangle_{\pi_0} = 0$. Observe that $\tilde{l}_2(S, \pi_0)$ is invariant with respect to the operator Q. We introduce a norm in $\tilde{l}_2(S, \pi_0)$ in the following way:

$$|||\psi||| = \max_{i=1,\dots,|s|-1} \{|c_i|\}, \quad (2.6)$$

where

$$\psi = \sum_{i=1}^{|s|-1} c_i e_i \in \tilde{l}_2(S, \pi_0).$$

It is obvious that for any function $\psi \in \tilde{l}_2(S, \pi_0)$ we have

$$|||Q\psi||| \leqslant |\mu_1| \cdot |||\psi|||. \tag{2.7}$$

For each $x \in \Pi(Y)$ consider the "vertical" line $h_x^T = \{(x, t), 0 \leqslant t \leqslant T\}$, and let

$$\tau_1(x) > \tau_2(x) > \cdots > \tau_k(x), \qquad k = k(x),$$

be the moments at which the particle arrives at the point $x \in \Pi(Y)$. For any specified path $Y = \{y_t\}_{t=0}^{t=T}$ we have

$$\prod_{t=1}^{T} e^{-\beta f(\xi_{y_t}(t))} = \alpha^{-T} \prod_{x \in \Pi(Y)} \prod_{i=1}^{k(x)} g(\xi_x(\tau_i(x))), \tag{2.8}$$

where $\alpha = (\langle e^{-\beta f}\rangle_{\pi_0})^{-1}$ and $g(s) = \alpha e^{-\beta f(s)}$.

Further, we obtain

$$\sum_{\{\xi_x(t),\, 0 \leqslant t \leqslant T,\, x \in \Pi(Y)\}} \prod_t e^{-\beta f(\xi_{y_t}(t))} \prod_{\substack{t \\ x \in \Pi(Y)}} q(\xi_x(t-1), \xi_x(t)) \tag{2.9}$$

$$= \alpha^{-T} \prod_{x \in \Pi(Y)} Q^{\tau_k(x)}(p_0 + p)\hat{g} Q^{\tau_{k-1}(x) - \tau_k(x)} \cdots Q^{\tau_1(x) - \tau_2(x)}(p_0 + p) g,$$

where we denote by $\hat{g}$ the operator of multiplication by the function g in the space $l_2(S, \pi_0)$. Removing the brackets in the product of operators (2.9), for each fixed x we obtain a sum of terms of the form

$$k_{[\tau_{i_1}(x), \tau_{i_1'}(x)]} \cdots k_{[\tau_{i_s}(x), \tau_{i_s'}(x)]} \tag{2.10}$$

or, in equivalent form, $k_{[\tau_{i_1}(x), \tau_{i_1'}(x)]} \cdots k^{\text{in}}_{[\tau_{i_s}(x), 0]}(\xi_x(0))$, where

$$1 \leqslant i_1 < i_1' < i_2 < i_2' < \cdots < i_s < i_s' \leqslant k(x). \tag{2.11}$$

We have

$$k_{[\tau_i, \tau_{i'}]} = \langle \hat{g} Q^{\tau_{i'-1} - \tau_{i'}} p \hat{g} Q^{\tau_{i'-2} - \tau_{i'-1}} \cdots Q^{\tau_i - \tau_{i+1}} p g \rangle_{\pi_0} \tag{2.12}$$

for $i' \leqslant k(x)$ and

$$k^{\text{in}}_{[\tau_i, 0]}(\xi_x(0)) = (Q^{\tau_k} p \hat{g} \cdots p \hat{g} Q^{\tau_i - \tau_{i+1}} p g)(\xi_x(0)). \tag{2.12a}$$

The intervals $[\tau_i(x), \tau_{i'}(x)] \equiv m_{i,i'}(x, Y)$, $i < i'$, will be called *elementary segments*, and the interval $[\tau_i(x), 0] \equiv m_i^{\text{in}}(x, Y)$ the *initial elementary segment*. The set of elementary segments

$$\{m_{i_1, i_1'}(x_1, Y), \ldots, m_{i_l, i_l'}(x_l, Y)\}$$

(possibly including the initial segment) will be called *connected* if the set of intervals on the one-dimensional lattice

$$[\tau_{i_1}, \tau_{i'_1}], \ \dots, \ [\tau_{i_l}, \tau_{i'_l}]$$

is connected. The maximal connected set of elementary segments

$$G = \{m_{i_1, i'_1}(x_1, Y), \dots, m_{i_l, i'_l}(x_l, Y)\} \tag{2.13}$$

together with part of the path $Y_g = \{y_t, \tau_{\min}(G) \leqslant t \leqslant \tau_{\max}(G)\}$ generate the cluster $\{G, Y_G\}$ (note that the cluster does not depend on the behavior of the path Y outside the interval $[\tau_{\min}(G), \tau_{\max}(G)]$). Observe also that

$$\tau_{\max}(G) = \max_k \tau_{i_k}, \qquad \tau_{\min}(G) = \min_k \tau_{i'_k}. \tag{2.14}$$

The point of the path $y_{\tau_{\max}}(G) = y_{\max}(G)$ will be called the *upper end* and the point $y_{\tau_{\min}}(G) = y_{\min}(G)$ the *lower end* of the cluster $\{G, Y_G\}$. The quantity

$$K(G, Y_G) = \prod_{m_{j,j'} \in G} K_{m_{j,j'}} \prod_{t=\tau_{\min}(G)}^{\tau_{\max}(G)-1} p_0(y_{t+1} - y_t) \tag{2.15}$$

will be called the *weight of the cluster* $\{G, Y_G\}$. The cluster will be called *initial* if it contains at least one initial segment $m^{\text{in}}_{i,0}$. Thus we obtain the following representation of the partition function:

$$\begin{aligned} &Z_T(\xi(0), y_0) \\ &= \alpha^{-T}\Bigg[\sum'_{(G_1, Y_{G_1}) < \cdots < (G_s, Y_{G_s})} \prod_{i=1}^{s} K(G_i, Y_{G_i}) p_0^{\min \tau(G_1)}(y_{\min}(G_1) - y_0) \\ &\qquad \times \prod_{k=2}^{s} p_0^{\tau_{\min}(G_k) - \tau_{\max}(G_{k-1})}(y_{\min}(G_k) - y_{\max}(G_{k-1})) \\ &\qquad + \sum''_{(G_0^{\text{in}}, Y_{G_0^{\text{in}}}) < \cdots < (G_s^{\text{in}}, Y_{G_s^{\text{in}}})} K(G_0^{\text{in}}, Y_{G_0^{\text{in}}}) \prod_{i=1}^{s} K(G_i, Y_{G_i}) \\ &\qquad \times \prod_{k=1}^{s} p_0^{\tau_{\min}(G_k) - \tau_{\max}(G_{k-1})}(y_{\min}(G_k) - y_{\max}(G_{k-1}))\Bigg]. \end{aligned} \tag{2.16}$$

The inequality $(G, Y_G) < (G', Y_{G'})$ implies that $\tau_{\max}(G) < \tau_{\min}(G')$. Note also the relation

$$p_0^n(\xi) = (\underbrace{p_0 * p_0 * \cdots * p_0}_{(n)})(\xi), \tag{2.17}$$

where $*$ is the convolution of functions on the lattice $\mathbb{Z}^\nu$. The sum $\sum'$ is taken over all collections of clusters that do not include the initial cluster, and the sum $\sum''$ involves the collections containing the initial cluster. If we denote the first sum $\sum'$ by $\widetilde{Z}^T$ (as is easily seen, it does not depend on $\xi(0)$ and y_0), then it follows from the expansion (2.16) that

$$Z_T(\xi(0), y_0) = \alpha^{-T}\left[\widetilde{Z}_T + \sum_{(G^{\text{in}}, Y_{G^{\text{in}}})} K(G^{\text{in}}, Y_{G^{\text{in}}})(\xi_x(0))\widetilde{Z}_{T-\tau_{\max}}(G^{\text{in}})\right], \quad (2.18)$$

where the sum $\sum_{(G^{\text{in}}, Y_{G^{\text{in}}})}$ is taken over all initial clusters. To conclude the proof of the lemma, we need one more lemma.

LEMMA 2. *The following relation holds:*

$$\widetilde{Z}_T = \exp\{\gamma_0 T + \gamma_1 + \delta(T)\}, \quad (2.19)$$

where $\gamma_0 > 0$ *and* γ_1 *are constants independent of* T *and*

$$|\delta(T)| < C\lambda^T, \quad (2.20)$$

where $C > 0$ *and* $0 < \lambda < 1$ *do not depend on* T.

Expressions (2.18) and (2.20) lead directly to formula (2.3), in which

$$K(\xi(0), y_0) = 1 + \sum_{(G^{\text{in}}, Y_{G^{\text{in}}})} K(G^{\text{in}}, Y_{G^{\text{in}}})(\xi_x(0)) \exp\{-\gamma_0\tau_{\max}(G^{\text{in}})\} \quad (2.21)$$

and $\widehat{\gamma} = \alpha^{-1}e^{\gamma_0}$. The convergence of the series (2.21) follows from the estimates given below.

PROOF OF LEMMA 2. We shall specify the collection of segments

$$m_1 = [t_1, t_1'], \dots, m_s = [t_s, t_s'], \qquad 0 \leqslant t_1 < t_1' < t_2 < t_2' < \cdots < t_s < t_s' \leqslant T,$$

and consider the sum

$$\sum_{(G_1, Y_{G_1}) < \cdots < (G_s, Y_{G_s})} \prod_{i=1}^{s} K(G_i, Y_{G_i}) p_0^{t_1}(y_{\min}(G_1) - y_0)$$
$$\times \prod_{i=2}^{s} p_0^{t_i - t_{i-1}'}(y_{\min}(G_i) - y_{\max}(G_{i-1})) = \prod_{i=1}^{s} e(t_i' - t_i), \quad (2.22)$$
$$\tau_{\max}(G_i) = t_i', \quad \tau_{\min}(G_i) = t_i, \quad i = 1, \dots, s,$$

where

$$e(u) = \sum_{\substack{(G, Y_G): \\ \tau_{\min}(G)=0,\, \tau_{\max}(G)=u}} K(G, Y_G). \quad (2.23)$$

It follows from (2.7) that for each elementary segment $m[t(x), t'(x)]$ the following inequality holds:

$$|K_m| < (C\beta)^{p-1}|\mu_1|^{|t-t'|-1}, \quad (2.24)$$

where p is the number of times the path Y passes through the point x in the time interval $[t(x), t'(x)]$. Therefore, since the cluster G is connected, for sufficiently small β we have

$$\prod_{\mu \in G} |K_m| < |\mu_1|^{\tau_{\max}(G) - \tau_{\min}(G)}.$$

Since for any specified "duration" $t = \tau_{\max}(G) - \tau_{\min}(G)$ of the cluster G and for a specified path Y_G there exist no more than 2^t clusters, the following estimate holds:

$$|e(t)| < (2|\mu_1|)^t. \tag{2.25}$$

Let us now consider an ensemble of collections

$$\mathfrak{M} = \{m_1, \dots, m_s\}, \qquad m_i \subseteq [0, T], \quad |m_i| > 1,$$

of pairwise nonintersecting segments, the weight of each collection m being equal to $\prod_{i=1}^{s} e(|m_i|)$ ($|m_i|$ is the length of the segment m_i). It is readily seen that

$$\widetilde{Z}_T = \sum_{\mathfrak{M}} \prod_{m_i \in \mathfrak{M}} e(|m_i|). \tag{2.26}$$

Given that β and μ_1 are sufficiently small, some general cluster expansions can be applied to the ensemble of segments (see, for example, [10]). Using these methods, we obtain

$$\ln \widetilde{Z}_T = \sum_{m \subseteq [0, T], |m| > 1} s(m), \tag{2.27}$$

where $s(m)$ is a function that is defined on the set of finite segments, depends only on the length of the segment: $s(m) = s(|m|)$, and satisfies the estimate

$$|s(m)| < C\lambda^{|m|}, \tag{2.28}$$

where $C > 0$ and $0 \leqslant \lambda \leqslant 1$ are constants, the constant λ being small simultaneously with β and μ_1. The representation (2.27) implies the representation (2.19), in which

$$\gamma_0 = \sum_{m = [0, t], t > 0} s(m), \quad \gamma_1 = \sum_{m : 0 \in m} s(m), \quad \delta(T) = \sum_{m : [0, T] \subset m} s(m). \tag{2.29}$$

The estimate (2.28) leads to the convergence of the series in (2.29), the estimate (2.20), and the following estimate:

$$|\gamma_0| < C_1 |\mu_1|, \tag{2.30}$$

where C_1 is an absolute constant. Since $\widetilde{Z}_T > 1$ for all T, we obtain $\gamma_0 > 0$. The proof of Lemma 2 and, consequently, that of Lemma 1 is complete. It is readily seen that the proof of Theorem 1 follows immediately from these lemmas.

COROLLARIES. I. The estimates given above imply that

$$\left| \sum_{(G^{\text{in}}, Y_{G^{\text{in}}}), \tau_{\max}(G^{\text{in}})=t} K(G^{\text{in}}, Y_{G^{\text{in}}}) \right| < C(2|\mu_1|)^t. \tag{2.31}$$

It follows that

$$\left| \sum_{(G^{\text{in}}, Y_{G^{\text{in}}})} K(G^{\text{in}}, Y_{G^{\text{in}}}) \exp\{-\gamma_0 \tau_{\max}(G^{\text{in}})\} \right| < C_1|\mu_1| \tag{2.32}$$

(C_1 being an absolute constant), and thus the function $[K(\xi(0), y_0)]^{-1}$ is bounded.

II. Since the sum

$$\sum_{(G^{\text{in}}, Y_{G^{\text{in}}}), \tau_{\max}(G^{\text{in}})=t} K(G^{\text{in}}, Y_{G^{\text{in}}})$$

depends only on the field values $\xi(0)$ at the points x that are within $R_0 t$ distance of y_0, R_0 being the value of the maximum jump of the particle in one step, it follows that the function $K(\xi(0), y_0)$ depends "exponentially weakly" on the field values at points far removed from y_0:

$$|K(\xi(0), y_0) - K(\xi'(0), y_0)| < C(2|\mu_1|)^{b(\xi(0), \xi'(0), y_0)/R_0}, \tag{2.33}$$

where $\xi(0)$ and $\xi'(0)$ are two field configurations, and

$$b(\xi(0), \xi'(0), y_0) = \min_{x:\, \xi_x(0) \neq \xi'_x(0)} |x - y_0|.$$

§3. Proof of Theorems 2 and 3

As we mentioned in the Introduction, the proof of the local limit theorem concerning the particle's position and also the deduction of the correlation asymptotics follows the same arguments as in [8, 9]. We shall introduce the transfer matrix (stochastic operator) of the limit chain under consideration

$$\begin{aligned}(\mathscr{T}\Phi)(\xi, y) &= \int \Phi(\eta, z)\, dQ(\eta, z \mid \xi, y) \\ &= \hat{\gamma} \int \Phi(\eta, z) e^{-\beta f(\eta_z)} K(\eta, z)(K(\xi, y))^{-1} dQ^0(\eta, z \mid \xi, y).\end{aligned} \tag{3.1}$$

In this formula Φ is an arbitrary (measurable) bounded function on the space $\Sigma \times \mathbb{Z}^\nu$. As is easily verified, the operator $\mathscr{T}$ is defined on all functions $\Phi \in L_2(\Sigma, \Pi_0) \otimes l_2(\mathbb{Z}^\nu) = \mathscr{H}$ and is a bounded operator in this space ($\Pi_0 = (\pi_0)^{\times \mathbb{Z}^\nu}$ is the stationary distribution of the unperturbed chain, which is equal to a countable product of the measures π_0). Observe that

$$\Pr(y_t = y \mid y_{t=0} = y_0) = \int_\Sigma (\mathscr{T}^t \Phi_y)(\xi, y)\, d\Pi_0(\xi), \tag{3.2}$$

where $\Phi_y(\xi, y) = \delta_{y,z}$.

Now we choose the following basis in the space $\mathscr{H}$:

$$\Phi_{\Gamma, z}(\xi, y) = \frac{1}{K(\xi, y)} \prod_x e_{\gamma(x)}(\xi_x) \delta_{y, z}, \tag{3.3}$$

where $\Gamma = \{\gamma(x), x \in \mathbb{Z}^\nu\}$ is a multi-index, i.e., a finite function on $\mathbb{Z}^\nu$ with values $\gamma(x) = 0, \dots, |s| - 1$, and e_γ are eigenvectors of the operator Q. As in [8], it is easily verified that $\{\Phi_{\Gamma, z}\}$ is a basis in $\mathscr{H}$, i.e., any function $\Phi \in \mathscr{H}$ can be uniquely expanded in the series

$$\Phi = \sum_{(\Gamma, z)} C_{(\Gamma, z)} \Phi_{(\Gamma, z)}. \tag{3.4}$$

A simple computation shows that

$$\mathscr{T} \Phi_{(\Gamma, z)} = \sum_{(\Gamma', z')} R_{(\Gamma, z), (\Gamma', z')} \Phi_{(\Gamma', z')}, \tag{3.5}$$

where

$$R_{(\Gamma, z), (\Gamma', z')} = \mu_{\Gamma_z} \delta_{\Gamma_z, \Gamma'_z} L(\gamma(z), \gamma'(z), z - z'). \tag{3.6}$$

In this formula the multi-index $\Gamma_z = \{\gamma_z(x), x \in \mathbb{Z}^\nu\}$ is defined by

$$\gamma_z(x) = \begin{cases} \gamma(x), & \text{for } x \neq z, \\ 0, & \text{for } x = z, \end{cases} \tag{3.7}$$

where $\gamma(x)$ are the values of the multi-index Γ,

$$\mu_\Gamma = \prod_x \mu_{\gamma(x)},$$

$$\begin{aligned} L(j, j', u) &= e^{-\gamma_0} \mu_j \delta_{jj'} p_0(u) + \widehat{\gamma} \mu_{j'} \sum_{k \neq 0} \varkappa_k b^{j'}_{k, j} p_0(u) \\ &= \mu_j \delta_{jj'} p_0(u) + \widehat{\gamma} \mu_{j'} \sum_{k \neq 0} \varkappa_k b^{j'}_{k, j} p_0(u) + (e^{-\gamma_0} - 1) \mu_\gamma \delta_{jj'} p_0(u). \end{aligned} \tag{3.8}$$

Note that the numbers $\varkappa_k$ and $b^{j'}_{k, j}$ are defined by the expansions

$$e^{-\beta f(s)} = \sum_{k=0}^{|s|-1} \varkappa_k e_k(s), \quad s \in S \ (\varkappa_0 = \alpha); \tag{3.9}$$

$$e_k(s) e_j(s) = \sum_{j'} b^{j'}_{k, j} e_{j'}(s), \quad s \in S, \ k, j = 0, \dots, |s| - 1. \tag{3.10}$$

By (2.30) we have

$$|e^{-\gamma_0} - 1| < C_1 |\mu_1|$$

and also

$$|\varkappa_k| < C_2 |\beta|, \qquad k \neq 0, \tag{3.11}$$

where C_1 and C_2 are absolute constants. Further steps in proving Theorems 2 and 3 follow the same arguments as in [8, 9], and so they will not be repeated here.

REFERENCES

1. V. V. Anshelevich, K. M. Khanin, and Ya. G. Sinai, *Symmetric random walks in random environments*, Comm. Math. Phys. **85** (1982), 449–470.
2. G. A. Ritter, *Random walk in a random environment*, Ph.D. Thesis, Cornell University, Ithaca, NY, 1976.
3. J. Z. Imbrie and T. Spencer, *Diffusion of directed polymers in a random environment*, J. Statist. Phys. **52** (1988), no. 3 (4), 609–626.
4. Ya. G. Sinai, *Random walk under a random potential*, Teor. Veroyatnost. i Primenen. **38** (1993), 457–460; English transl. in Theory Probab. Appl. **38** (1993).
5. J. Glimm and A. Jaffe, *Quantum physics. A functional integral point of view*, Springer-Verlag, New York, 1981.
6. B. Simon, *Functional integration and quantum physics*, Academic Press, New York, 1979.
7. C. Boldrighini, I. A. Ignatyuk, V. A. Malyshev, and A. Pellegrinotti, *Random walk in a dynamic environment with mutual influence*, Stochastic Processes Appl. **41** (1992), 157–177.
8. C. Boldrighini, R. A. Minlos, and A. Pellegrinotti, *Central limit theorem for the random walk of one and two particles in a random environment, with mutual interaction*, in this volume.
9. ———, *Interacting random walk in a dynamical random environment. I. Decay of correlation (one particle)*, Ann. Inst. H. Poincaré (to appear).
10. V. A. Malyshev and R. A. Minlos, *Gibbs random fields*, Kluwer Academic Publishers, Dordrecht, 1991.
11. I. I. Gikhman and A. V. Skorokhod, *Theory of random processes*. I, "Nauka", Moscow, 1971; English transl., Springer-Verlag, New York, 1974.

MOSCOW STATE UNIVERSITY, MECHANICS AND MATHEMATICS DEPARTMENT, MOSCOW 119899, RUSSIA

Translated by N. K. KULMAN

ADVANCES IN SOVIET MATHEMATICS
Volume 20, 1994

Large and Moderate Deviations in the Ising Model

ROLAND L. DOBRUSHIN AND SENYA B. SHLOSMAN

CONTENTS

1. Introduction and Statements of Main Results

§1.1. Introduction

The theory of large and moderate deviations of sums of random variables is now a wide and fastly growing branch of probability theory (see references in §1.9). It was created initially in the framework of the theory of sums of independent identically distributed random variables and then extended to a wide class of random processes, i.e., random functions in one variable, with some general conditions of weak dependence traditional for the theory of random processes. It was found that in these cases the asymptotic structure of probabilities of deviations is completely similar to the situation for the case of independent variables. We shall call this case a classical behavior of large and moderate deviations.

In the last few years probability theory has turned to the study of random fields, i.e., random functions in several variables. It became clear that the main difference of this multidimensional case is the possibility of a phase transition, which occurs even in the simplest typical situations. It turns out that the existence of a phase transition can radically change the asymptotic

1991 *Mathematics Subject Classification.* Primary 60F10, 60K35, 82B20.

behavior of large and moderate deviations. The aim of this paper is to study these new effects and to compare them with the classical behavior. It turns out also that this behavior is closely related to the physical phenomenon of condensation of droplets of one phase inside another phase. To simplify the exposition we will present the results for the simplest nontrivial Gibbs field, namely the ferromagnetic Ising model. Possible generalizations are discussed in §1.9.

§1.2. Classical behavior of moderate and large deviations

In this section we will describe the classical behavior of the large and moderate deviations for sums of random variables. The description we give differs somewhat from the standard one, and contains in fact more information than usual, so we call it the *Classical Large Deviation Principle in the Strong Form.* Since we are Ising model oriented, we consider random variables taking the values $+1$ and -1 only.

Consider a random variable

$$S_N = \xi_1^N + \cdots + \xi_N^N , \tag{1.2.1}$$

where the random variables ξ_i^N take the values $+1$ and -1. Let

$$P_N(b) = \mathbf{P}\{S_N = b\} , \tag{1.2.2}$$

where b are integers such that

$$(b - N) \equiv 0 \pmod 2. \tag{1.2.3}$$

(For all other values of b the probabilities (1.2.2) vanish.) We consider the logarithmic generating function

$$H_N(z) = \ln \sum_b P_N(b) e^{zb} \tag{1.2.4}$$

and suppose that for each real $\lambda \in \mathbb{R}^1$ there exist a complex neighborhood $\mathscr{O}_\lambda$ of λ in $\mathbb{C}^1$ and a constant C_λ such that for all large enough N

$$|H_N(z)| \le NC_\lambda \quad \text{if } z \in \mathscr{O}_\lambda . \tag{1.2.5}$$

We suppose also that the limit

$$\lim_{N\to\infty} \frac{1}{N} H_N(z) = H(z) \tag{1.2.6}$$

exists in the sense of uniform convergence in $\mathscr{O}_\lambda$ for all λ. This limit is a holomorphic function of $z \in \mathscr{O}_\lambda$.

For any $\lambda \in \mathbb{R}^1$ denote

$$G^n_{N,\lambda} = \left. \frac{d^n H_N(z)}{dz^n} \right|_{z=\lambda} . \tag{1.2.7}$$

For $\lambda = 0$ these quantities are called the semiinvariants of order n of the variables (1.2.1). It follows from (1.2.5) and well-known properties of holomorphic functions that the following limits exist:

$$G_\lambda^n = \lim_{N\to\infty} \frac{1}{N} G_{N,\lambda}^n = \left.\frac{d^n H(z)}{dz^n}\right|_{z=\lambda}. \tag{1.2.8}$$

In the case $\lambda = 0$ we shall omit the index λ in the notation $G_{N,\lambda}^n$. Also,

$$E_N = G_N^1, \qquad E = G^1, \qquad D_N = G_N^2, \qquad D = G^2. \tag{1.2.9}$$

The quantities E_N and D_N are, of course, the mean value and the variance of S_N. It is easy to see that $G_{N,\lambda}^2 \geq 0$ for all $\lambda \in \mathbb{R}^1$. So the functions $H_N(\lambda)$ and $H(\lambda)$ are concave functions of $\lambda \in \mathbb{R}^1$. We suppose additionally that $H_N(\lambda)$ and $H(\lambda)$ are strictly concave. (For $H_N(\lambda)$ this property holds under the condition that the probability distribution $P_N(b)$ is not concentrated at a single point; for the limiting function $H(\lambda)$ it is an essential new restriction.) This assumption implies in particular that the limiting variance D is positive. Our hypothesis about the asymptotic behavior of the functions $\frac{1}{N}H_N(\lambda)$ and $H(\lambda)$ at $\lambda = \pm\infty$ is that their derivatives $\frac{1}{N}G_{N,\lambda}^1$ and G_λ^1 tend to ± 1 as $\lambda \to \pm\infty$.

We define the *action functions*

$$I_N(b) = \sup_{\lambda\in\mathbb{R}^1}(\lambda b - H_N(\lambda)), \qquad I(\hat{b}) = \sup_{\lambda\in\mathbb{R}^1}(\lambda\hat{b} - H(\lambda)). \tag{1.2.10}$$

Under the strict concavity hypothesis introduced above, there exist unique values $\lambda_N(b)$, $b \in (-N, N)$, and $\lambda(\hat{b})$, $\hat{b} \in (-1, 1)$, such that

$$b = G_{N,\lambda}^1 = \left.\frac{dH_N(\lambda)}{d\lambda}\right|_{\lambda=\lambda_N(b)}, \qquad \hat{b} = \left.\frac{dH(\lambda)}{d\lambda}\right|_{\lambda=\lambda(\hat{b})} \tag{1.2.11}$$

and so

$$I_N(b) = \lambda_N(b)b - H_N(\lambda_N(b)), \qquad I(\hat{b}) = \lambda(\hat{b})\hat{b} - H(\lambda(\hat{b})), \tag{1.2.12}$$

and

$$I(\hat{b}) = \lim_{N\to\infty} \frac{1}{N} I_N(N\hat{b}). \tag{1.2.13}$$

So the implicit function theorem implies that $I_N(b)$ and $I(\hat{b})$ can be extended to holomorphic functions in a neighborhood of $\mathbb{R}^1$. It is easy to see that

$$I_N(E_N) = I(E) = 0, \qquad \left.\frac{dI_N(b)}{db}\right|_{b=E_N} = \left.\frac{dI(\hat{b})}{d\hat{b}}\right|_{\hat{b}=E} = 0. \tag{1.2.14}$$

So the Taylor expansions of $I_N(b)$ and $I(\hat{b})$ at points E_N and E respectively have the following form:

$$I_N(b) = \frac{(b-E_N)^2}{2D_N} + \sum_{j=3}^{\infty} \frac{K_N^j}{j!}\left(\frac{(b-E_N)}{D_N}\right)^j, \tag{1.2.15}$$

$$I(\hat{b}) = \frac{(\hat{b}-E)^2}{2D} + \sum_{j=3}^{\infty} \frac{K^j}{j!}\left(\frac{(\hat{b}-E)}{D}\right)^j, \tag{1.2.16}$$

where

$$K^j = \lim_{N\to\infty} \frac{1}{N} K_N^j. \tag{1.2.17}$$

It is easy to see that the coefficients K_N^j are given by the formula

$$K_N^j = D_N Q_j\left(\frac{G_N^3}{D_N}, \dots, \frac{G_N^j}{D_N}\right), \qquad j = 3, 4, \dots, \tag{1.2.18}$$

where Q_j is a polynomial in $j-2$ variables. To find the polynomials Q_j we observe first that

$$\frac{dI_N(b)}{db} = \lambda_N(b). \tag{1.2.19}$$

So if we substitute the expansion

$$b - E_N = \frac{dH_N(\lambda_N(b))}{d\lambda} - E_N = D_N\lambda_N(b) + \sum_{j=3}^{\infty} \frac{G_N^j(\lambda_N(b))^{j-1}}{(j-1)!} \tag{1.2.20}$$

in the expansion

$$\lambda_N(b) = \frac{dI_N(b)}{db} = \frac{(b-E_N)}{D_N} + \sum_{j=3}^{\infty} \frac{K_N^j}{(j-1)!}\frac{(b-E_N)^{j-1}}{(D_N)^j}, \tag{1.2.21}$$

and take into account that the coefficients at all terms $\lambda_N(b)^j$, $j = 2, 3, \dots$, must vanish, we get the equations

$$\frac{K_N^j}{D_N} = \hat{Q}_j\left(\frac{G_N^3}{D_N}, \dots, \frac{G_N^j}{D_N}, \frac{K_N^3}{D_N}, \dots, \frac{K_N^{j-1}}{D_N}\right), \tag{1.2.22}$$

where the $\hat{Q}_j$ are polynomials that can be written down explicitly. So one can find the polynomials Q_j recurrently. In particular,

$$K_N^3 = -G_N^3, \qquad K_N^4 = -G_N^4 + 3(G_N^3)^2(D_N)^{-1}, \tag{1.2.23}$$

and so on. Similar relations hold between the limiting coefficients K^j and the limiting semiinvariants G^j.

Observe that it follows from (1.2.11), (1.2.19) that the derivatives

$$\frac{d^2I_N(b)}{db^2} = \left(\frac{dH_N(\lambda_N(b))}{d\lambda}\right)^{-1}, \qquad \frac{d^2I(\hat{b})}{d\hat{b}^2} = \left(\frac{dH(\lambda(\hat{b}))}{d\lambda}\right)^{-1}. \tag{1.2.24}$$

Because we have assumed that the functions $H_N(\lambda)$ and $H(\lambda)$ are strictly concave, the relation (1.2.24) implies that the same holds for the functions $I_N(b)$ and $I(\hat{b})$.

We say that the *Classical Large Deviation Principle in the Strong Form* holds for the sequence S_N and the sequence b_N, $N = 1, 2, 3, \ldots$, satisfying the conditions

$$(b_N - N) \equiv 0 \pmod 2 \tag{1.2.25}$$

and

$$\limsup_{N\to\infty} \frac{|b_N|}{N} < 1 \tag{1.2.26}$$

if the following relation holds:

$$P_N(b_N) = 2(2\pi G^2_{N,\lambda_N(b_N)})^{-1/2} \exp\{-I_N(b_N)\}(1 + o_N(1)), \tag{1.2.27}$$

where $o_N(1) \to 0$ as $N \to \infty$. It follows from (1.2.27) and (1.2.12) that if the sequence b_N satisfies the condition (1.2.25) and the condition

$$\lim_{N\to\infty} \frac{b_N}{N} = \hat{b}, \tag{1.2.28}$$

where $|\hat{b}| < 1$, then

$$\lim_{N\to\infty} \frac{1}{N} \ln P_N(b_N) = -I(\hat{b}). \tag{1.2.29}$$

This is a familiar form of the *Large Deviation Principle* as it usually appears in the literature. Of course, it is weaker than (1.2.27).

On the other hand, we can consider the case when

$$\lim_{N\to\infty} \frac{|b_N - E_N|}{N} = 0, \tag{1.2.30}$$

which is called the *Moderate Deviations Case*. Then $\lambda_N(b_N) \to 0$ as $N \to \infty$, so that the Classical Large Deviation Principle in the Strong Form implies

$$P_N(b_N) = 2(2\pi D_N)^{-1/2} \exp\{-I_N(b_N)\}(1 + o_N(1)). \tag{1.2.31}$$

If an even stronger condition holds, namely

$$\lim_{N\to\infty} \frac{|b_N - E_N|}{N^{k/(k+1)}} = 0 \tag{1.2.32}$$

for some $k = 2, 3, \ldots$, then

$$P_N(b_N) = q_N(b_N) \exp\left\{ -\sum_{j=3}^{k} \frac{K_N^j}{(j)!} \left(\frac{b_N - E_N}{D_N}\right)^j \right\}(1 + o_N(1)), \tag{1.2.33}$$

where

$$q_N(b_N) = 2(2\pi D_N)^{-1/2} \exp\left\{ -\frac{1}{2} \frac{(b_N - E_N)^2}{D_N} \right\} \tag{1.2.34}$$

is the usual normal approximation for the probabilities $P_N(b_N)$. In the case $k=2$, called the *Small Deviation Case*, it implies that

$$P_N(b_N) = q_N(b_N)(1+o_N(1)) \tag{1.2.35}$$

if

$$\lim_{N\to\infty} \frac{|b_N - E_N|}{N^{2/3}} = 0, \tag{1.2.36}$$

and this is a natural region of validity of the local central limit theorem. The equations (1.2.29), (1.2.15) and the positivity of the limiting variance D imply a normal approximation in the domain of moderate deviations (1.2.30):

$$\ln P_N(b_N) = \ln q_N(b_N)(1+o_N(1)) = -\frac{1}{2}\frac{(b_N - E_N)^2}{D_N}(1+o_N(1)). \tag{1.2.37}$$

In the domain of large deviations (1.2.28) with $b \neq 0$ we can reduce (1.2.31) further to

$$\ln(-\ln P_N(b_N)) = \ln(-\ln q_N(b_N))(1+o_N(1)) = \ln N(1+o_N(1)). \tag{1.2.38}$$

So the normal approximation correctly predicts the order of the exponential asymptotics of the probabilities of large deviations.

§1.3. Ising model

In this section we will introduce certain notation and recall results about the Ising model; these results will be used later in this paper. Let $\mathbb{Z}^\nu$ be a ν-dimensional integer lattice with sites $t = (t^1, \dots, t^\nu)$, $t^i \in \mathbb{Z}$, $i = 1, \dots, \nu$, and the norm

$$|t| = |t^1| + \dots + |t^\nu|, \qquad t \in \mathbb{Z}^\nu, \tag{1.3.1}$$

which define a metric in $\mathbb{Z}^\nu$. For any $V \subseteq \mathbb{Z}^\nu$ we denote by Ω_V the set of all configurations $\{\sigma_t, t \in V\}$, $\sigma_t = \pm 1$. We reserve the notation Ω for the case when $V = \mathbb{Z}^\nu$. If $V' \subset V$, we use $\sigma_{V'} = \{\sigma_t, t \in V'\}$ to denote the restriction of the configuration $\sigma \in \Omega_V$ to V'.

Let $V \subset \mathbb{Z}^\nu$ be a finite subset of $\mathbb{Z}^\nu$, $\overline{\sigma} = \{\sigma_t, t \in \mathbb{Z}^\nu\}$ be a fixed configuration, and $\beta > 0$ and h be two real constants. In a usual way we define the energy function

$$H_V^h(\sigma|\overline{\sigma}) = -\frac{1}{2} \sum_{\substack{s \neq t: \\ s, t \in V, |s-t|=1}} \sigma_s\sigma_t - \sum_{\substack{s \in V, t \in V^c: \\ |s-t|=1}} \sigma_s\sigma_t - h\sum_{s\in V}\sigma_s, \quad \sigma \in \Omega_V, \tag{1.3.2}$$

and the Gibbs probabilities

$$P_V^{\beta,h}(\sigma|\overline{\sigma}) = (Z_V(\beta, h, \overline{\sigma}))^{-1}\exp\{-\beta H_V^h(\sigma|\overline{\sigma})\}, \quad \sigma \in \Omega_V, \tag{1.3.3}$$

where the partition function $Z_V(\beta, h, \overline{\sigma})$ is given by

$$Z_V(\beta, h, \overline{\sigma}) = \sum_{\sigma\in\Omega_V} \exp\{-\beta H_V^h(\sigma|\overline{\sigma})\}. \tag{1.3.4}$$

The probability distribution (1.3.3) is called the *Ising model distribution in the volume* V *with the boundary condition* $\overline{\sigma}$ *in magnetic field* h *at inverse temperature* β. Two particular boundary conditions will be mainly used:

$$\overline{\sigma}^{+} = (\sigma_t \equiv 1\,,\, t \in \mathbb{Z}^{\nu}) \quad \text{and} \quad \overline{\sigma}^{-} = (\sigma_t \equiv -1\,,\, t \in \mathbb{Z}^{\nu}). \tag{1.3.5}$$

The boundary conditions $\overline{\sigma}^{\pm}$ are called the $\pm$*-boundary conditions.* The corresponding Gibbs probabilities and the partition functions will be denoted by $P_V^{\pm,\beta,h}(\sigma)$ and $Z_V^{\pm}(\beta\,,\,h)$.

We will consider also the free boundary conditions, which correspond to the energy function

$$H_V^{\text{free},h}(\sigma) = -\frac{1}{2} \sum_{\substack{s \neq t:\\ s,t \in V,\, |s-t|=1}} \sigma_s \sigma_t - h \sum_{s \in V} \sigma_s\,, \qquad \sigma \in \Omega_V\,, \tag{1.3.6}$$

and the Gibbs probabilities $P_V^{\text{free},\beta,h}(\sigma)$ and partition function s $Z_V^{\text{free}}(\beta\,,\,h)$ which are defined by the relations (1.3.3), (1.3.4), where $H_V^h(\sigma|\overline{\sigma})$ is changed to $H_V^{h,\text{free}}(\sigma)$.

The periodic boundary conditions correspond to the case when instead of a box V we consider the discrete torus $\mathbf{T}_N = \mathbb{Z}^{\nu}/N\mathbb{Z}^{\nu}$ with the side N and the obvious norm $|t|$. We define

$$H_{\mathbf{T}_N}^{\text{per},h}(\sigma) = -\frac{1}{2} \sum_{\substack{s \neq t:\\ s,t \in \mathbf{T}_N,\, |s-t|=1}} \sigma_s \sigma_t - h \sum_{s \in \mathbf{T}_N} \sigma_s\,, \qquad \sigma \in \Omega_{\mathbf{T}_N}. \tag{1.3.7}$$

Again, using $\langle f\rangle_{W,V}^{+,\beta,h,\omega}$ we define the probabilities $P_{\mathbf{T}_N}^{\text{per},\beta,h}(\sigma)$ and partition functions $Z_{\mathbf{T}_N}^{\text{per}}(\beta\,,\,h)$ by relations similat to the relations (1.3.3), (1.3.4).

With a slight abuse of notation we shall use the same notation $P_V^{\beta,h}(*\mid\overline{\sigma})$, $P_V^{\pm,\beta,h}(*)$, $P_V^{\text{free},\beta,h}(*)$, $P_{\mathbf{T}_N}^{\text{per},\beta,h}(*)$ for the corresponding probability measures on both Ω_V and $\Omega_{\mathbf{T}_N}$. For any function $f(\sigma)$, $\sigma \in \Omega_W$, where $W \subseteq V$, we denote by

$$\langle f\rangle_V^{+,\beta,h} = \sum_{\sigma \in \Omega_V} f(\sigma_W) P_V^{+,\beta,h}(\sigma) \tag{1.3.8}$$

its Ising model mean value in the volume V with $+$-boundary conditions. In a similar way we define the mean values $\langle f\rangle_V^{-,\beta,h}$, $\langle f\rangle_V^{\text{free},\beta,h}$, $\langle f\rangle_{\mathbf{T}_N}^{\text{per},\beta,h}$. We shall add similiar indices and arguments to other quantities and in particular to quantities defined in §1.2 corresponding to the probability distributions introduced above. Sometimes we shall omit indices and the arguments h and β.

In what follows we assume that the dimension ν satisfies the bound $\nu \geq 2$, since otherwise there is no phase transition.

It is well known (see, e.g., [R]), that for any function $f(\sigma)$, $\sigma \in \Omega_W$, where $|W| < \infty$, and any sequence of volumes $V_1 \subset V_2 \subset \cdots$ such that $\bigcup_N V_N = \mathbb{Z}^\nu$, the limits

$$\langle f\rangle^{\pm} = \langle f\rangle^{\pm,\beta,h} = \lim_{N\to\infty} \langle f\rangle^{\pm}_{V_N} \tag{1.3.9}$$

exist and do not depend on a choice of the sequence V_N. These collections of mean values define probability measures $P^{\pm} = P^{\pm,\beta,h}$ on $\Omega_{\mathbb{Z}^\nu}$, which describe stationary ergodic random fields and are called the *limit pure ±-Gibbs states with parameters* β, h. Another well-known fact is the existence of the critical values β_ν^{crit} such that $0 < \beta_\nu^{\text{crit}} < \infty$ and the limit Gibbs states $P^{+,\beta,h}$ and $P^{-,\beta,h}$ coincide if $h \neq 0$ or if $h = 0$ and $\beta \leq \beta_\nu^{\text{crit}}$. In this case for any function f we have

$$\lim_{N\to\infty} \langle f\rangle^{\text{free}}_{V_N} = \lim_{N\to\infty} \langle f\rangle^{\text{per}}_{\mathbf{T}_N} = \langle f\rangle^{+,\beta,h} = \langle f\rangle^{-,\beta,h}. \tag{1.3.10}$$

(Here for large enough N we identify a finite subvolume $W \subset \mathbb{Z}^\nu$ with a subset of $\mathbf{T}_N$ in a natural way.) This case corresponds to the physical situation of the absence of a phase transition. In case $h = 0$, $\beta > \beta_\nu^{\text{crit}}$, when the phase transition does take place, the limit states $P^{\pm,\beta,0}$ are distinct. The corresponding mean values are connected by the symmetry relation:

$$\langle f(*)\rangle^{+,\beta,0} = \langle f(-*)\rangle^{-,\beta,0}. \tag{1.3.11}$$

In this case

$$\lim_{N\to\infty} \langle f\rangle^{\text{free}}_{V_N} = \lim_{N\to\infty} \langle f\rangle^{\text{per}}_{\mathbf{T}_N} = \frac{1}{2}(\langle f\rangle^{+,\beta,0} + \langle f\rangle^{-,\beta,0}), \tag{1.3.12}$$

and so the corresponding Gibbs states, being mixtures, are nonergodic random fields.

Finally, we define magnetizations

$$m^+(\beta, h) = \langle \sigma_t\rangle^{+,\beta,h}, \qquad m^-(\beta, h) = \langle \sigma_t\rangle^{-,\beta,h}. \tag{1.3.13}$$

For $\beta \leq \beta_\nu^{\text{crit}}$ or for $\beta > \beta_\nu^{\text{crit}}$, $h \neq 0$, the functions $m^+(\beta, h) = m^-(\beta, h)$ are monotone continuous functions of h, such that $|m^+(\beta, h)| < 1$ and $m^+(\beta, h) \to \pm 1$ as $h \to \pm\infty$. For $\beta > \beta_\nu^{\text{crit}}$, as $h \to 0$ we have a jump in magnetization:

$$\begin{aligned} \lim_{h\to 0+} m^{\pm}(\beta, h) &= m^+(\beta, 0) > 0, \\ \lim_{h\to 0-} m^{\pm}(\beta, h) &= m^-(\beta, 0). \end{aligned} \tag{1.3.14}$$

We define

$$m(\beta) = m^+(\beta, 0) = -m^-(\beta, 0). \tag{1.3.15}$$

§1.4. The classical behavior of the probabilities of deviations for the Ising model

We fix a sequence of finite volumes $V_1 \subset V_2 \subset \cdots \subset \mathbb{Z}^\nu$ such that $\bigcup_N V_N = \mathbb{Z}^\nu$ and the following Van Hove condition holds:

$$\lim_{N\to\infty} \frac{|\partial V_N|}{|V_N|} = 0. \tag{1.4.1}$$

Here the boundary of a set $V \subset \mathbb{Z}^\nu$ is

$$\partial V = \{t \in V^c : \operatorname{dist}(t, V) = 1\} \tag{1.4.2}$$

and $|V|$ is the number of points in V. For $\sigma = (\sigma_t,\ t \in V_N) \in \Omega_N$ we consider

$$S_N = S_N(\sigma) = \sum_{t\in V_N} \sigma_t \tag{1.4.3}$$

and we will study the large deviation behavior of this random variable, governed by one of the distributions defined in §1.3. A small difference between the setting of §1.2 and the rest of the paper is that the sum S_N now has $|V_N|$ terms. So we must use the factor $|V_N|$ instead of N. We assume that

$$(b_N - |V_N|) \equiv 0 \pmod 2, \quad \limsup_{N\to\infty} \frac{|b_N|}{|V_N|} < \begin{cases} 1, & \text{if } \beta \le 1, \\ \tanh 2C\beta, & \text{if } \beta > 1, \end{cases} \tag{1.4.4}$$

where $C > 0$ is any constant. (Since $(1-\tanh 2C\beta)/(1-m(\beta)) \to 0$ when $\beta \to \infty$ for any sufficiently large C, this implies that the fraction of values of b_N excluded by the second line in (1.4.4) is negligibly small.)

The following probability distributions for $\sigma \in \Omega_{V_N}$ in (1.4.3) will be considered: the Ising model distribution in V_N with arbitrary boundary conditions $\overline{\sigma} = (\sigma_t,\, t \in \mathbf{Z}^\nu)$ and in particular with +-boundary conditions $\overline{\sigma}^+ = (\sigma_t \equiv +1)$ (see (1.3.3)), the free boundary conditions (see (1.3.6)), and the periodic boundary conditions (see (1.3.7)); in the last case it is assumed that $V_N = \mathbf{T}_N$. For these cases we shall use all the subscript-N notation of §1.2 supplemented also with the superscripts $\overline{\sigma}$, $+$, free and per to distinguish among them. We shall also consider the case of the limit pure +-Gibbs state (see (1.3.9)) when the probability distribution for $\sigma \in \Omega_{V_N}$ is obtained by projecting the probability distribution P^+ to Ω_{V_N}. This case will have $+, \infty$ superscripts. The $-$ and $-, \infty$ cases will not be considered since they are identical to the $+$ cases via the symmetry. In particular, we denote

$$P_N^{+,h}(b_N) = P_{V_N}^{+,\beta,h}(\sigma : S_N(\sigma) = b_N). \tag{1.4.5}$$

Similar notation will be used for other boundary conditions also.

THEOREM 1.4.1. *Suppose that the Van Hove condition* (1.4.1) *and the conditions* (1.4.4) *are satisfied. There exists a value $\beta_0 > 0$ such that for the Ising model with the inverse temperature $\beta \le \beta_0$, any value of the magnetic field h,* *and all types of Gibbs distributions for* $\sigma \in \Omega_{V_N}$ *discussed above, the Classical Large Deviation Principle in the Strong Form* (*see* §1.2) *holds together with its implications* (1.2.29), (1.2.31), (1.2.33), (1.2.35), (1.2.37), (1.2.38). *The limit action function does not depend on the type of the Gibbs distribution and on the sequence of volumes* V_N:

$$\begin{aligned} I(\hat{b}) &= \lim_{N\to\infty} \frac{1}{|V_N|} I_N^{\bar{\sigma}}(|V_N|\hat{b}) = \lim_{N\to\infty} \frac{1}{|V_N|} I_N^{+}(|V_N|\hat{b}) \\ &= \lim_{N\to\infty} \frac{1}{|V_N|} I_N^{+,\infty}(|V_N|\hat{b}) = \lim_{N\to\infty} \frac{1}{|V_N|} I_N^{\mathrm{free}}(|V_N|\hat{b}) \\ &= \lim_{N\to\infty} \frac{1}{|V_N|} I_N^{\mathrm{per}}(|V_N|\hat{b}). \end{aligned} \tag{1.4.6}$$

This limit action function is a strictly concave holomorphic function of $\hat{b}$.

We emphasize that although the limit action function does not depend on boundary conditions and shapes of volumes—so the same holds for the asymptotics of usual large deviations (1.2.27)—the semiinvariants do depend on them essentially; so it is impossible to neglect asymptotically this dependence in the study of moderate deviations (cf. (1.2.33), (1.2.35)).

The proof of this theorem is given in §2.3.

§1.5. The behavior of deviations in the phase transition regime

As we have already mentioned, the behavior of large deviations in the phase transition regime is not classical. In particular, it is known (Isakov [I]) that the function $H(z)$ is not holomorphic at $z = 0$ if $h = 0$ and β is large enough, contrary to the functions $H_N(z)$, which are holomorphic in some neighborhoods around zero; however, these neighborhoods shrink to zero as $N \to \infty$. The same holds for the functions $I(z)$ and $I_N(z)$. Still, we shall use the coefficients K_N^j defined from the semiinvariants G_N^j using the relations (1.2.18), (1.2.22), (1.2.23) even in the phase transition region.

The statements of the present section use the notion of a *contractible volume.* By a contractible volume we mean any finite volume $V \subset \mathbb{Z}^\nu$ such that the complement V^c is connected. We will sometimes use the following additional conditions on a sequence of finite volumes V_N, which will be called the *thickness condition*: There exists a constant $L < \infty$ such that for any N and any integer $K \le |V_N|$ there exists a contractible subvolume $V_N' \subset V_N$ such that $|V_N'| = K$ and

$$|\partial V_N'| \le L |V_N'|^{(\nu-1)/\nu}. \tag{1.5.1}$$

This restriction is rather weak. For example, if V_N is a sequence of parallelepipeds $\{t = (t_1, \dots t_\nu) : 0 \le t_i \le l_i,\ i = 1, \dots, \nu\}$, then the thickness condition (and also the Van Hove condition and the contractibility condition) is fulfilled provided all ratios l_i/l_j, $i, j = 1, \dots, \nu$, are uniformly bounded.

THEOREM 1.5.1. *Suppose that* $h = 0$ *is a value of the magnetic field. Suppose that a sequence of contractible volumes* V_N *and a sequence of integers* b_N *are given such that the Van Hove condition* (1.4.1) *and the conditions* (1.4.4) *are fulfilled. The following statements hold for all* $\beta > \beta_1 = \beta_1(C)$, *where* β_1 *is some finite inverse temperature depending on the parameter* A *from* (1.4.4).

Consider the $+$*-boundary conditions. Let*

$$b_N - E_N^+ \geq 0, \tag{1.5.2}$$

where

$$E_N^+ = \langle S_N \rangle_{V_N}^{+,\beta,0} \tag{1.5.3}$$

is the mean value of S_N *with respect to the Gibbs distribution considered* (*cf. notations* (1.2.9) *and* (1.3.8)). *Then for the random variable* S_N *defined by* (1.4.3) *and any sequence* b_N *the Classical Large Deviation Principle in the Strong Form holds, i.e. the relation* (1.2.27) *holds, together with* (1.2.29), (1.2.31), (1.2.33), (1.2.35), (1.2.37), (1.2.38) (*which in the situation of* §1.2 *were its corollaries*). *The limit action function* $I^0(\hat{b})$ (*see* (1.2.13)) *defined for the corresponding Gibbs distribution with* $h = 0$ *is a holomorphic strictly concave function of* $\hat{b}$ *in the interval*

$$\hat{b} > m^+(\beta). \tag{1.5.4}$$

In the opposite case

$$b_N - E_N^+ < 0 \tag{1.5.5}$$

for any integer k *such that*

$$k \leq \nu, \tag{1.5.6}$$

and for values b_N *satisfying*

$$\lim_{N\to\infty} \frac{|b_N - E_N^+|}{|V_N|^{k/(k+1)}} = 0 \tag{1.5.7}$$

the equality (1.2.33) *and its implication* (1.2.35) *hold.*

Suppose now additionally that the thickness condition (1.5.1) *is satisfied. Then in the complementary region*

$$\limsup_{N\to\infty} \frac{b_N - E_N^+}{|V_N|^{\nu/(\nu+1)}} < 0 \tag{1.5.8}$$

provided that for some constant $Q > 0$

$$b_N \geq -E_N^+ - Q|V_N|^{(2\nu-1)/2\nu} \tag{1.5.9}$$

there exist constants $\underline{c} > 0$, $\overline{c} < \infty$ *depending only on* Q, *on a sequence of volumes* V_N, *and on the value of inverse temperature* β *such that for all* N

$$-\underline{c}(E_N^+ - b_N)^{(\nu-1)/\nu} \leq \ln P_N^{+,0}(b_N) \leq -\overline{c}(E_N^+ - b_N)^{(\nu-1)/\nu}. \tag{1.5.10}$$

Finally, if

$$\limsup_{N\to\infty} \frac{b_N + E_N^+}{|V_N|^{(2\nu-1)/2\nu}} = -\infty, \tag{1.5.11}$$

$\beta > \beta_1(C)$, *where* β_1 *is a finite unverse temperature depending on the parameter* C *from* (1.4.4), *and* (*cf.* (1.4.4))

$$\liminf_{N\to\infty} \frac{b_N}{|V_N|} > -\tanh C\beta \tag{1.5.12}$$

then

$$\lim_{N\to\infty} \frac{-\ln P_N^{+,0}(b_N)}{I_N^0(-b_N)} = 1. \tag{1.5.13}$$

where $I_N^0(b)$ *is the action function defined by the relation* (1.2.12) *for the considered probability distribution.*

Similar statements hold for the case of the limit pure +-state. The limit action functions for the +-boundary conditions and for the limit pure +-state coincide and do not depend on a choice of the sequences of volumes V_N (*cf.* (1.4.5)).

To compare the classical and nonclassical behavior of the deviations described above we suppose that for some $1 \geq \varkappa \geq \frac{1}{2}$ and some $a \neq 0$

$$\lim_{N\to\infty} \frac{b_N - E_N^+}{|V_N|^{\varkappa}} = a. \tag{1.5.14}$$

Then it follows from the previous theorem that

$$-\ln P_N^{+,0}(b_N) \asymp \begin{cases} |V_N|^{2\varkappa-1}, & \text{if } a > 0,\ \varkappa < 1, \text{ or} \\ & 1 - m^+(\beta, 0) > a > 0,\ \varkappa = 1, \text{ or} \\ & a < 0,\ \varkappa < \frac{\nu}{\nu+1}, \\ |V_N|^{\varkappa(\nu-1)/\nu}, & \text{if } a < 0,\ \frac{\nu}{\nu+1} < \varkappa < 1, \text{ or} \\ & -2m^+(\beta, 0) \leq a < 0,\ \varkappa = 1, \\ |V_N|, & \text{if } -1 - m^+(\beta, 0) < a \\ & < -2m^+(\beta, 0),\ \varkappa = 1, \end{cases} \tag{1.5.15}$$

where $\asymp$ means that the ratio of the corresponding quantities is bounded from both sides uniformly with respect to N. This relation illustrates a difference between the asymptotic behavior of large deviations in the region of phase transitions and in the region without phase transitions. Another interesting picture occurs if we assume that for some $0 < \varkappa \leq 1$, $a > 0$

$$\lim_{N\to\infty} \frac{b_N + E_N^+}{|V_N|^{\varkappa}} = a. \tag{1.5.16}$$

Then using the statements (1.5.10) and (1.5.13) and the normal approximation for the action function $I_N^0(b)$ for $b = -b_N \approx E_N^+$ (cf. (1.2.37)) we see that

$$-\ln P_N^{+,0}(b_N) \asymp \begin{cases} |V_N|^{(\nu-1))/\nu}, & \text{if } \varkappa \le (2\nu-1)/2\nu, \\ |V_N|^{2\varkappa-1}, & \text{if } \varkappa \ge (2\nu-1)/2\nu. \end{cases} \tag{1.5.17}$$

Finally, it follows from (1.5.13) that if

$$-\tanh C\beta < \lim_{N\to\infty} \frac{b_N}{|V_N|} = \hat{b} < -m^+(\beta, 0), \tag{1.5.18}$$

then we know a more exact asymptotics and can claim that

$$\lim_{N\to\infty} \frac{-\ln P_N^{+,0}(b_N)}{|V_N|} = I_N^0(-\hat{b}). \tag{1.5.19}$$

THEOREM 1.5.2. *Suppose that $h > 0$ is the value of the magnetic field. Suppose that the same restrictions on the sequences V_N and b_N as in Theorem 1.5.1 are fulfilled. Suppose that for some $\beta_1 = \beta_1(h) < \infty$ the inverse temperature satisfies $\beta > \beta_1$ and the condition* (1.5.2) *is fulfilled. Then again the Classical Large Deviations Principle in the Strong Form (see* (1.2.27)*) together with the relations* (1.2.29), (1.2.31), (1.2.33), (1.2.35), (1.2.37), (1.2.38) *hold for any sequence b_N. The limit action function $I^h(\hat{b})$ defined for the corresponding Gibbs distribution with the value h of the magnetic field is a holomorphic strictly concave function in the interval*

$$\hat{b} > m^+(\beta, 0). \tag{1.5.20}$$

If the thickness condition (1.5.1) *is also satisfied and*

$$\lim_{N\to\infty} \frac{b_N}{|V_N|} = \hat{b}, \tag{1.5.21}$$

where

$$m^+(\beta, 0) \ge \hat{b}, \tag{1.5.22}$$

then

$$\lim_{N\to\infty} -\frac{\ln P_N^{+,h}(b_N)}{|V_N|} = \tilde{I}^h(\hat{b}) \tag{1.5.23}$$

where

$$\tilde{I}^h(\hat{b}) = \begin{cases} I^h(m^+(\beta, 0)) + \beta h(m^+(\beta, 0) - \hat{b}), & \text{if } \hat{b} > -m^+(\beta, 0) \\ I^h(m^+(\beta, 0)) + \beta h(m^+(\beta, 0) - \hat{b}) + I^0(-\hat{b}), & \\ & \text{if } -m^+(\beta, 0) \ge \hat{b} > -1. \end{cases} \tag{1.5.24}$$

Similar statements hold for the limit pure $+$-state. The limit action functions for the $+$-boundary conditions and for the limit $+$-pure state coincide on the interval (1.5.22) *and do not depend on the sequence V_N.*

Observe that the function (1.5.24) is a continuous extension of the limit action function $I^h(\hat{b})$ to the region (1.5.22). However, it does not possess the strict concavity property of $I^h(\hat{b})$. This is again a reflection of the existence of the phase transition in the Ising model.

Theorems 1.5.1 and 1.5.2 are proved in §4.6 for the cases when the Classical Large Deviations Principle in the Strong Form is valid and in §5.2 for the cases of nonstandard phase transition structure of deviations.

§1.6. The behavior of deviations in the phase transition domain. Symmetric case

The behavior of deviations in the cases of periodic and free boundary conditions differs slightly from the previous case. In the case $h = 0$ the symmetry with respect to change of variables $\sigma \to -\sigma$ implies that in the limit $V_N \to \infty$ these phases are mixtures. Since

$$P_N^{\mathrm{per},0}(b_N) = P_N^{\mathrm{per},0}(-b_N), \tag{1.6.1}$$

we can assume without loss of generality that in this case

$$b_N \geq 0. \tag{1.6.2}$$

Let

$$P_N^{+,\mathrm{per},h}(b) = P_{\mathbf{T}_N}^{\mathrm{per},\beta,h}(\sigma \in \Omega_{\mathbf{T}_N} : S_{\mathbf{T}_N}(\sigma) = b | S_{\mathbf{T}_N}(\sigma) \geq 0) \tag{1.6.3}$$

be the conditional distribution induced by the probability distribution with the periodic boundary condition under the condition $S_{\mathbf{T}_N}(\sigma) \geq 0$. Consider also the logarithmic generating function (cf. (1.2.4))

$$H_N^{+,\mathrm{per},h}(z) = \ln \sum_{b \geq 0} P_N^{+,\mathrm{per},h}(b) e^{zb}. \tag{1.6.4}$$

In the case $h = 0$ the symmetry (1.6.1) implies that

$$H_N^{\mathrm{per},h,+}(z) = \ln \left(2 \sum_{b>0} P_N(b) e^{zb} \right). \tag{1.6.5}$$

(For definiteness we assume here that $|V_N|$ is even.)

THEOREM 1.6.1. *Suppose that* $h > 0$ *or* $h = 0$ *and* $b_N \geq 0$, *and in all the definitions of* §1.2 *we use the function* (1.6.4) *instead of* (1.2.4). *Then for the case of periodic boundary condition all the statements of Theorems* 1.5.1 *and* 1.5.2 *are valid.*

A similar result holds for the case of the free boundary condition if we suppose that all the volumes V_N *are contractible and convex* (*in the sense of the natural group structure in* $\mathbb{Z}^\nu$).

NOTE. The limit action functions defined by (1.2.13) coincide for all types of boundary conditions considered (see the end of §4.7). So for a familiar rough form (1.2.29) of the Large Deviation Principle the modification (1.6.4) of the logarithmic generating function introduced above is not essential.

§1.7. Large deviations and shape of droplets

The relation (1.5.10) gives much rougher information compared with the precise knowledge of the large deviation behavior one gets in the classical regime. So it is natural to try to get at least the value of the constant c which describes the true asymptotic in (1.5.10):

$$\lim_{N\to\infty} \frac{-\ln P_N^+(b_N)}{(E_N^+ - b_N)^{(\nu-1)/\nu}} = c. \tag{1.7.1}$$

It turns out that this is a difficult question, and the answer is closely connected with an important physical theory which describes the typical shape of a droplet of a phase floating in the opposite phase. In the recent book [DKS] the question of the typical shape was rigorously studied in the simplest nontrivial situation of the two-dimensional Ising model with periodic boundary conditions, and this is about the only case when we can prove the existence of the limit (1.7.1) and exhibit a construction which produces the value of this constant c. For this we have to repeat some notions discussed in detail in [DKS].

Let

$$W_N = \{t = (t^1, t^2) \in \mathbb{Z}^2 : -N \le t^i \le N, i = 1, 2\} \tag{1.7.2}$$

be a square. For any direction $n \in \mathbb{S}^1$ (where $\mathbb{S}^1 \subset \mathbb{R}^2$ is the unit circle centered at the origin) we introduce the boundary conditions $\overline{\sigma}^n$ such that

$$\overline{\sigma}_t^n = \begin{cases} +1 & \text{if } (t, n) > 0, \\ -1 & \text{if } (t, n) \le 0. \end{cases} \tag{1.7.3}$$

Suppose that the magnetic field $h = 0$ and the inverse temperature β is large enough. We introduce the surface tension in the direction n as the limit

$$\tau_\beta(n) = -\lim_{N\to\infty} \frac{1}{\beta d(N, n)} \ln \frac{Z_{W_N}(\beta, 0, \overline{\sigma}^n)}{Z_{W_N}(\beta, 0, \overline{\sigma}^+)}, \tag{1.7.4}$$

where $\overline{\sigma}^+$ are $+$-boundary conditions (see (1.3.5)) and $d(N, n)$ is the length of the segment

$$\{t \in \mathbb{R}^2 : (t, n) = 0, t^1 \in [-N, N]\}. \tag{1.7.5}$$

The existence of the limit (1.7.4) for the case of large enough inverse temperatures is proved in [DKS].

Let γ be a closed nonselfintersecting smooth curve in $\mathbb{R}^2$. Let

$$\mathscr{W}_\beta(\gamma) = \int_\gamma \tau_\beta(n_s)\, ds, \tag{1.7.6}$$

where $n_s \in \mathbb{S}^1$ is the normal to the curve γ at the point s and ds is the differential of the length. Let

$$F_\beta = \inf \mathscr{W}_\beta(\gamma) \tag{1.7.7}$$

where the infimum is taken over all curves γ enclosing a region of area 1.

THEOREM 1.7.1. *There exists a value $\beta_1 < \infty$ such that for the two-dimensional Ising model with the inverse temperature $\beta > \beta_1$ and the magnetic field $h = 0$ and for any sequence of nonnegative integers b_N such that the conditions* (1.4.4) *hold and*

$$\liminf_{N\to\infty} \frac{b_N - E_{\mathbf{T}_N}^{\mathrm{per},\beta,0,+}}{|\mathbf{T}_N|^{2/3}} < 0, \tag{1.7.8}$$

where $E_{\mathbf{T}_N}^{\mathrm{per},\beta,0,+}$ is the mean value defined in §1.6, *the following equality holds*:

$$\lim_{N\to\infty} \frac{-\ln P_{\mathbf{T}_N}^{\mathrm{per},\beta,0}(b_N)}{w_N(b_N)} = 1, \tag{1.7.9}$$

where

$$w_N(b_N) = \begin{cases} \beta F_\beta\left(\frac{E_{\mathbf{T}_N}^{\mathrm{per},\beta,0,+} - b_N}{2m(\beta)}\right)^{1/2} & \text{if } F_\beta\left(\frac{E_{\mathbf{T}_N}^{\mathrm{per},\beta,0,+} - b_N}{2m(\beta)}\right)^{1/2} \le 2\beta N \tau_\beta(n_0), \\ 2\beta N\tau_\beta(n_0) & \text{otherwise,} \end{cases} \tag{1.7.10}$$

and $n_0 = (1, 0)$ is the unit lattice vector.

The case of positive b_N can be reduced to the case of negative b_N with the help of the symmetry.

Part of Theorem 1.7.1 is proved in [DKS]. This part corresponds to the case of exponent 1 instead of $2/3$ in (1.7.8) and to the region specified in the first line of (1.7.10). The remaining part of the proof will appear elsewhere.

§1.8. The contour point of view and a physical interpretation of the results

To explain the mystery of nonclassical behavior of large deviations in the phase transition domain we use the contour technique, which forms the basis of the mathematical theory of first order phase transitions.

In the case of small β a typical configuration of the Ising model is a chaotic mixture of pluses and minuses. In the limit case $\beta = 0$ one gets just the realization of a Bernoulli field. With the growth of β, a clustering of pluses and minuses begins. For β large, when we have distinct plus- and minus-phases, a typical configuration of a realization of the plus-phase looks like a sea of pluses with islands of minuses floating inside it (some of these islands have lakes of pluses inside and so on). In the opposite phase a typical realization is a continent of minuses with lakes of pluses inside it (and again with islands of minuses inside some of these lakes and so on.) To formalize this visual picture we shall use the notion of contours of a configuration. To be specific, we consider configurations in a finite volume with + boundary conditions.

Let a finite volume $V \subset \mathbb{Z}^\nu$ be given. We define the boundary of a configuration $\sigma \in \Omega_V$ to be the set of all unit plaquettes of the dual lattice $\mathbb{Z}^{\nu *} = \mathbb{Z}^\nu + (1/2, \dots, 1/2) \subset \mathbb{R}^\nu$ that cross any of the bonds (t_1, t_2) of the initial lattice such that $\sigma_{t_1} \neq \sigma_{t_2}$. (If t lies outside V we suppose that $\sigma_t = +1$.) The boundary of a configuration σ splits into connected components that are disjoint polyhedrons; they will be called the *contours of the configuration* σ (see §3.1 for details). The family of all the contours of a configuration σ will be denoted by $\mathcal{G}(\sigma)$. For any contour $\Gamma \in \mathcal{G}(\sigma)$ we denote by $|\Gamma|$ the number of its ($(\nu - 1)$-dimensional) faces, by $\operatorname{Int}\Gamma$ the set of lattice sites inside the corresponding polyhedron, and by $|\operatorname{Int}\Gamma|$ the number of points in this set. The interiors of external contours (i.e., contours that do not lie inside other contours) can be interpreted as droplets of the minus-phase inside the plus-phase.

We shall describe a typical large deviation configuration in the volume V_N by using the conditional probabilities (see the notation (1.4.5))

$$P_N^{+,h,b_N}(\sigma) = \frac{P_N^{+,h}(\sigma)}{\sum_{\sigma \in \Omega_V \,:\, S_N(\sigma) = b_N} P_N^{+,h}(\sigma)} \tag{1.8.1}$$

induced by the condition $S_N(\sigma) = b_N$, where b_N satisfies the parity condition (1.4.4). The following theorem describes some typical properties of such configurations.

THEOREM 1.8. *Assume that sequences of contractible volumes V_N and of integers b_N satisfying the conditions* (1.4.1) *and* (1.4.4) *are given. Suppose that $\beta > \beta_1$ (see Theorem* 1.5.1), $h \geq 0$, *and either the condition* (1.5.2) *is satisfied or the condition* (1.5.7) *with $k = \nu$ is satisfied. Then there exists a constant $C < \infty$ such that for the events*

$$A_N = \{\sigma \in \Omega_{V_N} : \max_{\Gamma \in \mathcal{G}(\sigma)} |\Gamma| \leq C \ln |V_N|\} \tag{1.8.2}$$

we have

$$\lim_{N \to \infty} \sum_{\sigma \in A_N} P_N^{+,h,b_N}(\sigma) = 1. \tag{1.8.3}$$

On the other hand, assume that $\beta > \beta_1$, $h \geq 0$, and the conditions (1.5.1), (1.5.8), (1.5.9) *are satisfied. Then there exist constants $C < \infty$, $c > 0$ such that for the event*

$$\begin{aligned} B_N = \{\sigma \in \Omega_{V_N} :\ & \text{there exists a contour } \Gamma_0 \in \mathcal{G}(\sigma) \text{ of measure} \\ & |\Gamma_0| > c(E_N^+ - b_N)^{(\nu-1)/\nu} \text{ and volume } |\operatorname{Int}\Gamma_0| > c(E_N^+ - b_N), \\ & \text{while any other contour } \Gamma \in \mathcal{G}(\sigma),\ \Gamma \neq \Gamma_0, \\ & \text{has the measure } |\Gamma| \leq C \ln |V_N|\} \end{aligned} \tag{1.8.4}$$

we have

$$\lim_{N \to \infty} \sum_{\sigma \in B_N} P_N^{+,h,b_N}(\sigma) = 1. \tag{1.8.5}$$

Finally, suppose that $\beta > \beta_1$, $h \geq 0$, *and the conditions* (1.5.1), (1.5.11), *and* (1.5.12) *are valid. Then there exist constants* $C < \infty$, $c > 0$ *such that for the event*

$$\begin{aligned} B'_N = \{\sigma \in \Omega_{V_N} &: \textit{ there exists a contour } \Gamma_0 \in \mathcal{G}(\sigma) \textit{ with} \\ &|\operatorname{Int}(\Gamma_0)| > |V_N| - c|\partial V_N|, \textit{ while any other contour} \\ &\Gamma \in \mathcal{G}(\sigma),\ \Gamma \neq \Gamma_0, \textit{ has the length } |\Gamma| \leq C \ln |V_N|\} \end{aligned} \tag{1.8.6}$$

we have

$$\lim_{N\to\infty} \sum_{\sigma \in B'_N} P_N^{+,h,b_N}(\sigma) = 1. \tag{1.8.7}$$

Similar results are true for all other boundary conditions introduced in §1.3.

This theorem can be interpreted in the following way. We discuss first the case $h = 0$. If a value of S_N is close to its mean value E_N^+, then all droplets of the opposite minus-phase have microscopic sizes (they are smaller than $C \ln |V_N|$). If we reduce the number of minus-particles (by considering the values $b_N > E_N^+$), these droplets further shrink in size and decrease in number, but there will be no qualitative change in a picture. The total deviation of S_N is caused by deviations of the sizes of separate droplets, which are almost independent, and so the probability of the deviation behaves in the same way as in the classical case of independent variables. If we increase the number of minus-particles (the case $b_N < E_N^+$), the same situation holds, provided that the order of the number of additional minus-particles is smaller than $|V_N|^{(\nu-1)/\nu}$. But after we pass this threshold, the microscopic droplets can no longer absorb all the additional minus-particles, so a large minus-droplet of a macroscopic size is formed, which contains most of these particles. It is the moment of the condensation of the minus-phase. One can give nonrigorous arguments explaining why the scale of the order $|V_N|^{(\nu-1)/\nu}$ is critical. If the number of additional minus-particles $E_N^+ - b_N$ is approximately equal to $|V_N|^{\varkappa}$, then the classical mechanism related to the collective behavior of many small droplets predicts the deviation probability to be of the order of

$$\exp\left\{ -\frac{(E_N^+ - b_N)^2}{2D_N^+} \right\} \approx \exp\{-\bar{c}|V_N|^{2\varkappa-1}\}, \tag{1.8.8}$$

where D_N^+ is the variance of the random variable S_N. On the other hand, the large droplet containing approximately $|V_N|^{\varkappa}$ particles the surface of order $|V_N|^{\varkappa(\nu-1)/\nu}$, and so its probability is of order

$$\exp\{-\tilde{c}|V_N|^{\varkappa(\nu-1)/\nu}\}. \tag{1.8.9}$$

The exponents in (1.8.8) and (1.8.9) are equal if $\varkappa = \nu/(\nu+1)$.

At the level $b_N \approx -E_N^+$ we have a new change of the picture. The volume of the large contour containing the minus-phase reaches its maximal possible value $|V_N|$ and so cannot grow further. So the droplets of the plus-phase inside the minus-phase begin to shrink. Typical configurations inside the minus-phase become symmetric to typical configurations of the plus-phase via the transformation $b_N \to -b_N$. This explains the appearence of the denominator $I_N^0(-b_N)$ in (1.5.13).

The situation in the case $h > 0$ is closely connected with the case $h = 0$. Definition (1.3.3) implies that for any h we have

$$P_N^{+,h}(\sigma) = P_N^{+,0}(\sigma)\exp\{\beta h S_N(\sigma)\}\frac{Z_{V_N}^+(\beta, 0)}{Z_{V_N}^+(\beta, h)}. \tag{1.8.10}$$

So the conditional distribution under the condition that b_N is fixed does not depend on h, and typical configurations have the same structure as in the case $h = 0$. It is possible to explain the expression (1.5.24) for the function $\tilde{I}^h(\hat{b})$ if we observe that

$$\tilde{I}^h(\hat{b}) = \tilde{I}^0(\hat{b}) + I_\Delta^h(\hat{b}), \tag{1.8.11}$$

where the term

$$\tilde{I}^0(\hat{b}) = \begin{cases} I^0(m^+(\beta, 0)) = 0 & \text{if } \hat{b} > -m^+(\beta, 0), \\ I^0(-\hat{b}) & \text{if } -m^+(\beta'0) \geq \hat{b} > -1. \end{cases} \tag{1.8.12}$$

comes from the statements (1.5.10), (1.5.13) and the term

$$I_\Delta^h(\hat{b}) = I^h(m^+(\beta, 0)) + \beta h(m^+(\beta, 0) - \hat{b}) \tag{1.8.13}$$

describes the asymptotics of the product of two last factors in (1.8.10).

The reason why it is possible to find a more precise asymptotics for the probabilities of large deviations in the case $\nu = 2$ with periodic boundary conditions (see §1.7) lies in the possibility of finding in this case a limiting shape of the large droplet as $N \to \infty$. This shape is given by the curve γ which minimizes the functional $\mathscr{W}_\beta(\gamma)$ in (1.7.6) (see [DKS]) in the case defined by the first line in (1.7.10). This shape is called *Wulff shape* in the physical literature (see a detailed discussion in [DKS]) and describes a typical shape of a droplet of a phase inside another phase.

A seemingly strange behavior of the final answer (the second line in (1.7.10)) is due to the geometry of the discrete torus. If the droplet is too large, its optimal shape would be that of a strip on the torus between two parallels or meridians. In other words, if one sought the minima of the functional (1.7.6) over the boundaries of all the domains on the torus with a given area, the minimum is attained on the Wulff shape if this area is small enough and on the strip shape if the area exceeds a certain value.

§1.9. References and generalizations

1.9.1. In the exposition of the classical theory of deviations in §1.2 we followed essentially the pioneering papers of Khinchin [Kh], Smirnov [Sm], Cramer [Cr], who laid the foundations of the theory (see also books by Ibragimov and Linnik [IL], Petrov [P], and Saulis and Statulevičius [SS]). Instead of the local variant of the theory used in §1.2, its integral variant is discussed more often. The local approach was used in papers [Ri, CS1, CS2] and in the book [B]. In the integral theory the asymptotics of sums

$$P_N = \sum_{b \in B_N} P_N(b), \tag{1.9.1}$$

where $B_N \subset \mathbb{R}^1$ is a sequence of subsets, is studied. Since the main estimate (1.2.27) is true for any sequence b_N, we can estimate the sum (1.9.1) termwise uniformly with respect to N, so an estimate for the sum can be obtained as an easy implication of the local theory. The possibility of obtaining results about both large and moderate deviations as consequences of a unique estimate (1.2.27), discussed in §1.2, might be a methodological innovation. The elegant relation (1.2.29) became a starting point of deep generalizations to the cases of dependent variables and of variables with values in general functional spaces, to a study of empirical distributions, and so on (see the books [DeSt, DZ, V, VF]). In §2.1 we prove a general result which shows that the Classical Large Deviations Principle in the Strong Form follows easily from some analyticity properties of the logarithmic generating function. This result is influenced by similar results of Saulis and Statulevičius [SS].

1.9.2. Theorem 1.4.1 is a simple consequence of the results of §2.1 mentioned above. The analyticity condition which one has to check was intensively studied in the literature on rigorous results in statistical physics (see, for example, [DS1, DS2, MM]) and so this theorem can be extended to a wide class of situations. On the other hand, it is natural to expect that Theorem 1.4.1 remains true for all β in the domain $\beta_0 < \beta < \beta^{\nu}_{\mathrm{crit}}$, but at present there is no known method to check the corresponding analyticity properties. Some result of the similar type were obtained by Čepulēnas (see [Č]).

Recently several authors have developed a general theory of large deviations for empirical distributions of Gibbs fields (see [B, BD, Co, FO1, G1, G2, GZ, O]). This theory implies, in particular, the existence of the limit (1.2.29) and can be applied to a wide class of Gibbs fields including the phase transition region. This does not contradict a special behavior of large deviations in the phase transition regime described in §1.5: the general theory does not exclude the possibility that the action function $I(b)$ vanishes on some interval, and this is a typical situation in the case of phase transitions of the first order. In such a situation the equality (1.2.29) does not reveal the true asymptotics of the probabilities.

1.9.3. The main estimate (1.5.10) for the case of deviations proportional to the volume was proved also in the recent papers of Föllmer and Ort [FO2] and Schonmann [Sch]. They obtained for this case explicit estimates of the constants $\underline{c}$ and $\overline{c}$ in (1.5.10). Their papers are based on methods different from the one used in the present paper. Without being explicitly stated, such results essentially already were presented in an early paper of Minlos and

Sinaĭ [MS]. The results of our paper about moderate deviations seem to be new. The results of §§1.5–1.8 can be extended to a wide class of Gibbs fields undergoing phase transitions, for which it is possible to have a full control of probabilistic properties of contour systems (such as in the Pirogov-Sinaĭ situation, see [S]). Again, it is natural to expect that the theorems of §§1.5, 1.6 are valid for all values of $\beta > \beta^{\nu}_{\text{crit}}$, but at present there are no methods to prove it.

1.9.4. The results of §1.8 are connected very closely with the main result of the book [DKS] about the shape of the droplet. Their proofs are generalizations of the proof in [DKS] and will appear elsewhere. Other sections of this paper are independent of [DKS]. Since we want to use the constructions of [DKS] directly, we restrict ourselves in §1.8 to the case of two-dimensional Ising models with periodic boundary conditions. Generalizations to a wider class of two-dimensional Gibbs models and to other boundary conditions seem to be possible. A study of the three-dimensional case meets very serious mathematical difficulties (see the discussion in §1 of [DKS]).

1.9.5. The book [DKS] contains some of the results of the present paper for the two-dimensional case, which were necessary to obtain the main result of [DKS]. Some other results were formulated in the papers [Sh, DS2].

2. Classical Theory of Large Deviations

In this chapter we prove certain general facts about the behavior of large deviations and then apply them to Gibbs fields with small value of the parameter β.

§2.1. Logarithmic generating function

Consider a sequence of random variables S_N taking values in the sublattice $\mathbb{Z}_d = \{k \in \mathbb{Z}^1 : k \equiv 0 \pmod{d}\}$ with a period $d \in \mathbb{Z}^1$. So the probability distribution of S_N is defined by the probabilities

$$P_N(b) = \mathbf{P}\{S_N = b\}, \qquad b \in \mathbb{Z}_d. \tag{2.1.1}$$

We suppose that this probability distribution is nondegenerate in the sense that its support contains more than one point. Let

$$B_N^- = \inf\{b : P_N(b) > 0\}, \qquad B_N^+ = \sup\{b : P_N(b) > 0\} \tag{2.1.2}$$

be the minimal and the maximal values of the random variable S_N. (We do not exclude the cases $B_N^- = -\infty$ or $B_N^+ = \infty$.) We assume that for any real $\lambda \in \mathbb{R}^1$,

$$\sum_{b\in\mathbb{Z}_d} P_N(b)e^{\lambda b} < \infty. \tag{2.1.3}$$

In that case the generating function

$$F_N(z) = \sum_{b\in\mathbb{Z}_d} P_N(b)e^{zb}, \qquad z \in \mathbb{C}^1, \tag{2.1.4}$$

is defined for all z and is a holomorphic function. The logarithmic generating function

$$H_N(z) = \ln F_N(z) \tag{2.1.5}$$

is defined for all z with $F_N(z) \neq 0$ and is holomorphic in its domain of definition. We define

$$G^n_{N,z} = \frac{d^n H_N(z)}{dz^n}. \tag{2.1.6}$$

In particular,

$$E_{N,z} = G^1_{N,z} = (F_N(z))^{-1} \sum_{b\in\mathbb{Z}_d} P_N(b)be^{zb} \tag{2.1.7}$$

and

$$D_{N,z} = G^2_{N,z} = (F_N(z))^{-1} \sum_{b\in\mathbb{Z}_d} P_N(b)(b - E_{N,z})^2 e^{zb}. \tag{2.1.8}$$

Sometimes it is convenient to consider a new probability distribution

$$P_{N,\lambda}(b) = (F_N(\lambda))^{-1} e^{\lambda b} P_N(b), \qquad b \in \mathbb{Z}_d, \tag{2.1.9}$$

defined for all real $\lambda \in \mathbb{R}^1$. This probability distribution is called *λ-tilted distribution.* Then $E_{N,\lambda}$ is the mean value, $D_{N,\lambda}$ is the variance, and $G^n_{N,\lambda}$ is the kth semiinvariant of a random variable with this λ-tilted distribution.

From the nondegeneracy of $P_N(b)$ it follows that

$$D_{N,\lambda} > 0, \qquad \lambda \in \mathbb{R}^1. \tag{2.1.10}$$

So $E_{N,\lambda}$ is strictly increasing in $\lambda \in \mathbb{R}^1$ and $H_N(\lambda)$ is a strictly convex function of $\lambda \in \mathbb{R}^1$. The function $E_{N,\lambda}$ has limits $B^\pm_N$ as $\lambda \to \pm\infty$. Hence for any $b \in \mathbb{Z}_d$ such that $B^-_N < b < B^+_N$ there exists a unique value $\lambda = \lambda_N(b) \in \mathbb{R}^1$ such that

$$E_{N,\lambda_N(b)} = b. \tag{2.1.11}$$

The function

$$I_N(b) = \sup_{\lambda\in\mathbb{R}^1}(\lambda b - H_N(\lambda)) = \lambda_N(b)b - H_N(\lambda_N(b)), \quad B^-_N < b < B^+_N, \tag{2.1.12}$$

is called the *action function.* If B^+_N is finite, we can define the value $I_N(B^+_N)$ by

$$I_N(B^+_N) = \sup_{\lambda\in\mathbb{R}^1}(\lambda B^+_N - H_N(\lambda)) = \lim_{\lambda\to+\infty}(\lambda B^+_N - H_N(\lambda)) = -\ln P_N(B^+_N). \tag{2.1.13}$$

Similar definitions can be made for B^-_N.

In the theorem that will be formulated below we claim that certain analyticity properties of the logarithmic generating function imply classical large deviation behavior. We now introduce the corresponding notation. We say that the function $H_N(z)$ satisfy the *analyticity condition* in a complex neighborhood $\mathscr{O}_N(\bar{\lambda}^1, \bar{\lambda}^2) \subset \mathbb{C}^1$ of a segment $[\bar{\lambda}^1, \bar{\lambda}^2] \subset \mathbb{R}^1$ with a constant $Q_N(\bar{\lambda}^1, \bar{\lambda}^2) \in \mathbb{R}^1$ if

$$|H_N(z)| \le Q_N(\bar{\lambda}^1, \bar{\lambda}^2) D_{N, \mathrm{Re}\, z} \quad \text{for } z \in \mathscr{O}_N(\bar{\lambda}^1, \bar{\lambda}^2). \tag{2.1.14}$$

(The variables $\bar{\lambda}^1$, $\bar{\lambda}^2$ need not be finite.) For any $\lambda \in \mathbb{R}^1$ we denote by $R_N(\lambda) = R_{N, \mathscr{O}_N(\bar{\lambda}^1, \bar{\lambda}^2)}(\lambda)$ the radius of the maximal circle in $O_N(\bar{\lambda}^1, \bar{\lambda}^2)$ centered at λ.

We need also a locality condition. Consider the characteristic function

$$\varphi_{N,\lambda}(t) = \sum_{b \in \mathbb{Z}_d} P_N(b) e^{itb} = \frac{F_N(\lambda + it)}{F_N(\lambda)}. \tag{2.1.15}$$

Let $b_N \in \mathbb{Z}_d$, $N = 1, 2, \ldots$, be a given sequence such that $B_N^- < b_N < B_N^+$,

$$\hat{D}_N = D_{N, \lambda_N(b_N)}, \qquad \hat{\varphi}_N(t) = \varphi_{N, \lambda_N(b_N)}(t), \tag{2.1.16}$$

and

$$\gamma_N = K(\hat{D}_N)^{-\frac{1}{2}} \ln(\hat{D}_N)^{\frac{1}{2}}. \tag{2.1.17}$$

We say that the *locality condition* holds for $\hat{\varphi}_N(t)$ if there exists a sufficiently large K such that

$$\lim_{N \to \infty} (\hat{D}_N)^{\frac{1}{2}} \int_{\gamma_N < |t| < \pi/d} |\hat{\varphi}_N(t)|\, dt = 0. \tag{2.1.18}$$

Let us comment on the last condition. Note that $|\hat{\varphi}_N(t)| \le 1$, $\hat{\varphi}_N(0) = 1$, and $\hat{\varphi}_N(t)$ is periodic in t with period $2\pi/d$. If the $\lambda_N(b)$-tilted distribution is concentrated on a sublattice $\mathbb{Z}_{d'}$ with $d' > d$, then $\hat{\varphi}_N(2\pi/d') = 1$, so there is no hope that the condition (2.1.18) holds. So this condition rules out situations when a probability distribution is concentrated on a sublattice or is close enough to such a distribution.

In many cases it is possible to verify the following simpler condition, which implies locality. It states that for some positive constants $\varkappa = \varkappa(\bar{\lambda}^1, \bar{\lambda}^2)$ and for all $\lambda \in [\bar{\lambda}^1, \bar{\lambda}^2]$ we have

$$|\varphi_{N,\lambda}(t)| \le \exp\{-\varkappa t^2 D_{N,\lambda}\} \quad \text{for } |t| < \pi/d. \tag{2.1.19}$$

The Taylor expansion around t

$$\varphi_{N,\lambda}(t) = \exp\{itE_{N,\lambda} - \frac{t^2}{2} D_{N,\lambda} + O(|t|^3)\} \tag{2.1.20}$$

suggests that the condition (2.1.19) should hold in many cases. We will show in §§2.2, 2.3 that this condition indeed holds for independent variables and for the ordinary Ising model. But in §4.1 we are dealing with cut-off Ising models, where we "exclude large contours", and there we can check only the weaker condition (2.1.18).

THEOREM 2.1. *Let a sequence* $b_N \in \mathbb{Z}_d$, $N = 1, 2, \dots$, *be given such that* $B_N^- \le b_N \le B_N^+$ *for all* N. *Then*

$$P_N(b_N) \le \exp\{-I_N(b_N)\}, \qquad N = 1, 2, \dots. \tag{2.1.21}$$

Further, assume that $B_N^- < b_N < B_N^+$ *and that for some finite* $\bar{\lambda}^1$, $\bar{\lambda}^2$ *and for all* N *we have*

$$\bar{\lambda}^1 \le \lambda_N(b_N) \le \bar{\lambda}^2. \tag{2.1.22}$$

Let

$$\hat{E}_N = E_{N, \lambda_N(b_N)}, \qquad \hat{D}_N = D_{N, \lambda_N(b_N)}, \qquad \hat{G}_N^n = G_{N, \lambda_N(b_N)}^n. \tag{2.1.23}$$

We assume that $\hat{D}_N \to \infty$ *when* $N \to \infty$. *Let the analyticity condition* (2.1.14) *hold, and let*

$$\hat{R}_N = R_{N, \mathscr{O}_N(\bar{\lambda}^1, \bar{\lambda}^2)}(\lambda_N(b_N)), \qquad \hat{Q}_N = Q_N(\bar{\lambda}^1, \bar{\lambda}^2). \tag{2.1.24}$$

Assume further that for some $n \ge 3$ *and all* k, $3 \le k \le n-1$, *we have*

$$\lim_{N\to\infty} \frac{(\ln \hat{D}_N)^{k/2}}{(\hat{D}_N)^{k/2}} \hat{G}_N^k = 0 \tag{2.1.25}$$

and

$$\lim_{N\to\infty} \frac{(\ln \hat{D}_N)^{n/2} \hat{Q}_N}{(\hat{D}_N)^{n/2}} (\hat{R}_N)^{-n} = 0. \tag{2.1.26}$$

Finally, let the locality condition (2.1.18) *hold.*

Then

$$P_N(b_N) = d(2\pi \hat{D}_N)^{-1/2} \exp\{-I_N(b_N)\}(1 + o_N(1)), \tag{2.1.27}$$

where $o_N(1) \to 0$ *as* $N \to \infty$.

REMARK. In many cases (see the following corollary and its applications to the cases of independent random variables in §2.2 and of the high temperature Ising model in §2.3) it is enough to use the above theorem for $n = 3$. The general form is necessary for the low-temperature Ising model.

PROOF. Let us consider the characteristic function of the λ-tilted distribution $\varphi_{N,\lambda}(t)$, introduced in (2.1.15). The usual inversion formula for the Fourier transform and the definition (2.1.9) imply that

$$\begin{aligned} P_N(b_N) &= F_N(\lambda) e^{-\lambda b_N} \frac{d}{2\pi} \int_{-\pi/d}^{\pi/d} e^{-itb_N} \varphi_{N,\lambda}(t)\, dt \\ &= \exp\{-(\lambda b_N - H_N(\lambda))\} \frac{d}{2\pi} \int_{-\pi/d}^{\pi/d} e^{-itb_N} \varphi_{N,\lambda}(t)\, dt. \end{aligned} \tag{2.1.28}$$

Since $|\varphi_{N,\lambda}(t)| \le 1$, it follows from (2.1.28) that

$$P_N(b_N) \le \exp\{-\sup_{\lambda}(\lambda b_N - H_N(\lambda))\}, \tag{2.1.29}$$

where the supremum is taken over all $\lambda \in \mathbb{R}^1$. So the estimate (2.1.21) follows from (2.1.29) and the definitions (2.1.12), (2.1.13).

The proof of the asymptotic equality (2.1.27) is, naturally, not that simple. We rewrite first (2.1.28) by using (2.1.12) and (2.1.16) as

$$P_N(b_N) = \exp\{-I_N(b_N)\}\frac{d}{2\pi}\int_{-\pi/d}^{\pi/d} e^{-itb_N}\hat{\varphi}_N(t)\,dt. \tag{2.1.30}$$

It follows from the Taylor expansion of $H_N(z)$ at $z = \lambda_N(b_N)$ and from the Cauchy formula that for $|t| \le \hat{R}_N/2$ we have

$$\ln \hat{\varphi}_N(t) = it\hat{E}_N - \frac{t^2}{2}\hat{D}_N + \sum_{k=3}^{n-1}\frac{(it)^k}{k!}\hat{G}_N^k + t^n(\hat{R}_N)^{-n}\hat{Q}_N\hat{D}_N\theta_N(t), \tag{2.1.31}$$

where the functions $\theta_N(t)$ are uniformly bounded with respect to t and N.

Let γ_N be defined by the relation (2.1.17). Then it follows from (2.1.30), (2.1.31) and conditions (2.1.25), (2.1.26) that

$$\int_{|t|\le\gamma_N} e^{-itb_N}\hat{\varphi}_N(t)\,dt = \int_{|t|\le\gamma_N}\exp\{-itb_N + it\hat{E}_N - \frac{t^2}{2}\hat{D}_N\}\,dt\,(1+o_N(1)). \tag{2.1.32}$$

Using the definitions (2.1.23) and (2.1.11), we can rewrite (2.1.32) further as follows:

$$\begin{aligned}\int_{|t|\le\gamma_N} e^{-itb_N}\hat{\varphi}_N(t)\,dt &= \int_{|t|\le\gamma_N}\exp\left\{-\frac{t^2}{2}\hat{D}_N\right\}dt\,(1+o_N(1))\\ &= (\hat{D}_N)^{-1/2}\int_{|t|\le\gamma_N(\hat{D}_N)^{1/2}}\exp\left\{-\frac{t^2}{2}\right\}dt\,(1+o_N(1))\\ &= (\hat{D}_N)^{-1/2}(2\pi)^{1/2}(1+o_N(1)).\end{aligned} \tag{2.1.33}$$

Using the locality condition (2.1.18) we see that

$$\int_{\gamma_N<|t|<2\pi/d} e^{-itb_N}\hat{\varphi}_N(t)\,dt = (\hat{D}_N)^{-1/2}o_N(1). \tag{2.1.34}$$

Relations (2.1.30), (2.1.33), and (2.1.34) prove the main statement (2.1.27) of the theorem.

COROLLARY. *Suppose that the analyticity condition* (2.1.14) *holds in a domain* $\mathscr{O}_N(\bar{\lambda}^1, \bar{\lambda}^2) = \mathscr{O}$ *with a constant* $Q_N(\bar{\lambda}^1, \bar{\lambda}^2) = Q$ *that does not depend on* N. *Assume that for some sequence* $v_N \in [0, \infty)$, $v_N \to \infty$ *as* $N \to \infty$, *there exists the limit*

$$\lim_{N\to\infty} \frac{1}{v_N} H_N(z) = H(z) \tag{2.1.35}$$

in the sense of uniform convergence in $\mathscr{O}$, *with* $H(\lambda)$ *strictly concave for* $\lambda \in [\bar{\lambda}^1, \bar{\lambda}^2]$. *Denoting by* B_N^i *the (unique) solutions of*

$$\lambda_N(B_N^i) = \bar{\lambda}^i, \qquad i = 1, 2, \tag{2.1.36}$$

we assume that the limits

$$B^i = \lim_{N\to\infty} \frac{B_N^i}{v_N}, \qquad i = 1, 2, \tag{2.1.37}$$

exist. In the case $\bar{\lambda}^1 = -\infty$ *we assume similarly that the limit*

$$B^- = \lim_{N\to\infty} \frac{B_N^-}{v_N} \tag{2.1.38}$$

exists and, moreover, coincides with the limit of

$$E_\lambda = \frac{dH(\lambda)}{d\lambda} \tag{2.1.39}$$

as $\lambda \to -\infty$; *we assume that the same is true for the case* $\bar{\lambda}^2 = +\infty$. *Finally, we assume that the locality condition* (2.1.18) *holds.*

Then for any sequence $b_N \in \mathbb{Z}_d$, $N = 1, 2, \dots$, *such that*

$$\liminf_{N\to\infty} \frac{b_N}{v_N} = b^1 > B^1, \qquad \limsup_{N\to\infty} \frac{b_N}{v_N} = b^2 < B^2, \tag{2.1.40}$$

the statement (2.1.27) *of Theorem* 2.1.1 *holds.*

PROOF. It suffices to verify the growth condition for the sequence $\hat{D}_N = D_{N,\lambda_N(b_N)}$, the conditions (2.1.25) and (2.1.26) of the theorem which turn out to be valid even for $n = 3$, and the boundedness of the sequence $\lambda_N(b_N)$ (if $\bar{\lambda}^1$ or $\bar{\lambda}^2$ is infinite). Note that in our case the function $R_N(\lambda)$ does not depend on N. Our conditions imply that the function E_λ is strictly monotone and

$$\lim_{N\to\infty} \frac{E_{N,\lambda}}{v_N} = E_\lambda. \tag{2.1.41}$$

So, if μ^1, μ^2 are the solutions of the equations

$$E_{\mu^1} = b^1, \qquad E_{\mu^2} = b^2, \tag{2.1.42}$$

then (2.1.41) implies that

$$\liminf_{N\to\infty} \lambda_N(b_N) \ge \mu^1, \qquad \limsup_{N\to\infty} \lambda_N(b_N) \le \mu^2, \tag{2.1.43}$$

so that $\lambda_N(b_N)$ is uniformly bounded. Similarly,

$$\lim_{N\to\infty} \frac{D_{N,\lambda}}{v_N} = D_\lambda = \frac{d^2 H_N(\lambda)}{d\lambda^2}, \qquad \lim_{N\to\infty} \frac{G^k_{N,\lambda}}{v_N} = G^k_\lambda = \frac{d^k H_N(\lambda)}{d\lambda^k}, \quad (2.1.44)$$

where D_λ is a strictly positive function, and so the sequence $\hat{D}_N = D_{N,\lambda_N(b_N)}$ is such that both sequences $\hat{D}_N/v_N$ and $v_N/\hat{D}_N$, as well as the sequences $\hat{G}^k_N/v_N$, $k = 3, \dots, n-1$, are bounded. Therefore the conditions (2.1.25) and (2.1.26) of the theorem are satisfied for any $n \geq 3$.

Applying the above statements to the setting of §1.2 for the case $S_N = \xi_1^N + \dots + \xi_N^N$, we put $v_N = N$ and the sublattice spacing $d = 2$. We then get the statement that the Classical Large Deviation Principle in the Strong Form is valid here. In the case of Gibbs distributions in a volume V_N one should use the normalization $v_N = |V_N|$.

§2.2. Sums of independent variables

To illustrate the results of the previous section we shall consider sums of independent identically distributed (i.i.d.) random variables.

THEOREM 2.2. *Let*

$$S_N = \xi_1 + \dots + \xi_N, \quad (2.2.1)$$

where the random variables ξ_i *are i.i.d. with values in a sublattice* $\mathbb{Z}_d$. *Assume that the distribution of these variables is nondegenerate and is not concentrated on a proper sublattice of* $\mathbb{Z}_d$. *Assume also that the generating function of* ξ_i,

$$F(z) = \sum_{b\in\mathbb{Z}_d} \mathbf{P}\{\xi_i = b\} e^{zb}, \quad (2.2.2)$$

is finite for all $z \in \mathbb{C}^1$. *Let*

$$B^- = \inf\{b : \mathbf{P}\{\xi_i = b\} > 0\}, \qquad B^+ = \sup\{b : \mathbf{P}\{\xi_i = b\} > 0\}, \quad (2.2.3)$$

and

$$H(\lambda) = \ln F(\lambda), \qquad E_\lambda = \frac{dH(\lambda)}{d\lambda}, \qquad D_\lambda = \frac{d^2 H(\lambda)}{d\lambda^2}. \quad (2.2.4)$$

For any b *in the interval* (B_-, B_+) *denote by* $\lambda(b)$ *the unique solution of the equation*

$$b = E_{\lambda(b)}. \quad (2.2.5)$$

Let finally

$$I(b) = \sup_\lambda(\lambda b - H(\lambda)) = \lambda(b)b - H(\lambda(b)), \qquad B^- < b < B^+, \quad (2.2.6)$$

and

$$\hat{D}(b) = D_{\lambda(b)}. \quad (2.2.7)$$

Then for any sequence b_N *such that*

$$B^- < \liminf_{N\to\infty} \frac{b_N}{N} \le \limsup_{N\to\infty} \frac{b_N}{N} < B^+ \tag{2.2.8}$$

the following equality holds asymptotically as $N \to \infty$:

$$\mathbf{P}\{S_N = b_N\} = d(2\pi N\hat{D}(b_N/N))^{-1/2} \exp\{-NI(b_N/N)\}(1 + o_N(1)). \tag{2.2.9}$$

PROOF. It suffices to note (see the notation in §2.1) that in our case

$$H_N(\lambda) = NH(\lambda), \qquad E_{N,\lambda} = NE_\lambda, \qquad D_{N,\lambda} = ND_\lambda, \tag{2.2.10}$$

so that

$$\lambda_N(b_N) = \lambda(b_N/N), \qquad I_N(b_N) = NI(b_N/N), \qquad \hat{D}_N = N\hat{D}(b). \tag{2.2.11}$$

Therefore, it remains only to check the conditions of Corollary 2.1 (for $v_N = N$ and $\overline{\lambda}_1 = -\infty$, $\overline{\lambda}_2 = \infty$). The analiticity condition evidently holds. It is easy to see that $E_\lambda \to B^\pm$ as $\lambda \to \pm\infty$. To verify the locality condition (2.1.18) we note that the assumption of the theorem that the probability distribution is not concentrated on a proper sublattice implies that for any finite λ_1, λ_2 the characteristic function

$$\varphi_\lambda(t) = \frac{F(\lambda + it)}{F(\lambda)} \tag{2.2.12}$$

of the tilted distribution for ξ_i satisfies the estimate

$$\inf_{\lambda : \lambda_1 < \lambda < \lambda_2} (1 - |\varphi_\lambda(t)|) > 0 \quad \text{for any } t \ne 0,\ |t| \le \pi/d. \tag{2.2.13}$$

Also

$$\ln|\varphi_\lambda(t)| = -\frac{t^2}{2} D(\lambda) + O_\lambda(t^3), \tag{2.2.14}$$

where $|O_\lambda(t^3)| < Ct^3$ and $(D(\lambda))^{-1} < C$ for a constant $C < \infty$ unformly with respect to $\lambda \in [\lambda_1, \lambda_2]$, which implies that for some $\gamma = \gamma(\lambda_1, \lambda_2)$ we have

$$|\varphi_\lambda(t)| \le \exp\{-\gamma t^2\} \quad \text{for any } |t| \le \pi/d,\ \lambda \in [\lambda_1, \lambda_2]. \tag{2.2.15}$$

Hence

$$|\varphi_{N,\lambda}(t)| = \frac{|F_N(\lambda + it)|}{F_N(\lambda)} = |\varphi_\lambda(t)|^N \le \exp\{-\gamma N t^2\}, \tag{2.2.16}$$

Since $\overline{D}_N = ND_\lambda(b_N/N)$ and since the condition (2.2.8) implies that the sequence $\lambda(b_N/N)$ lies in a finite interval $[\lambda_1, \lambda_2]$, the estimate (2.2.16) gives the condition (2.1.18).

§2.3. High-temperature Ising model

In this section we show that Theorem 1.4.1 about the behavior of deviation for the Ising model with sufficiently small values of β can be obtained as a corollary of Theorem 2.1.

LEMMA 2.3.1. *Under the conditions of Theorem* 1.4.1 *for all types of Gibbs distributions considered there, there exists a neighborhood* $\mathscr{O} \subset \mathbb{C}$ *of the real line and numbers* $Q_{\overline{\lambda}}$, $\overline{\lambda} \in (0, \infty)$, *such that for any* N *we have*

$$|H_N(z)| \le Q_{\overline{\lambda}}|V_N| \quad \textit{for } z \in \mathscr{O} \cap \{|z| < \overline{\lambda}\} \tag{2.3.1}$$

(*see the notation in* §2.1). *The limit*

$$H(z) = \lim_{N\to\infty} \frac{1}{|V_N|} H_N(z) \tag{2.3.2}$$

exists in the sense of uniform convergence on the sets $\mathscr{O} \cap \{|z| < \overline{\lambda}\}$ *for each* $\overline{\lambda} > 0$. *This limit does not depend on the choice of the sequence of the volumes* V_N *satisfying the Van Hove condition* (1.4.1) *or on the type of the Gibbs distribution. The restriction of* $H(\lambda)$ *to* $\lambda \in \mathbb{R}^1$ *is a strictly concave function of* $\lambda \in \mathbb{R}^1$. *Its derivative is given by*

$$\frac{dH(\lambda)}{d\lambda} = m^+(\beta, \beta^{-1}\lambda + h). \tag{2.3.3}$$

PROOF. Denote by $Z_V(z \mid \overline{\sigma})$ the partition function for the boundary conditions $\overline{\sigma}$ defined by the relation (1.3.2) with the parameter h changed to a complex number z. Then the generating function (see (1.3.3) and (2.1.4)) is given by

$$F_N(z) = \frac{Z_V(\beta^{-1}z + h \mid \overline{\sigma})}{Z_V(h \mid \overline{\sigma})}. \tag{2.3.4}$$

It is well known in the theory of Gibbs fields (see, for example, [MM, DS1]) that for a sufficiently small β and for any sequences of boundary conditions σ_N and of volumes V_N with the Van Hove condition (1.4.1), the inequality

$$|\ln Z_{V_N}(z|\overline{\sigma}_N)| \le Q_h|V_N|, \qquad z \in \mathscr{O}_h, \tag{2.3.5}$$

holds in a complex neighborhood $\mathscr{O}_h$ of any real h with some constant Q_h. Furthermore, the limit

$$f(z) = \lim_{N\to\infty} \frac{\ln Z_{V_N}(z \mid \overline{\sigma}_N)}{|V_N|} \tag{2.3.6}$$

exists in the sense of the uniform convergence in $\mathscr{O}_h$. This limit is the same for all Van Hove sequences V_N. These facts together with the relation (2.3.4) imply statements (2.3.1), (2.3.2) for the case of Gibbs distributions with $\pm$-boundary conditions.

In the cases of free and of periodic boundary conditions, the situation is completely similar. Although we cannot give a precise reference to a book or a paper where the estimate (2.3.5) is explicily described for free or periodic boundary conditions, all known methods of deriving of the estimate (2.3.5) for $\pm$-boundary conditions can be easily extended to these cases.

The case of the limit +-pure boundary conditions requires additional comments. It follows from the Markov properly of Gibbs fields that in this case the generating function is given by

$$F_N(z) = \left\langle \frac{Z_{V_N}(\beta^{-1}z + h \mid \cdot)}{Z_{V_N}(h \mid \cdot)} \right\rangle^+, \tag{2.3.7}$$

where we average the ratio of conditional partition functions, cosidered as functions of the conditions, with respect to the +-pure limit Gibbs state. When z is real, so that the partition functions are positive, relations (2.3.7) and (2.3.5), (2.3.6) imply the estimate (2.3.1) and the existence of the limit (2.3.2) coinciding with similar limits for other boundary conditions. But if z is complex, it is necessary to be more careful because the possible cancellations of the terms can alter the picture. To avoid this difficulty we need another representation for $F_N(z)$. It follows from the definitions (1.3.9) and (2.1.4) that

$$F_N(z) = \lim_{n\to\infty} \left\langle \exp\left\{ z \sum_{t\in V_N} \sigma_t \right\} \right\rangle^+_{W_n}, \tag{2.3.8}$$

where the W_n are cubes centered at 0 with diameters tending to ∞. Let the energy (see (1.3.2)) be given by

$$H^{z,N}_{W_n}(\sigma \mid \overline{\sigma}^+) = H_{W_n}(\sigma \mid \overline{\sigma}^+) - \beta^{-1} z \sum_{t\in V_N} \sigma_t, \tag{2.3.9}$$

so that the corresponding partition function is

$$Z^{z,N,+}_{W_n} = \sum_{\sigma\in\Omega_V} \exp\{-\beta H^{z,N}_{W_n}(\sigma \mid \overline{\sigma}^+)\}. \tag{2.3.10}$$

Then

$$\left\langle \exp\left\{ z \sum_{t\in V_N} \sigma_t \right\} \right\rangle^+_{W_n} = Z^{z,N,+}_{W_n} / Z^+_{W_n}. \tag{2.3.11}$$

The difference

$$H^{z,N}_{W_n}(\sigma \mid \overline{\sigma}^+) - H_{W_n}(\sigma \mid \overline{\sigma}^+) \tag{2.3.12}$$

of energies defining the numerator and the denominator in (2.3.11) depends only on σ_t with $t \in V_N$. Therefore for some constant $Q = Q_h$, all sufficiently small z, and all n and N we have

$$|\ln Z^{z,N,+}_{W_n} - \ln Z^+_{W_n}| \le Q|V_N|. \tag{2.3.13}$$

This statement can be proven by a term-by-term comparison of the cluster expansions for $\ln Z^{z,N,+}_{W_n}$ and for $\ln Z^+_{W_n}$, or obtained as a direct consequence of the estimate (**2.8**) of the paper [DS1]. The estimate (2.3.13) together with relations (2.3.11) and (2.3.8) implies (2.3.1). The estimate (2.3.1) means that the sequence of functions $(V_N)^{-1} H_N(z)$ satisfies the well-known criterion of compactness in the space of analytic functions. So the convergence in (2.3.2) for the real z proved above implies the convergence for complex z also. The case of the limit $-$-pure boundary condition is similar.

The concavity of the limit function $H(\lambda)$, $\lambda \in \mathbb{R}^1$, follows from the concavity of the functions $H_N(\lambda)$ mentioned in §2.1. However, its strict concavity is a deeper fact. It is proved in [DN]. Nevertheless, we reproduce the proof here because it can help to understand similar but more complicated constructions in §4.1. Recall (see §2.1) that the second derivative $D_{N,\lambda} = d^2H_N(\lambda)/d^2\lambda$ is the variance of the λ-tilted distribution. It suffices to verify the existence of positive constants $c = c_{\bar\lambda}$ such that for all N we have

$$D_{N,\lambda} \geq c|V_N| \quad \text{for } |\lambda| \leq \bar\lambda. \tag{2.3.14}$$

We use the following general fact. For any random variable ξ and any partition $\mathscr{B} = \{B_1, \cdots, B_s\}$ of the space of its realization the variance $D\xi$ is given by

$$D\xi = \sum_{k=1}^{s} \mathbf{P}(B_k)D(\xi \mid B_k) + \sum_{k=1}^{s} \mathbf{P}(B_k)(E(\xi \mid B_k) - E\xi)^2, \tag{2.3.15}$$

where $E(\xi|B_k)$ and $D(\xi|B_k)$ are conditional mean values and variances of ξ under conditions B_k. Since the second term in (2.3.15) is nonnegative, we obtain the following lower bounds for variances:

$$D\xi \geq \sum_{k=1}^{s} \mathbf{P}(B_k)D(\xi \mid B_k). \tag{2.3.16}$$

Now we note that

$$D_N(\lambda) = DS_N, \tag{2.3.17}$$

where S_N is the sum (1.4.3) and the variance is taken with respect to the λ-tilted distribution, which coincides with the Gibbs distribution with the parameter h changed to $h + \beta^{-1}\lambda$. Now we introduce the following construction. Let $V_N^e \subset V_N$ be the set of all points $t \in V_N$ such that all components t_i, $i = 1, \ldots, \nu$, of t are even and $V_N^c = V_N \setminus V_N^e$. Let the events B_k correspond to all the possible ways to fix values of σ_t at points of the set V_N^c. Then under each condition B_k the variables σ_s, $s \in V_N^c$, are fixed and because of the Markov property of Gibbs fields the variables σ_s, $s \in V_N^e$, are conditionally independent. So

$$D(S_N \mid B_k) = \sum_{s \in V_N^e} D(\sigma_s \mid B_k). \tag{2.3.18}$$

The terms in (2.3.18) are positive uniformly with respect to t, N, and k, and their number is asymptotically proportional to $|V_N|$. So (2.3.18) together with (2.3.16) imply the desired estimate (2.3.14). It is known that

$$m^+(\beta, h) = \lim_{N\to\infty} \frac{1}{|V_N|}\langle S_N\rangle^{+,\beta,h} \tag{2.3.19}$$

(see [R] or a similiar statement proved in §3.19 for the more complicated case of large β). The comparison of this relation with (2.1.7) proves (2.3.3).

LEMMA 2.3.2. *Under the conditions of Theorem* 1.4.1 *for all types of Gibbs distributions considered there, all conditions of Corollary* 2.1 *are valid for* $v_N = |V_N|$.

PROOF. We can let the interval $[\lambda_1, \lambda_2]$ introduced in Corollary 2.1 be equal to $[-\bar{\lambda}, \bar{\lambda}]$. Then the condition (2.1.35) of this corollary follows from Lemma 2.3.1. Comparing relations (2.1.11), (2.1.6), (2.3.3) with the properties of the function $m^+(\beta, h)$ formulated in §1.3, we see that limits (2.1.37) exist and are equal to $m^+(\beta, \beta^{-1}\lambda - \bar{\lambda})$ and $m^+(\beta, \beta^{-1}\lambda + \bar{\lambda})$ respectively. Therefore, the condition (1.4.4) of Theorem 1.4.1 implies that the conditions (2.1.40) hold for an appropriate choice of $\bar{\lambda}$.

However, the verification of the locality condition (2.1.18) requires a special construction. In the case under consideration, (2.3.1), (2.1.6), and the Cauchy theorem imply that the variance satisfies the bound

$$D_N(\lambda) = \frac{d^2 H_N(\lambda)}{d\lambda^2} \le C_{\bar{\lambda}}|V_N| \quad \text{if } |\lambda| \le \bar{\lambda}, \tag{2.3.20}$$

where $\bar{\lambda}$ is a constant. So the condition (2.1.18) will be verified if we show that the characteristic function $\varphi_{\lambda, N}(t)$ (see (2.1.15)) satisfies the estimate

$$|\varphi_{N,\lambda}(t)| \le \exp\{-\varkappa|V_N|t^2\} \quad \text{if } |t| \le \pi/2 \tag{2.3.21}$$

(in our case $d = 2$). To show this we use a construction similar to that used in [DT]. Let $\mathscr{B} = \{B_1, \dots, B_k\}$ be the same partition as in the proof of the previous lemma. Then

$$\varphi_{N,\lambda}(t) = \sum_k \mathbf{P}(B_k)E(e^{itS_N} \mid B_k), \tag{2.3.22}$$

where $\mathbf{P}(\cdot)$ and the conditional expectations $E(\cdot \mid B_k)$ are induced by the Ising model Gibbs distribution with the parameter $h + \beta^{-1}\lambda$. So

$$|\varphi_{N,\lambda}(t)| \le \sum_k \mathbf{P}(B_k)|E(e^{itS_N} \mid B_k)|. \tag{2.3.23}$$

Similarly to (2.3.18) we find that

$$E(e^{itS_N} \mid B_k) = \left|\exp\left\{it \sum_{s\in V_N^c} \sigma_s^k\right\}\right| \prod_{s\in V_N^e} E(e^{it\sigma_s} \mid B_k), \tag{2.3.24}$$

where σ_s^k are the values of σ_s, $s \in V_N^c$, defined by the condition B_k. Each of the factors in (2.3.24) satisfies the estimate

$$|E(e^{it\sigma_s}|B_k)| \le e^{-\tilde{\varkappa}t^2} \quad \text{for } |t| < \pi/2, \tag{2.3.25}$$

where $\tilde{\varkappa} > 0$ is a constant that does not depend on t, B_k, and N. This estimate follows from the equality

$$E(e^{it\sigma_s}|B_k) = e^{it}\mathbf{P}(\sigma_s = 1) + e^{-it}\mathbf{P}(\sigma_s = -1), \tag{2.3.26}$$

where both probabilities $\mathbf{P}(\sigma_s = 1)$ and $\mathbf{P}(\sigma_s = -1)$ are uniformly separated from zero. Since the number of factors in (2.3.24) is asymptotically proportional to $|V_N|$, the estimates (2.3.23)–(2.3.25) imply the desired estimate (2.3.20).

Theorem 1.4.1 follows from Corollary 2.1 and Lemma 2.3.2.

3. Contour Ensembles

A straightforward application of the results of §2.1 to Gibbs fields in the phase transition regime is not possible, since zeroes of the generating functions typically tend to the real axis as the volume increase, so the parameters $\hat{R}_N$ used in Theorem 2.1 go to zero as $N \to \infty$. To cope with this obstacle we introduce auxiliary cutoff contour ensembles, that are obtained by excluding all configurations with too large contours. For these ensembles we can control the rate at which $\hat{R}_N \to 0$; at the same time we can compare the distributions of the sums S_N for the initial and the cutoff ensembles. In the present section we introduce the cutoff ensembles and study their properties.

§3.1. Contours

We shall treat the ν-dimensional integer lattice $\mathbb{Z}^\nu$ as a graph whose set of edges is the set of all pairs (s, t), $s, t \in \mathbb{Z}^\nu$, such that $|s - t| = 1$. We shall sometimes treat the lattice $\mathbb{Z}^\nu$ as a subset of the Euclidean space $\mathbb{R}^\nu$. The closed cube $W_t \subset \mathbb{R}^\nu$ of size 1 centered at a point $t \in \mathbb{Z}^\nu$ and oriented along the axes of $\mathbb{Z}^\nu$ is called the *(unit) cell centered at* t. Vertices of these cells form the *dual lattice* $\widehat{\mathbb{Z}^\nu}$, while the $(\nu - 1)$-dimensional closed faces of these cells will be called *faces of the dual lattice* $\widehat{\mathbb{Z}^\nu}$. The faces of the cell W_t will be called faces *adjacent to* t. Each face has two sites of $\mathbb{Z}^\nu$ to which it is adjacent. We say that this face *separates* those sites. For any set $V \subset \mathbb{Z}^\nu$ we denote by $\hat{V}$ the set of all faces of $\widehat{\mathbb{Z}^\nu}$ which are adjacent to at least one site of V. We denote by ∂V the set of all faces in $\widehat{\mathbb{Z}^\nu}$ separating points of the set V and of its complement V^c.

Let $\sigma \in \Omega$ be a configuration on $\mathbb{Z}^\nu$. The set of all faces of $\widehat{\mathbb{Z}^\nu}$ such that $\sigma_s \neq \sigma_t$ for a pair (s, t) of points adjacent to it is called the *boundary of the configuration* σ. For a finite volume $V \subset \mathbb{Z}^\nu$ and a configuration $\sigma \in \Omega_V$ we define its (plus) boundary (for the case of +-boundary conditions) as a boundary of the configuration $\sigma^+ \in \Omega$, which is the extension of σ to $\mathbb{Z}^\nu$ by $+1$ outside V. The connected components of this boundary (treated as subsets of $\mathbb{R}^\nu$) are called *contours*, and the collection of these contours is denoted by $\mathscr{G}^+(\sigma)$ or $\mathscr{G}(\sigma)$. We shall write $\mathscr{G}(\sigma) = \{\Gamma_1, \dots, \Gamma_k\}$, where $\Gamma_1, \dots, \Gamma_k$ are all the contours of σ. The boundary for the case of $-$-boundary conditions is defined in a similar way and is denoted by $\mathscr{G}^-(\sigma)$.

Let G_V be the set of all possible contours of configurations $\sigma \in \Omega_V$. It is easy to see that the transformation $\sigma \in \Omega_V \to \mathscr{G}(\sigma) = \{\Gamma_1, \dots, \Gamma_k\}$ is a one-to-one correspondence between the set Ω_V and the set of all the systems $\{\Gamma_1, \dots, \Gamma_k\}$, $k = 0, 1, \dots$, of contours $\Gamma_i \in G_V$ such that the contours Γ_i are mutually disjoint (as subsets of $\mathbb{R}^\nu$). The set

$$G = \bigcup_{V : |V| < \infty} G_V \tag{3.1.1}$$

is called the *set of contours in* $\mathbb{Z}^\nu$.

Let $\overline{V}$ be the set of all faces of $\widehat{\mathbb{Z}^\nu}$ such that they are adjacent to at least one site of V. It is clear that $G_V = \{\Gamma \in G : \Gamma \subset \overline{V}\}$. For $\Gamma \in G$ we introduce the sets

$$\begin{aligned} \operatorname{Int}\Gamma &= \{t \in \mathbb{Z}^\nu : t \text{ belongs to a finite} \\ &\qquad\qquad \text{connected component of } \mathbb{R}^\nu \setminus \Gamma\}, \\ \operatorname{Ext}\Gamma &= \{t \in \mathbb{Z}^\nu : t \text{ belongs to the infinite} \\ &\qquad\qquad \text{connected component of } \mathbb{R}^\nu \setminus \Gamma\}, \\ \Delta(\Gamma) &= \{t \in \mathbb{Z}^\nu : t \text{ is adjacent to at least one face of } \Gamma\}, \\ \Delta^I(\Gamma) &= \Delta(\Gamma) \cap \operatorname{Int}\Gamma, \\ \Delta^E(\Gamma) &= \Delta(\Gamma) \cap \operatorname{Ext}\Gamma, \\ \operatorname{Int}^0\Gamma &= \operatorname{Int}\Gamma \setminus \Delta^I(\Gamma). \end{aligned} \tag{3.1.2}$$

If a configuration $\sigma \in \Omega_V$ has a contour Γ, then $\sigma_t = +1$ for $t \in \Delta^E(\Gamma)$ and $\sigma_t = -1$ for $t \in \Delta^I(\Gamma)$ provided the number of contours $\Gamma_i \in \mathscr{G}(\sigma)$ such that $\operatorname{Int}\Gamma \subset \operatorname{Int}\Gamma_i$ is even. In the opposite case one should take the opposite signs. For other points $t \in V \setminus \Delta(\Gamma)$ the values of σ are not specified by the existence of the contour Γ in the configuration σ.

It is clear that the number of points $|\operatorname{Int}\Gamma|$ in $\operatorname{Int}\Gamma$ is always finite, and that if $\Gamma_1, \Gamma_2 \in \mathscr{G}(\sigma)$ for some $\sigma \in \Omega_V$ and $\operatorname{Int}\Gamma_2 \subseteq \operatorname{Int}\Gamma_1$, then $\operatorname{Int}\Gamma_2 \subseteq \operatorname{Int}^0\Gamma_1$.

We define the *size of the contour* Γ to be

$$q(\Gamma) = \frac{1}{2\nu}|\operatorname{Int}\Gamma|^{1/\nu}. \tag{3.1.3}$$

Elementary geometrical constructions show that if $|\Gamma|$ is the number of faces in Γ, then

$$|\operatorname{Int}\Gamma| \le |\Gamma| q(\Gamma), \tag{3.1.4}$$

which implies that

$$q(\Gamma)^{\nu-1} \le \frac{1}{(2\nu)^\nu}|\Gamma|. \tag{3.1.5}$$

The equality is achieved when Γ is the boundary of a cube.

§3.2. Contour ensembles

Let V be fixed finite volume. A contour $\Gamma \in G_V$ is called an *inner contour of* V if $\operatorname{Int}\Gamma \subset V$. Let $\mathscr{L} \subset G_V$ be a fixed subset of inner contours of V.

The most important case is when

$$\mathscr{L} = G_V^{\omega} = \{\Gamma \in G_V : \Gamma \text{ is an inner contour of } V\ ,\ q(\Gamma) \leq \omega\} \tag{3.2.1}$$

for some integer $\omega > 1$ (or $\omega = \infty$), which is called the *cutoff level.* A more general case, which will be used later in the study of restrictions of Gibbs distributions to W, is the case when for some set $W \subset V$

$$\mathscr{L} = G_{V,W}^{\omega} = \{\Gamma \in G_V : \Gamma \text{ is an inner contour of } V\ , \\ q(\Gamma) \leq \omega \text{ provided } \operatorname{Int}\Gamma \cap W \neq \varnothing\}. \tag{3.2.2}$$

Clearly, $G_{V,V}^{\omega} = G_V^{\omega}$. Below we shall mainly consider sequences of volumes V_N, W_N and of cutoff levels ω_N such that $|V_N| \to \infty$, $|W_N| \to \infty$, $\omega_N \to \infty$. The appearance of the parameter ω in any formula would automatically mean that we are dealing with one of the cases (3.2.1) or (3.2.2).

For any $W \subset V$ let $G_W^{\mathscr{L}} \subset \mathscr{L}$ be the family of all contours $\Gamma \in \mathscr{L}$ such that $\operatorname{Int}\Gamma \subset W$. Let $\Omega_V^{\mathscr{L}}$ be the set of all configurations $\sigma \in \Omega_V$ such that any contour $\Gamma \in \mathscr{G}(\sigma)$ lies in the set $\mathscr{L}$. Similarly, let $\Omega_W^{\mathscr{L}}$ be $\Omega_W^{\mathscr{L}(W)}$. In the case $\mathscr{L} = G_V^{\omega}$ we write Ω_V^{ω} instead of $\Omega_V^{\mathscr{L}}$. It is easy to see that a configuration satisfies $\sigma \in \Omega_V^{\mathscr{L}}$ if and only if the system of contours satisfies $\mathscr{G}(\sigma) \subset \mathscr{L}$. The set of all such systems of contours will be denoted by $D_V^{\mathscr{L}}$ (or D_V^{ω} in case $\mathscr{L} = G_V^{\omega}$ or simply D_V if $\mathscr{L} = G_V^{\infty}$ is the set of all contours inner to V). Finally, let $D_W^{\mathscr{L}} = D^{G_W^{\mathscr{L}}}$.

A contour $\Gamma \in \mathscr{G}(\sigma)$ is called an *exterior contour of the configuration* $\sigma \in \Omega_V$ if for any other contour $\Gamma' \in \mathscr{G}(\sigma)$ either $\operatorname{Int}\Gamma' \subset \operatorname{Int}\Gamma$ or $\operatorname{Int}\Gamma \cap \operatorname{Int}\Gamma' = \varnothing$. The set of all systems of contours $\partial \in D_V^{\mathscr{L}}$ such that all contours of the corresponding configuration are exterior contours is denoted by $\overline{D}_V^{\mathscr{L}}$. The set of all exterior contours of a configuration $\sigma \in \Omega_V$ is denoted by $\mathscr{G}^E(\sigma)$. It is evident that for any configuration $\sigma \in \Omega_V$ the subsystem $\partial \subset \mathscr{G}(\sigma)$ of all contours $\Gamma \in \mathscr{G}(\sigma)$ that are external contours of this configuration belongs to $\overline{D}_V^{\mathscr{L}}$. The sets of external contours $\overline{D}_V^{\omega}$ and $\overline{D}_V$ are defined in a similar way.

Let $h = (h_t, t \in V)$ be a set of real numbers, $h_t \in \mathbb{R}^1$. The $\mathscr{L}$-cutoff Gibbs distribution in V with $+$-boundary conditions and parameters β, h is defined by the relation

$$P_V^{+,\beta,h,\mathscr{L}}(\sigma) = \begin{cases} \exp\{-\beta H_h(\sigma)\} Z(V, +, \beta, h, \mathscr{L})^{-1}, & \text{if } \sigma \in \Omega_V^{\mathscr{L}}, \\ 0, & \text{if } \sigma \in \Omega_V \setminus \Omega_V^{\mathscr{L}}, \end{cases} \tag{3.2.3}$$

where

$$H_h(\sigma) = 2 \sum_{\Gamma \in \mathscr{G}(\sigma)} |\Gamma| - \sum_{t \in V} h_t \sigma_t \tag{3.2.4}$$

and $Z(V, +, \beta, h, \mathscr{L})$ is the partition function

$$Z(V, +, \beta, h, \mathscr{L}) = \sum_{\sigma \in \Omega_V^{\mathscr{L}}} \exp\{-\beta H_h(\sigma)\}. \tag{3.2.5}$$

For any set $A \subset \Omega_V$ we set

$$P_V^{+, \beta, h, \mathscr{L}}(A) = \sum_{\sigma \in A} P_V^{+, \beta, h, \mathscr{L}}(\sigma), \tag{3.2.6}$$

and for any function $f(\sigma)$ of $\sigma \in \Omega_V$,

$$\langle f \rangle_V^{+, \beta, h, \mathscr{L}} = \sum_{\sigma \in \Omega_V^{\mathscr{L}}} f(\sigma) P_V^{+, \beta, h, \mathscr{L}}(\sigma). \tag{3.2.7}$$

For any $W \subset V$, let

$$S_W(\sigma) = \sum_{t \in W} \sigma_t, \quad \sigma \in \Omega_V, \tag{3.2.8}$$

and let

$$G^n(W, V, +, \beta, h, \mathscr{L}) = \left. \frac{\partial^n \ln \langle \exp\{\lambda S_W(\sigma)\}\rangle_V^{+, \beta, h, \mathscr{L}}}{\partial \lambda^n} \right|_{\lambda=0} \tag{3.2.9}$$

be the semiinvariant of order n of the variable S_W in the volume W. In particular, we have the mean value

$$E(W, V, +, \beta, h, \mathscr{L}) = G^1(W, V, +, \beta, h, \mathscr{L}) = \langle S_W \rangle_V^{+, \beta, h, \mathscr{L}}, \tag{3.2.10}$$

and the variance

$$\begin{aligned} D(W, V, +, \beta, h, \mathscr{L}) &= G^2(W, V, +, \beta, h, \mathscr{L}) \\ &= \langle (S_W)^2 \rangle_V^{+, \beta, h, \mathscr{L}} - (E(W, V, +, \beta, h, \mathscr{L}))^2. \end{aligned} \tag{3.2.11}$$

Below we omit some of the indices $W, V, +, \beta, h, \mathscr{L}$ in the notation when it will not lead to ambiguities; we also use index ω instead of $\mathscr{L}$ if $\mathscr{L} = G_V^{\omega}$. The notation (3.2.4), (3.2.5) will be used also for complex values of fields h_t. In the case when $\mathscr{L} = G_V^{\infty}$ and V is contractible, the probability distribution (3.2.3) consides with the main probability distribution $P_V^{+, \beta, h}$ introduced in §1.3. The partition function (3.2.5) is connected with the partition function $Z_V^+(\beta, h)$ by the relation

$$Z_V^+(\beta, h) = \exp\{-\beta b(V)\} Z(V, +, \beta, h, \infty), \tag{3.2.12}$$

where $b(V)$ is the number of faces adjacent to sites of V. We shall use also $\mathscr{L}$-cutoff Gibbs distributions with $-$-boundary conditions, which are defined in a completely analogous way with $\mathscr{G}^+(\sigma)$ replaced by $\mathscr{G}^-(\sigma)$ in the definition (3.2.4) of the Hamiltonian. In this case we use " $-$ " instead of " $+$ " in all the previous notation.

In the following sections of this chapter we will study the properties of the cutoff and the original Gibbs ensembles. We will assume in all the considerations that the inverse temperature satisfies the condition $\beta > \beta_\nu$, where β_ν is a sufficiently large constant depending on the dimension ν only. All the constants which will appear in the following estimates do not depend on $\beta > \beta_\nu$.

§3.3. Estimates and expansions for partition functions in complex domains. Formulations

As it is clear from the construction of Chapter 2 for Gibbs fields, the generating functions are closely connected with the partition functions. So it is very important for our purposes to have a good control over the partition functions for complex values of the parameters h_t. It might be that the results formulated below can be obtained as an implication of constructions used in earlier papers (see [GKK, Za, BI]). But since these constructions are very complicated and the results we need are not formulated explicitly in these papers, we present here more transparent and direct proofs.

PROPOSITION 3.3. *Let a number γ, $0 < \gamma < 1$, a set of complex valued parameters $h = (h_t, t \in \mathbb{Z}^\nu)$, and a finite volume $V \subset \mathbb{Z}^\nu$ be given. Let*

$$\begin{aligned} V^+ &= V^+(h) = \{t \in V : |\operatorname{Im}(h_t)| \le 2(1-\gamma)\operatorname{Re}(h_t)\}, \\ V^0 &= V^0(h) = V \setminus V^+(h). \end{aligned} \tag{3.3.1}$$

Suppose that a system $\mathscr{L}$ of inner contours of V is such that for each contour $\Gamma \in \mathscr{L}$

$$\sum_{t \in \operatorname{Int}\Gamma \cap V^0} |h_t| \le |\Gamma|(1-\gamma). \tag{3.3.2}$$

By (3.1.4), *the condition* (3.3.2) *holds in the case $\mathscr{L} = G_V^\omega$ if*

$$|h_t| \le (1-\gamma)\omega^{-1}, \qquad t \in V^0(h). \tag{3.3.3}$$

(Here $\omega^{-1} = 0$ if $\omega = \infty$.) Then the following statements hold:

First, for some $\beta_0 = \beta_0(\nu, \gamma)$ depending on the dimension ν and the parameter γ only, for some $K = K(\nu, \gamma)$, and for all $\beta \ge \beta_0$ and all V, h, $\mathscr{L}$, the partition function $Z(V, +, \beta, h, \mathscr{L})$ (see (3.2.5)) *does not vanish and*

$$|\ln Z(V, +, \beta, h, \mathscr{L})| \le K|V|. \tag{3.3.4}$$

Second, for any contour $\Gamma \in \mathscr{L}$ we have (see (3.1.2) *for notation)*

$$\left| \frac{Z(\operatorname{Int}^0\Gamma, -, \mathscr{L})}{Z(\operatorname{Int}^0\Gamma, +, \mathscr{L})} \right| \le 2\exp\left\{ (2\beta + \gamma/2) \sum_{t \in (\operatorname{Int}^0\Gamma)^0} |h_t| \right\}. \tag{3.3.5}$$

Finally, if we define for any nonempty finite $\Lambda \subset \mathbb{Z}^\nu$

$$\Phi(\Lambda) = \Phi(\Lambda, +, \beta, h, \mathscr{L}) = \sum_{\Lambda' \subset \Lambda, \Lambda' \neq \varnothing} (-1)^{|\Lambda \setminus \Lambda'|} \ln Z(\Lambda', +, \beta, h, \mathscr{L}), \tag{3.3.6}$$

then for any finite $V \subset \mathbb{Z}^\nu$

$$\ln Z(V, +, \beta, h, \mathscr{L}) = \sum_{\Lambda \subset V} \Phi(\Lambda). \tag{3.3.7}$$

For any fixed β, $\mathscr{L}$ *and* γ *the functions* $\Phi(\Lambda)$ *depend on* h_t, $t \in \Lambda$, *only and are holomorphic functions of* h_t *in the domain* (3.3.1). *They have the following properties: for any* $\mathscr{L}$, $\mathscr{L}'$ *with* $G_\Lambda^{\mathscr{L}} = G_\Lambda^{\mathscr{L}'}$ *(see* §3.2 *for notation)*

$$\Phi(\Lambda, +, \beta, h, \mathscr{L}) = \Phi(\Lambda, +, \beta, h, \mathscr{L}'). \tag{3.3.8}$$

In particular, if $\mathscr{L} = G_V^\omega$, $\mathscr{L}' = G_V^{\omega'}$ *with* $\omega' \geq \omega$, *then* (3.3.8) *holds provided the volume* Λ *satisfies*

$$|\Lambda| \leq (2\nu\omega)^\nu. \tag{3.3.9}$$

Next, for any Λ *and* $\Lambda' = \Lambda + s$, $s \in \mathbb{Z}^\nu$, *and any* $h = (h_t, t \in \mathbb{Z}^\nu)$ *and* $h' = (h_{t+s}, t \in \mathbb{Z}^\nu)$ *satisfying* (3.3.1) *we have*

$$\Phi(\Lambda, +, \beta, h, \mathscr{L}) = \Phi(\Lambda', +, \beta, h', \mathscr{L}), \tag{3.3.10}$$

provided that $\mathscr{L}(\Lambda') = \{\Gamma + s : \Gamma \in \mathscr{L}(\Lambda)\}$. *The last (and the most important) property is*

$$\begin{aligned} |\Phi(\Lambda)| &\leq \exp\{-2\gamma(\beta - \beta_0) d(\Lambda)\}, & |\Lambda| > 1, \\ |\Phi(\{t\}) - \beta h_t| &\leq \exp\{-2\gamma(\beta - \beta_0) 2\nu\}, & |\lambda| = 1, \Lambda = \{t\}. \end{aligned} \tag{3.3.11}$$

Here $d(\Lambda)$ *is defined by*

$$d(\Lambda) = \min |B|, \tag{3.3.12}$$

where the minimum is taken over all connected sets B *of faces (see* §3.1) *such that* $\partial\Lambda \subset B$. *(If the complement* Λ^c *is connected, then* $d(\Lambda) = |\partial\Lambda|$. *If there are no such connected sets* B, *we let* $d(\Lambda) = \infty$, *and it means that* $\Phi(\Lambda) = 0$.)

The statement (3.3.7) follows from (3.3.6) and is a reformulation of a well-known Mobius inversion relation (see, for example, [MM]). Relations (3.3.8), (3.3.10) are evident consequences of the definition (3.3.6). The nontrivial parts of the proposition, which will be proved in §§3.7–3.17, are the statements about the nonvanishing of the partition functions and the estimates (3.3.4), (3.3.5), and (3.3.11).

§3.4. An estimate for combinatoric sums involving $d(\Lambda)$

In using the estimates of the type of (3.3.11), we will find the following result helpful.

LEMMA 3.4. *There exists a constant* $\alpha_0 = \alpha_0(\nu)$ *such that for any* $\alpha > \alpha_0$, $t \in \mathbb{Z}^\nu$, *and any integer* N

$$\sum_{\Lambda : t\in\Lambda, \operatorname{diam}(\Lambda)\geq N} \exp\{-\alpha d(\Lambda)\} \leq \exp\{-(\alpha-\alpha_0)(N+1)\}. \tag{3.4.1}$$

PROOF. Note that the number of different connected sets B of faces such that $|B| = n$ and a fixed face Q is in B does not exceed C^n with a constant $C = C(\nu)$. For a given B, the number of different sets $\partial\Lambda \subset B$ certainly does not exceed 2^n. Since the boundary $\partial\Lambda$ uniquely determines the set Λ, the number of different sets Λ with $d(\Lambda) = n$ and with $Q \in \partial\Lambda$ does not exceed $(2C)^n$. If $t \in \Lambda$, then the set $\partial\Lambda$ contains a face distanced from t by at most $d(\Lambda)$. The number of such faces is at most $C'(d(\Lambda))^\nu$, where $C' = C'(\nu)$ is another constant. Finally, the relation $d(\Lambda) \geq \operatorname{diam}(\Lambda) + 1$ is evident. Hence,

$$\sum_{\Lambda : t\in\Lambda, \operatorname{diam}(\Lambda)\geq N} \exp\{-\alpha d(\Lambda)\} \leq \sum_{n\geq N+1} C' n^\nu \exp\{-\alpha n\}(2C)^n. \tag{3.4.2}$$

The estimate (3.4.1) follows from (3.4.2).

§3.5. Polymer representation for partition functions

The main technique in the proof of Proposition 3.3 is the use of so-called polymer representations of partition functions (see [KP, MM]).

Let $V \subset \mathbb{Z}^\nu$ be a finite volume and let a complex-valued function $f(\Gamma)$, $\Gamma \in G_V$, be defined on the set of all contours in the volume V. This function will be called a *contour functional.* For a nonemply volume $W \subset V$, the quantity

$$Z_f(W) = \sum_{C\in D_W} \prod_{\Gamma\in C} f(\Gamma) \tag{3.5.1}$$

will be called the *contour polymer partition function in the volume* W defined by the contour functional f. Here D_W is the set of all systems of mutually disjoint contours in the volume W (see §3.2), including the emply system $C = \varnothing$ for which we let the product in (3.5.1) be equal to 1.

In the following proposition we use the notation from §§3.1, 3.2 for fixed β and h, therefore omitting these parameters in the notation.

PROPOSITION 3.5. *Suppose that all partition functions satisfy*

$$Z(W, +, \mathscr{L}) \neq 0, \qquad W \subset V. \tag{3.5.2}$$

Then for any $W \subset V$,

$$Z(W, +, \mathscr{L}) = \exp\left\{\beta \sum_{t\in W} h_t\right\} Z_w(W), \tag{3.5.3}$$

where the contour functional is given by

$$w(\Gamma) = w_{\beta, h, \mathscr{L}}(\Gamma) = \begin{cases} \exp\left\{-2\beta\left(|\Gamma| + \sum_{t\in\Delta^I(\Gamma)} h_t\right)\right\} d\dfrac{Z(\mathrm{Int}^0\Gamma, -, \mathscr{L})}{Z(\mathrm{Int}^0\Gamma, +, \mathscr{L})}, & \text{if } \Gamma \in \mathscr{L}, \\ 0, & \text{if } \Gamma \notin \mathscr{L}. \end{cases} \tag{3.5.4}$$

(Here and later we set $Z(W, \pm, \mathscr{L}) = 0$ *for an empty* W.)

PROOF. We shall obtain the representation (3.5.3) by induction on $|W|$. Let $\overline{D}_W^{\mathscr{L}}$ be the set of all systems of external contours of configurations $\sigma \in \Omega_W^{\mathscr{L}}$ (see §3.2). Then, using (3.2.5), we see that

$$\begin{aligned} Z(W, +, \mathscr{L}) &= \sum_{\partial\in\overline{D}_W^{\mathscr{L}}} \exp\left\{\beta \sum_{t\in W\setminus\bigcup_{\Gamma\in\partial}\mathrm{Int}\,\Gamma} h_t\right\} \\ &\quad\times \left(\prod_{\Gamma\in\partial} \exp\left\{-2\beta|\Gamma| - \beta\sum_{t\in\Delta^I(\Gamma)} h_t\right\} Z(\mathrm{Int}^0\Gamma, -, \mathscr{L})\right) \\ &= \sum_{\partial\in\overline{D}_W^{\mathscr{L}}} \exp\left\{\beta \sum_{t\in W\setminus\bigcup_{\Gamma\in\partial}\mathrm{Int}^0\Gamma} h_t\right\} \prod_{\Gamma\in\partial} w(\Gamma) Z(\mathrm{Int}^0\Gamma, +, \mathscr{L}). \end{aligned} \tag{3.5.5}$$

By the induction hypothesis, the partition functions $Z(\mathrm{Int}^0\Gamma, +, \mathscr{L})$ have the representation (3.5.3). Since any contour system $\partial \in D_W^{\mathscr{L}}$ can be represented as

$$\partial = \overline{\partial} \cup \left(\bigcup_{\Gamma\in\overline{\partial}} \partial_\Gamma\right), \tag{3.5.6}$$

where $\overline{\partial} \in \overline{D}_W^{\mathscr{L}}$, $\partial_\Gamma \in D_{\mathrm{Int}^0\Gamma}^{\mathscr{L}}$, the equality (3.5.3) is proved.

§3.6. Estimates of polymer partition functions

Contour polymer partition functions form a special case of a more general class of polymer partition functions intensively studied in the literature on Gibbs fields. Here we formulate some properties of polymer partition functions in the form which is convenient for their subsequent use.

PROPOSITION 3.6. *Let* $\alpha > 0$ *be a constant. Suppose that a contour functional* f *is such that for any face* $Q \in \widehat{\mathbb{Z}^\nu}$ *(see* §3.1)

$$\sum_{\Gamma:\,\Gamma\in G_V,\, Q\subset\Gamma} |f(\Gamma)| \exp\{(1+\alpha)|\Gamma|\} \le 1. \tag{3.6.1}$$

Then for any $W \subset V$ *the partition function* $Z_f(W)$ *satisfies*

$$Z_f(W) \ne 0. \tag{3.6.2}$$

For any $\Lambda \subset V$, $\Lambda \neq \varnothing$, *let*

$$\Phi_f(\Lambda) = \sum_{\Lambda' \subset \Lambda, \Lambda' \neq \varnothing} (-1)^{|\Lambda \setminus \Lambda'|} \ln Z_f(\Lambda'). \tag{3.6.3}$$

Then for any $W \subset V$,

$$\ln Z_f(W) = \sum_{\Lambda \subset W, \Lambda \neq \varnothing} \Phi_f(\Lambda) \tag{3.6.4}$$

and the functions $\Phi_f(\Lambda)$ *satisfy the condition*

$$|\Phi_f(\Lambda)| \leq 2\nu \exp\{-\alpha d(\Lambda)\}, \tag{3.6.5}$$

where $d(\Lambda)$ *was defined in Proposition* 3.3. *The partition function* $Z_f(W)$ *satisfies the inequality*

$$|\ln Z_f(W)| \leq K|W|, \tag{3.6.6}$$

and for any $\Gamma \in G_V$ *the following estimate for the partial derivative is valid:*

$$\left| \frac{\partial \ln Z_f(W)}{\partial f(\Gamma)} \right| \leq |\Gamma| \exp\{-2\nu\alpha\}. \tag{3.6.7}$$

PROOF. Estimates of this type can be obtained from several versions of cluster expansions described in the literature on Gibbs fields (see, for example, [Z, MM]). We think that the shortest way is to use the technique of Kotecký and Preiss [KP]. A special case of a more general result proved in [KP] states that if $\mathscr{C} \subset D_V$ is any set of contours, and the partition function is defined by the relation

$$Z_f(\mathscr{C}) = \sum_{C \in \mathscr{C}} \prod_{\Gamma \in C} f(\Gamma), \tag{3.6.8}$$

then $Z_f(\mathscr{C}) \neq 0$ and the following representation for $\ln Z_f(\mathscr{C})$ holds. There exists a function $\Phi_f(\mathscr{C})$ defined on all nonempty systems of contours $\mathscr{C} \subset D_V$ such that

$$\ln Z_f(\mathscr{C}) = \sum_{\substack{\mathscr{C}' \subset \mathscr{C}, \\ \mathscr{C}' \text{ is connected}}} \Phi_f(\mathscr{C}'), \tag{3.6.9}$$

where the sum is taken over all connected nonempty subsystems $\mathscr{C}'$ of contours of $\mathscr{C}$. Recall that $\mathscr{C}'$ is called *connected* if it cannot be represented as $\mathscr{C}' = \mathscr{C}_1' \cup \mathscr{C}_2'$, where $\mathscr{C}_1'$, $\mathscr{C}_2'$ are nonempty set of contours such that any two contours $\Gamma_1 \in \mathscr{C}_1'$, $\Gamma_2 \in \mathscr{C}_2'$ are disjoint (as sets in $\mathbb{R}^\nu$). The function $\Phi_f(\mathscr{C})$ satisfies the following estimate: for any contour $\Gamma \in \mathscr{L}$

$$\sum_{\substack{\mathscr{C} : (\cup_{\bar{\Gamma} \in \mathscr{C}} \bar{\Gamma}) \cap \Gamma \neq \varnothing, \\ \mathscr{C} \text{ is connected}}} |\Phi_f(\mathscr{C})| \exp\left\{\alpha \sum_{\bar{\Gamma} \in \mathscr{C}} |\bar{\Gamma}|\right\} \leq |\Gamma|, \tag{3.6.10}$$

where the sum is taken over all connected systems of contours $\mathscr{C}$ that contain a contour $\bar{\Gamma} \in \mathscr{C}$ with $\bar{\Gamma} \cap \Gamma \neq \varnothing$. To obtain this statement from the Kotecký-Preiss theorem [KP], we must treat contours $\Gamma \in G_V$ as polymers, saying that contours Γ_1, Γ_2 are incompatible if they have a nonempty (as sets in $\mathbb{R}^\nu$) intersection, and defining

$$a(\Gamma) = |\Gamma|, \qquad d(\Gamma) = \alpha|\Gamma|. \tag{3.6.11}$$

We can obtain the desired representation (3.6.4) for any $W \subset V$ from the representation (3.6.9) if we set $\mathscr{C} = \mathscr{L}(W)$ and

$$\Phi_f(\Lambda) = \sum_{\substack{\mathscr{C}' : \bigcup_{\Gamma\in\mathscr{C}'} \operatorname{Int}\Gamma = \Lambda, \\ \mathscr{C}' \text{ is connected}}} \Phi_f(\mathscr{C}'). \tag{3.6.12}$$

The relation (3.6.4) for the function $\Phi_f(\Lambda)$ defined by (3.6.12) follows from (3.6.9), since any term $\Phi_f(\mathscr{C}')$ in the sum (3.6.9) occurs in the sum (3.6.12) for exactly one Λ. To obtain the estimate (3.6.5) we must note that any connected system of contours with $\bigcup_{\Gamma\in\mathscr{C}'} \operatorname{Int}\Gamma = \Lambda$ contains all faces of the boundary $\partial\Lambda$ of the set Λ, so that for such a system we have

$$d(\Lambda) \le \sum_{\Gamma\in\mathscr{C}'} |\Gamma|. \tag{3.6.13}$$

Then we can apply the estimate (3.6.10) for the case when Γ is one of contours that contains a face $Q \in \partial\Lambda$ and has the minimal length 2ν. If such a connected set of contours does not exist, we have $d(\Lambda) = \infty$ and $\Phi_f(\Lambda) = 0$.

Finally, we must prove the estimate (3.6.7). Using the definition (3.5.1) and the notation (3.6.8) we see that

$$\frac{\partial \ln Z_f(W)}{\partial f(\Gamma)} = (Z_f(W))^{-1} \sum_{\mathscr{C}\in D_W : \Gamma\in\mathscr{C}} \prod_{\bar\Gamma\in\mathscr{C}\setminus\Gamma} f(\bar\Gamma) = (Z_f(W))^{-1} Z_f(\mathscr{C}_\Gamma), \tag{3.6.14}$$

where $\mathscr{C}_\Gamma \subset D_W$ is the set of all contours $\bar\Gamma \in G_W$ that are disjoint from the contour Γ. Now we apply the expansions (3.6.9) and (3.6.4) to the numerator and the denominator in (3.6.14). We see that

$$\frac{\partial \ln Z_f(W)}{\partial f(\Gamma)} = \exp\Big\{ - \sum_{\substack{\mathscr{C}'\in D_W : (\bigcup_{\bar\Gamma\in\mathscr{C}'} \bar\Gamma)\cap\Gamma\neq\varnothing, \\ \mathscr{C}' \text{ is connected}}} \Phi_f(\mathscr{C}') \Big\}. \tag{3.6.15}$$

Since for any nonemply system of contours $\mathscr{C}'$ we have

$$\sum_{\Gamma\in\mathscr{C}'} |\Gamma| \ge 2\nu, \tag{3.6.16}$$

the estimate (3.6.7) follows from (3.6.15) and (3.6.10).

§3.7. The main estimate

The aim of the subsequent sections 3.8–3.11 is to prove the following estimate.

PROPOSITION 3.7. *Under the conditions of Proposition* 3.3, *there exists a constant* $\beta_0 = \beta_0(\nu, \gamma)$ *such that for all contours* $\Gamma \in \mathscr{L}$ *the denominator in the definition* (3.5.4) *of the contour functional does not vanish and the functional satisfies the estimate*

$$|w(\Gamma)| \le \exp\{-2(\beta - \beta_0)\gamma|\Gamma|\}, \quad \Gamma \in \mathscr{L}. \tag{3.7.1}$$

The estimates (3.3.4) *and* (3.3.5) *also hold.*

Here we show that the main Proposition 3.3 is an easy consequence of Propositions 3.6 and 3.7.

DERIVATION OF PROPOSITION 3.3 FROM PROPOSITIONS 3.6 AND 3.7. The number of different contours of a fixed length $|\Gamma| = n$ containing a fixed face $Q \in \widehat{\mathbb{Z}^\nu}$ does not exceed C^n, where the constant $C = C(\nu)$ depends on the dimension ν only. So the main condition (3.6.1) for the functional $w(\Gamma)$ follows from (3.7.1) if $\beta \ge \beta_0$ and

$$\alpha = 2\gamma\beta - \beta_0', \tag{3.7.2}$$

where β_0 and $\beta_0' > 0$ are sufficiently large. The expansion (3.6.4) applied to the case $f = w$ and the relations (3.5.2) (they follow from (3.6.2)) and (3.5.3) imply the expansion (3.3.7) with

$$\Phi(\Lambda) = \begin{cases} \Phi_w(\Lambda) & \text{if } |\Lambda| > 1, \\ \Phi_w(\Lambda) + \beta h_t & \text{if } \Lambda = \{t\}, \quad t \in \mathbb{Z}^\nu. \end{cases} \tag{3.7.3}$$

and so the estimate (3.6.5) will imply the main estimate (3.3.11) of Proposition 3.3 if β_0 is sufficiently large. Properties (3.3.6), (3.3.8), and (3.3.10) follow from the definition (3.6.3) and the relation (3.5.3).

We will use the following notations (cf. (3.3.1)):

$$\begin{aligned} W^+ &= W^+(h) = V^+ \cap W = \{t \in W : |\operatorname{Im}(h_t)| \le 2(1-\gamma)\operatorname{Re}(h_t)\}, \\ W^0 &= W^0(h) = V^0 \cap W = W \setminus W^+(h), \quad W \subset V. \end{aligned} \tag{3.7.4}$$

PLAN OF THE PROOF OF PROPOSITION 3.7. The estimate (3.7.1) will be proved by induction on the number $|\operatorname{Int}\Gamma|$ of points in the set $\operatorname{Int}\Gamma$. It will be convenient to prove together with (3.7.1) the following three estimates (under the same conditions).

First, the polymer partition function $Z_w(W)$ defined by relation (3.5.1) with the contour functional (3.5.4) does not vanish, and its derivatives with respect to the parameters h_t satisfy for all h_t the estimates

$$\left|\frac{\partial \ln Z_w(W)}{\partial h_t}\right| \le \gamma/2, \qquad W \subset \operatorname{Int}\Gamma, \quad t \in W, \quad \Gamma \in \mathscr{L}. \tag{3.7.5}$$

Second, the partition functions $Z(W, \pm, \mathscr{L})$ defined in §3.2 do not vanish, and for some $\beta_0 = \beta_0(\nu, \gamma)$ and all $\beta \ge \beta_0$ we have

$$\left|\frac{Z(W, -, \mathscr{L})}{Z(W, +, \mathscr{L})}\right| \le 2\exp\left\{(2\beta + \gamma/2)\sum_{t \in W^0} |h_t|\right\}, \tag{3.7.6}$$

$$W \subset \operatorname{Int}\Gamma, \quad \Gamma \in \mathscr{L}.$$

Third, for some new $\beta_0 = \beta_0(\nu, \gamma)$ and all $\beta \geq \beta_0$ we have

$$\left|\frac{\partial}{\partial h_t}\left(\frac{Z(W, -, \mathscr{L})}{Z(W, +, \mathscr{L})}\right)\right| \leq 8\beta \exp\left\{\left(2\beta + \frac{\gamma}{2}\right) \sum_{t \in W^0} |h_t|\right\}, \tag{3.7.7}$$

$$W \subset \operatorname{Int}\Gamma, \quad t \in W, \quad \Gamma \in \mathscr{L}.$$

The estimates (3.7.1), (3.7.5), (3.7.6), (3.7.7) follow easily from the main conditions (3.3.1), (3.3.2) in the case when Γ is a minimal contour with $|\operatorname{Int}\Gamma| = 1$ and $W = \operatorname{Int}\Gamma$. The induction proceeds in the following way: first we assume that for some integer N the estimate (3.7.1) is valid if $\Gamma \in \mathscr{L}$, $|\operatorname{Int}\Gamma| < N$, and that the estimate (3.7.5) is valid if $|W| < N$, $W \subset \operatorname{Int}\Gamma$, and prove (see §3.14) that (3.7.1) is also valid if $|\operatorname{Int}\Gamma| = N$.

Then we assume that (3.7.1) is true if $|\operatorname{Int}\Gamma| = N$ and (3.7.7) is true if $|W| < N$, and prove (3.7.5) for $|W| = N$ (see §3.15). After that we assume that (3.7.1) holds if $|\operatorname{Int}\Gamma| = N$ and (3.7.5) holds if $|W| \leq N$, and prove that (3.7.6) is true if $|W| = N$ (see §3.16). At the last step we assume that (3.7.1) is true if $|\operatorname{Int}\Gamma| = N$, (3.7.5) and (3.7.6) are true if $|W| \leq N$, and prove that (3.7.7) is true for $|W| = N$ (see §3.17). This will give the proof of all the estimates (3.7.1), (3.7.5), (3.7.6), (3.7.7) for all Γ and W. In all cases when we state that some of these estimates are true we mean that they are true for all values of h_t specified in Proposition 3.3. The estimate (3.3.4) follows from the estimate (3.3.6) for $f = w$. The estimate (3.3.5) follows immediately from the estimate (3.7.6).

In §§3.10–3.14 we prove estimates for the ratio of minus and plus partition functions which are essentially used in the proof of the induction statements.

§3.8. The main decomposition of the minus partition function

In treating the Ising ensembles it is necessary to apply different approaches when the h_t are close to zero and far from zero. Since the parameters h_t vary from point to point, we must find some compromise. So we break all contours into two classes: the small ones, for which the ratio of the sum of h_t over its interior to the size of this interior remains bounded, and the large ones, for which this is not the case.

In the inductive proof of the estimates (3.7.1), (3.7.5), (3.7.6), (3.7.7) we must treat small and large contours of the configuration differently.

We fix a set of parameters $h = \{h_t, t \in \mathbb{Z}^\nu\}$ for which the conditions (3.3.2) of Proposition 3.3 are valid. We say that a contour $\Gamma \in \mathscr{L}$ is *small* if (in the notation (3.3.1))

$$\sum_{t \in \operatorname{Int}\Gamma \cap V^+} \operatorname{Re}(h_t) \leq 1, \tag{3.8.1}$$

and that it is *large* if

$$\sum_{t \in \operatorname{Int}\Gamma \cap V^+} \operatorname{Re}(h_t) > 1. \tag{3.8.2}$$

We denote the sets of all small and of all large contours in the volume $W \subset \mathbb{Z}^\nu$ by G_W^s and by G_W^l respectively, and set also $\mathscr{L}^s(W) = \mathscr{L} \cap G_W^s$, $\mathscr{L}^l(W) = \mathscr{L} \cap G_W^l$.

Let $\overline{D}_W^{\mathscr{L}}$ be the set of all families of external contours of configurations $\sigma \in \Omega_W^{\mathscr{L}}$ with the $+$-boundary conditions (see §3.2). It is evident that this set coincides with the similir set defined for the $-$-boundary conditions. Let also $\overline{D}_W^{\mathscr{L}^l} \subset \overline{D}_W^{\mathscr{L}}$ be a subset containing only large contours. Similiarly to (3.5.5), we have

$$Z(W, -, \mathscr{L}) = \sum_{\partial \in \overline{D}_W^{\mathscr{L}}} \exp\Big\{ -\beta \sum_{t \in W \setminus \bigcup_{\Gamma \in \partial} \operatorname{Int}\Gamma} h_t \Big\} \times \prod_{\Gamma \in \partial} \Big(\exp\Big\{ -2\beta|\Gamma| + \beta \sum_{t \in \Delta^I(\Gamma)} h_t \Big\} Z(\operatorname{Int}^0\Gamma, +, \mathscr{L}) \Big). \tag{3.8.3}$$

Summing in (3.8.3) over all systems $\partial \in \overline{D}_W^{\mathscr{L}}$ such that the subsystem $\partial^l \subset \partial$ consisting of all large contours of the system ∂ is fixed, we arrive at the relation

$$Z(W, -, \mathscr{L}) = \sum_{\partial^l \in \overline{D}_W^{\mathscr{L}^l}} Z(W(\partial^l), -, \mathscr{L}^s) \exp\Big\{ -\beta \sum_{t \in \Delta^E(\partial^l)} h_t \Big\} \times \prod_{\Gamma \in \partial^l} \Big(\exp\Big\{ -2\beta|\Gamma| + \beta \sum_{t \in \Delta^I(\Gamma)} h_t \Big\} Z(\operatorname{Int}^0\Gamma, +, \mathscr{L}) \Big). \tag{3.8.4}$$

Here (see (3.1.2))

$$\Delta^E(\partial^l) = \bigcup_{\Gamma \in \partial^l} \Delta^E(\Gamma), \tag{3.8.5}$$

and

$$W(\partial^l) = W \setminus (\operatorname{Int}\partial^l \cup \Delta^E(\partial^l)), \tag{3.8.6}$$

where for any $\partial \in \overline{D}_W^{\mathscr{L}}$ we set

$$\operatorname{Int}\partial = \bigcup_{\Gamma \in \partial} \operatorname{Int}\Gamma. \tag{3.8.7}$$

The volume $W(\partial^l)$ is the set of points on which the configuration can take arbitrary values with the only restriction that there are no large outer contours. The partition function $Z(W, -, \mathscr{L}^s)$, called the *small contour partition function,* is, by definition of $\mathscr{L}^s$, calculated on the set of all configurations $\sigma \in \Omega_W^{\mathscr{L}}$ such that all contours $\Gamma \in \mathscr{G}^-(\sigma)$ are small. (The important point here is that our definition (3.8.1) implies that if a contour Γ is small, then any contour $\Gamma' \in G_{\operatorname{Int}\Gamma}$ is also small.) In §3.9 we obtain an upper bound for the small contour partition function. The analysis of large contour partition functions will be performed in §§3.10, 3.11.

§3.9. Estimates on the small contour partition functions

In this section we prove the following result.

LEMMA 3.9. *Let us fix a set* $W \subset V$. *Suppose that for any contour* $\Gamma \in \mathscr{L}(W)$ *the estimate* (3.7.1) *is valid and that the estimate* (3.7.5) *holds for the set* W. *Then for all* $\beta \geq \beta_0$, *where* $\beta_0 = \beta_0(\nu, \gamma)$ *is sufficiently large, the partition function* $Z(W, +, \mathscr{L})$ *does not vanish and* (*see notation* (3.7.4))

$$\left|\frac{Z(W, -, \mathscr{L}^s)}{Z(W, +, \mathscr{L})}\right| \leq 2\exp\left\{(-\gamma\beta + \gamma(1-\gamma)) \sum_{t\in W^+} \operatorname{Re}(h_t) + \left(2\beta + \frac{\gamma}{2}\right) \sum_{t\in W^0} |h_t|\right\} \tag{3.9.1}$$

PROOF. Consider new real-valued Hamiltonians

$$\overline{H}_h^{\pm}(\sigma) = \sum_{\Gamma\in\mathscr{G}^{\pm}(\sigma)} |\Gamma| - \sum_{t\in W^+} \operatorname{Re}(h_t)\sigma_t. \tag{3.9.2}$$

and the corresponding partition functions

$$\overline{Z}(W, \pm, \mathscr{L}) = \sum_{\sigma\in\Omega_W^{\mathscr{L}}} \exp\{-\beta\overline{H}_h^{\pm}(\sigma)\}, \tag{3.9.3}$$

$$\overline{Z}(W, \pm, \mathscr{L}^s) = \sum_{\sigma\in\Omega_W^{\mathscr{L}^s}} \exp\{-\beta\overline{H}_h^{\pm}(\sigma)\}. \tag{3.9.4}$$

It is essential that the Hamiltonians $\overline{H}_h^{\pm}$ take real values and depend only on those h_t that satisfy the condition (3.3.1). Evidently,

$$\frac{Z(W, -, \mathscr{L}^s)}{Z(W, +, \mathscr{L})} = \frac{Z(W, -, \mathscr{L}^s)}{\overline{Z}(W, -, \mathscr{L}^s)} \frac{\overline{Z}(W, -, \mathscr{L}^s)}{\overline{Z}(W, +, \mathscr{L})} \frac{\overline{Z}(W, +, \mathscr{L})}{Z(W, +, \mathscr{L})}, \tag{3.9.5}$$

and we estimate separately the three factors in (3.9.5).

The estimation of the first factor is simple. Comparing (3.9.2) and (3.2.4) we see that

$$\operatorname{Re}(H_h^-(\sigma)) \leq \overline{H}_h^-(\sigma) + \sum_{t\in W^0} |h_t|, \qquad \sigma \in \Omega_W^{\mathscr{L}}. \tag{3.9.6}$$

Therefore,

$$\left|\frac{Z(W, -, \mathscr{L}^s)}{\overline{Z}(W, -, \mathscr{L}^s)}\right| \leq \exp\left\{\beta \sum_{t\in W^0} |h_t|\right\}. \tag{3.9.7}$$

The estimation of the third factor in (3.9.5) needs a special approach because here we have a complex-valued partition function in the denominator. We introduce a family $h^\lambda = \{h_t^\lambda, t \in W\}$, $0 \leq \lambda \leq 1$, of magnetic fields such that

$$h_t^\lambda = \begin{cases} \operatorname{Re}(h_t) + i\lambda \operatorname{Im}(h_t), & \text{if } h_t \in W^+, \\ \lambda h_t, & \text{if } h_t \in W^0. \end{cases} \tag{3.9.8}$$

Then (using the notation (3.2.5)) we have

$$\ln \frac{\overline{Z}(W, +, \mathscr{L})}{Z(W, +, \mathscr{L})} = \int_0^1 \frac{\partial}{\partial\lambda} \ln Z(W, +, \beta, h^\lambda, \mathscr{L})\, d\lambda. \tag{3.9.9}$$

Note that the main condition (3.3.2) remains valid for all fields h^λ. So we can use the inductive estimate (3.7.5) with h replaced by h^λ. Together with the relation (3.5.3) it shows that

$$\int_0^1 \left|\frac{\partial}{\partial\lambda} \ln Z(W, +, \beta, h^\lambda, \mathscr{L})\right| d\lambda \le (\beta + \gamma/2) \left(\sum_{t\in W^+} |\operatorname{Im}(h_t)| + \sum_{t\in W^0} |h_t|\right). \tag{3.9.10}$$

Using the main condition (3.3.1), we see from (3.9.9) and (3.9.10) that for a sufficiently large $\beta_0 = \beta_0(\nu, \gamma)$ and all $\beta \ge \beta_0(\nu, \gamma)$ the partition function $Z(W, +, \mathscr{L})$ does not vanish and

$$\left|\frac{\overline{Z}(W, +, \mathscr{L})}{Z(W, +, \mathscr{L})}\right| \le \exp\left\{(\beta + \gamma/2)\left(2(1-\gamma) \sum_{t\in W^+} \operatorname{Re}(h_t) + \sum_{t\in W^0} |h_t|\right)\right\}. \tag{3.9.11}$$

Since the Hamiltonians in the second factor in (3.9.5) are real-valued, we can try to estimate them using probabilistic arguments. For any $W \subset V$ define the subset of configurations $Q^\pm$ by the formula

$$Q^\pm = \left\{\sigma \in \Omega_W^{\mathscr{L}^s} : \sum_{t\in W^+ : \sigma_t = \pm 1} \operatorname{Re}(h_t) \ge \left(1 - \frac{\gamma}{4}\right) \sum_{t\in W^+} \operatorname{Re}(h_t)\right\}, \tag{3.9.12}$$

and introduce new partition functions

$$\overline{Z}(W, \pm, Q^\pm) = \sum_{\sigma\in Q^\pm} \exp\{-\beta \overline{H}_h^\pm(\sigma)\}. \tag{3.9.13}$$

Note that for any $\sigma \in \Omega_W$ we have $\mathscr{G}^+(\sigma) = \mathscr{G}^-(-\sigma)$ and so for $\sigma \in Q^+$ the difference $\overline{H}_h^+(\sigma) - \overline{H}_h^-(-\sigma)$ can be estimated as follows:

$$\overline{H}_h^+(\sigma) - \overline{H}_h^-(-\sigma) \le -2\left(1 - \frac{\gamma}{4}\right) \sum_{t\in W^+} \operatorname{Re}(h_t) \le -2\left(1 - \frac{\gamma}{2}\right) \sum_{t\in W^+} \operatorname{Re}(h_t). \tag{3.9.14}$$

Since $Q^- = \{\sigma : -\sigma \in Q^+\}$, we have

$$\frac{\overline{Z}(W, -, Q^-)}{\overline{Z}(W, +, Q^+)} \le \exp\left\{-2\beta\left(1 - \frac{\gamma}{2}\right) \sum_{t\in W^+} \operatorname{Re}(h_t)\right\}, \tag{3.9.15}$$

and we shall try to reduce the estimate of the second factor in (3.9.5) to the estimate (3.9.15). First we observe that since the Hamiltonian $\overline{H}_h^+(\sigma)$ takes real values,

$$\overline{Z}(W, +, \mathscr{L}) \ge \overline{Z}(W, +, Q^+). \tag{3.9.16}$$

The opposite estimates for the numerator are not so evident. We claim that for $\beta \geq \beta_0$, where $\beta_0 = \beta_0(\gamma)$ is sufficiently large, we have

$$\overline{Z}(W, -, \mathscr{L}^s) \leq 2\overline{Z}(W, -, Q^-). \tag{3.9.17}$$

To show this we introduce the Gibbs distribution corresponding to the partition function $\overline{Z}(W, -, \mathscr{L}^s)$, and for any function $f(\sigma)$, $\sigma \in \Omega_W^{\mathscr{L}^s}$, define

$$\langle f \rangle_W^{-, \beta, \mathscr{L}^s} = (\overline{Z}(W, -, \mathscr{L}^s))^{-1} \sum_{\sigma \in \Omega_W^{\mathscr{L}^s}} f(\sigma) \exp\{-\beta \overline{H}_h^-(\sigma)\}. \tag{3.9.18}$$

Then the desired inequality (3.9.17) can be rewritten as

$$\langle \chi_{Q^-} \rangle_W^{-, \beta, \mathscr{L}^s} \geq 1/2, \tag{3.9.19}$$

where χ_{Q^-} is the indicator of the event Q^-. The definition (3.9.12) can be rewritten in the form

$$Q^- = \left\{ \sigma \in \Omega_W^{\mathscr{L}^s} : \sum_{t \in W^+} (1 + \sigma_t) \operatorname{Re}(h_t) \leq (\gamma/2) \sum_{t \in W^+} \operatorname{Re}(h_t) \right\}. \tag{3.9.20}$$

So we consider the mean value

$$m^- = \left\langle \sum_{t \in W^+} (1 + \sigma_t) \operatorname{Re}(h_t) \right\rangle_W^{-, \beta, \mathscr{L}^s} \sum_{t \in W^+} \operatorname{Re}(h_t) \langle 1 + \sigma_t \rangle_W^{-, \beta, \mathscr{L}^s}, \tag{3.9.21}$$

and use the Chebyshev inequality to observe that

$$\langle \chi_{Q^-} \rangle_W^{-, \beta, \mathscr{L}^s} \geq 1 - \frac{2}{\gamma} \frac{m^-}{\sum_{t \in W^+} \operatorname{Re}(h_t)}, \tag{3.9.22}$$

since $\sum_{t \in W^+} (1 + \sigma_t) \operatorname{Re}(h_t) \geq 0$. It is clear that $\sigma_t = 1$ only if there exists an external contour Γ of the configuration σ such that $t \in \operatorname{Int} \Gamma$. In the ensemble under consideration this contour has to be small. So

$$m^- \leq 2 \sum_{t \in W^+} \sum_{\substack{\Gamma \in \mathscr{L}^s(W), t \in \operatorname{Int} \Gamma, \\ \Gamma \text{ is external}}} \operatorname{Re}(h_t) \langle \chi_\Gamma \rangle_W^{-, \beta, \mathscr{L}^s}, \tag{3.9.23}$$

where χ_Γ is the indicator of the set of configurations $\sigma \in \Omega_W^{\mathscr{L}}$ such that $\Gamma \in \mathscr{G}^-(\sigma)$. To evaluate the mean value of χ_Γ we use the Peierls transformation: to any configuration $\sigma \in \mathscr{L}$ such that $\Gamma \in \mathscr{G}^-(\sigma)$ we assign a configuration $\sigma' \in \mathscr{L}$ having the contour system $\mathscr{G}^-(\sigma') = \mathscr{G}^-(\sigma) \setminus \Gamma$. It follows from the definition (3.9.2) of the Hamiltonian and the definition (3.8.1) of a small contour that since $|\Gamma| \geq 2\nu$, we have

$$\overline{H}_h^-(\sigma) - \overline{H}_h^-(-\sigma') \geq 2|\Gamma| - 2 \sum_{t \in W^+ \cap \operatorname{Int} \Gamma} \operatorname{Re}(h_t) \geq |\Gamma|. \tag{3.9.24}$$

So

$$\langle \chi_\Gamma \rangle_W^{-,\beta,\mathscr{L}^s} = \Big(\sum_{\sigma \in \Omega_W^{\mathscr{L}^s} : \Gamma \in \mathscr{G}(\sigma)} \exp\{-\beta \overline{H}_h^-(\sigma)\} \Big) \Big(\sum_{\sigma \in \Omega_W^{\mathscr{L}^s}} \exp\{-\beta \overline{H}_h^-(\sigma)\} \Big)^{-1}$$
$$\leq \exp\{-\beta|\Gamma|\}. \qquad (3.9.25)$$

The number of different contours Γ with $t \in \operatorname{Int}\Gamma$ and $|\Gamma| = N$ does not exceed C^N, where the constant C depends on the dimension ν only. So if $\beta \geq \beta_0$, where $\beta_0 = \beta_0(\nu, \gamma)$ depends only on the constant γ in the condition of Proposition 3.3 and the dimension ν, it follows from (3.9.23) and (3.9.25) that

$$m^- \leq (\gamma/4) \sum_{t \in W^+} \operatorname{Re}(h_t), \qquad (3.9.26)$$

and the desired estimates (3.9.19) and (3.9.17) follow from (3.9.22). Now we can combine the estimates (3.9.16), (3.9.17), and (3.9.15) to see that

$$\frac{\overline{Z}(W, -, \mathscr{L}^s)}{\overline{Z}(W, +, \mathscr{L})} \leq 2\exp\Big\{ -2\beta\Big(1 - \frac{\gamma}{2}\Big) \sum_{t \in W^+} \operatorname{Re}(h_t)\Big\}, \qquad (3.9.27)$$

and this is the necessary estimate of the second factor in (3.9.5). Now we see that the desired estimate (3.9.1) follows from the equality (3.9.5) and the estimates (3.9.7), (3.9.11), and (3.9.27).

§3.10. Reduction to the estimate of the contour partition functions

In this and the following sections we prove several lemmas that enable us to find an estimate for the polymer weights $w(\Gamma)$.

LEMMA 3.10. *Under the same conditions as in Lemma* 3.9 *for* $\beta \geq \beta_0$, *where* $\beta_0 = \beta_0(\nu, \gamma)$ *is sufficiently large, we have*

$$\left|\frac{Z(W, -, \mathscr{L})}{Z(W, +, \mathscr{L})}\right| \leq \sum_{\partial^l \in \overline{D}_W^{\mathscr{L}^l}} \Big(2\exp\Big\{ (-\gamma\beta + \gamma(1-\gamma)) \sum_{t \in (W \setminus \operatorname{Int}\partial)^+} \operatorname{Re}(h_t)$$
$$+ \Big(2\beta + \frac{\gamma}{2}\Big) \sum_{t \in (W\setminus \operatorname{Int}\partial)^0} |h_t| \Big\} \prod_{\Gamma \in \partial} q_\Gamma \Big); \qquad (3.10.1)$$

here the real numbers $q_\Gamma = q_\Gamma(\partial)$ *satisfy*

$$q_\Gamma \leq \exp\{-(2\beta - \gamma/4)|\Gamma|\}. \qquad (3.10.2)$$

PROOF. Using (3.8.4) we see that

$$\frac{Z(W, -, \mathscr{L})}{Z(W, +, \mathscr{L})} = \sum_{\partial^l \in \overline{D}_W^{\mathscr{L}^l}} \frac{Z(W(\partial^l), -, \mathscr{L}^s)}{Z(W(\partial^l), +, \mathscr{L})}$$
$$\times \exp\Big\{ -2\beta \sum_{t \in \Delta^E(\partial^l)} h_t \Big\} R(\partial^l) \prod_{\Gamma \in \partial^l} \exp\{-2\beta|\Gamma|\}, \qquad (3.10.3)$$

where

$$R(\partial^l) = \frac{Z(W(\partial^l), +, \mathscr{L})}{Z(W, +, \mathscr{L})} \exp\Big\{\beta \sum_{t\in\Delta^E(\partial^l)} h_t\Big\} \times \prod_{\Gamma\in\partial^l} \Big[Z(\mathrm{Int}^0\Gamma, +, \mathscr{L}) \exp\Big\{\beta \sum_{t\in\Delta^I(\Gamma)} h_t\Big\}\Big]. \tag{3.10.4}$$

To estimate the first factor in (3.10.3) we can use Lemma 3.9 with W changed to $W(\partial^l)$. We see that

$$\begin{aligned}\Big|\frac{Z(W(\partial^l), -, \mathscr{L}^s)}{Z(W(\partial^l), +, \mathscr{L})} &\exp\Big\{-2\beta \sum_{t\in\Delta^E(\partial^l)} h_t\Big\}\Big| \\ &\le 2\exp\Big\{(-\gamma\beta + \gamma(1-\gamma)) \sum_{t\in W(\partial^l)^+} \mathrm{Re}(h_t) + (2\beta+\gamma/2) \\ &\qquad\times \sum_{t\in W(\partial^l)^0} |h_t| - 2\beta \sum_{t\in\Delta^E(\partial^l)} h_t\Big\} \\ &\le 2\exp\Big\{(-\gamma\beta + \gamma(1-\gamma)) \sum_{t\in(W\setminus\mathrm{Int}\,\partial^l)^+} \mathrm{Re}(h_t) + (2\beta+\gamma/2) \\ &\qquad\times \sum_{t\in(W\setminus\mathrm{Int}\,\partial^l)^0} |h_t|\Big\}.\end{aligned} \tag{3.10.5}$$

To obtain the desired estimate (3.10.1) we must estimate (3.10.4). Applying the representation (3.5.3) to each factor in (3.10.4) we see that

$$R(\partial^l) = \frac{Z_w(W(\partial^l))}{Z_w(W)} \prod_{\Gamma\in\partial^l} Z_w(\mathrm{Int}^0\Gamma), \tag{3.10.6}$$

where the polymer partition functions $Z_w(*, \mathscr{L})$ are defined by the polymer weights (3.5.4). The partition functions used in (3.10.6) include only weights $w(\Gamma')$ such that $\mathrm{Int}\,\Gamma' \subset W$. So the induction hypothesis implies that the estimate (3.7.1) is valid for these partition fucntions. So we can check that the condition (3.6.1) is true for $f = w$ and $V = W$ with the constant α defined by the relation (3.7.2). Using now the results of §3.6, we get that for the set W we have

$$Z_w(W) = \exp\Big\{\sum_{\Lambda\subset W, \Lambda\neq\varnothing} \Phi_w(\Lambda)\Big\}, \tag{3.10.7}$$

and the same holds for the sets $W^0(\partial^l)$ and $\mathrm{Int}^0\Gamma$, where the functions $\Phi_w(\Lambda)$ are defined by (3.6.3) for $f = w$, and (what is the only essential property) it follows from (3.6.5) that

$$|\Phi_w(\Lambda)| \le 2\nu \exp\{-\alpha d(\Lambda)\}. \tag{3.10.8}$$

Observe that if a set $\Lambda \subset W$ with $d(\Lambda) \neq \infty$ is not contained in $W(\partial^l)$ or in $\mathrm{Int}^0 \Gamma$ for some $\Gamma \in \partial^l$, then it has an nonempty intersection with some of the sets $\Delta(\Gamma)$, $\Gamma \in \partial^l$. So we can rewrite (3.10.6) as

$$R(\partial^l) = \prod_{\Gamma \in \partial^l} r_\Gamma \tag{3.10.9}$$

with

$$r_\Gamma = r_\Gamma(\partial^l) = \exp\left\{ - \sum_{\Lambda \subset \mathrm{Int}^0 \Gamma \,:\, \Lambda \cap \Delta(\Gamma) \neq \varnothing} (k(\Lambda))^{-1} \Phi_w(\Lambda) \right\}, \tag{3.10.10}$$

where $k(\Lambda)$ is the number of different contours $\Gamma \in \partial^l$ such that $\Lambda \cap \Delta(\Gamma) \neq \varnothing$. It follows from (3.10.8) that

$$r_\Gamma \leq \exp\left\{ 2\nu \sum_{\Lambda \subset \mathbb{Z}^\nu \,:\, \Lambda \cap \Delta(\Gamma) \neq \varnothing} \exp\{-\alpha d(\Lambda)\} \right\}. \tag{3.10.11}$$

It is clear that $|\Delta(\Gamma)| \leq C|\Gamma|$, where the constant C depends only on the dimension ν. So Lemma 3.4 used with $N = 0$ implies that for some constant α' we have

$$r_\Gamma \leq \exp\{2\nu C|\Gamma| \exp\{-(\alpha - \alpha')\}\} \tag{3.10.12}$$

Let

$$q_\Gamma = r_\Gamma \exp\{-2\beta|\Gamma|\}. \tag{3.10.13}$$

Then relations (3.10.3), (3.10.5), (3.10.9), and (3.10.13) imply the statement of the proposition.

§3.11. Estimates for partition functions on graphs

The next sections are devoted to estimates of sums over systems ∂ of external contours, like the sum in (3.10.1). For that, we need a general result about partition functions on directed graphs.

Let F be a finite set on which a structure of a directed graph is defined. This means that for each $f \in F$ a system of subsets $Q(f) \subset F \setminus \{f\}$ is given, called the set of vertices adjacent to the vertex f. A path in the graph starting at a vertex f and ending at a vertex f' is a sequence $\{f_0 = f, f_1, \dots, f_k = f'\}$ such that $f_i \in Q(f_{i-1})$, $i = 1, \dots, k-1$. A set $F' \subset F$ is called a connected subset of F with the initial point f if for any $f' \in F'$ there is a path $\{f_0 = f, f_1, \dots, f_k = f'\}$ in F such that $f_0, f_1, \dots, f_k \in F'$. The set of all connected subsets $F' \subset F$ with the initial point f_0 is denoted by $\mathscr{F}(f_0, F)$.

Suppose that a real-valued function $a(f)$, $f \in F$, is given. The partition function is defined by

$$Z_{f_0}(F) = \sum_{F' \in \mathscr{F}(f_0, F)} \exp\left\{ - \sum_{f \in F'} a(f) \right\}. \tag{3.11.1}$$

LEMMA 3.11. *For some* $\varkappa$, $0 < \varkappa < 1$, *and for all* $f \in F$ *let*

$$\sum_{g \in Q(f)} \exp\{-(1-\varkappa)a(g)\} \leq \varkappa a(f). \tag{3.11.2}$$

Then for any $f_0 \in F$,

$$Z_{f_0}(F) \leq \exp\{-(1-\varkappa)a(f_0)\}. \tag{3.11.3}$$

PROOF. We prove the lemma by induction on the number of vertices $|F|$. So we fix $f_0 \in F$ and consider the graph $\overline{F} = F \setminus \{f_0\}$. We treat $\overline{F}$ as a directed graph with sets of adjacent points $\overline{Q}(f) = Q(f) \setminus \{f_0\}$, $f \in \overline{F}$. By the induction hypothesis we can claim that for any $f \in \overline{F}$,

$$Z_f(\overline{F}) \leq \exp\{-(1-\varkappa)a(f)\}. \tag{3.11.4}$$

Any connected set $F' \in \mathscr{F}(f_0, F)$ can be represented as

$$F' = \left(\bigcup_{g \in Q(f_0)} F_g\right) \cup \{f_0\}, \tag{3.11.5}$$

where either $F_g \in \mathscr{F}(g, \overline{F})$ is a connected set in the graph $\overline{F}$ or F_g is empty. So using (3.11.5) together with the induction hypothesis (3.11.4), the inequality $1 + b \leq e^b$, $b \geq 0$, and the main condition (3.11.2) we see that

$$\begin{aligned} Z_{f_0}(F) &\leq \prod_{g \in Q(f_0)} (1 + Z_g(\overline{F}))e^{-a(f_0)} \leq \exp\left\{\sum_{g \in Q(f_0)} Z_g(\overline{F}) - a(f_0)\right\} \qquad (3.11.6) \\ &\leq \exp\left\{\sum_{g \in Q(f_0)} \exp\{-(1-\varkappa)a(g)\} - a(f_0)\right\} \leq \exp\{-(1-\varkappa)a(f_0)\}, \end{aligned}$$

and this is the desired estimate (3.11.2).

§3.12. Clans of large contours

In the following sections, where we prove the main estimate for polymer contour weight $w(\Gamma)$ and the ratio $Z(W, -, \mathscr{L})/Z(W, +, \mathscr{L})$, we shall use some constructions related to a special clustering of large contours.

Let $\overline{\Gamma} \in G_W^l$, $W \subset V$, be a large coutour (see (3.8.2)). The set $\varkappa(\overline{\Gamma}) \subset G_W^l$ is called the *clan of large contours with the ancestor* $\overline{\Gamma}$ if it consists of all contours $\tilde{\Gamma} \in G_W^l$ such that

$$\operatorname{Int} \tilde{\Gamma} \cap \operatorname{Int} \overline{\Gamma} \neq \varnothing \tag{3.12.1}$$

and

$$|\tilde{\Gamma}| \geq |\overline{\Gamma}|. \tag{3.12.2}$$

We say that large contours $\overline{\Gamma}_1, \dots, \overline{\Gamma}_k \in G_W^l$ generate a *basis system of clans in* W *with the ancestors* $\overline{\Gamma}_1, \dots, \overline{\Gamma}_k$ if the following two statements hold:

$$\operatorname{Int} \overline{\Gamma}_i \cap \operatorname{Int} \overline{\Gamma}_j = \varnothing, \qquad i \neq j, \quad i, j = 1, \dots, k \tag{3.12.3}$$

and

$$\bigcup_{i=1}^{k} \varkappa(\overline{\Gamma}_i) = G_W^l. \tag{3.12.4}$$

(We emphasise that in general different clans $\varkappa(\overline{\Gamma}_i)$ can have nonempty intersection.)

LEMMA 3.12.1. *Basis systems of clans in W exist.*

PROOF. We choose ancestors $\overline{\Gamma}_1, \overline{\Gamma}_2, \dots$ inductively in such a way that the condition (3.12.3) holds. For $\overline{\Gamma}_1$ we take any large contour $\overline{\Gamma} \in G_W^l$ with the minimal length $|\overline{\Gamma}|$ among all contours in G_W^l. If ancestors $\overline{\Gamma}_1, \dots, \overline{\Gamma}_j$ are already chosen, we take for $\overline{\Gamma}_{j+1}$ any contour $\overline{\Gamma} \in G_W^l$ that satisfies the condition

$$\operatorname{Int}\overline{\Gamma} \cap \operatorname{Int}\overline{\Gamma}_i = \varnothing, \qquad i = 1, \dots, j, \tag{3.12.5}$$

and has the minimal length $|\overline{\Gamma}|$ among large contours satisfying (3.12.5). The process terminates when for some $j = k$ there are no contours $\overline{\Gamma}$ satisfying (3.12.5). The property (3.12.3) is evident. To prove the property (3.12.4), we use the induction on j to verify the following statement: any large contour $\overline{\Gamma}$ such that

$$\overline{\Gamma} \in G_W^l \setminus \bigcup_{i=1}^{j} \varkappa(\overline{\Gamma}_i) \tag{3.12.6}$$

satisfies the condition (3.12.5). Indeed, let a contour $\overline{\Gamma}$ satisfy the condition (3.12.6). By the induction hypothesis we know that

$$\operatorname{Int}\overline{\Gamma} \cap \operatorname{Int}\overline{\Gamma}_i = \varnothing, \qquad i = 1, \dots, j-1. \tag{3.12.7}$$

So this contour $\overline{\Gamma}$ was eligible to participate in the choice of the ancestor $\overline{\Gamma}_j$, hence $|\overline{\Gamma}| \ge |\overline{\Gamma}_j|$. So either $\operatorname{Int}\overline{\Gamma} \cap \operatorname{Int}\overline{\Gamma}_j \ne \varnothing$ and then $\overline{\Gamma} \in \varkappa(\overline{\Gamma}_j)$, or $\operatorname{Int}\overline{\Gamma} \cap \operatorname{Int}\overline{\Gamma}_j = \varnothing$ and then the condition (3.12.5) is valid. Applying (3.12.5) for $j = k$ we obtain the relation (3.12.4).

Below we assume that a basis system $\{\varkappa(\overline{\Gamma}_1), \dots, \varkappa(\overline{\Gamma}_k)\}$ of clans in W is fixed. We assume also that these clans are enumerated in such a way that

$$|\overline{\Gamma}_1| \le |\overline{\Gamma}_2| \le \cdots \le |\overline{\Gamma}_k|. \tag{3.12.8}$$

We say that a large contour $\tilde{\Gamma}_1 \in G_W^l$ is a *parent* of a large contour $\tilde{\Gamma}_2 \in G_W^l$ if for some $j = 1, \dots, k$

$$\operatorname{Int}\tilde{\Gamma}_1 \cap \operatorname{Int}\overline{\Gamma}_j \ne \varnothing, \qquad (W \setminus \operatorname{Int}\tilde{\Gamma}_1) \cap \operatorname{Int}\overline{\Gamma}_j \ne \varnothing, \tag{3.12.9}$$

and $\tilde{\Gamma}_2 \in \varkappa(\overline{\Gamma}_j)$. We regard the set of all large contours G_W^l as a directed graph (see §3.11) with the set of vertices $Q(\tilde{\Gamma}_1)$ adjacent to $\tilde{\Gamma}_1$ to be the set of all contours $\tilde{\Gamma}_2 \in G_W^l$ for which $\tilde{\Gamma}_1$ is a parent. Any set of large contours $\mathcal{F} \subset G_W^l$ which is a connected subset of this graph with an initial point $\tilde{\Gamma}$ is called a family of large contours generated by the ancestor $\tilde{\Gamma}$.

LEMMA 3.12.2. *Let* $\overline{\mathscr{F}}_j$ *be the set of all families of large contours with ancestors* $\tilde{\Gamma} \in \varkappa(\overline{\Gamma}_j)$. *Then for any* $\varkappa$, $0 < \varkappa < 1$, *there exists a constant* $\alpha_0 = \alpha_0(\varkappa, \nu)$, *depending on* $\varkappa$ *and the dimension* ν *only, such that*

$$\sum_{F=\{\tilde{\Gamma}_1, \dots, \tilde{\Gamma}_s\} \in \overline{\mathscr{F}}_j} \exp\{-\alpha(|\tilde{\Gamma}_1| + \cdots + |\tilde{\Gamma}_s|)\} \leq \exp\{-(\alpha(1-\varkappa) - \alpha_0)|\overline{\Gamma}_j|\} \tag{3.12.10}$$

for $\alpha \geq \alpha_0$ *and* $j = 1, \dots, k$.

PROOF. To prove this lemma we apply the general estimate of Lemma 3.11 to the graph described above. Let $F = G_W^l$ and for some $\alpha > 0$ let

$$a(\tilde{\Gamma}) = \alpha|\tilde{\Gamma}|, \qquad \tilde{\Gamma} \in G_W^l. \tag{3.12.11}$$

The main point is to verify the condition (3.11.2), and to do this we use some auxiliary estimates.

First of all, we fix a point $t \in W$ and prove that for some $\alpha_0' = \alpha_0'(\nu)$, all $\alpha \geq \alpha_0'$, and all integers N,

$$\sum_{\tilde{\Gamma} \in G_W^l \,:\, |\tilde{\Gamma}| \geq N,\, t \in \operatorname{Int} \tilde{\Gamma}} e^{-\alpha|\tilde{\Gamma}|} \leq e^{-N(\alpha - \alpha_0')}. \tag{3.12.12}$$

Indeed, it is clear that for some constant $C = C(\nu)$ the number of different contours $\tilde{\Gamma}$ with $|\tilde{\Gamma}| = m$ that contain a fixed face does not exceed C^m. If a contour $\tilde{\Gamma}$, $|\tilde{\Gamma}| = m$, is such that $t \in \operatorname{Int} \tilde{\Gamma}$, it must contain a face distanced from t by at most m, and the number of such faces does not exceed $C_1 m^\nu$, where $C_1 = C_1(\nu)$ depends on the dimension. Therefore,

$$\sum_{\tilde{\Gamma} \in G_W^l \,:\, |\tilde{\Gamma}| \geq N,\, t \in \operatorname{Int} \tilde{\Gamma}} e^{-\alpha|\tilde{\Gamma}|} \leq \sum_{m=N}^{\infty} C_1 m^\nu C^m e^{-\alpha m}, \tag{3.12.13}$$

and this implies (3.12.12) for an appropriate $\alpha_0'(\nu)$. The next step is to prove that for some $\alpha_0'' = \alpha_0''(\nu)$, all $\alpha \geq \alpha_0''$, and all $j = 1, \dots, k$,

$$\sum_{\tilde{\Gamma} \in \varkappa(\overline{\Gamma}_j)} e^{-\alpha|\tilde{\Gamma}|} \leq e^{-(\alpha - \alpha_0'')|\overline{\Gamma}_j|}. \tag{3.12.14}$$

Indeed, it follows from (3.12.12) and conditions (3.12.1), (3.12.2) in the definition of a clan that

$$\begin{aligned} \sum_{\tilde{\Gamma} \in \varkappa(\overline{\Gamma}_j)} e^{-\alpha|\tilde{\Gamma}|} &\leq \sum_{t \in \operatorname{Int} \overline{\Gamma}_j} \left(\sum_{\tilde{\Gamma} \in G_W^l \,:\, |\tilde{\Gamma}| \geq |\overline{\Gamma}_j|,\, t \in \operatorname{Int} \tilde{\Gamma}} e^{-\alpha|\tilde{\Gamma}|} \right) \\ &\leq |\operatorname{Int} \overline{\Gamma}_j| e^{-|\overline{\Gamma}_j|(\alpha - \alpha_0')}, \end{aligned} \tag{3.12.15}$$

Since $|\operatorname{Int} \overline{\Gamma}_j| \leq K|\overline{\Gamma}_j|^{\nu/(\nu-1)}$, where the constant $K = K(\nu)$ depends on the dimension ν only, the estimate (3.12.14) is valid for an appropriate $\alpha_0'' \geq \alpha_0'$.

Now we can proceed directly to the verification of the condition (3.11.2). We state that in our case for any $\varkappa$, $0 < \varkappa < 1$, all $\tilde{\Gamma}_1 \in G_W^l$, and all $\alpha \geq \alpha_0(\varkappa, \nu)$, where $\alpha_0(\varkappa, \nu)$ is a constant depending on $\varkappa$ and ν only, we have

$$\sum_{\tilde{\Gamma}_2 \in Q(\tilde{\Gamma}_1)} \exp\{-\alpha(1-\varkappa)|\tilde{\Gamma}_2|\} \leq \varkappa\alpha|\tilde{\Gamma}_1|. \tag{3.12.16}$$

Observe that by (3.12.1), the number of different indexes $j = 1, \dots, k$ such that the condition (3.12.9) is valid for a fixed contour $\tilde{\Gamma}_1$ does not exceed $K|\tilde{\Gamma}_1|$, where the constant $K = K(\nu)$ depends on the dimension ν only. So from the definition of $Q(\tilde{\Gamma}_1)$, the condition (3.12.2), and the estimate (3.12.1) we derive that

$$\sum_{\tilde{\Gamma}_2 \in Q(\tilde{\Gamma}_1)} \exp\{-\alpha(1-\varkappa)|\tilde{\Gamma}_2|\} \leq K|\tilde{\Gamma}_1|e^{\alpha_0''|\tilde{\Gamma}_1|} \exp\{-\alpha(1-\varkappa)|\Gamma_1|\}. \tag{3.12.17}$$

Since $|\tilde{\Gamma}_1| \geq 2\nu$ (this is true for any contour), we find that the inequality (3.12.7) implies the estimate (3.12.16) if $\alpha \geq \alpha_0$ for a sufficiently large $\alpha_0 = \alpha_0(\varkappa, \nu)$.

Now Lemma 3.11 implies that for any $\tilde{\Gamma} \in \varkappa(\overline{\Gamma}_j)$ and all $\alpha \geq \alpha_0(\varkappa, \nu)$,

$$\sum_{F=\{\tilde{\Gamma}_1, \dots, \tilde{\Gamma}_s\} \in \mathscr{F}(\tilde{\Gamma})} \exp\{-\alpha(|\tilde{\Gamma}_1| + \dots + |\tilde{\Gamma}_s|)\} \leq \exp\{-(1-\varkappa)\alpha|\tilde{\Gamma}|\}, \tag{3.12.18}$$

where $\mathscr{F}(\tilde{\Gamma})$ is the set of all the families of large coutours with a fixed ancestor $\tilde{\Gamma}$. Now we can apply again the estimate (3.12.14) for the sum of the right-hand sides of the estimates (3.12.18) to obtain the desired estimate (3.12.10).

§3.13. An estimate of the ratio of minus- and plus-partition functions

In this section we put together the results of §§3.8–3.12 to obtain a final estimate of a ratio of minus- and plus-partition functions in the same volume. We shall see (in §§3.14–3.17) that all four main induction statements formulated in §3.7 are easy consequences of the following lemma.

LEMMA 3.13. *Under the conditions of Lemma* 3.9 *with some* $\beta_0 = \beta_0(\nu, \gamma)$, *the partition function* $Z(W, +, \mathscr{L})$ *does not vanish for all* $\beta \geq \beta_0$ *and*

$$\left|\frac{Z(W, -, \mathscr{L})}{Z(W, +, \mathscr{L})}\right| \leq 2\exp\left\{\left(2\beta + \frac{\gamma}{2}\right)\sum_{t \in W^0} |h_t|\right\} \tag{3.13.1}$$

PROOF. Comparing the estimate (3.13.1) with the estimate (3.10.1) proved earlier, we see that it suffices to establish the inequality

$$\sum_{\partial^l \in \overline{D}_W^{\mathscr{L}^l}} \exp\left\{(-\gamma\beta + \gamma(1-\gamma)) \sum_{t \in (W \setminus \operatorname{Int} \partial^l)^+} \operatorname{Re}(h_t)\right\} \prod_{\Gamma \in \partial^l} q_\Gamma \leq 1. \tag{3.13.2}$$

Fixing a large contour system $\partial^l \in \overline{D}_W^{\mathscr{L}^l}$ (see §3.7) and following the construction of §3.12, we denote by $B(\partial^l)$ the set of all indexes $j = 1, \dots, k$ such that $\operatorname{Int}\Gamma \cap \operatorname{Int}\overline{\Gamma}_j = \varnothing$ for any contour $\Gamma \in \partial^l$. From the definition (3.8.7) it is clear that $\operatorname{Int}\overline{\Gamma}_j \cap V^+ \subset (W \setminus \operatorname{Int}\partial^l)^+$ for $j \in B(\partial^l)$, hence it follows from (3.12.3) and the definition (3.8.2) of large contours that

$$\sum_{t \in (W \setminus \operatorname{Int}\partial^l)^+} \operatorname{Re}(h_t) \geq |B(\partial^l)|. \tag{3.13.3}$$

Therefore, it suffices to prove that

$$\sum_{\partial^l \in \overline{D}_W^{\mathscr{L}^l}} \exp\{(-\gamma\beta + \gamma(1-\gamma))|B(\partial^l)|\} \prod_{\Gamma \in \partial^l} q_\Gamma \leq 1, \tag{3.13.4}$$

for $\beta \geq \beta_0(\nu, \gamma)$, where the constant $\beta_0(\nu, \gamma)$ is sufficiently large.

We prove the estimate (3.13.4) by induction, and it will be convenient to prove a more general estimate. Let $A \subset \{1, \dots, k\}$ and let $D(A) \subset \overline{D}_W^{\mathscr{L}^l}$ be the set of all families of large contours $\partial^l \in \overline{D}_W^{\mathscr{L}^l}$ such that the clans $\varkappa(\overline{\Gamma}_j)$ have empty intersections with ∂^l for $j \notin A$. Let

$$R(A) = \sum_{\partial^l \in D(A)} \exp\{(-\gamma\beta + \gamma(1-\gamma))|B(\partial^l) \cap A|\} \prod_{\Gamma \in \partial^l} q_\Gamma. \tag{3.13.5}$$

We claim that for $\beta \geq \beta_0(\nu, \gamma)$ and for all $A \subset \{1, \cdots, k\}$,

$$R(A) \leq 1. \tag{3.13.6}$$

For $A = \{1, \dots, k\}$, the estimate (3.13.6) is the desired estimate (3.13.4). We prove (3.13.6) by induction on $|A|$. So we assume that (3.13.6) holds for all $A' \subset A$, $A' \neq A$, and prove that it holds for A also. To do so we note first that large contours $\Gamma_1, \Gamma_2 \in \partial^l$, $\Gamma_1 \neq \Gamma_2$, where $\partial^l \in D(A)$, are necessarily external contours, so that $\operatorname{Int}\Gamma_1 \cap \operatorname{Int}\Gamma_2 = \varnothing$. This means that in such a case the condition

$$\operatorname{Int}\Gamma_1 \cap \operatorname{Int}\overline{\Gamma}_j \neq \varnothing, \qquad \operatorname{Int}\Gamma_2 \cap \operatorname{Int}\overline{\Gamma}_j \neq \varnothing \tag{3.13.7}$$

implies that also $(W \setminus \operatorname{Int}\Gamma_i) \cap \operatorname{Int}\overline{\Gamma}_j \neq \varnothing$, $i = 1, 2$ and so in the definition of a parent given in §3.12 we can use the condition (3.13.7) instead of (3.12.9). Now let

$$k(A) = \min_{j \in A} j \tag{3.13.8}$$

and for $\partial^l \in D(A)$ denote by $\partial_0^l \subset \partial^l$ any maximal (with respect to the number of elements) family of large contours with an ancestor from the clan $\varkappa(\overline{\Gamma}_{k(A)})$. Let $I(\partial^l) = I(\partial_0^l)$ be the set of all indices $j \in A$ such that

$$\operatorname{Int}\tilde{\Gamma} \cap \operatorname{Int}\overline{\Gamma}_j \neq \varnothing \tag{3.13.9}$$

for some $\tilde{\Gamma} \in \partial_0^l$. Using the condition (3.13.7) in the definition of a parent and the condition (3.12.1) in the definition of a clan we see that

$$\partial^l \cap \varkappa(\overline{\Gamma}_j) \subset \partial_0^l \quad \text{if } j \in I(\partial^l). \tag{3.13.10}$$

So,

$$\partial_1^l = \partial^l \setminus \partial_0^l \in D(A \setminus I(\partial^l)). \tag{3.13.11}$$

In particular, it follows from (3.3.10) that

$$\partial^l \cap \varkappa(\overline{\Gamma}_{k(A)}) \subset \partial_0^l \quad \text{if } \partial_0^l \text{ is nonempty.} \tag{3.13.12}$$

So

$$\partial_1^l = \partial^l \setminus \partial_0^l \in D(A \setminus I(\partial^l)),$$

and (cf. (3.13.5))

$$\begin{aligned} &\sum_{\partial^l \in D(A)\,:\,\partial_0^l \neq \varnothing} \exp\{(-\gamma\beta + \gamma(1-\gamma))|B(\partial^l) \cap A|\} \prod_{\Gamma \in \partial^l} q_\Gamma \\ &\quad \leq \sum_{\partial_0^l\,:\,\partial_0^l \neq \varnothing} \prod_{\tilde{\Gamma} \in \partial_0^l} q_{\tilde{\Gamma}} R_{A \setminus \{k(A)\}}. \end{aligned} \tag{3.13.13}$$

In the case when ∂_0^l is empty, $\partial^l \in D(A \setminus \{k(A)\})$, so that $k(A) \in B(\partial^l)$. We see that

$$\begin{aligned} &\sum_{\partial^l \in D(A)\,:\,\partial_0^l = \varnothing} \exp\{(-\gamma\beta + \gamma(1-\gamma))|B(\partial^l) \cap A|\} \prod_{\Gamma \in \partial^l} q_\Gamma \\ &\quad \leq e^{(-\gamma\beta + \gamma(1-\gamma))} R_{A \setminus I(\partial_0^l)}. \end{aligned} \tag{3.13.14}$$

Using (3.13.6) as the induction hypothesis, we derive from (3.13.5), (3.13.13), and (3.13.14) that

$$R(A) \leq \sum_{\partial_0^l\,:\,\partial_0^l \neq \varnothing} \prod_{\tilde{\Gamma} \in \partial_0^l} q_{\tilde{\Gamma}} + e^{(-\gamma\beta + \gamma(1-\gamma))}. \tag{3.13.15}$$

But if $\partial_0^l \neq \varnothing$, then $\partial_0^l \in \overline{\mathscr{F}}_{k(1)}$, where $\overline{\mathscr{F}}_j$ are sets of families considered in Lemma 3.11. Recalling the estimate (3.10.2), we can apply (3.12.10) with $\alpha = 2\beta - \gamma/4$ and $\varkappa = 1/2$ (for example). It shows that

$$\sum_{\partial_0^l\,:\,\partial_0^l \neq \varnothing} \prod_{\tilde{\Gamma} \in \partial_0^l} q_{\tilde{\Gamma}} \leq \exp\{-(\beta - \gamma/8) - \alpha_0)|\overline{\Gamma}_{k(A)}|\}, \tag{3.13.16}$$

and if $\beta \geq \beta_0$ with a sufficiently large $\beta_0 = \beta_0(\nu, \gamma)$, the estimates (3.13.14) and (3.13.15) imply the desired estimate (3.13.6). As we have mentioned before, this implies the statement of the lemma.

§3.14. Inductive estimate of polymer weights

LEMMA 3.14. *Let a contour* $\Gamma \in \mathscr{L}$ *be fixed and* $\beta \geq \beta_0'$. *Assume that for any contour* $\tilde{\Gamma}$ *such that* $|\operatorname{Int}\tilde{\Gamma}| < |\operatorname{Int}\Gamma|$ *the estimate* (3.7.1) *is valid and that the estimate* (3.7.5) *is valid for any* W *with* $|W| < |\operatorname{Int}\Gamma|$. *Then the estimate* (3.7.1) *is valid for the contour* Γ *itself provided that* $\beta_0' = \beta_0'(\nu, \gamma)$ *is sufficiently large.*

PROOF. Recall the definition (3.5.4). Since $|\operatorname{Int}^0\Gamma| < |\operatorname{Int}\Gamma|$, the conditions of Lemma 3.9 are satisfied for $W = \operatorname{Int}^0\Gamma$ and we can apply the estimate (3.13.1). We see that

$$\left|\frac{Z(\operatorname{Int}^0\Gamma, -, \mathscr{L})}{Z(\operatorname{Int}^0\Gamma, +, \mathscr{L})}\right| \leq 2\exp\left\{\left(2\beta + \frac{\gamma}{2}\right) \sum_{t\in(\operatorname{Int}^0\Gamma)^0} |h_t|\right\}. \tag{3.14.1}$$

Note that, by the definition (3.3.1), $\operatorname{Re} h_t \geq 0$ for V^+, so that

$$\left|\exp\left\{-2\beta \sum_{t\in\Delta^l(\Gamma)} h_t\right\}\right| \leq \exp\left\{2\beta \sum_{t\in(\Delta^l(\Gamma))^0} |h_t|\right\}. \tag{3.14.2}$$

Therefore, applying (3.5.4) and then (3.3.2), we find that

$$\begin{aligned} |w(\Gamma)| &\leq 2\exp\{-2\beta|\Gamma|\}\exp\left\{\left(2\beta + \frac{\gamma}{2}\right) \sum_{t\in(\operatorname{Int}\Gamma)^0} |h_t|\right\} \qquad (3.14.3)\\ &\leq 2\exp\{-2\beta|\Gamma|\}\exp\{(2\beta + \gamma/2)|\Gamma|(1-\gamma)\} \leq \exp\{-2(\beta - \beta_0)\gamma|\Gamma|\}, \end{aligned}$$

which is the desired estimate.

§3.15. Inductive estimate of the derivate of the polymer partition function

LEMMA 3.15. *Suppose that* $\beta \geq \beta_0(\nu, \gamma)$ *for a sufficienlty large* $\beta_0(\nu, \gamma)$. *Let for some fixed volume* $W \subset V$ *and any contour* $\Gamma \in G_W^{\mathscr{L}}$ *the estimate* (3.7.1) *hold and for any set* $W' \subset W$, $|W'| < |W|$, *the estimate* (3.7.7) *hold. Then the estimate* (3.7.5) *is valid for the set* W.

PROOF. We must apply the general estimate (3.6.7) to the case $f = w$. Indeed, using the induction hypothesis about the estimate (3.7.1) of contour weights, we see again that the main condition (3.6.1) of Proposition 3.6 is satisfied with α given by (3.7.2). So we obtain from (3.6.7) that

$$\left|\frac{\partial \ln Z_w(W)}{\partial w(\Gamma)}\right| \leq |\Gamma|\exp\{-2\nu(2\gamma\beta - \beta_0')\} \tag{3.15.1}$$

for any $\Gamma \in G_W^{\mathscr{L}}$. It is clear that

$$\frac{\partial \ln Z_w(W)}{\partial h_t} = \sum_{\Gamma\in G_W^{\mathscr{L}}} \frac{\partial \ln Z_w(W)}{\partial w(\Gamma)} \frac{\partial w(\Gamma)}{\partial h_t}, \qquad t \in W. \tag{3.15.2}$$

Now we recall the definition (3.5.4), which shows that

$$\frac{\partial w(\Gamma)}{\partial h_t} = \begin{cases} \exp\{-2\beta(|\Gamma| + \sum_{t'\in\Delta^I(\Gamma)} h_{t'})\} \\ \qquad \times \dfrac{\partial}{\partial h_t}\left(\dfrac{Z(\mathrm{Int}^0\Gamma, -, \mathscr{L})}{Z(\mathrm{Int}^0\Gamma, +, \mathscr{L})}\right), & \text{if } t \in \mathrm{Int}^0\Gamma, \\ -2\beta w(\Gamma), & \text{if } t \in \Delta^I(\Gamma), \\ 0, & \text{if } t \notin \mathrm{Int}\,\Gamma. \end{cases} \tag{3.15.3}$$

Using (3.7.7) as an induction hypothesis (this is possible since $|\mathrm{Int}^0\Gamma| < |W|$ for $\Gamma \in G_W^{\mathscr{L}}$), we see that

$$\left|\frac{\partial}{\partial h_t}\frac{Z(\mathrm{Int}^0\Gamma, -, \mathscr{L})}{Z(\mathrm{Int}^0\Gamma, +, \mathscr{L})}\right| \le 8\beta \exp\left\{\left(2\beta + \frac{\gamma}{2}\right) \sum_{t\in(\mathrm{Int}^0\Gamma)^0} |h_t|\right\}. \tag{3.15.4}$$

Definition (3.3.1) implies that

$$\left|\exp\left\{-2\beta \sum_{t\in\Delta^I(\Gamma)} h_t\right\}\right| \le \exp\left\{2\beta \sum_{t\in(\Delta^I(\Gamma))^0} |h_t|\right\}. \tag{3.15.5}$$

Using the estimate (3.3.2) again, we obtain from (3.5.4), (3.15.4), and (3.15.5) that

$$\begin{aligned} \left|\frac{\partial w(\Gamma)}{\partial h_t}\right| &\le 8\beta \exp\left\{-2\beta|\Gamma| + \left(2\beta + \frac{\gamma}{2}\right) \sum_{t\in(\mathrm{Int}\,\Gamma)^0} |h_t|\right\} \\ &\le \exp\{-2(\beta - \beta_0')\gamma|\Gamma|\} \end{aligned} \tag{3.15.6}$$

for $t \in \mathrm{Int}^0\Gamma$ (we killed the factor 8β in (3.15.6) by choosing $\beta_0' = \beta_0'(\nu, \gamma)$ sufficiently large and using the inequality $|\Gamma| \ge 2\nu$). Using the induction estimate (3.7.1) we obtain also from (3.15.4) that

$$|\partial w(\Gamma)/\partial h_t| \le \exp\{-2(\beta - \beta_0'')\gamma|\Gamma|\} \quad \text{if } t \in \Delta^I(\Gamma) \tag{3.15.7}$$

for a sufficiently large $\beta_0''(\nu, \gamma)$. So it follows from (3.15.1)–(3.15.7) that for a new $\beta_0''' = \beta_0'''(\nu, \gamma)$

$$\left|\frac{\partial \ln Z_w(W)}{\partial h_t}\right| \le \sum_{\Gamma\in G_W^{\mathscr{L}}\,:\, t\in\mathrm{Int}\,\Gamma} |\Gamma| \exp\{-2(\beta - \beta_0''')\gamma|\Gamma|\}. \tag{3.15.8}$$

Since the number of different contours Γ such that a fixed point t belongs to $\operatorname{Int}\Gamma$ and $|\Gamma| = N$ does not exceed C^N for some constant $C = C(\nu)$, the estimate (3.7.5) follows from (3.15.9) if we choose β_0 to be sufficiently large.

§3.16. Inductive estimate of the ratio of minus- and plus-partition functions

LEMMA 3.16. *Suppose that for some* $\beta \geq \beta_0(\nu, \gamma)$ *and a fixed volume* W *the estimate* (3.7.1) *holds for* $\Gamma \in G_W^{\mathscr{L}}$ *and the estimate* (3.7.5) *holds for* W. *Then the estimate* (3.7.6) *is also true for* W *provided that the constant* $\beta_0(\nu, \gamma)$ *is sufficiently large.*

PROOF. This is a reformulation of Lemma 3.13.

§3.17. Inductive estimate for the derivative of the ratio of minus- and plus-partition functions

LEMMA 3.17. *Suppose that for some* $\beta \geq \beta_0(\nu, \gamma)$ *and a fixed volume* W *the estimate* (3.7.1) *holds for any contour* $\Gamma \in G_W^{\mathscr{L}}$ *and the estimates* (3.7.5), (3.7.6) *hold for any subset* $W' \subset W$ *(including* $W' = W$*). Then the estimate* (3.7.7) *holds for* W.

PROOF. Observe that

$$\begin{aligned}\frac{\partial}{\partial h_t}\left(\frac{Z(W, -, \mathscr{L})}{Z(W, +, \mathscr{L})}\right) \\ &= (Z(W, +, \mathscr{L}))^{-1}\frac{\partial Z(W, -, \mathscr{L})}{\partial h_t} \\ &- \frac{Z(W, -, \mathscr{L})}{(Z(W, +, \mathscr{L}))^2}\frac{\partial}{\partial h_t}Z(W, +, \mathscr{L}).\end{aligned} \tag{3.17.1}$$

We begin with the estimation of the first term in (3.17.1). We use the representation (3.8.3). For each system $\partial \in \overline{D}_W^{\mathscr{L}}$ of external contours we denote by $\Gamma_t(\partial)$ the contour $\Gamma \in \partial$ such that $t \in \operatorname{Int}\Gamma$. If there are no such contours $\Gamma \in \partial$ we set $\Gamma_t(\partial) = t$. Fixing $t \in W$ and summing over all systems $\partial \in \overline{D}_W^{\mathscr{L}}$ with a given value of $\Gamma_t(\partial)$ we can rewrite (3.8.3) as (cf. (3.8.4))

$$\begin{aligned}Z(W, -, \mathscr{L}) = &\sum_{\Gamma \in G_W^{\mathscr{L}}\,:\, t \in \operatorname{Int}\Gamma} Z(W(\Gamma), -, \mathscr{L}) \\ &\times \exp\Bigg\{-\beta \sum_{t' \in \Delta^E(\Gamma)} h_{t'} - 2\beta|\Gamma| + \beta \sum_{t' \in \Delta^I(\Gamma)} h_{t'}\Bigg\} Z(\operatorname{Int}^0\Gamma, +, \mathscr{L}) \\ &+ Z(W \setminus \{t\}, -, \mathscr{L})e^{-\beta h_t},\end{aligned} \tag{3.17.2}$$

where $W(\Gamma) = W \setminus (\operatorname{Int}\Gamma \cup \Delta^E(\Gamma))$. Therefore,

$$\begin{aligned}
&(Z(W, +, \mathscr{L}))^{-1}\frac{\partial Z(W, -, \mathscr{L})}{\partial h_t} \\
&= \sum_{\Gamma\in G_W^{\mathscr{L}}\,:\,t\in\operatorname{Int}^0\Gamma} \frac{Z(W(\Gamma), -, \mathscr{L})}{Z(W(\Gamma), +, \mathscr{L})}\exp\Big\{-2\beta|\Gamma| - 2\beta\sum_{t'\in\Delta^E(\Gamma)} h_{t'}\Big\} \\
&\quad\times (Z(\operatorname{Int}^0\Gamma, +, \mathscr{L}))^{-1}\frac{\partial Z(\operatorname{Int}^0\Gamma, +, \mathscr{L})}{\partial h_t}Q_\Gamma \qquad (3.17.3)\\
&\quad+ \sum_{\Gamma\in G_W^{\mathscr{L}}\,:\,t\in\Delta^I(\Gamma)} \beta\frac{Z(W(\Gamma), -, \mathscr{L})}{Z(W(\Gamma), +, \mathscr{L})}\exp\Big\{-2\beta|\Gamma| - 2\beta\sum_{t'\in\Delta^E(\Gamma)} h_{t'}\Big\}Q_\Gamma \\
&\quad+ (-\beta)\frac{Z(W\setminus\{t\}, -, \mathscr{L})}{Z(W\setminus\{t\}, +, \mathscr{L})}e^{-2\beta h_t}Q_t,
\end{aligned}$$

where

$$Q_\Gamma = \frac{Z(W(\Gamma), +, \mathscr{L})Z(\operatorname{Int}^0\Gamma, +, \mathscr{L})}{Z(W, +, \mathscr{L})}\exp\Big\{\beta\sum_{t'\in\Delta(\Gamma)} h_{t'}\Big\}, \quad \Gamma\in G_W^{\mathscr{L}} \qquad (3.17.4)$$

and

$$Q_t = \frac{Z(W\setminus\{t\}, +, \mathscr{L})}{Z(W, +, \mathscr{L})}e^{\beta h_t}. \qquad (3.17.5)$$

(The fact that all the plus-partition functions appearing above do not vanish is a part of the induction hypothesis (3.7.6).) The factors Q_Γ, Q_t can be estimated in the same way as the factor $R(\partial^l)$ (see (3.10.4)). Using the representation (3.5.3) we find (cf. (3.10.7)) that

$$\begin{aligned}
Q_\Gamma &= (Z_w(W))^{-1}Z_w(W(\Gamma))Z_w(\operatorname{Int}^0\Gamma),\\
Q &= (Z_w(W))^{-1}Z_w(W\setminus\{t\}).
\end{aligned} \qquad (3.17.6)$$

Since the estimate (3.7.1) is valid for any contour $\Gamma\in G_W^{\mathscr{L}}$, we can apply (cf. (3.10.8)) the representation (3.6.4) (with $f = w$) to all sets W, $W(\Gamma)$, $\operatorname{Int}^0\Gamma$, and $W\setminus\{t\}$ that occur in (3.17.6). Proceeding similarly to the derivation of the estimate (3.10.2) (see (3.10.7)–(3.10.12) in §3.10), we arrive at the estimates

$$|Q_\Gamma| \le \exp\{\gamma|\Gamma|/4\}, \qquad |Q_0| \le \exp\{\gamma/4\}, \qquad (3.17.7)$$

if $\beta \ge \beta_0(\nu, \gamma)$ for a sufficiently large $\beta_0(\nu, \gamma)$. Using the relation (3.5.3) we see that

$$\begin{aligned}
&Z(\operatorname{Int}^0\Gamma, +, \mathscr{L})^{-1}\frac{\partial Z(\operatorname{Int}^0\Gamma, +, \mathscr{L})}{\partial h_t}\\
&\quad= \frac{\partial \ln Z(\operatorname{Int}^0\Gamma, +, \mathscr{L})}{\partial h_t} = \beta + \frac{\partial \ln Z_w(W)}{\partial h_t},
\end{aligned} \qquad (3.17.8)$$

and so the induction estimate (3.7.5) implies that

$$\left| Z(\mathrm{Int}^0\Gamma, +, \mathscr{L})^{-1} \frac{\partial Z(\mathrm{Int}^0\Gamma, +, \mathscr{L})}{\partial h_t} \right| \le \beta + \frac{\gamma}{2}. \tag{3.17.9}$$

The induction estimate (3.7.6), together with the main definition (3.3.1), implies that

$$\left| \frac{Z(W(\Gamma), -, \mathscr{L})}{Z(W(\Gamma), +, \mathscr{L})} \exp\left\{ -2\beta \sum_{t' \in \Delta^E(\Gamma)} h_{t'} \right\} \right| \le 2\exp\left\{ \left(2\beta + \frac{\gamma}{2}\right) \sum_{t' \in W^0} |h_{t'}| \right\}, \tag{3.17.10}$$

and similarly

$$\left| \frac{Z(W \setminus \{t\}, -, \mathscr{L})}{Z(W \setminus \{t\}, +, \mathscr{L})} e^{-2\beta h_t} \right| \le 2\exp\left\{ \left(2\beta + \frac{\gamma}{2}\right) \sum_{t' \in W^0} |h_{t'}| \right\}. \tag{3.17.11}$$

So we obtain from (3.17.3) and (3.17.7), (3.17.9), (3.17.10), (3.17.11) that

$$\begin{aligned} \left| Z(W, +, \mathscr{L})^{-1} \frac{\partial Z(W, -, \mathscr{L})}{\partial h_t} \right| &\le 2\left(\beta + \frac{\gamma}{2}\right) \exp\left\{ \left(2\beta + \frac{\gamma}{2}\right) \sum_{t' \in W^0} |h_{t'}| \right\} \\ &\quad \times \left(\sum_{\Gamma \in G_W^{\mathscr{L}} :\, t \in \mathrm{Int}\,\Gamma} \exp\{(-2\beta + \gamma/4)|\Gamma|\} + e^{\gamma/4} \right) \\ &\le 4\beta \exp\left\{ \left(2\beta + \frac{\gamma}{2}\right) \sum_{t' \in W^0} |h_{t'}| \right\}. \end{aligned} \tag{3.17.12}$$

In the last inequality we use the fact that the number of different contours Γ such that $|\Gamma| = N$, $t \in \mathrm{Int}\,\Gamma$, does not exceed C^N, where $C = C(\nu)$ is a contant depending on the dimension ν only, and assume again that $\beta \ge \beta_0(\nu, \gamma)$ for a sufficiently large $\beta_0(\nu, \gamma)$.

To estimate the second term in (3.17.1) is much simpler. Similarly to (3.17.9), using the induction hypothesis (3.7.5) for the set W we find that

$$\left| Z(W, +, \mathscr{L})^{-1} \frac{\partial}{\partial h_t} Z(W, +, \mathscr{L}) \right| \le \beta + \frac{\gamma}{2}. \tag{3.17.13}$$

Applying the induction hypothesis (3.7.6) to the set W we see that

$$\left| \frac{Z(W, -, \mathscr{L}) \frac{\partial}{\partial h_t} Z(W, +, \mathscr{L})}{(Z(W, +, \mathscr{L}))^2} \right| \le 4\beta \exp\left\{ \left(2\beta + \frac{\gamma}{2}\right) \sum_{t' \in W^0} |h_{t'}| \right\}, \tag{3.17.14}$$

again if β is sufficiently large. So the relation (3.17.1) and the estimates (3.17.14) and (3.17.12) imply the desired estimate (3.7.7).

Lemmas 3.14–3.17 prove all the induction statements formulated in §3.7, so the proof of the main Proposition 3.3 is completed (at last!).

§3.18. Domain of analyticity

We apply Proposition 3.3 to the case when the cutoff set $\mathscr{L}$ is $\mathscr{L} = G^{\omega}_{V, W}$ (see (3.2.2)), where $W \subset V$ and $h = \{h_t, t \in V\}$,

$$h_t = \begin{cases} h^0 & \text{if } t \in V \setminus W, \\ h^1 + z, \; z \in \mathbb{C} & \text{if } t \in W. \end{cases} \tag{3.18.1}$$

We assume that h^0, h^1 are real and

$$h^0 \geq 0, \qquad h^1 > -\omega^{-1}. \tag{3.18.2}$$

(Here and below, $\omega^{-1} = 0$ for $\omega = \infty$.)

LEMMA 3.18. *Suppose that for some* r, $0 < r < 1$,

$$|z| \leq \max\left(\frac{2r}{\sqrt{5}}h^1, \, r\omega^{-1} - |h^1|\right). \tag{3.18.3}$$

Then there exists a number $\gamma = \gamma(r)$, $0 < \gamma < 1$, *such that the condition* (3.3.2) *of Proposition* 3.3 *holds, and therefore all its implications are valid.*

PROOF. We must show that conditions (3.18.1), (3.18.2), (3.18.3) imply the estimate (3.3.2) for an appropriate $\gamma = \gamma(r)$. Observe that the condition

$$|z| \leq \frac{2}{\sqrt{5}}h^1 \tag{3.18.4}$$

implies that

$$|\operatorname{Im}(h^1 + z)| \leq 2\operatorname{Re}(h^1 + z). \tag{3.18.5}$$

(The distance from the point $h^1 \in \mathbb{R} \subset \mathbb{C}$ to each of the rays $\operatorname{Im} z = \pm 2 \operatorname{Re} z$ is equal to $2h^1/\sqrt{5}$.) So if $|z| \leq 2r/\sqrt{5}h^1$, then for a sufficiently small $\gamma = \gamma(r)$ the set V^0 is empty (see (3.3.1)) and the condition (3.3.2) is trivially valid. Now let

$$|z| < r\omega^{-1} - |h^1|. \tag{3.18.6}$$

Then

$$|h^1 + z| < r\omega^{-1}. \tag{3.18.7}$$

Since $h^0 \geq 0$, we have $V^0 \subset W$. Therefore (see (3.3.2) and (3.1.4))

$$\sum_{t \in \operatorname{Int}\Gamma \cap V^0} |h_t| \leq |\operatorname{Int}\Gamma| r\omega^{-1} \leq |\Gamma| r q(\Gamma)\omega^{-1} \leq |\Gamma|(1-\gamma) \tag{3.18.8}$$

if $\gamma < 1 - r$, verifying the condition (3.3.2) in this second case.

§3.19. Estimates of semiinvariants

In this section we assume that $\mathscr{L} = G^{\omega}_{V,W}$, where $W \subset V$ (see §3.2) and that $h = \{h_t, t \in V\}$, where

$$h_t = \begin{cases} h^0, & t \in V \setminus W, \\ h^1, & t \in W. \end{cases} \tag{3.19.1}$$

By (3.2.9), the semiinvariants G^n are given by

$$G^n(W, V, +, \beta, h, \omega) = (\beta)^{-n} \frac{\partial^n \ln Z(V, +, \beta, h, \omega)}{\partial (h^1)^n}, \qquad n = 1, 2, \ldots. \tag{3.19.2}$$

We also define

$$G^0(W, V, +, \beta, h, \omega) = \ln \frac{Z(V, +, \beta, h, \omega)}{Z(V, +, \beta, \overline{h}^0, \omega)}, \tag{3.19.3}$$

where $\overline{h}^0 = \{h_t \equiv h^0, t \in V\}$. The quantities G^0 can be estimated similarly to the semiinvariants G^n, $n = 1, 2, \ldots$.

PROPOSITION 3.19. *Let (cf. (3.18.2))*

$$h^0 \geq 0, \qquad h^1 \geq -c\omega^{-1}, \tag{3.19.4}$$

for some c, $0 < c < 1$. *Then there exist positive constants* $K_n = K_n(\nu, c)$, $n = 0, 1, \ldots$, $\beta_0 = \beta_0(\nu, c)$, $\gamma = \gamma(\nu, c)$, $\rho = \rho(\nu, c)$, *depending on* c *and on the dimension* ν *only, such that for any* $\beta \geq \beta_0$ *and any* ω, h *satisfying the conditions* (3.19.1), (3.19.4), *the semiinvariant of order* n *can be extended to a holomorphic function of* h^1 *in a disk* $\mathscr{O}_{h^1, \omega}$ *of radius* $\rho \min(1, h^1)$ *if* $h^1 > \rho\omega^{-1}$ *and of radius* $\rho\omega^{-1}$ *if* $h^1 < \rho\omega^{-1}$, *with the center at the point* h^1, *and in this disk we have*

$$|G^n(W, V, +, \beta, h, \omega)| \leq K_n e^{-\beta\gamma} |W|, \qquad n = 2, 3, \ldots, \tag{3.19.5}$$

and

$$\begin{aligned} &|G^1(W, V, +, \beta, h, \omega)| \leq K_1 |W|, \\ &|G^0(W, V, +, \beta, h, \omega)| \leq (|h^1|\beta + K_0)|W|. \end{aligned} \tag{3.19.5a}$$

Furthermore, let another cutoff level $\tilde{\omega}$, *volumes* $\tilde{W} \subset \tilde{V}$, *parameters* $\tilde{h}^0$, $\tilde{h}^1$, *and magnetic fields* $\tilde{h}_t$ *defined by the relations similar to* (3.19.1) *be given. Again, assume that* $\tilde{h}^0 \geq 0$, $\tilde{h}^1 \geq -c\tilde{\omega}^{-1}$ (*cf.* (3.19.4)). *Let* $\tilde{h} = \{\tilde{h}_t, t \in V\}$. *Then for some* $\alpha = \alpha(\nu, c)$ *and all* $n = 2, 3, \ldots$ *we have*

$$\begin{aligned} &|G^n(W, V, +, \beta, h, \omega) - G^n(\tilde{W}, \tilde{V}, +, \beta, \tilde{h}, \tilde{\omega})| \\ &\leq K_n e^{-\beta\gamma} \Big[|W \Delta \tilde{W}| + |V \Delta \tilde{V}| \exp\{-\alpha\beta \min(\operatorname{dist}(V\Delta\tilde{V}, W), \operatorname{dist}(V\Delta\tilde{V}, \tilde{W}))\} \\ &\qquad + \max(|\partial W|, |\partial \tilde{W}|)|h^0 - \tilde{h}^0| + \max(|W|, |\tilde{W}|)|h^1 - \tilde{h}^1| \\ &\qquad + \exp\{-\alpha\beta(\min(\omega, \tilde{\omega}))^{\nu-1}\} \max(|W|, |\tilde{W}|) \Big], \end{aligned} \tag{3.19.6}$$

and

$$|G^1(W, V, +, \beta, h, \omega) - G^1(\tilde{W}, \tilde{V}, +, \beta, \tilde{h}, \tilde{\omega})|$$
$$\le K_1 \Big[|W\Delta\tilde{W}| + |V\Delta\tilde{V}| \exp\{-\alpha\beta \min(\operatorname{dist}(V\Delta\tilde{V}, W), \operatorname{dist}(V\Delta\tilde{V}, \tilde{W}))\}$$
$$+ e^{-\beta\gamma} \max(|\partial W|, |\partial\tilde{W}|)|h^0 - \tilde{h}^0| + \max(|W|, |\tilde{W}|)|h^1 - \tilde{h}^1|$$
$$+ \exp\{-\alpha\beta(\min(\omega, \tilde{\omega}))^{\nu-1}\} \max(|W|, |\tilde{W}|)\Big],$$

$$|G^0(W, V, +, \beta, h, \omega) - G^0(\tilde{W}, \tilde{V}, +, \beta, \tilde{h}, \tilde{\omega})|$$
$$\le K_0 \Big[\beta|h^1||W\Delta\tilde{W}|$$
$$+ |V\Delta\tilde{V}| \exp\{-\alpha\beta \min(\operatorname{dist}(V\Delta\tilde{V}, W), \operatorname{dist}(V\Delta\tilde{V}, \tilde{W}))\}$$
$$+ \beta \max(|\partial W|, |\partial\tilde{W}|)|h^0 - \tilde{h}^0| + \beta \max(|W|, |\tilde{W}|)|h^1 - \tilde{h}^1|$$
$$+ \exp\{-\alpha\beta(\min(\omega, \tilde{\omega}))^{\nu-1}\} \max(|W|, |\tilde{W}|)\Big], \tag{3.19.7}$$

where Δ *denotes the symmetric difference of two sets.*

Let $W_N \subset V_N$ *be sequences of sets such that*

$$\lim_{N\to\infty} \frac{|\partial W_N|}{|W_N|} = 0. \tag{3.19.8}$$

(*This is the Van Hove condition* (1.4.1).) *Then for* $\beta \ge \beta_0$ *the limits*

$$\overline{G}^n(\beta, h^1, \omega) = \lim_{N\to\infty} \frac{G^n(V_N, W_N, +, \beta, h, \omega)}{|W_N|}, \quad n = 0, 1, \dots, \tag{3.19.9}$$

exist and do not depend on the choice of h^0. *For all* $W \subset V$ *and all* $n = 0, 1, \dots$ *we have*

$$|G^n(W, V, +, \beta, h, \omega) - |W|\overline{G}^n(\beta, h^1, \omega)| \le K_n e^{-\beta\gamma}|\partial W|. \tag{3.19.10}$$

For all $n = 0, 1, \dots$ *the functions* $\overline{G}^n$ *are holomorphic function in* h^1 *in the same disks* $\mathscr{O}_{h^1,\omega}$ *as above, and*

$$\frac{d\overline{G}^n(\beta, h^1, \omega)}{dh^1} = \beta^{-1}\overline{G}^{(n+1)}(\beta, h^1, \omega). \tag{3.19.11}$$

PROOF. The representation (3.3.7) and the relation (3.19.2) imply that

$$G^n(W, V, +, \beta, h, \omega) = \beta^{-n} \sum_{\Lambda\subset V} \Phi^n(\Lambda, h, \omega), \qquad n = 1, 2, \dots, \tag{3.19.12}$$

where

$$\Phi^n(\Lambda, h, \omega) = \frac{\partial^n}{\partial(h^1)^n}\Phi(\Lambda, +, \beta, h, \omega). \tag{3.19.13}$$

It follows from the definitions (3.3.6), (3.19.1) and Proposition 3.3 that $\Phi(\Lambda, \beta, h, +, \omega)$ does not depend on h^1 if $\Lambda \cap W = \varnothing$. Therefore,

$$\Phi^n(\Lambda, h, \omega) = 0 \quad \text{if } \Lambda \cap W = \varnothing, \ n = 1, 2, \dots, \tag{3.19.14}$$

so that we can rewrite (3.19.13) as

$$G^n(W, V, +, \beta, h, \omega) = \beta^{-n} \sum_{\Lambda \subset V : \Lambda \cap W \neq \varnothing} \Phi^n(\Lambda, h, \omega). \qquad (3.19.15)$$

Applying the representation (3.3.7) to each of two partition functions used in the definition (3.19.3), we see that

$$\begin{aligned} G^0(W, V, +, \beta, h, \omega) &= \sum_{\Lambda \subset V} (\Phi(\Lambda, h, \omega) - \Phi(\Lambda, \overline{h}^0, \omega)) \qquad (3.19.16) \\ &= \sum_{\Lambda \subset V : \Lambda \cap W \neq \varnothing} (\Phi(\Lambda, h, \omega) - \Phi(\Lambda, \overline{h}^0, \omega)), \end{aligned}$$

since it follows from Proposition 3.3 that $\Phi(\Lambda, h, \omega) = \Phi(\Lambda, \overline{h}^0, \omega)$ when $\Lambda \cap W = \varnothing$.

Proposition 3.3 and Lemma 3.18 for some r, $c < r < 1$, imply that $\Phi(\Lambda, +, \beta, h, \omega)$ is a holomorphic function of z in the disk (3.18.3). Differentiating the estimate (3.3.11) and applying the Cauchy formula in this domain, we find the following estimates in the disk $\mathscr{O}_{h_1, \omega}$ for the case $|\Lambda| > 1$:

$$|\Phi^n(\Lambda, h, \omega)| \leq \overline{K}((2r/\sqrt{5} - \rho)h^1)^{-n} \exp\{-2\beta\gamma d(\Lambda)\}, \quad \text{if } h_1 > \rho\omega^{-1}, \qquad (3.19.17)$$

and

$$|\Phi^n(\Lambda, h, \omega)| \leq \overline{K}((r - c - \rho)\omega^{-1})^{-n} \exp\{-2\beta\gamma d(\Lambda)\}, \quad \text{if } h_1 \leq \rho\omega^{-1}, \qquad (3.19.18)$$

where we assume that $\beta \geq \beta_0$, ρ is sufficiently small, and $\overline{K} = \overline{K}(\nu, c)$, $\beta_0 = \beta_0(\nu, c)$ are some constants. We can improve the estimate (3.19.18) if we observe (cf. (3.1.3), (3.1.5), and (3.3.12)) that for any contour $\Gamma \in G_\Lambda$

$$q(\Gamma) \leq (1/2\nu)|\Lambda|^{1/\nu} \leq |\partial\Lambda|^{1/(\nu-1)} \leq d(\Lambda)^{1/(\nu-1)}, \qquad (3.19.19)$$

and so

$$\Phi^n(\Lambda, h, \omega) = \Phi^n(\Lambda, h, d(\Lambda)^{1/(\nu-1)}), \quad \text{if } \omega > d(\Lambda)^{1/(\nu-1)}. \qquad (3.19.20)$$

Now it follows from this remark and the estimates (3.9.17), (3.19.18) applied for $\omega = d(\Lambda)^{1/(\nu-1)}$ that for some new constants $\overline{K}_n = \overline{K}_n(\nu, c)$, $n = 1, 2, \dots$, we have

$$\begin{aligned} |\Phi^n(\Lambda, h, \omega)| \leq \overline{K}_n &\max\{(r - c - \rho), (2r/\sqrt{5} - \rho)\rho\}^{-n} d(\Lambda)^{n/(\nu-1)} \\ &\times \exp\{-2\beta\gamma(d(\Lambda)\} \leq \overline{K}_n \exp\{-\beta\gamma d(\Lambda)\} \end{aligned} \qquad (3.19.21)$$

if $\omega > d(\Lambda)^{1/(\nu-1)}$.

It follows directly from (3.19.17), (3.9.18) that

$$|\Phi^n(\Lambda, h, \omega)| \leq \overline{K}_n \exp\{-\beta\gamma d(\Lambda)\}, \qquad (3.19.22)$$

if $\omega \le d(\Lambda)^{1/(\nu-1)}$ (with a new constant $\overline{K}_n = \overline{K}_n(\nu, c)$). So we can claim that for some $K'_n = K'_n(\nu, c)$, $\beta'_0 = \beta'_0(\nu, c)$, all $n = 1, 2, \ldots$, and all $\beta \ge \beta'_0$

$$|\beta^{-n}\Phi^n(\Lambda, h, \omega)| \le K'_n \exp\{-\beta\gamma d(\Lambda)\} \text{ for } |\Lambda| > 1 \tag{3.19.23}$$

in the entire domain (3.19.4). It follows from the explicit formula (3.3.6) that for $\Lambda = \{t\}$ we have

$$\begin{aligned} |\beta^{-n}\Phi^n(\{t\}, \overline{h}, \omega)| &\le K'_n \exp\{-\beta\gamma\}, \quad n = 2, 3, \ldots, \\ |\beta^{-1}\Phi^1(\{t\}, \overline{h}, \omega) - 1| &\le K'_1 \exp\{-\beta\gamma\}, \quad \text{if } |\Lambda| = 1, \end{aligned} \tag{3.19.24}$$

where the constants K'_n, β'_0 are chosen in an appropriate way.

Now Lemma 3.4 immediately implies the estimates (3.19.5), (3.19.5a) for appropriate constants K_n. In proving (3.19.6) it suffices to consider the case when we change only one of parameters V, W, h_0, h_1, ω. So observe first that it follows from relations (3.19.15), (3.19.23), (3.19.24) together with Lemma 3.4 that if $W \subset V$, $\tilde{W} \subset V$, then

$$\begin{aligned} &|G^n(W, V, +, \beta, h, \omega) - G^n(V, \tilde{W}, +, \beta, h, \omega)| \\ &\quad \le \beta^{-n} \sum_{\Lambda : \Lambda \cap (W \Delta \tilde{W}) \neq \varnothing,\ \Lambda \subset V} |\Phi^n(\Lambda, h, \omega)| \le K_n |W \Delta \tilde{W}|, \qquad n > 1, \end{aligned} \tag{3.19.25}$$

if we choose K_n is an appropriate way. Similarly, if $W \subset V$, $W \subset \tilde{V}$, then

$$\begin{aligned} &|G^n(W, V, +, \beta, h, \omega) - G^n(W, \tilde{V}, +, \beta, h, \omega)| \\ &\quad \le \beta^{-n} \sum_{\substack{\Lambda : \Lambda \cap W \neq \varnothing, \\ \Lambda \cap (V \Delta \tilde{V}) \neq \varnothing}} |\Phi^n(\Lambda, h, \omega)| \\ &\quad \le \overline{K}_n \sum_{\substack{\Lambda : \Lambda \cap (V \Delta \tilde{V}) \neq \varnothing, \\ d(\Lambda) \ge \operatorname{dist}(W, V \Delta \tilde{V})}} \exp\{-\beta\gamma d(\Lambda)\} \\ &\quad \le K_n \exp\{-\alpha\beta \operatorname{dist}(W, V \Delta \tilde{V})\}, \qquad n > 1, \end{aligned} \tag{3.19.26}$$

for an appropriate choice of constants α, K_n. To evaluate the dependence on h^0 we observe that

$$\frac{\partial \Phi^n(\Lambda, h, \omega)}{\partial h^0} = \frac{\partial^{n+1}}{(\partial h^1)^n \partial h^0} \Phi(\Lambda, +, \beta, h, \omega). \tag{3.19.27}$$

Considerations completely similar to the derivation of (3.19.23) show that

$$\begin{aligned} &\left| \frac{\partial^{n+1}}{(\partial h^1)^n \partial h^0} \Phi(\Lambda, +, \beta, \overline{h}, \omega) \right| \\ &\quad \le \overline{K}^0_n \exp\{-\beta\gamma d(\Lambda)\} \quad \text{for } |\Lambda| > 1, \ n > 1 \end{aligned} \tag{3.19.28}$$

for an appropriate constant $\overline{K}^0_n$. Observe also that the derivatives (3.19.28) vanish if $\Lambda \cap W^c = \varnothing$ or $\Lambda \cap W = \varnothing$. Therefore, since we can assume that

Λ is a connected set, they do not vanish only if $\Lambda \cap \partial W \neq \varnothing$. So if $h(h'^0)$ is given by (3.19.1) with h^0 changed to h'^0, we can use Lemma 3.4 to find that for $\tilde{h}^0 > h^0$

$$\begin{aligned} &|G^n(W, V, +, \beta, h(\tilde{h}^0), \omega) - G^n(W, V, +, \beta, h, \omega)| \\ &\quad \leq \int_{h^0}^{\tilde{h}^0} \sum_{\Lambda : \Lambda \cap \partial W \neq \varnothing} \left| \frac{\partial \Phi^n(\Lambda, h(h'^0), \omega)}{\partial h'^0} \right| dh'^0 \\ &\quad \leq K_n |h^0 - \tilde{h}^0| |\partial W|, \qquad n > 1, \end{aligned} \tag{3.19.29}$$

for a corresponding choice of the constant K_n. In a completely similar way we consider the case when $h(h'^1)$ is given by (3.19.1) with h^1 changed to h'^1. We find that

$$|G^n(W, V, +, \beta, h(\tilde{h}^1), \omega) - G^n(W, V, +, \beta, h, \omega)| \leq K_n |h^1 - \tilde{h}^1| |W|. \tag{3.19.30}$$

(In this case we can restrict the summation range to include only the sets Λ with $\Lambda \cap W \neq \varnothing$.) Finally, if $\tilde{\omega} \geq \omega$, it follows from (3.3.8), (3.19.23), (3.19.24), and Lemma 3.4 that for an appropriate constant K_n

$$\begin{aligned} &|G^n(W, V, +, \beta, h, \omega) - G^n(W, V, +, \beta, h, \tilde{\omega})| \\ &\quad \leq \sum_{\Lambda : |\Lambda| > (2\nu\omega)^\nu} (|\Phi^n(\Lambda, h, \omega)| + |\Phi^n(\Lambda, h, \tilde{\omega})|) \\ &\quad \leq K_n \beta^{-n} \exp\{-\alpha\beta\omega^{\nu-1}\}, \qquad n > 1. \end{aligned} \tag{3.19.31}$$

(We use here that $d(\Lambda) \geq (2\nu)|\Lambda|^{(\nu-1)/\nu}$.) The statement (3.19.6) of the proposition follows from (3.19.25), (3.19.26), (3.19.29), (3.19.30), (3.19.31). The statements (3.19.7) and (3.19.5a) for $n = 0$ are proved in a similar way (in the case $n = 0$ the estimate (3.3.11) and the definition (3.3.6) for $|\Lambda| = 1$ are used directly).

To prove the existence of the limit (3.19.9) we observe that for $\Lambda \subset W$ the function $\Phi^n(\Lambda, h, \omega)$ depends on h^1 only, and denote it by $\Phi^n(\Lambda, h^1, \omega)$. The function $\Phi^n(\Lambda, h^1, \omega)$ is translation-invariant, and so (3.19.15) can be rewritten as

$$\begin{aligned} &G^n(W, V, +, \beta, h, \omega) \\ &\quad = \beta^{-n} \Bigg(\sum_{\Lambda : 0 \in \Lambda} \Phi^n(\Lambda, h^1, \omega) |\Lambda|^{-1} |\{t \in \mathbb{Z}^\nu : \Lambda + t \subset W\}| \\ &\qquad + \sum_{\Lambda \subset V : \Lambda \cap W \neq \varnothing, \Lambda \cap (V \setminus W) \neq \varnothing} \Phi^n(\Lambda, h, \omega) \Bigg), \qquad n > 1. \end{aligned} \tag{3.19.32}$$

It is easy to verify that

$$\big| |\{t \in \mathbb{Z}^\nu : \Lambda + t \subset W\}| - |W| \big| \leq (\operatorname{diam} \Lambda + 1) |\partial W| \tag{3.19.33}$$

and that (3.19.23) implies that

$$\left|\beta^{-n} \sum_{\Lambda\subset V\,:\,\Lambda\cap W\neq\varnothing\,,\,\Lambda\cap(V\setminus W)\neq\varnothing} \Phi^n(\Lambda\,,\,h\,,\,\omega)\right| \le K_n e^{-\beta\gamma} \tag{3.19.34}$$

for appropriate constants K_n. So using relations (3.19.32)–(3.19.34) together with the Van Hove condition (3.9.18) we see that the limit (3.19.9) exists and

$$\overline{G}^n(\beta\,,\,h^1\,,\,\omega) = \beta^{-n} \sum_{\Lambda\,:\,0\in\Lambda} |\Lambda|^{-1}\Phi^n(\Lambda\,,\,h^1\,,\,\omega). \tag{3.19.35}$$

The convergence of the series (3.19.35) follows from the estimate (3.19.23) and Lemma 3.4. The estimate (3.19.10) can be proved in a similiar way. The case $n = 0$ is also similiar. The equality (3.19.11) follows from (3.19.12) and (3.19.13).

REMARK 1. Similar considerations permit us to verify that the constants K_n can be bounded from above, $K_n \le K(Q(h^1))^n$. However, $Q(h^1) \to \infty$ as $h_1 \to 0$. This looks natural because in the case $h^1 = 0$ estimates for semiinvariants that are exponential in n cannot be valid since that would imply the analyticity of magnetization (see [I]). Our main goal is to obtain estimates of semiinvariants that are uniform in the entire domain (3.19.1).

REMARK 2. Comparing the estimates (3.19.7) and (3.19.9) with the definitions (1.3.13), we see that

$$\overline{G}^1(\beta\,,\,h\,,\,\infty) = m^+(\beta\,,\,h). \tag{3.19.36}$$

§3.20. Estimate for probabilities of systems of external contours

Again, as in §3.19, we assume that a cutoff set of contours is $\mathscr{L} = G^{\omega}_{V\,,\,W}$ and that the fields $h = \{h_t\,,\,t \in V\}$ satisfies the conditions (3.19.1), (3.19.4).

PROPOSITION 3.20. *There exist constants $\gamma = \gamma(\nu\,,\,c) \ge 0$, $\beta_0 = \beta(\nu\,,\,c) < \infty$ such that for $\beta \ge \beta_0$ and any system of contours $\mathscr{G} \subset \mathscr{L}$ we have*

$$P_V^{+\,,\,\beta\,,\,h\,,\,\mathscr{L}}(\sigma : \mathscr{G} \subset \mathscr{G}^E(\sigma)) \le \exp\{-\sum_{\Gamma\in\mathscr{G}} 2(\beta-\beta_0)\gamma|\Gamma|\}\,, \tag{3.20.1}$$

where $\mathscr{G}^E(\sigma)$ is the set of all external contours of the configuration σ (see §3.2).

PROOF. In the special case $h_t \equiv 0$, $t \in V$, the estimate (3.20.1) is reduced to the well-known Peierls estimate. To prove it in the general case we must use the main construction of this chapter. Let

$$\begin{aligned}
\operatorname{Int}^0 \mathscr{G} &= \left\{ t \in V : t \in \bigcup_{\Gamma \in \mathscr{G}} (\operatorname{Int} \Gamma \setminus \Delta(\Gamma)) \right\}, \\
\operatorname{Ext}^0 \mathscr{G} &= \left\{ t \in V : t \in \bigcup_{\Gamma \in \mathscr{G}} (\operatorname{Ext} \Gamma \setminus \Delta(\Gamma)) \right\}, \\
\Delta^I(\mathscr{G}) &= \left\{ t \in V : t \in \bigcup_{\Gamma \in \mathscr{G}} (\operatorname{Int} \Gamma \cap \Delta(\Gamma)) \right\}, \\
\Delta^E(\mathscr{G}) &= \left\{ t \in V : t \in \bigcup_{\Gamma \in \mathscr{G}} (\operatorname{Ext} \Gamma \cap \Delta(\Gamma)) \right\}
\end{aligned} \tag{3.20.2}$$

(see the notation in §3.1). Then

$$\begin{aligned}
P_V^{+,\beta,h,\mathscr{L}}(\sigma : \mathscr{G} \subset \mathscr{G}^E(\sigma)) &= \exp\left\{ \beta\left(\sum_{t \in \Delta^E(\mathscr{G})} h_t - \sum_{t \in \Delta^I(\mathscr{G})} h_t - 2\sum_{\Gamma \in \mathscr{G}} |\Gamma| \right) \right\} \\
&\quad \times \frac{Z(\operatorname{Ext}^0 \mathscr{G}, +, \beta, h, \mathscr{L}) Z(\operatorname{Int}^0 \mathscr{G}, -, \beta, h, \mathscr{L})}{Z(V, +, \beta, h, \mathscr{L})}.
\end{aligned} \tag{3.20.3}$$

(It is essential here that $\mathscr{G}$ is a system of external contours. Otherwise the restriction of σ to $\operatorname{Ext}^0 \mathscr{G}$ might contain contours that are not inner in $\operatorname{Ext}^0 \mathscr{G}$.) It is clear that

$$Z(\operatorname{Int}^0 \mathscr{G}, \pm, \beta, h, \mathscr{L}) = \prod_{\Gamma \in \mathscr{G}} Z(\operatorname{Int}^0 \Gamma, \pm, \beta, h, \mathscr{L}). \tag{3.20.4}$$

So using the estimate (3.3.5) we see that for a sufficiently large $\beta_0 < \infty$

$$\begin{aligned}
P_V^{+,\beta,h,\mathscr{L}}(\sigma : \mathscr{G} \subset \mathscr{G}_E(\sigma)) &\le \exp\left\{ \beta\left(\sum_{t \in \Delta^E(\mathscr{G})} h_t - \sum_{t \in \Delta^I(\mathscr{G})} h_t - 2\sum_{\Gamma \in \mathscr{G}} |\Gamma| \right) \right\} \\
&\quad \times \prod_{\Gamma \in \mathscr{G}} 2 \exp\left\{ (2\beta + \gamma/2) \sum_{t \in (\operatorname{Int}^0 \Gamma)^0} |h_t| \right\} \\
&\quad \times \frac{Z(\operatorname{Ext}^0 \mathscr{G}, +, \beta, h, \mathscr{L}) Z(\operatorname{Int}^0 \mathscr{G}, +, \beta, h, \mathscr{L})}{Z(V, +, \beta, h, \mathscr{L})}.
\end{aligned} \tag{3.20.5}$$

Now we want to apply the expansion (3.3.7) to the partition functions in (3.20.5). Since $\Phi(\Lambda)$ vanishes when Λ is not connected, we can consider connected sets Λ only. But any connected set Λ such that $\Lambda \cap \operatorname{Int}^0 \mathscr{G} \neq \varnothing$ and $\Lambda \cap \operatorname{Ext}^0 \mathscr{G} \neq \varnothing$ has a nonempty intersection with the set $\Delta(\mathscr{G}) =$

$\bigcup_{\Gamma\in\mathscr{G}}\Delta(\Gamma)$. So we have

$$\frac{Z(\mathrm{Ext}^0\mathscr{G},+,\beta,h,\mathscr{L})Z(\mathrm{Int}^0\mathscr{G},+,\beta,h,\mathscr{L})}{Z(V,+,\beta,h,\mathscr{L})} = \exp\Big\{-\sum_{\Lambda:\Lambda\cap\Delta(\mathscr{G})\neq\varnothing,\Lambda\subset V}\Phi(\Lambda)\Big\}. \tag{3.20.6}$$

Now it follows from the estimate (3.3.11) and Lemma 3.4 that for appropriate constants β_0', γ' we have

$$\sum_{\Lambda:\Lambda\cap\Delta(\mathscr{G})\neq\varnothing,|\Lambda|>1}|\Phi(\Lambda)|\le\exp\Big\{-2\gamma'(\beta-\beta_0')\sum_{\Gamma\in\mathscr{G}}|\Gamma|\Big\}. \tag{3.20.7}$$

Furthermore, (3.3.11) implies that

$$\Big|\beta\Big(\sum_{t\in\Delta^E(\mathscr{G})}h_t-\sum_{t\in\Delta^I(\mathscr{G})}h_t\Big)-\sum_{t\in\Delta(\mathscr{G})}\Phi(\{t\})\Big|$$
$$\le -2\beta\sum_{t\in\Delta^I(\mathscr{G})}h_t+C|\Delta(\mathscr{G})|\le 2\beta\sum_{(\Delta^I(\mathscr{G}))^0}|h_t|+C|\Delta(\mathscr{G})|. \tag{3.20.8}$$

for some constant $C=C(\nu,c)$. The desired estimate (3.20.1) follows from relations (3.20.5)–(3.20.8) and the main condition (3.3.2).

4. Classical Behavior of Deviations

In this chapter we will prove part of the results formulated in §§1.5, 1.6, and 1.8 about the classical behavior of large and moderate deviations for sufficiently large values of β. We use the general statements proved in Chapters 2 and 3. First we consider the cases of +-boundary conditions and the limit Gibbs +-state. In §4.7 we extend the results to periodic and free boundary conditions.

§4.1. The verification of the locality condition and the lower bound for the variance

In this section we shall verify that the locality condition (2.1.18) used in §2.1 in the derivation of asymptotics for probabilities of large deviations holds, and the variance grows linearly with the volume for the cutoff Ising distribution with sufficiently large β.

PROPOSITION 4.1. *We assume (using the notation of* §3.2) *that the cutoff set is* $\mathscr{L}=G^{\omega}_{W,V}$, *where* $W\subset V\subset\mathbb{Z}^\nu$ *are finite sets, and that (cf.* (3.19.1)) $h=\{h_t,\ t\in V\}$, *where*

$$h_t=\begin{cases}h^0\ge 0, & t\in V\setminus W,\\ h^1, & t\in W.\end{cases} \tag{4.1.1}$$

Assume also that a cutoff level ω *satisfies* $\omega > 2\nu$ *and for some* c, $0 < c < 1$, $H < \infty$ *(cf.* (3.19.4)) *we have*

$$H > h^1 > -c\omega^{-1}. \tag{4.1.2}$$

We denote by $P_{W,V}^{+,\beta,h,\omega}$ *the probability distributions given by* (3.2.3) *for this case and by* $\langle f\rangle_{W,V}^{+,\beta,h,\omega}$ *the mean value of a function* f *with respect to this distribution.*

Then there exist positive numbers $\beta_0 = \beta_0(c, H, \nu)$, $\varkappa = \varkappa(\beta, c, H, \nu)$, *and* $\alpha = \alpha(\beta, c, H, \nu)$ *such that for all* V, W, h^0, h^1 *and* $\beta \geq \beta_0(c, H, \nu)$

$$|\varphi_W(t)| \leq \exp\{-\varkappa D(W)t^2\} + \exp\{-\varkappa |W|^{(\nu-1)/\nu}\}, \quad 0 \leq |t| < \pi/2 \tag{4.1.3}$$

where

$$\varphi_W(t) = \varphi_{W,V,\beta,h,\omega}(t) = \langle \exp\{itS_W\}\rangle_{W,V}^{+,\beta,h,\omega} \tag{4.1.4}$$

is the characteristic function of the random variable S_W *with the distribution* $P_{W,V}^{+,\beta,h,\omega}$ *and (see* (3.2.11))

$$D(W) = D(W, V, +, \beta, h, \omega) \tag{4.1.5}$$

is the variance of this random variable. Finally,

$$D(W) \geq \alpha|W|. \tag{4.1.6}$$

PROOF. In the course of this proof we will omit $V, \beta, +, h, \omega$ from the notation for probabilities and mean values. Both statements (4.1.3) and (4.1.6) will be proved using the construction generalizing the construction used in the proof of the similar facts for the ensemble without a cutoff in §2.3. Similarly to §2.3, let $W^e \subset V$ be a maximal set of points $t \in W$ such that all the distances $|s - t|$, $s, t \in W^e$, $s \neq t$, are not smaller than 2, and $V^c = V \setminus W^e$. The number of points $|W^e| \gtrsim 2^{-\nu}|W|$ as $|W| \to \infty$.

We shall use the method of §2.3, which is based on the estimation of conditional variances and conditional characteristic functions. However, the present situation has additional difficulty: in the case of a cutoff ensemble the field $\{\sigma_t, t \in V\}$ is not a Markov field any more, so that the variables σ_s, $s \in W^e$, are not conditionally independent under the condition that the values σ_t, $t \in V^c$, are fixed. The following construction is used to resolve this difficulty.

Let G^0 be the set of all elementary contours Γ_t having 2ν faces and surrounding points $t \in W^e$. Let us be given a system of contours $G \subset G_{W,V}^{\omega} \setminus G^0$ (see §3.2) and let $\Omega(G)$ be the set of all configurations $\sigma \in \Omega_{W,V}^{\omega}$ such that the corresponding set of contours $\mathscr{G}(\sigma)$ satisfies $\mathscr{G}(\sigma) \setminus G^0 = G$ (so to fix G means to fix all the contours of a configuration except the elementary contours surrounding points $t \in W^e$). Let $Q(G)$ be the collection of points $t \in W^e$ such that the elementary contours Γ_t have empty intersections with all the contours from G. It is easy to see that for any $\sigma \in \Omega(G)$ the corresponding contour system $\mathcal{G}(\sigma)$ can be represented as

$$\mathcal{G}(\sigma) = G \cup \Big(\bigcup_{t \in Q(\sigma)} \Gamma_t \Big), \tag{4.1.7}$$

where $Q(\sigma)$ is a subset of the set $Q(G)$ and the correspondence between such subsets and the configurations $\sigma \in \Omega(G)$ is one-to-one. Let $Q^{\pm}(G)$ be a collection of sites $t \in Q(G)$ such that t belongs to an even (odd) number of sets $\operatorname{Int}\Gamma$, $\Gamma \in G$. If $t \in Q^{+}(G)$ and $\sigma \in \Omega(G)$, then $\sigma_t = +1 \Longleftrightarrow \sigma_t = +1 \Longleftrightarrow t \notin Q(\sigma)$ (or $\Longleftrightarrow t \in Q(\sigma)$ in the opposite case $t \in Q^{-}(G)$). So the conditional probablilities are given by

$$P(\sigma_t, t \in Q(G) \mid \Omega(G)) = \prod_{t \in Q^{+}(G)} p_{+}(\sigma_t) \prod_{t \in Q^{-}(G)} p_{-}(\sigma_t), \tag{4.1.8}$$

where

$$p_{\pm}(\sigma) = \frac{\exp\{\sigma\beta(h^1 \pm 2\nu)\}}{\exp\{\beta(h^1 \pm 2\nu)\} + \exp\{-\beta(h^1 \pm 2\nu)\}}. \tag{4.1.9}$$

The condition $\sigma \in \Omega(G)$ fixes the values of σ_t, $t \notin Q(\sigma)$. Repeating the considerations of §2.3, we find (cf. (2.3.15)) that

$$D(S_W) \geq \sum_{G \subset G^{\omega}_{W,\nu} \setminus G^0} P(\sigma : \mathcal{G}(\sigma) \setminus G^0 = G) D(S_W \mid \{\sigma : \mathcal{G}(\sigma) \setminus G^0 = G\}). \tag{4.1.10}$$

Under the condition (4.1.2), for each fixed β the variances of random variables having the probability distributions $p_{\pm}$ are uniformly separated from zero, and since the variance of the sum of independent variables is equal to the sum of their variances, (4.1.8) implies that the conditional variance satisfies

$$D(S_W \mid \{\sigma : \mathcal{G}(\sigma) \setminus G^0 = G\}) \geq q|Q(G)|, \tag{4.1.11}$$

where $q = q(\beta, c, H, \nu)$ is a positive constant, and so

$$D(S_W) \geq q\langle |Q(\mathcal{G}(\sigma) \setminus G^0)| \rangle. \tag{4.1.12}$$

In a similar way we can repeat the construction of §2.3 and find (cf. (2.3.21)) that

$$|\varphi_W(t)| \leq \sum_{G \subset G^{\omega}_{W,\nu} \setminus G^0} P(\sigma : \mathcal{G}(\sigma) \setminus G^0 = G) |\langle e^{itS_W} \mid \{\sigma : \mathcal{G}(\sigma) \setminus G^0 = G\} \rangle|. \tag{4.1.13}$$

Formula (4.1.9) implies that for some $\gamma = \gamma(\beta, c, H, \nu)$

$$|\langle e^{it\sigma} \rangle_{\pm}| \leq e^{-\gamma t^2}, \tag{4.1.14}$$

(where we average with respect to the probability distributions $p^{\pm}(\sigma)$). Since the characteristic function of the sum of independent variables is equal to the product of the chacteristic functions of summands, we see that

$$|\varphi_W(t)| \leq \sum_{G\subset G^{\omega}_{W,\nu}\setminus G_0} P(\sigma : \mathcal{G}(\sigma)\setminus G^0 = G)\exp\{-\gamma t^2|Q(G)|\}. \qquad (4.1.15)$$

We will prove below that for an appropiate choice of the constants $\gamma = \gamma(\nu, \beta, c, H)$ and $\beta_0 = \beta_0(\nu, c, H)$

$$\sum_{\substack{G\subset G^{\omega}_{W,\nu}\setminus G_0 : \\ |Q(G)|\leq 2^{-(\nu+1)}|W|}} P(\sigma : \mathcal{G}(\sigma)\setminus G_0 = G) \leq \exp\{-\gamma|W|^{(\nu-1)/\nu}\} \qquad (4.1.16)$$

for $\beta > \beta_0$. It is clear that from (4.1.16) and (4.1.12) follows the desired estimate (4.1.6) for an appropriate value of $\alpha = \alpha(\beta, c, H, \nu)$. Then using (3.19.5) with $n = 2$ we derive the estimate (4.1.3) from (4.1.15) and (4.1.16) for an appropriate $\varkappa = \varkappa(\nu, \beta, c, H)$.

To prove the estimate (4.1.16) we observe that

$$\operatorname{dist}\left(t, \bigcup_{\Gamma\in\mathcal{G}^E(\sigma)} \operatorname{Int}\Gamma\right) \leq \frac{1}{2}\nu^{1/2} \quad \text{if } t\in W^e\setminus Q(\mathcal{G}(\sigma)), \qquad (4.1.17)$$

where $\mathcal{G}^E(\sigma)$ is the set of all external contours of the configuration σ (see §3.2). So for a positive constant $d_1 = d_1(\nu)$, depending on the dimension ν only,

$$\sum_{\Gamma\in\mathcal{G}^E(\sigma)} |\operatorname{Int}\Gamma| \geq d_1|W^e\setminus Q(\mathcal{G}(\sigma))|. \qquad (4.1.18)$$

It is easy to see that there exists a constant $d_2 = d_2(\nu)$ such that for any system of contours G

$$\sum_{\Gamma\in G}|\Gamma| \geq d_2\left(\sum_{\Gamma\in G}|\operatorname{Int}\Gamma|\right)^{(\nu-1)/\nu}. \qquad (4.1.19)$$

If $|Q(G)| \leq 2^{-(\nu+1)}|W|$, then $|W^e\setminus Q(G)| \gtrsim 2^{-(\nu+1)}|W|$ since $|W^e| \gtrsim 2^{\nu}|W|$, so that (4.1.18) and (4.1.19) imply that, for a new positive constant $d_3 = d_3(\nu)$,

$$\sum_{\Gamma\in\mathcal{G}^E(\sigma)} |\Gamma| \geq d_3|W|^{(\nu-1)/\nu} \quad \text{if } |Q(\mathcal{G}(\sigma))| \leq 2^{-(\nu+1)}|W|. \qquad (4.1.20)$$

Now the desired estimate (4.1.16) follows from Proposition 3.20.

§4.2. Large deviations in the case of cutoff ensembles

PROPOSITION 4.2. *Let sequences $W_N \subset V_N$ of finite volumes in $\mathbb{Z}^{\nu}$ and a sequence $\omega_N \to \infty$ of cutoff levels be given. Denote by h_N the set of the magnetic fields given by* (4.1.1) *with the condition* (4.1.2) *for $V = V_N$, $W = W_N$, by P_N^+ the probability distribution $P^{+,\beta,h_N,\omega_N}_{W_N,V_N}$ used in Proposition* 4.1, *and* *by* $\langle\cdot\rangle_N^+$ *the expectation with respect to this probability distribution. Suppose that a sequence of integers* b_N *is given such that*

$$(b_N - |W_N|) = 0 \pmod 2, \qquad |b_N| \leq |W_N|. \tag{4.2.1}$$

Let

$$E_N(\overline{h}) = E(W_N, V_N, +, \beta, \hat{h}_N(\overline{h}), \omega_N) \tag{4.2.2}$$

be the mean value of the variable S_{W_N} *(see* (3.2.10)*) defined for the fields* $\hat{h}_N(\overline{h}) = (\hat{h}_t, t \in V_N)$ *such that*

$$\hat{h}_t = \begin{cases} h^0, & t \in V_N \setminus W_N, \\ \overline{h}, & t \in W_N, \end{cases} \tag{4.2.3}$$

where $\overline{h}$ *is a constant. Let* $\overline{h}_N(b)$ *be the (unique) solution of the equation*

$$E_N(\overline{h}_N(b)) = b. \tag{4.2.4}$$

We suppose that the sequence $\overline{h}_N(b_N)$ *is bounded from above uniformly with respect to* β *and for some* $\overline{r}$, $c < \overline{r} < 1$, *(see* (4.1.2)*)*

$$\overline{h}_N(b_N) > -\overline{r}\omega_N^{-1}. \tag{4.2.5}$$

We suppose also that for some $\varepsilon > 0$

$$\lim_{N\to\infty} \max\{\overline{h}_N(b_N), \omega_N^{-1}\}|W_N|^{\frac{1}{2}-\varepsilon} = \infty. \tag{4.2.6}$$

Then for some $\beta_0 = \beta_0(\nu, c, \overline{r}, \varepsilon)$ *and all* $\beta \geq \beta_0$

$$\begin{aligned} P_N^+(b_N) &= P_N^+(\sigma : S_{W_N}(\sigma) = b_N) \\ &= \frac{2}{\sqrt{2\pi D_N(\overline{h}_N(b_N))}} \exp\{-I_N(b_N)\}(1 + o_N(1)), \end{aligned} \tag{4.2.7}$$

where $o_N(1) \to 0$, *as* $N \to \infty$,

$$I_N(b) = \beta(\overline{h}_N(b) - h^1)b - \overline{H}_N(\overline{h}_N(b)), \tag{4.2.8}$$

$$\overline{H}_N(\overline{h}) = \ln\left(\frac{Z(V_N, +, \beta, \hat{h}_N(\overline{h}), G^{\omega_N}_{W_N, V_N})}{Z(V_N, +, \beta, h_N, G^{\omega_N}_{W_N, V_N})}\right) \tag{4.2.9}$$

with the partition functions defined by (3.2.5), *and the variance (see* (3.2.11)*) is given by*

$$D_N(\overline{h}) = D(W_N, V_N, +, \beta, \hat{h}_N(\overline{h}), \omega_N). \tag{4.2.10}$$

The inequality

$$P_N^+(b_N) \leq \exp\{-I_N(b_N)\} \tag{4.2.11}$$

is valid even if we drop the assumption that the sequence $\overline{h}_N(b_N)$ *is bounded from above and the conditions* (4.2.5), (4.2.6) *hold.*

PROOF. We apply Theorem 2.1 to the sequence of random variables

$$S_N = S_{W_N}(\sigma), \tag{4.2.12}$$

where configurations σ have the probability distribution P_N^+. Then the function $H_N(\lambda)$ introduced in (2.1.4), (2.1.5) is equal to

$$H_N(\lambda) = \ln\left(\frac{Z(V_N, +, \beta, h_N^\lambda, G_{W_N, V_N}^{\omega_N})}{Z(V_N, +, \beta, h_N, G_{W_N, V_N}^{\omega_N})}\right), \tag{4.2.13}$$

where $h_N^\lambda = \hat{h}_N(h^1 + \lambda\beta^{-1}) = \{h_t^\lambda, t \in V_N\}$ and

$$h_t^\lambda = \begin{cases} h^0, & t \in V_N \setminus W_N, \\ h^1 + \lambda\beta^{-1}, & t \in W_N. \end{cases} \tag{4.2.14}$$

The statement (4.2.7) coincides with the main statement (2.1.27) of Theorem 2.1, and the estimate (4.2.11) coincides with the estimate (2.1.21).

We must verify the condition of Theorem 2.1 used in the proof of the relation (4.2.7). The condition (2.1.22) follows from (4.2.5) and the boundedness of $\overline{h}_N(b_N)$. It follows from the estimate (4.1.6) of Theorem 4.1 that

$$\hat{D}_N \equiv D_N(\overline{h}_N(b_N)) \geq \alpha|W_N|, \tag{4.2.15}$$

and so $\hat{D}_N \to \infty$ as $N \to \infty$. The estimate (4.1.3) of Theorem 4.1 together with (4.2.15) shows that if

$$\hat{\varphi}_N(t) = \langle \exp\{itS_{W_N}\}\rangle^{+, \beta, \hat{h}_N(\overline{h}_N(b_N)), \omega_N} \tag{4.2.16}$$

is the characteristic function of the variable S_{W_N} with respect to the tilted distribution, then for a constant $K = K(\varkappa)$ and the number γ_N defined in (2.1.17) we have

$$|W_N|^{1/2} \int_{\gamma_N < |t| < \pi} |\hat{\varphi}_N(t)|\, dt$$
$$\leq 2|W_N|^{1/2}\left(\int_{\gamma_N}^{\infty} e^{-\varkappa \hat{D}_N t^2}\, dt + \pi e^{-\varkappa|W_N|^{(\nu-1)/\nu}}\right) \to 0 \tag{4.2.17}$$

as $N \to \infty$, and by (4.2.15) the locality condition (2.1.18) holds.

Finally, let a number r satisfy $\overline{r} < r < 1$. Then we use Lemma 3.18 to observe that the main condition (3.3.2) of Proposition 3.3 is fulfilled with a constant $\gamma = \gamma(r)$ for fields $\hat{h}_N(\overline{h})$, $\overline{h} \in \mathscr{O}_N$, where

$$\mathscr{O}_N = \{\overline{h} \in \mathbb{C} : |\overline{h}| \leq r(\omega_N)^{-1}\} \cup \{\overline{h} \in \mathbb{C} : |\operatorname{Im} \overline{h}| \leq 2r \operatorname{Re} \overline{h}\}. \tag{4.2.18}$$

This condition is applicable both to fields h_N^λ, where $h^1 + \lambda\beta^{-1} \in \mathscr{O}_N$, and to fields h_N. So we can apply the main expansion (3.3.1) to each partition function in (4.2.13). Since $\Phi(\Lambda)$ depends only on h_t, $t \in \Lambda$, we see that

$$H_N(\lambda) = \exp\left\{ \sum_{\Lambda : \Lambda \cap W_N \neq \varnothing} (\Phi(\Lambda, +, \beta, h_N^\lambda, G_{W_N, V_N}^{\omega_N}) \right.$$
$$\left. - \Phi(\Lambda, +, \beta, h_N, G_{W_N, V_N}^{\omega_N}))\right\}. \tag{4.2.19}$$

So it follows from the estimate (3.3.11) together with Lemma 3.4 that for a sufficiently large β_0, $\beta \geq \beta_0$, and some constant $C = C(\nu, c, \bar{r})$

$$|\ln H_N(\lambda)| \leq C|W_N| \tag{4.2.20}$$

if $h^1 + \lambda\beta^{-1}$ lies in the domain (4.2.18). Using the estimate (4.2.15) we see that the analyticity condition (2.1.14) is satisfied with a constant $Q_N \equiv Q(\nu, \beta, H, c, \bar{r})$ which does not depend on N. We can take (see (2.1.24), (2.1.14), and (3.18.3))

$$\hat{R}_N = \beta \max\left\{\frac{2r}{\sqrt{5}} h_N(b_N),\ (r - \bar{r})(\omega_N)^{-1}\right\}. \tag{4.2.21}$$

The last thing we need is a bound on the semiinvariants. In our case (cf. (3.2.9) and (2.1.23))

$$\hat{G}_N^k = G^k(W_N, V_N, +, \beta, \hat{h}_N(\bar{h}_N(b_N)), \omega_N), \tag{4.2.22}$$

and these semiinvariants were estimated in Proposition 3.19, which is applicable because of (4.2.5). The inequality (3.19.5) implies that for $\beta \geq \beta_0(\nu, c, \bar{r})$

$$|\hat{G}_N^k| \leq K_k|W_N|, \qquad k = 3, 4, \ldots, \tag{4.2.23}$$

where $K_k = K_k(\nu, c, \bar{r})$ are certain constants.

Now we derive from the estimates (4.2.23) and (4.2.15) that the condition (2.1.25) is valid for any $k \geq 3$. It follows from (4.2.21), (4.2.23), (4.2.15), and the main condition (4.2.6) that

$$\lim_{N\to\infty} \frac{(\ln \hat{D}_N)^{\frac{n}{2}}}{(\hat{D}_N)^{\frac{n}{2}-1}} (\hat{R}_N)^{-n} = 0 \quad \text{if } n > 1/\varepsilon, \tag{4.2.24}$$

and this is the condition (2.1.26). So all the conditions of Theorem 2.1 are fulfilled for an appropriate n, and the proof is completed.

§4.3. A refinement of the previous proposition

In §4.2 we formulated the theorem about large deviations using the number $\bar{h}_N(b_N)$, which is a solution of the transcendental equation (4.2.4). Now we want to state this condition more explicitly.

PROPOSITION 4.3. *Let the assumptions of Proposition* 4.2 (*except* (4.2.5), (4.2.6)) *be satisfied. For any* $C \leq \infty$ *there exist* $\beta_0 = \beta_0(\nu, C)$, $c_0 = c_0(\nu, \beta)$ *such that if* $\beta \geq \beta_0$ *and* (*see the notation* (4.2.2))

$$\tanh(C\beta)|W_N| \geq b_N \geq E_N(0) - c_0\omega_N^{-1}|W_N|, \tag{4.3.1}$$

then the condition (4.2.5) *of Proposition* 4.2 *is valid for some* $\bar{r} = \bar{r}(\nu, \beta, c_0) > 0$. *If*

$$b_N \leq \tanh(C\beta)|W_N|, \tag{4.3.2}$$

then the sequence $\overline{h}_N(b_N)$ *is bounded from above by a constant which does not depend on* β. *If for some* $\varepsilon > 0$ *either*

$$\lim_{N\to\infty}(b_N - E_N(0))|W_N|^{-(1/2-\varepsilon)} = \infty \quad \textit{and} \quad b_N \geq E_N(0) \tag{4.3.3}$$

or

$$\lim_{N\to\infty}\omega_N|W_N|^{-(1/2-\varepsilon)} = 0, \tag{4.3.4}$$

then the condition (4.2.6) *of Proposition* 4.2 *holds.*

PROOF. It is clear from the definition (3.2.9)–(3.2.11) that (see notations (4.2.2) and (4.2.10))

$$\frac{dE_N(\overline{h})}{d\overline{h}} = \beta D_N(\overline{h}). \tag{4.3.5}$$

It follows from (3.19.5) and (4.1.6) that for some $\alpha = \alpha(\nu, \beta, c, C) > 0$, $K = K(\nu, c) < \infty$ and all sufficiently large β,

$$\alpha|W_N| \leq \frac{dE_N(\overline{h})}{dh} \leq K|W_N| \tag{4.3.6}$$

if

$$C \geq \overline{h} \geq -c\omega_N^{-1} \tag{4.3.7}$$

for some positive $c < 1$, $C < \infty$. So if the second inequality in (4.3.1) is valid with $c_0 = \overline{r}\alpha$, then the condition (4.2.5) must hold. If the conditions (4.3.3) and (4.3.1) are satisfied, then (4.3.6) implies that

$$\lim_{N\to\infty}\overline{h}_N(b_N)|W_N|^{\frac{1}{2}-\varepsilon} = \infty, \tag{4.3.8}$$

and the condition (4.2.6) is fulfilled. It is clear that (4.2.6) follows also from (4.3.4).

Now we want to prove that the condition (4.3.2) implies that the sequence h_N is bounded. It suffices to prove that uniformly with respect to N

$$E_N(\overline{h}) = E(W_N, V_N, +, \beta, \hat{h}_N(\overline{h}), \omega_N) \geq \tanh(C\beta)|W_N|, \tag{4.3.9}$$

provided that $\overline{h} \geq \overline{h}(C)$. Using the definition (4.2.2) and (3.2.10), we note that

$$E_N(\overline{h}) = \sum_{t\in W_N}\langle\sigma_t\rangle_{W_N, V_N}^{+, \beta, \hat{h}_N(\overline{h}), \omega_N}. \tag{4.3.10}$$

If $\omega_N \equiv \infty$ (the ensemble without a cutoff), we can rewrite the expectations in the above sum as linear combinations of conditional expectations when the configurations are given on $V_N \setminus \{t\}$ and use explicit formulas for these conditional expectations (using the Markov property of the field). This implies (4.3.9) for the case $\omega_N \equiv \infty$. The field with a cutoff has no Markov property, but we can use the estimate (3.19.6) to compare the quantities $E_N(\overline{h})$

in the cases $\omega_N \equiv \infty$ and $\omega_N \to \infty$, and this comparison proves the desired statement (4.3.9).

§4.4. The absence of large contours

The condition (4.3.3) of Theorem 4.3 can be used in the case of the ensemble without cutoff, i.e., in the case when $\omega_N \equiv \infty$. We shall see in §4.6 that this permits us to prove Theorems 1.5.1 and 1.5.2 directly in the case when the ratio $b_N/|V_N|$ is positive and does not tend to zero. But for the case when $b_N/|V_N|$ is close to zero we must use the condition (4.3.4), which can be used only for cutoff ensembles. To overcome this difficulty we prove here a result which shows that the influence of the cutoff is asymptotically negligible. This result is very close to the first statement of Theorem 1.8, which was discussed in details in §1.8 (see also the proof of Theorem 1.8 in §4.5).

PROPOSITION 4.4. *Let sequences of contractible* (*see* §1.5) *finite volumes* $W_N \subset V_N \subset \mathbb{Z}^\nu$ *be given such that* $|W_N| \to \infty$ *and* W_N *satisfies the Van Hove condition* (1.4.1). *Let* $h^0 \geq 0$, $h^1 \geq 0$. *Let a cutoff level* ω_N *be such that for some* $\varepsilon > 0$

$$|W_N|^{1/(\nu+1)} \geq \omega_N \geq (\ln|W_N|)^{1/(\nu-1)}. \tag{4.4.1}$$

Suppose that a sequence b_N *of integers is given such that the condition* (4.2.1) *holds, and for some* $C < \infty$, $c > 0$

$$\tanh(C\beta)|W_N| \geq b_N \geq E_N(0) - c|W_N|^{\nu/(\nu+1)}, \tag{4.4.2}$$

where $E_N(0)$ *is defined by the relation* (4.2.2). *Denote by* $P_N^{+,h}$ *the probabilities* $P_{W_N, V_N}^{+,\beta,h,\infty}$, *where* h *is defined by* (4.1.1). *Then for a sufficiently large* $\beta_0 = \beta_0(\nu, C)$, *all* $\beta \geq \beta_0$, *and a sufficiently small* $c = c(\beta, \nu)$ *we have*

$$\frac{P_N^{+,h}\left(\sigma \in \Omega_{V_N} : S_{W_N}(\sigma) = b_N,\ \max_{\Gamma \in \mathscr{G}(\sigma)\,:\,\mathrm{Int}\,\Gamma \cap W_N \neq \varnothing} q(\Gamma) > \omega_N\right)}{P_N^{+,h}(\sigma \in \Omega_{V_N} : S_{W_N}(\sigma) = b_N)}$$
$$\leq e^{-2(\beta-\beta_0)(\omega_N)^{\nu-1}}. \tag{4.4.3}$$

PROOF. It follows immediately from definitions (3.2.3)–(3.2.5) that the ratio considered in (4.4.3) does not depend on the choice of the field h^1. So without loss of generality we can consider the case $h = \{h_t,\ t \in V_N\}$, where

$$h_t = \begin{cases} h^0, & \text{if } t \in V_N \setminus W_N, \\ 0, & \text{if } t \in W_N. \end{cases} \tag{4.4.4}$$

We need a special construction that specifies a subsystem of contours of a configuration $\sigma \in \Omega_{V_N}$. Let $\mathscr{G}^L_{W_N, \omega_N}(\sigma) = \mathscr{G}^L(\sigma)$ be a subsystem of all large contours of the configuration σ having nonempty intersection with W_N, i.e., the system of all contours $\Gamma \in \mathscr{G}(\sigma)$ such that

$$q(\Gamma) > \omega_N, \qquad \mathrm{Int}\,\Gamma \cap W_N \neq \varnothing. \tag{4.4.5}$$

(Note that the above definition is different from the definition of large contours used in Chapter 3.) Let D_N^L be the set of all systems of contours ∂^L such that $\partial^L = \mathcal{G}^L(\sigma)$ for some $\sigma \in \Omega_{V_N}$. For each $\partial^L \in D_N^L$ there exists a unique configuration $\sigma(\partial^L) \in \Omega_{V_N}$ such that its system of contours is $\mathcal{G}(\sigma(\partial^L)) = \mathcal{G}^L(\sigma(\partial^L)) = \partial^L$ (with $+$-boundary condition outside V_N). Let $U^+(\partial^L) = \{t \in V_N : \sigma_t(\partial^L) = +1\}$, $U^-(\partial^L) = \{t \in V_N : \sigma_t(\partial^L) = -1\}$. The sets $U^\pm(\partial^L)$ will be called the domains of $\pm$-phases defined by the system ∂^L. Let $U_W^\pm(\partial^L) = U^\pm(\partial^L) \cap W_N$ and $U^\pm(\sigma) = U^\pm(\partial^L(\sigma))$, $U_W^\pm(\sigma) = U_W^\pm(\partial^L(\sigma))$. Let $\mathcal{G}^{E,-}(\sigma)$, $\sigma \in \Omega_{V_N}$, be the system of all external contours Γ of the restriction $\sigma_{U^-(\sigma)}$ of the configuration σ to the $-$-phase domain $U^-(\sigma)$, having points of W_N inside them, i.e. the set of all external contours Γ of the family of contours $\mathcal{G}(\sigma) \setminus \mathcal{G}^L(\sigma)$ such that

$$\operatorname{Int}\Gamma \cap W_N \neq \varnothing, \qquad \operatorname{Int}\Gamma \subset U^-(\sigma). \tag{4.4.6}$$

Let $\mathcal{G}_s^{E,-}(\sigma) \subset \mathcal{G}^{E,-}(\sigma)$ be the set of all contours with s faces belonging to $\mathcal{G}^{E,-}(\sigma)$ and $k_s(\sigma)$ be the number of these contour. Finally, for a constant $q > 0$ we define

$$\hat{\mathcal{G}}(\sigma) = \bigcup_{s\,:\,k_s(\sigma) > (qs^4)^{-1}|U_W^-(\sigma)|} \mathcal{G}_s^{E,-}(\sigma). \tag{4.4.7}$$

(So we include contours with s faces into $\hat{\mathcal{G}}(\sigma)$ only if there are sufficiently many of them.) We put further

$$\overline{\mathcal{G}}(\sigma) = \mathcal{G}^L(\sigma) \cup \hat{\mathcal{G}}(\sigma). \tag{4.4.8}$$

We denote by D_N the set of possible systems of contours $\partial = \overline{\mathcal{G}}(\sigma)$ for $\sigma \in \Omega_{V_N}$ and by $\Omega_\partial \subset \Omega_{V_N}$, $\partial \in D_N$, the set of all configurations σ such that $\partial = \overline{\mathcal{G}}(\sigma)$. It is clear that $\partial^L \subset \partial$ is the subsystem of all large contours in ∂. For any $\partial^L \in D_N^L$ let $D_N(\partial^L)$ be the set of all possible systems of contours g such that $g = \partial \setminus \partial^L$ for some $\partial \in D_N$. For $\sigma \in \Omega_\partial$ we let

$$\begin{aligned} &\mathcal{G}^L(\partial) = \mathcal{G}^L(\sigma), &&\hat{\mathcal{G}}(\partial) = \hat{\mathcal{G}}(\sigma), \\ &U^\pm(\partial) = U^\pm(\partial^L) = U^\pm(\sigma), &&U_W^\pm(\partial) = U_W^\pm(\partial^L) = U_W^\pm(\sigma). \end{aligned} \tag{4.4.9}$$

(It is clear that these objects do not depend on $\sigma \in \Omega_\partial$.) Finally, we set

$$W^-(\partial) = U_W^-(\partial) \setminus \bigcup_{\Gamma \in \hat{\mathcal{G}}(\partial)} \operatorname{Int}\Gamma. \tag{4.4.10}$$

To justify the introduction of such an enigmatic construction we prove two of its properties. These properties are the only features used below.

(i) For any configuration $\sigma \in \Omega_\partial$, $\partial \in D_N$, we have

$$S_{W^-(\partial)} \leq -(1-\lambda)|W^-(\partial)|, \tag{4.4.11}$$

where the positive constant $\lambda = \lambda(q, \nu)$ is such that $\lambda(q, \nu) \to 0$ as $q \to \infty$. Indeed, a configuration $\sigma \in \Omega_\partial$ can take positive values $\sigma_t = +1$ for $t \in W^-(\partial)$ only if $t \in \operatorname{Int}\Gamma'$ for some contour $\Gamma' \in \mathscr{G}^{E,-}(\sigma)\backslash\hat{\mathscr{G}}(\partial)$. Hence,

$$\begin{aligned} S_{W^-(\partial)} &\leq \sum_{\Gamma'\in\mathscr{G}^{E,-}(\sigma)\backslash\hat{\mathscr{G}}(\partial)} |\operatorname{Int}\Gamma'| - \Big(|W^-(\partial)| - \sum_{\Gamma'\in\mathscr{G}^{E,-}(\sigma)\backslash\hat{\mathscr{G}}(\partial)} |\operatorname{Int}\Gamma'|\Big) \\ &= -|W^-(\partial)| + 2 \sum_{\Gamma'\in\mathscr{G}^{E,-}(\sigma)\backslash\hat{\mathscr{G}}(\partial)} |\operatorname{Int}\Gamma'|. \end{aligned} \tag{4.4.12}$$

If $k_s(\sigma) \leq (qs^4)^{-1}|U_W^-(\partial)|$ there are $k_s(\sigma)$ contours $\Gamma' \in \mathscr{G}^{E,-}(\sigma) \setminus \hat{\mathscr{G}}(\partial)$ with s faces, and there are no such contours if $k_s(\sigma) > (qs^4)^{-1}|U_W^-(\partial)|$. So

$$\sum_{\Gamma'\in\mathscr{G}^{E,-}(\sigma)\backslash\hat{\mathscr{G}}(\partial)} |\operatorname{Int}\Gamma'| \leq \sum_{s=1}^{\infty}(qs^4)^{-1}|U_W^-(\partial)|c_\nu s^{\nu/(\nu-1)}, \tag{4.4.13}$$

where c_ν is a constant such that $|\operatorname{Int}\Gamma| \leq c_\nu|\Gamma|^{\nu/(\nu-1)}$ for any contour Γ. If we choose the constant q to be sufficiently large, the sum in (4.4.13) can be made smaller than $(\lambda/2)|U_W^-(\partial)|$, and this implies the estimate (4.4.11).

(ii) Let us fix a constant $\gamma > 0$ and define Σ_N^γ by

$$\Sigma_N^\gamma = \sum_{\partial\in D_N\,:\,\partial\neq\varnothing} \exp\Big\{-\gamma\sum_{\Gamma\in\partial}|\Gamma|\Big\}. \tag{4.4.14}$$

We claim that for some $\gamma_0 = \gamma_0(\nu) > 0$, all $\gamma \geq \gamma_0$, and any fixed $q > 0$

$$\lim_{N\to\infty} \Sigma_N^\gamma = 0. \tag{4.4.15}$$

To see this we make the following construction. We enumerate points of the set W_N in an arbitrary way. For each contour Γ such that $\operatorname{Int}\Gamma \cap W_N \neq \varnothing$ we choose the first point lying in $\operatorname{Int}\Gamma \cap W_N$, denote it by $t(\Gamma)$, and call it the *anchor* of the contour Γ. The usual combinatorial considerations show that there exist constants $R = R(\nu)$, $\varkappa = \varkappa(\nu) > 0$, depending only on the dimension ν, such that for any $t \in W_N$ the number of all possible contours with s faces and with the anchor at t satisfies the condition

$$|\{\Gamma \in G : |\Gamma| = s,\ t(\Gamma) = t\}| \leq Re^{\varkappa s}. \tag{4.4.16}$$

It is clear from the definition (4.4.14) that

$$\Sigma_N^\gamma = \sum_{\partial^L\in D_N^L,\,\partial^L\neq\varnothing} \exp\Big\{-\gamma\sum_{\Gamma\in\partial^L}|\Gamma|\Big\}\Big(\sum_{g\in D_N(\partial^L)} \exp\Big\{-\gamma\sum_{\Gamma\in g}|\Gamma|\Big\}\Big). \tag{4.4.17}$$

First we estimate the second factor. If a system g contains $\overline{k}_s(g)$ contours of area s, then

$$\exp\left\{-\gamma\sum_{\Gamma\in g}|\Gamma|\right\} = \prod_s e^{-\gamma s\overline{k}_s(g)}. \tag{4.4.18}$$

It follows from the main construction that if $g \in D(\partial^L)$, then

$$\overline{k}_s(g) > (qs^4)^{-1}|U_W^-(\partial^L)|, \quad \text{if } \overline{k}_s(g) \neq 0. \tag{4.4.19}$$

Let $N(k, s)$ be the number of different sets of k contours of area s that can enter in a system $g \in D_N(\partial^L)$. The anchors of such contours are all different (for a fixed g the interiors of contours $\Gamma \in g$ are disjoint). So the number of ways these contours can be anchored is estimated by a binomial coefficient and the number of contours with a fixed anchor is estimated by (4.4.16). We see that

$$N(k, s) \leq \binom{|U_W^-(\partial^L)|}{k} Re^{\varkappa s} \leq R_1\left(\frac{|U_W^-(\partial^L)|}{k}\right)^k e^{k+\varkappa s}, \tag{4.4.20}$$

where $R_1 = R_1(\nu)$ is another constant and where we have used an elementary estimate for the binomial coefficient obtained from the Stirling formula. Using (4.4.18), (4.4.19) we obtain that

$$\sum_{g\in D_N(\partial^L)} \exp\{-\gamma\sum_{\Gamma\in g}|\Gamma|\} \leq \prod_s\left(\sum_{k>(gs^4)^{-1}|U_W^-(\partial^L)|} N(k, s)e^{-\gamma sk}+1\right). \tag{4.4.21}$$

Using (4.4.20) we see that for any s

$$\begin{aligned}
&\sum_{k\,:\,k>(qs^4)^{-1}|U_W^-(\partial^L)|} N(k, s)e^{-\gamma sk} \\
&\quad\leq R_1 \sum_{k\,:\,k>(qs^4)^{-1}|U_W^-(\partial^L|} \exp\left\{-k\left((\gamma-\varkappa)s - \ln\left(\frac{|U_W^-(\partial^L)|}{k}\right) - 1\right)\right\} \\
&\quad\leq R_1\sum_{k=1}^{\infty}\exp\{-k((\gamma-\varkappa)s - 4\ln s - \ln q - 1)\} \leq R_2e^{-s\varkappa},
\end{aligned} \tag{4.4.22}$$

where we suppose γ is so large that $\gamma \geq 3\varkappa$, and where $R_2 = R_2(\nu, q)$ is a new constant. Using (4.4.21) and (4.4.22), we find that

$$\sum_{g\in D_N(\partial^L)} \exp\left\{-\gamma\sum_{\Gamma\in g}|\Gamma|\right\} \leq R_3, \tag{4.4.23}$$

where $R_3 = R_3(\nu, q)$ does not depend on ∂^L and N. Now in a similar way we can estimate the first sum in (4.4.17). Here we can use the estimate

$$N^L(k, s) \le Re^{\varkappa s}|W_N|^k \tag{4.4.24}$$

for the number $N^L(k, s)$ of possible choices of a system of k contours of length s in a system $\partial^L \in D_N^L$, which is more rough than (4.4.20). Using the main condition (4.4.1), relation (3.1.5), and the definition (4.4.5) we see that

$$N^L(k, s) = 0 \quad \text{if } s \le (2\nu)^\nu \ln|W_N| \le (2\nu)^\nu |\omega_N|^{\nu-1}, \tag{4.4.25}$$

and so

$$\begin{aligned} &\sum_{\partial^L \in D_N^L : \partial^L \neq \varnothing} \exp\Big\{-\gamma \sum_{\Gamma\in\partial^L} |\Gamma|\Big\} \\ &\le \prod_{s>(2\nu)^\nu \ln|W_N|} \Big(\sum_{k=1}^{\infty} N^L(k, s)e^{-\gamma ks} + 1\Big) - 1. \end{aligned} \tag{4.4.26}$$

(The additional term -1 arises because $\partial^L \neq \varnothing$). It follows from (4.4.24) that for $s > (2\nu)^\nu \ln|W_N|$ we have

$$N^L(k, s)e^{-\gamma ks} \le Re^{-s(k\gamma - \varkappa - k(2\nu)^{-\nu})} \le Re^{-\gamma ks/2}, \tag{4.4.27}$$

if $\gamma \ge \gamma_0(\nu)$, where $\gamma_0(\nu)$ is sufficiently large. Hence

$$\lim_{N\to\infty} \sum_{\partial^L \in D_N^L : \partial^L \neq \varnothing} \exp\Big\{-\gamma \sum_{\Gamma\in\partial^L} |\Gamma|\Big\} = 0, \tag{4.4.28}$$

and the desired (4.4.15) follows from (4.4.17), (4.4.23), and (4.4.28).

It is clear (see (4.4.11)) that

$$\begin{aligned} &P_N^{+,h}\Big(\sigma \in \Omega_{V_N} : S_{W_N}(\sigma) = b_N, \max_{\Gamma\in\mathscr{G}(\sigma) : \operatorname{Int}\Gamma\cap W_N \neq \varnothing} q(\Gamma) > \omega_N\Big) \\ &= \sum_{\partial\in D_N : \partial\neq\varnothing} P_N^{+,h}(\sigma \in \Omega_{V_N} : S_{W_N}(\sigma) = b_N, \overline{\mathscr{G}}(\sigma) = \partial) \\ &= \sum_{\partial\in D_N : \partial\neq\varnothing} \sum_{\bar b=0}^{[\lambda|W_N^-(\partial)|]} P_N^{+,h}(\sigma \in \Omega_{V_N} : S_{W_N}(\sigma) = b_N, \\ &\qquad\qquad S_{W^-(\partial)} = -|W^-(\partial)| + \bar b, \overline{\mathscr{G}}(\sigma) = \partial). \end{aligned} \tag{4.4.29}$$

Let (cf. §3.1 and (4.4.10)) for any $\partial \in D_N$

$$
\begin{aligned}
&V^-(\partial) = U^-(\partial) \setminus \bigcup_{\Gamma \in \hat{\mathscr{G}}(\partial)} \operatorname{Int}\Gamma \\
&V^+(\partial) = U^+(\partial) \bigcup_{\Gamma \in \hat{\mathscr{G}}(\partial)} \operatorname{Int}\Gamma = V_N \setminus V^-(\partial), \\
&W^+(\partial) = U_W^+(\partial) \bigcup_{\Gamma \in \hat{\mathscr{G}}(\partial)} (\operatorname{Int}\Gamma \cap W_N) = W_N \setminus W^-(\partial), \\
&\Delta(\partial) = \bigcup_{\Gamma \in \partial} \Delta(\Gamma), \\
&\Delta_W(\partial) = \Delta(\partial) \cap W_N, \\
&\Delta^\pm(\partial) = \Delta(\partial) \cap V^\pm(\partial), \\
&\Delta_W^\pm(\partial) = \Delta_W(\partial) \cap W^\pm(\partial), \\
&\overline{V}^\pm(\partial) = V^\pm(\partial) \setminus \Delta^\pm(\partial), \\
&\overline{W}^\pm(\partial) = W^\pm(\partial) \setminus \Delta_W^\pm(\partial), \\
&|\partial| = \sum_{\Gamma \in \partial} |\Gamma|.
\end{aligned}
\tag{4.4.30}
$$

Observe now that if $\overline{\mathscr{G}}(\sigma) = \partial$ then $\sigma_t = 1$ for $t \in \Delta^+(\partial)$ and $\sigma_t = -1$ for $t \in \Delta^-(\partial)$. In this case

$$
S_{W_N}(\sigma) = S_{\overline{W}^+(\partial)}(\sigma) + |\Delta_W^+(\partial)| - |W^-(\partial)| + \overline{b} \qquad \text{if } S_{W^-(\partial)} = -|W^-(\partial)| + \overline{b}, \tag{4.4.31}
$$

and so

$$
\begin{aligned}
&P_N^{+,h}(\sigma \in \Omega_{V_N} : S_{W_N}(\sigma) = b_N, S_{W^-(\partial)} = -|W^-(\partial)| + \overline{b}, \overline{\mathscr{G}}(\sigma) = \partial) \\
&\quad \le P_N^{+,h}(\sigma \in \Omega_{V_N} : S_{\overline{W}^+(\partial)}(\sigma) = b(\overline{b}, \partial), \overline{\mathscr{G}}(\sigma) = \partial),
\end{aligned}
\tag{4.4.32}
$$

where

$$
b(\overline{b}, \partial) = b_N + |W^-(\partial)| - |\Delta_W^+(\partial)| - \overline{b}. \tag{4.4.33}
$$

Note also that for $\overline{\mathscr{G}}(\sigma) = \partial$,

$$
q(\Gamma) \le \omega_N \quad \text{if } \Gamma \in \mathscr{G}(\sigma) \setminus \partial, \operatorname{Int}\Gamma \cap W_N \neq \varnothing, \tag{4.4.34}
$$

and

$$
\sigma_t = +1 \quad \text{if } t \in \Delta^+(\partial), \tag{4.4.35}
$$

and these two conditions are the only constrains imposed on the restriction $\sigma_{V^+(\partial)}$ of the configuration σ by the condition $\overline{\mathscr{G}}(\sigma) = \partial$. It follows from (3.1.3) that if a contour Γ' is inner to Γ, then $q(\Gamma) > q(\Gamma')$. Using these facts we find that the conditional distribution for $\sigma_{\overline{V}^+(\partial)}$ under the condition that $\overline{\mathscr{G}}(\sigma) = \partial$ coincides with the probability distribution

$P^{+,\beta,h^\partial,\mathscr{L}_N}_{\overline{V}^+(\partial)}$, where the family $\mathscr{L}_N = G^{\omega_N}_{\overline{W}^+(\partial),\overline{V}^+(\partial)}$ was introduced in §3.2 and where $h^\partial = \{h^\partial_t, t \in \overline{V}^+(\partial)\}$ with

$$h^\partial_t = \begin{cases} h^0, & \text{if } t \in \overline{V}^+(\partial) \setminus \overline{W}^+(\partial), \\ 0, & \text{if } t \in \overline{W}^+(\partial). \end{cases} \tag{4.4.36}$$

We denote this probability distribution by $P^{+,h}_{N,\partial}$. We have

$$\begin{aligned} &P^{+,h}_N(\sigma \in \Omega_{V_N} : S_{\overline{W}^+(\partial)}(\sigma) = b(\overline{b},\partial), \overline{\mathscr{G}}(\sigma) = \partial) \\ &\quad = P^{+,h}_N(\sigma \in \Omega_{V_N} : \overline{\mathscr{G}}(\sigma) = \partial) P^{+,h}_{N,\partial}(\sigma \in \Omega_{\overline{V}^+(\partial)} : S_{\overline{W}^+(\partial)}(\sigma) = b(\overline{b},\partial)). \end{aligned} \tag{4.4.37}$$

Now we want to estimate the first factor in (4.4.37). In addition to the properties of the conditional distribution under the condition $\overline{\mathscr{G}}(\sigma) = \partial$ stated above, this condition implies that $\sigma_t = -1$ if $t \in \Delta^-(\partial)$ and the restriction $\sigma_{\overline{V}^-(\partial)}$ of the configuration σ is such that its set of contours $\mathscr{G}^-(\sigma_{\overline{V}^-(\partial)}) \in G^{\omega_N}_{\overline{W}^-(\partial),\overline{V}^-(\partial)}$ (actually there are extra restrictions on this set of contours). So using the notation of §3.2 we find that

$$\begin{aligned} &P^{+,h}_N(\sigma \in \Omega_{V_N} : \overline{\mathscr{G}}(\sigma) = \partial) \\ &\quad \le \exp\{-2\beta|\partial| + \beta h^0(|\Delta^+(\partial) \setminus W_N| - |\Delta^-(\partial) \setminus W_N|)\} \\ &\quad\quad \times Z(\overline{V}^+(\partial), \beta, h^\partial, +, G^{\omega_N}_{\overline{W}^+(\partial),\overline{V}^+(\partial)}) \\ &\quad\quad \times Z(\overline{V}^-(\partial), \beta, h^\partial, -, G^{\omega_N}_{\overline{W}^-(\partial),\overline{V}^-(\partial)})(Z(W_N, V_N, \beta, h, +, \infty))^{-1} \end{aligned} \tag{4.4.38}$$

(the fields h^∂_t for $t \in \overline{V}^-(\partial)$ are defined by an analog of (4.4.36)). Using the estimate (3.3.5) (in our case all the sets $(\operatorname{Int}^0\Gamma)^0$ are empty since $h^\partial_t \ge 0$), we see that

$$Z(\overline{V}^-(\partial), \beta, h^\partial, -, G^{\omega_N}_{\overline{W}^-(\partial),\overline{V}^-(\partial)}) \le 2Z(\overline{V}^-(\partial), \beta, h^\partial, +, G^{\omega_N}_{\overline{W}^-(\partial),\overline{V}^-(\partial)}). \tag{4.4.39}$$

Now it is convenient to estimate the denominator in (4.4.3). We see (cf. (4.4.37)) that

$$\begin{aligned} P^{+,h}_N(\sigma \in \Omega_{V_N} : S_{W_N}(\sigma) = b_N) &\ge P^{+,h}_N(\sigma \in \Omega_{V_N} : \mathscr{G}^L(\sigma) = \varnothing, S_{W_N} = b_N) \\ &= P^{+,h}_N(\sigma \in \Omega_{V_N} : \mathscr{G}^L(\sigma) = \varnothing) P^{+,h}_{N,\varnothing}(\sigma \in \Omega_{V_N} : S_{W_N} = b_N). \end{aligned} \tag{4.4.40}$$

If $\partial = \varnothing$ is an empty system, then $\overline{V}^+(\varnothing) = V_N$, $\overline{W}^+(\varnothing) = W_N$, and so similarly to (4.4.38) we find that

$$P^{+,h}_N(\sigma \in \Omega_{V_N} : \mathscr{G}(\sigma) = \varnothing) = \frac{Z(V_N, \beta, h, +, G^{\omega_N}_{W_N,V_N})}{Z(W_N, V_N, \beta, h, +, \infty)}. \tag{4.4.41}$$

We find from (4.4.38), (4.4.39) and (4.4.41) that

$$\begin{aligned}\frac{P_N^{+,h}(\sigma \in \Omega_{V_N} : \overline{\mathscr{G}}(\sigma) = \partial)}{P_N^{+,h}(\sigma \in \Omega_{V_N} : \mathscr{G}^L(\sigma) = \varnothing)} &\\ \leq 2\exp\{-2\beta|\partial| + \beta h^0(|\Delta^+(\partial) \setminus W_N| - |\Delta^-(\partial) \setminus W_N|)\} & \qquad (4.4.42)\\ \times Z(\overline{V}^+(\partial), \beta, h^\partial, +, G^{\omega_N}_{\overline{W}^+(\partial), \overline{V}^+(\partial)}) &\\ \times Z(\overline{V}^-(\partial), \beta, h^\partial, +, G^{\omega_N}_{\overline{W}^-(\partial), \overline{V}^-(\partial)})(Z(V_N, \beta, h, +, G^{\omega_N}_{W_N, V_N}))^{-1}. &\end{aligned}$$

We apply the decomposition (3.3.7) to the partition functions in the right-hand side of (4.4.42). (It is applicable since in the considered case all the values h_t^∂ are nonnegative and so the condition (3.3.2) holds automatically.) We obtain

$$\begin{aligned}Z(\overline{V}^+(\partial), \beta, h^\partial, +, G^{\omega_N}_{\overline{W}^+(\partial), \overline{V}^+(\partial)})Z(\overline{V}^-(\partial), \beta, h^\partial, +, G^{\omega_N}_{\overline{W}^-(\partial), \overline{V}^-(\partial)}) &\\ \times (Z(V_N, \beta, h, +, G^{\omega_N}_{W_N, V_N}))^{-1} &\\ = \exp\Big\{\sum_{\Lambda \subset \overline{V}^+(\partial)} \Phi(\Lambda) + \sum_{\Lambda \subset \overline{V}^-(\partial)} \Phi(\Lambda) - \sum_{\Lambda \subset V_N} \Phi(\Lambda)\Big\} &\\ = \exp\Big\{-\sum_{t \in \Delta(\partial)} \Phi(\{t\}) - \sum_{\Lambda \subset V_N : \Lambda \cap \Delta(\partial) \neq \varnothing, |\Lambda| > 1} \Phi(\Lambda)\Big\}. &\end{aligned} \qquad (4.4.43)$$

Using the estimate (3.3.11) and Lemma 3.4, we see that for $\beta \geq \beta_0(\nu)$ and some constant $C = C(\nu)$ we have

$$\Big|\sum_{\Lambda \subset V_N : \Lambda \cap \Delta(\partial) \neq \varnothing, |\Lambda| > 1} \Phi(\Lambda)\Big| \leq C|\partial|. \qquad (4.4.44)$$

Using a similar construction and the condition $h^0 \geq 0$, we find that (cf. (3.3.11))

$$\beta h^0(|\Delta^+(\partial) \setminus W_N| - |\Delta^-(\partial) \setminus W_N|) - \sum_{t \in \Delta(\partial)} \Phi(\{t\}) \leq C|\partial|. \qquad (4.4.45)$$

Gathering estimates (4.4.42)–(4.4.45), we find that

$$\frac{P_N^{+,h}(\sigma \in \Omega_{V_N} : \overline{\mathscr{G}}(\sigma) = \partial)}{P_N^{+,h}(\sigma \in \Omega_{V_N} : \mathscr{G}^L(\sigma) = \varnothing)} \leq \exp\{-2(\beta - C)|\partial|\}. \qquad (4.4.46)$$

Now we can turn to the ratio in the main statement (4.4.3). Using (4.4.29) and the estimate (4.4.32), we get

$$P_N^{+,h}\left(\sigma \in \Omega_{V_N} : S_{W_N}(\sigma) = b_N , \max_{\Gamma\in\mathscr{G}(\sigma) : \operatorname{Int}\Gamma\cap W_N\neq\varnothing} q(\Gamma) > \omega_N\right)$$

$$\leq \sum_{\partial\in D_N : \partial\neq\varnothing} \sum_{\bar{b}=0}^{[\lambda|W^-(\partial)|]} P_N^{+,h}(\sigma \in \Omega_{V_N} : S_{\overline{W}^+(\partial)}(\sigma) = b(\bar{b}, \partial), \overline{\mathscr{G}}(\sigma) = \partial). \tag{4.4.47}$$

Now we obtain from (4.4.40), (4.4.46), (4.4.47), and (4.4.37) that the ratio (4.4.3) satisfies

$$\frac{P_N^{+,h}\left(\sigma \in \Omega_{V_N} : S_{W_N}(\sigma) = b_N , \max_{\Gamma\in\mathscr{G}(\sigma) : \operatorname{Int}\Gamma\cap W_N\neq\varnothing} q(\Gamma) > \omega_N\right)}{P_N^{+,h}(\sigma \in \Omega_{V_N} : S_{W_N}(\sigma) = b_N)}$$

$$\leq \sum_{\partial\in D_N : \partial\neq\varnothing} \exp\{-2(\beta - C)|\partial|\} \sum_{\bar{b}=0}^{[\lambda|W^-(\partial)|]} q_N(\bar{b}, \partial), \tag{4.4.48}$$

where we denote

$$q_N(\bar{b}, \partial) = \frac{P_{N,\partial}^{+,h}(\sigma \in \Omega_{\overline{V}^+(\partial)} : S_{\overline{W}^+(\partial)}(\sigma) = b(\bar{b}, \partial))}{P_{N,\varnothing}^{+,h}(\sigma \in \Omega_{V_N} : S_{W_N}(\sigma) = b_N)}. \tag{4.4.49}$$

This is the starting point for the following estimates, in which we shall use only the estimates (4.4.48) and (4.4.15) from the previous construction.

Now our main goal is to estimate the ratio (4.4.49). For it we need some estimates for mean values

$$E_N^{\partial}(\bar{h}) = E(\overline{W}^+(\partial), \overline{V}^+(\partial), +, \beta, \hat{h}^{\partial}(\bar{h}), G^{\omega_N}_{\overline{W}^+(\partial), \overline{V}^+(\partial)}), \tag{4.4.50}$$

where $\{\hat{h}^{\partial}(\bar{h}) = (\hat{h}_t^{\partial}(\bar{h}), t \in \overline{V}^+(\partial))\}$ is such that

$$\hat{h}_t^{\partial}(\bar{h}) = \begin{cases} h^0, & t \in \overline{V}^+(\partial) \setminus \overline{W}^+(\partial), \\ \bar{h}, & t \in \overline{W}^+(\partial). \end{cases} \tag{4.4.51}$$

For $\partial = \varnothing$ we have (cf. (4.2.2), (4.2.3))

$$E_N^{\varnothing}(\bar{h}) = E_N(\bar{h}). \tag{4.4.52}$$

We assume that for some $\bar{r} > 0$ (cf. (3.19.4))

$$\bar{h} \geq -\bar{r}(\omega_N)^{-1}. \tag{4.4.53}$$

We know that under this condition (see (3.19.15))

$$E_N^{\partial}(\bar{h}) = \beta^{-1} \sum_{\Lambda\subset\overline{V}^+(\partial) : \Lambda\cap\overline{W}^+(\partial)\neq\varnothing} \Phi^1(\Lambda, \bar{h}, \omega_N) \tag{4.4.54}$$

with (see (3.19.23), (3.19.24))

$$\begin{aligned} |\beta^{-1}\Phi^1(\Lambda, \bar{h}, \omega_N)| &\leq K \exp\{-\beta\gamma d(\Lambda)\}, \quad \text{if } |\Lambda| > 1, \\ |\beta^{-1}\Phi^1(\Lambda, \bar{h}, \omega_N) - 1| &\leq K \exp\{-\beta\gamma\}, \quad \text{if } |\Lambda| = 1,\ \Lambda \subset V_N, \end{aligned} \tag{4.4.55}$$

where $K = K(\nu, \bar{r})$, $\gamma = \gamma(\nu, \bar{r})$ are constants; the functions $\Phi^1(\Lambda, \bar{h}, \omega_N)$ vanish for disconnected sets Λ. It follows from (4.4.54) that

$$E_N^\partial(\bar{h}) = |\overline{W}^+(\partial)|(1 + \hat{\theta}\hat{c}e^{-\beta\gamma}). \tag{4.4.56}$$

where $\hat{c} = \hat{c}(\nu, \bar{r})$ is a constant and $\theta = \theta(\nu, \bar{r}, N, \partial)$ is such that $|\theta| \le 1$. It follows also from (4.4.54), (4.4.55) that

$$\begin{aligned} E_N(\bar{h}) - E_N^\partial(\bar{h}) &= \beta^{-1} \sum_{t \in W_N \setminus \overline{W}^+(\partial)} \Phi^1(\{t\}, \bar{h}, \omega_N) \\ &+ \beta^{-1} \sum_{\substack{\Lambda : \Lambda \subset V_N, \Lambda \cap W_N \ne \varnothing, \\ \Lambda \cap (V_N \setminus \overline{V}^+(\partial)) \ne \varnothing, |\Lambda| > 1}} \Phi^1(\Lambda, \bar{h}, \omega_N) \\ &+ \beta^{-1} \sum_{\substack{\Lambda : \Lambda \subset \overline{V}^+(\partial) \setminus \overline{W}^+(\partial), \\ \Lambda \cap W_N \ne \varnothing, |\Lambda| > 1}} \Phi^1(\Lambda, \bar{h}, \omega_N). \end{aligned} \tag{4.4.57}$$

Each connected set included in the second and third sums contains a point $t \in (W_N \setminus \overline{W}^+(\partial)) \cup \Delta(\partial)$. So we can estimate this sum using the estimates (4.4.55) and (3.4.1) and find that under the condition (4.4.53)

$$|E_N(\bar{h}) - E_N^\partial(\bar{h}) - |W_N \setminus \overline{W}^+(\partial)|| \le g(|W_N \setminus \overline{W}^+(\partial)| + |\partial|), \tag{4.4.58}$$

where $g = g(\nu, \beta)$ is such that $g(\nu, \beta) \le C \exp\{-\beta\gamma\}$.

Now we begin to estimate the ratio (4.4.49). First we observe that the conditions (4.4.1), (4.4.2) imply that the conditions (4.3.1), (4.3.2), and (4.3.4) of Proposition 4.3 are valid. So we can apply the statement (4.2.7) of Proposition 4.2, which shows that the denominator in (4.4.49) equals

$$P_{N,\varnothing}^{+,h}(\sigma \in \Omega_{V_N} : S_{W_N}(\sigma) = b_N) = \frac{2}{\sqrt{2\pi D_N(\bar{h}_N(b_N))}} \exp\{-I_N(b_N)\}(1 + o_N(1)). \tag{4.4.59}$$

In the estimate of the numerator of (4.4.49) we can use a less precise upper estimate (4.2.11) which is true without additional restriction to the value b_N of the random variable. Applying (4.2.11) to the case $W_N = \overline{W}^+(\partial)$, $V_N = \overline{V}^+(\partial)$, $b_N = b(\bar{b}, \partial)$ we see that

$$P_{N,\partial}^{+,h}(\sigma \in \Omega_{\overline{V}^+(\partial)} : S_{\overline{W}^+(\partial)}(\sigma) = b(\bar{b}, \partial)) \le \exp\{-I_N^\partial(b(\bar{b}, \partial))\}, \tag{4.4.60}$$

where (see (4.2.4), (4.2.8), (4.2.9), (4.4.36), (4.4.51))

$$I_N^\partial(b) = \beta\bar{h}_N^\partial(b)b - \ln \frac{Z(\overline{V}^+(\partial), +, \beta, \hat{h}^\partial(\bar{h}_N^\partial(b)), G^{\omega_N}_{\overline{W}^+(\partial), \overline{V}^+(\partial)})}{Z(\overline{V}^+(\partial), +, \beta, h^\partial, G^{\omega_N}_{\overline{W}^+(\partial), \overline{V}^+(\partial)})} \tag{4.4.61}$$

and where

$$E_N^\partial(\bar{h}_N^\partial(b)) = b. \tag{4.4.62}$$

Using (4.4.59), (4.4.60) and (4.4.49), we can write

$$q_N(\overline{b}, \partial) \le \frac{2}{\sqrt{2\pi D_N(\overline{h}_N(b_N))}} \exp\{I_N(b_N) - I_N^{\partial}(b(\overline{b}, \partial))\}(1 + o_N(1)). \tag{4.4.63}$$

We shall use also the following observation. Since $|S_{\overline{W}^+(\partial)}(\sigma)| \le |\overline{W}^+(\partial)|$, the relation (4.4.49) implies that

$$q_N(\overline{b}, \partial) = 0 \quad \text{if } b(\overline{b}, \partial) > |\overline{W}^+(\partial)|. \tag{4.4.64}$$

Now we need some simple geometric facts. Let $G_W(\partial)$ be the set of all faces of contours $\Gamma \in \partial$ adjacent to points in W_N. It is clear from the definition (4.4.30) that for a constant $K = K(\nu)$,

$$|\Delta_W^+(\partial)| \le K|G_W(\partial)|. \tag{4.4.65}$$

On the other hand, it follows from the definition (4.4.10) that each face from $G_W(\partial)$ is adjacent to a point $t \in W^-(\partial)$ or to a point $t \in \partial W_N^c$. So for some positive constant $k = k(\nu)$,

$$|W^-(\partial)| + |\partial W_N^c| \ge k|G_W(\partial)|. \tag{4.4.66}$$

It is convenient to assume that $k < K$. Formulas (4.4.65), (4.4.66) imply that

$$|\Delta_W^+(\partial)| \le (K/k)(|W^-(\partial)| + |\partial W_N^c|). \tag{4.4.67}$$

It follows from the condition (4.4.2) and the estimate (4.4.56) applied for $\partial = \varnothing$, when $\overline{W}^+(\partial) = W_N$ and $\overline{h} = 0$, that for sufficiently large β and N we have

$$b_N \ge \left(1 - \frac{k}{4K}\right)|W_N|. \tag{4.4.68}$$

Now we find from the definition (4.4.33) and the estimate (4.4.68) that for any $\overline{b}$ such that $0 \le \overline{b} \le [\lambda|W^-(\partial)|]$,

$$b(\overline{b}, \partial) \ge \left(1 - \frac{k}{4K}\right)|W_N| + |W^-(\partial)| - |\Delta_W^+(\partial)| - [\lambda|W^-(\partial)|]. \tag{4.4.69}$$

Suppose that we have chosen the constant $q = q(\nu)$ in (4.4.7) so large that $\lambda = \lambda(q, \nu) < k/(4K)$ (see (4.4.11)). Recall also that (see (4.4.30))

$$|\Delta_W^+(\partial)| = |W_N| - |\overline{W}^+(\partial)| - |W^-(\partial)|. \tag{4.4.70}$$

Using these facts and the inequality $|W^-(\partial)| \le |W_N|$, we can rewrite (4.4.69) as

$$b(\overline{b}, \partial) \ge |\overline{W}^+(\partial)| + 2|W^-(\partial)| - (k/(2K))|W_N|. \tag{4.4.71}$$

So it follows from (4.4.64) that

$$q_N(\overline{b}, \partial) = 0 \quad \text{if } |W^-(\partial)| > (k/(4K))|W_N|, \qquad 0 \le \overline{b} \le [\lambda|W^-(\partial)|]. \tag{4.4.72}$$

In this case the corresponding term in (4.4.48) vanishes and we shall consider in the following only $\partial \in D_N$ such that

$$|W^-(\partial)| \le (k/(4K))|W_N|. \tag{4.4.73}$$

Using also the estimate (4.4.67), the equality (4.4.70), and the Van Hove condition for W_N, we see that in the case (4.4.73) for N sufficiently large

$$|\overline{W}^+(\partial)| = |W_N| - |W^-(\partial)| - |\Delta_W^+(\partial)| \ge |W_N|/2. \tag{4.4.74}$$

Now we want to estimate the exponent in (4.4.63). We have to consider different cases separately. First we assume that

$$b_N > E_N(0) + R(\partial), \tag{4.4.75}$$

where we denote (cf. (4.4.58))

$$R(\partial) = 2gK|\partial|, \tag{4.4.76}$$

with the constant $K = K(\nu)$ chosen in such a way that $|\Delta(\partial)| \le K|\partial|$ for all contour systems ∂. Using the definitions (4.2.9), (4.2.2), and (3.19.2) we observe that

$$\frac{d\overline{H}_N(\overline{h})}{d\overline{h}} = \beta E_N(\overline{h}), \tag{4.4.77}$$

and so it follows from (4.2.8) and (4.2.4) (for $h^1 = 0$) that

$$\frac{dI_N(b)}{db} = \beta \overline{h}_N(b). \tag{4.4.78}$$

In a similar way (see (4.4.61))

$$\frac{dI_N^\partial(b)}{db} = \beta \overline{h}_N^\partial(b). \tag{4.4.79}$$

Finally, definition (4.2.8) and (4.4.61) show that

$$I_N(E_N(0)) = I_N^\partial(E_N^\partial(0)) = 0, \tag{4.4.80}$$

Now let (see (4.4.33) and (4.4.76))

$$d(\overline{b}, \partial) = b(\overline{b}, \partial) - b_N + R(\partial). \tag{4.4.81}$$

It follows from (4.4.78) and (4.4.80) that

$$I_N(b_N) = \beta \int_{E_N(0)}^{b_N} \overline{h}_N(b)\, db = \beta \left(\int_{E_N(0)}^{b_N - R(\partial)} \overline{h}_N(b)\, db + \int_{b_N - R(\partial)}^{b_N} \overline{h}_N(b)\, db \right). \tag{4.4.82}$$

and it follows from (4.4.79) and (4.4.80) that

$$\begin{aligned} I_N^\partial(b(\overline{b}, d)) &= \beta \int_{E_N^\partial(0)}^{b(\overline{b}, \partial)} \overline{h}_N^\partial(b) db \\ &= \beta \left(\int_{E_N^\partial(0)}^{E_N(0)+d(\overline{b}, \partial)} \overline{h}_N^\partial(b)\, db + \int_{E_N(0)+d(\overline{b}, \partial)}^{b(\overline{b}, \partial)} \overline{h}_N^\partial(b)\, db \right). \end{aligned} \tag{4.4.83}$$

By definition (4.4.81), $b(\overline{b}, \partial) = b_N - R(\partial) + d(\overline{b}, \partial)$, and so a change of variable shows that

$$\int_{E_N(0)+d(\overline{b},\partial)}^{b(\overline{b},\partial)} \overline{h}_N^{\partial}(b)db = \int_{E_N(0)}^{b_N-R(\partial)} (\overline{h}_N^{\partial}(b+d(\overline{b},\partial))\,db. \tag{4.4.84}$$

Relations (4.4.82), (4.4.83), (4.4.84) imply that

$$\begin{aligned} I_N(b_N) - I_N^{\partial}(b(\overline{b},\partial)) = \beta\Bigg(& \int_{E_N(0)}^{b_N-R(\partial)} (\overline{h}_N(b) - \overline{h}_N^{\partial}(b+d(\overline{b},\partial))\,db \\ & + \int_{b_N-R(\partial)}^{b_N} \overline{h}_N(b)\,db - \int_{E_N^{\partial}(0)}^{E_N(0)+d(\overline{b},\partial)} \overline{h}_N^{\partial}(b)\,db\Bigg). \end{aligned} \tag{4.4.85}$$

We must estimate the integrals on the right-hand side of this relation. The condition (4.4.2) implies the condition (4.3.2) of Proposition 4.3. Therefore, there exists a constant $H = H(C)$ such that $\overline{h}_N(b) \le H$ for $b_N \ge b \ge b_N - R(\partial)$. This implies (see (4.4.76)) that the second integral in (4.4.85) satisfies

$$\int_{b_N-R(\partial)}^{b_N} \overline{h}_N(b)\,db \le HR(\partial) \le |\partial|. \tag{4.4.86}$$

(We have used here the fact that $g(\beta,\nu) \to 0$ as $\beta \to \infty$, and assumed that β is so large that $2HKg \le 1$.) It follows from relations (4.4.33) and (4.4.70) that if $0 \le \overline{b} \le [\lambda|W^-(\partial)|]$ with $\lambda < 1$, then

$$b_N - b(\overline{b},\partial) \le |\Delta_W^+(\partial)| \le g|\Delta_W^+(\partial)| + (1-g)|W_N \setminus \overline{W}^+(\partial)|. \tag{4.4.87}$$

It follows from (4.4.87), (4.4.58), and (4.4.76) that for β sufficiently large

$$b_N - b(\overline{b},\partial) \le E_N(\overline{h}) - E_N^{\partial}(\overline{h}) + R(\partial) \quad \text{if } \overline{h} > -\overline{r}(\omega_N)^{-1}. \tag{4.4.88}$$

Using definition (4.4.81) and the previous estimate for $\overline{h} = 0$, we see that the difference of the limits of integration in the third integral in (4.4.85) is $E_N(0) + d(\overline{b},\partial) - E_N^{\partial}(0) \ge 0$, and since $\overline{h}_N^{\partial}(b) \ge 0$ for $b \ge E_N^{\partial}(0)$ this implies that

$$-\int_{E_N^{\partial}(0)}^{E_N(0)+d(\overline{b},\partial)} \overline{h}_N^{\partial}(b)\,db \le 0. \tag{4.4.89}$$

Using the estimate (4.4.88) and the definitions (4.2.4), (4.4.62), and (4.4.81) we find that, for $b_N \ge E_N(0)$,

$$\begin{aligned} E_N^{\partial}(\overline{h}_N(b)) &= E_N(\overline{h}_N(b)) - (E_N(\overline{h}_N(b)) - E_N^{\partial}(\overline{h}_N(b))) \\ &\le b - (b_N - b(\overline{b},\partial)) + R(\partial) = b + d(\overline{b},\partial) = E_N^{\partial}(\overline{h}_N^{\partial}(b+d(\overline{b},\partial))). \end{aligned} \tag{4.4.90}$$

(For $b \ge E_N(0)$ we have $\overline{h}_N(b) \ge 0$, so the condition (4.4.53) is satisfied.) Since $E_N^{\partial}(\overline{h})$ is monotone in $\overline{h}$, we see that

$$\overline{h}_N(b) - \overline{h}_N^{\partial}(b+d(\overline{b},\partial)) \le 0. \tag{4.4.91}$$

The condition (4.4.75) implies that the upper limit in the first integral in (4.4.85) is larger than its lower limit, and so it follows from (4.4.81) that

$$\int_{E_N(0)}^{b_N-R(\partial)} (\bar{h}_N(b) - \bar{h}_N^{\partial}(b + d(\bar{b}, \partial)))\, db \le 0. \tag{4.4.92}$$

Together with (4.4.85), (4.4.86), and (4.4.89) this implies that

$$I_N(b_N) - I_N^{\partial}(b(\bar{b}, \partial)) \le \beta|\partial| \tag{4.4.93}$$

in the considered case (4.4.75).

Now we consider the second case when (cf. (4.4.75))

$$E_N(0) - R(\partial) \le b_N \le E_N(0) + R(\partial). \tag{4.4.94}$$

In this case it suffices to observe, using (4.4.82) and the nonnegativity of the rate function, that

$$I_N(b_N) - I_N^{\partial}(b(\bar{b}, \partial)) \le I_N(b_N) = \beta \int_{E_N(0)}^{b_N} \bar{h}_N(b)\, db. \tag{4.4.95}$$

It follows from the condition (4.4.2) of the proposition and from Lemma 4.3 that there exists a constant $H' = H'(\nu, C, c)$ such that $\bar{h}_N(b) \le H'$ if b is between $E_N(0)$ and b_N. So it follows from (4.4.95) and (4.4.94) that

$$I_N(b_N) - I_N^{\partial}(b_N) \le \beta H'|b_N - E_N(0)| \le \beta H' R(\partial) \le \beta|\partial| \tag{4.4.96}$$

in the case (4.4.94) also. (In the last estimate we again assume that β is sufficiently large and so g is sufficiently small in comparison with H'; cf. (4.4.86).)

It remains to consider the last case when (cf. (4.4.2))

$$E_N(0) - R(\partial) > b_N \ge E_N(0) - c|W_N|^{\nu/(\nu+1)}. \tag{4.4.97}$$

This case requires a different approach. Using relations (4.2.4), (3.2.9)–(3.2.11), and (4.2.10), we observe that

$$\frac{d\bar{h}_N(b)}{db} = \left(\left. \frac{dE_N(\bar{h})}{d\bar{h}} \right|_{\bar{h}=\bar{h}_N(b)} \right)^{-1} = \beta^{-1}(D_N(\bar{h}_N(b)))^{-1}. \tag{4.4.98}$$

Since

$$I_N(E_N(0)) = \bar{h}_N(E_N(0)) = 0, \tag{4.4.99}$$

we obtain from (4.4.78) and (4.4.98) that

$$I_N(b_N) = \int_{b_N}^{E_N(0)} (D_N(\bar{h}_N(b)))^{-1}(b - b_N)\, db, \tag{4.4.100}$$

and similarly

$$I_N^{\partial}(b(\bar{b}, \partial)) = \int_{b(\bar{b}, \partial)}^{E_N^{\partial}(0)} (D_N^{\partial}(\bar{h}_N^{\partial}(b)))^{-1}(b - b(\bar{b}, \partial))\, db, \tag{4.4.101}$$

where the variance is given by (cf. (4.2.10) and (4.4.61))

$$D_N^\partial(\overline{h}) = D(\overline{W}^+(\partial), \overline{V}^+(\partial), +, \beta, \hat{h}^\partial(\overline{h}), \omega_N). \tag{4.4.102}$$

We shall consider two subcases separately. First we assume that for a sufficiently small constant $c_1 = c_1(\nu, \beta)$

$$\max(|W^-(\partial)|, |\partial|) \leq c_1 |W_N|^{\nu/(\nu+1)}. \tag{4.4.103}$$

Since the definition (4.4.10) of the set $W^-(\partial)$ implies that all the points of the set $W^-(\partial)$ lie inside the contours $\Gamma \in \partial$,

$$|W^-(\partial)|^{(\nu-1)/\nu} \leq L_1 |\partial| \tag{4.4.104}$$

for some constant $L_1 = L_1(\nu)$. So it follows from (4.4.103) that

$$|W^-(\partial)| \leq (c_1)^{1/\nu} L_1 |\partial| |W_N|^{1/(\nu+1)}. \tag{4.4.105}$$

So we find, using (4.4.33) and the evident relation $(c_1)^{1/\nu} L_1 |W_N|^{1/(\nu+1)} \geq 1$, valid for sufficiently large N, that for $0 \leq \overline{b} \leq [\lambda |W^-(\partial)|]$ (cf. (4.4.48)) and a new constant $L_2 = L_2(\nu)$

$$|b(\overline{b}, d) - b_N| \leq (1+\lambda)|W^-(\partial)| + |\Delta_W^+(\partial)| \leq (c_1)^{1/\nu} L_2 |\partial| |W_N|^{1/(\nu+1)}. \tag{4.4.106}$$

It follows from (4.4.70) that

$$|W_N \setminus \overline{W}^+(\partial)| = |W^-(\partial)| + |\Delta_W^+(\partial)|. \tag{4.4.107}$$

Applying the estimates (4.4.58), (4.4.107) and the condition (4.4.105) we see that for any $\overline{h}$ satisfying the condition (4.4.53)

$$\begin{aligned} |E_N(\overline{h}) - E_N^\partial(\overline{h})| &\leq (1+g)(|W^-(\partial)| + |\Delta_W^+(\partial)|) + g|\partial| \\ &\leq (c_1)^{1/\nu}((1+g)L_1 + g|W_N|^{-1/(\nu+1)})|\partial||W_N|^{1/(\nu+1)} \\ &\leq q|\partial||W_N|^{1/(\nu+1)}, \end{aligned} \tag{4.4.108}$$

where the constant $q = q_N(\nu, \beta, \overline{r}, c_1)$ is such that $q \to 0$ as $c_1 \to 0$, $N \to \infty$.

It follows from Proposition 4.1 that if a constant $\overline{h}$ is such that for some $H < \infty$, $\overline{r} \leq 1$

$$H \geq \overline{h} \geq -\overline{r}(\omega_N)^{-1}, \tag{4.4.109}$$

then for some constants $\alpha = \alpha(\nu, \beta, \overline{r}, H) > 0$ and $\beta_0 = \beta_0(\nu, \overline{r}, H)$ for $\beta \geq \beta_0$

$$D_N(\overline{h}) \geq \alpha |W_N|. \tag{4.4.110}$$

In a similar way, applying Proposition 4.1 with W_N, V_N replaced by $\overline{W}^+(\partial)$, $\overline{V}^+(\partial)$, we can evaluate $D_N^\partial(\overline{h})$. Therefore, using (4.4.103), (4.4.74), we can state that for a sufficiently large N

$$D_N^\partial(\overline{h}) \geq \alpha |\overline{W}^+(\partial)| \geq \frac{\alpha}{2} |W_N| \tag{4.4.111}$$

in the same domain (4.4.109). Suppose now that b is such that for some $C < \infty$ and sufficiently small $c_0 < \infty$,

$$\tanh(C\beta)|W_N| \geq b \geq E_N(0) - c_0|W_N|^{\nu/(\nu+1)}. \tag{4.4.112}$$

It follows from the condition (4.4.1) and Proposition 4.3 that $\overline{h} = \overline{h}_N(b)$ satisfies (4.4.109) for a suitable choice of $\overline{r} = \overline{r}(\nu, \beta, c_0, C)$, $H = H(C) < \infty$. Using the intermediate estimate in (4.4.108) and the condition (4.4.103), we find that

$$|E_N(\overline{h}) - E_N^{\partial}(\overline{h})| \leq q'|W_N|^{\nu/(\nu+1)} \tag{4.4.113}$$

where $q' = q'(\nu, \beta, \overline{r}, c_1)$ is a new constant such that $q' \to 0$ as $c_1 \to 0$. It follows from the relations (4.4.113), (4.4.107), and (4.4.103) that the condition (4.4.112) implies that for a sufficiently small c_1, sufficiently large N, and a corresponding choice of constants C', c_0'

$$\tanh(C'\beta)|\overline{W}^+(\partial)| \geq b \geq E_N^{\partial}(0) - c_0'|\overline{W}^+(\partial)|^{\nu/(\nu+1)}. \tag{4.4.114}$$

So we can apply Proposition 4.3 to $\overline{W}^+(\partial)$ instead of W_N and see that the condition (4.4.109) is valid also for $\overline{h} = \overline{h}_N^{\partial}(b)$. Therefore, this condition holds for all $\overline{h}$ in the interval between $\overline{h}_N(b)$ and $\overline{h}_N^{\partial}(b)$, and we can apply the estimate (4.4.110) for all these values of $\overline{h}$. Applying this estimate and the relation (3.19.2), we find that

$$\begin{aligned} |\overline{h}_N^{\partial}(b) - \overline{h}_N(b)| &\leq (\alpha|W_N|)^{-1}\left|\int_{\overline{h}_N(b)}^{\overline{h}_N^{\partial}(b)} D_N(\overline{h})\,d\overline{h}\right| \\ &= (\alpha|W_N|)^{-1}\beta^{-1}|E_N(\overline{h}_N(b)) - E_N(\overline{h}_N^{\partial}(b))| \\ &= (\alpha|W_N|)^{-1}\beta^{-1}|E_N^{\partial}(\overline{h}_N^{\partial}(b)) - E_N(\overline{h}_N^{\partial}(b))| \end{aligned} \tag{4.4.115}$$

In the last equality we used the relation

$$E_N(\overline{h}_N(b)) = E_N^{\partial}(\overline{h}_N^{\partial}(b)) = b \tag{4.4.116}$$

that follows from the definitions (4.2.4) and (4.4.62). So using the estimates (4.4.108) (the condition (4.4.53) follows from the condition (4.4.109), which we have verified above for $\overline{h}$ between $\overline{h}_N(b)$ and $\overline{h}_N^{\partial}(b)$), we see that for a sufficiently large β and a new constant $L_2 = L_2(\nu)$

$$|\overline{h}_N^{\partial}(b) - \overline{h}_N(b)| \leq (\alpha\beta)^{-1}L_2|\partial||W_N|^{-\nu/(\nu+1)}. \tag{4.4.117}$$

Hence, we can prove that for a constant $G = G(\nu, \beta, c_0, C)$ and all sufficiently large β and N

$$\begin{aligned} &|D_N(\overline{h}_N(b)) - D_N^{\partial}(\overline{h}_N^{\partial}(b))| \\ &\quad \leq |D_N(\overline{h}_N(b)) - D_N^{\partial}(\overline{h}_N(b))| + |D_N^{\partial}(\overline{h}_N(b)) - D_N^{\partial}(\overline{h}_N^{\partial}(b))| \\ &\quad \leq G|\partial||W_N|^{1/(\nu+1)}. \end{aligned} \tag{4.4.118}$$

The necessary estimate for the first summand in the middle part of this inequality can be obtained by a construction which is completely similar to the construction used in the derivation of the estimate (4.4.108) (cf. (4.4.53)–(4.4.58) and (4.4.103)–(4.4.108)). The second term can be estimated using the relations (3.19.6) and (4.4.117).

Now we can go to the main estimate of the difference between $I_N(b_N)$ and $I_N^\partial(b(\overline{b}, \partial))$. We see from (4.4.100), (4.4.101) that

$$\begin{aligned}|I_N(b_N) - I_N^\partial(b(\overline{b}, \partial))| \\ \leq \int_{b_N}^{E_N(0)} |(D_N(\overline{h}_N(b)))^{-1}(b - b_N) - (D_N^\partial(\overline{h}_N^\partial(b)))^{-1}(b - b(\overline{b}, \partial))|\, db \\ + \left| \int_{E_N(0)}^{E_N^\partial(0)} (D_N^\partial(\overline{h}_N^\partial(b)))^{-1}(b - b(\overline{b}, \partial))\, db \right| \\ + \left| \int_{b(\overline{b}, \partial)}^{b_N} (D_N^\partial(\overline{h}_N^\partial(b)))^{-1}(b - b(\overline{b}, \partial))\, db \right|. \qquad (4.4.119)\end{aligned}$$

The integrand in the first term can be bounded as follows:

$$\begin{aligned}|(D_N(\overline{h}_N(b)))^{-1}(b - b_N) - (D_N^\partial(\overline{h}_N^\partial(b)))^{-1}(b - b(\overline{b}, \partial))| \\ \leq \frac{|D_N(\overline{h}_N(b)) - D_N^\partial(\overline{h}_N^\partial(b))|}{D_N(\overline{h}_N(b))D_N^\partial(\overline{h}_N^\partial(b))}|b - b_N| + \frac{|b_N - b(\overline{b}, \partial)|}{D_N^\partial(\overline{h}_N^\partial(b))}. \qquad (4.4.120)\end{aligned}$$

It follows from (4.4.97) that the condition (4.4.112) is valid for $c_0 = c$ and b between b_N and $E_N(0)$. Therefore, to estimate this integrand we can apply the relations (4.4.118), (4.4.110), (4.4.111), and (4.4.106). We find that for $c \leq 1$ and some new constant C_1 depending on β, ν we have, for sufficiently large N,

$$\begin{aligned}\int_{b_N}^{E_N(0)} |(D_N(\overline{h}_N(b)))^{-1}(b - b_N) - (D_N^\partial(\overline{h}_N^\partial(b)))^{-1}(b - b(\overline{b}, \partial))|\, db \\ \leq C_1|\partial||W_N|^{1/(\nu+1)} \int_{b_N}^{E_N(0)} ((b - b_N)|W_N|^{-2} + |W_N|^{-1})\, db \\ \leq C_1|\partial||W_N|^{1/(\nu+1)}(1/2(E_N(0) - b_N)^2|W_N|^{-2} + |W_N|^{-1}(E_N(0) - b_N)) \\ \leq C_1|\partial||W_N|^{1/(\nu+1)}(1/2c^2|W_N|^{\frac{2\nu}{\nu+1}-2} + c|W_N|^{\frac{\nu}{\nu+1}-1}) \leq \frac{1}{3}|\partial|, \qquad (4.4.121)\end{aligned}$$

where in the last inequality we use the assumption that the constant $c = c(\beta, \nu)$ in the main condition (4.4.2) is sufficiently small compared to C_1. Using (4.4.106), (4.4.108), (4.4.103), and (4.4.97) we observe that for b between $E_N(0)$ and $E_N^\partial(0)$,

$$|b - b(\overline{b}, \partial)| \leq |b(\overline{b}, \partial) - b_N| + |b_N - E_N(0)| + |E_N(0) - E_N^\partial(0)| \leq C_2|W_N|^{\nu/(\nu+1)}, \qquad (4.4.122)$$

where $C_2 = C_2(\nu, \beta, c)$ is a constant. Using the estimate (4.4.122) for the integrand $|b - b(\overline{b}, \partial)|$ in the second term in (4.4.119) and the estimates (4.4.108) and (4.4.111), we see that

$$\left|\int_{E_N(0)}^{E_N^\partial(0)} (D_N^\partial(\overline{h}_N^\partial(b)))^{-1}(b - b(\overline{b}, \partial))\, db\right| \\ \leq q|\partial||W_N|^{1/(\nu+1)} C_2 |W_N|^{\nu/(\nu+1)}((\alpha/2)|W_N|)^{-1} \leq |\partial|/3, \tag{4.4.123}$$

if for a fixed β we choose the constant c_1 such that the constant q is sufficiently small compared to $C_2\alpha^{-1}$. Similarly to (4.4.123) we obtain that

$$\left|\int_{b(\overline{b},\partial)}^{b_N} (D_N^\partial(\overline{h}_N^\partial(b)))^{-1}(b - b(\overline{b}, \partial))\, db\right| \leq |\partial|/3. \tag{4.4.124}$$

It follows from the relations (4.4.119), (4.4.121), (4.4.123), (4.4.124) that

$$|I_N(b_N) - I_N^\partial(b(\overline{b}, \partial))| \leq |\partial| \tag{4.4.125}$$

in the case when the conditions (4.4.97) and (4.4.103) are satisfied.

In the last case, when (4.4.97) is true but instead of (4.4.103) we have the opposite condition

$$\max(|W^-(\partial)|, |\partial|) > c_1|W_N|^{\nu/(\nu+1)}, \tag{4.4.126}$$

we can use less precise estimates. Using (4.4.100) and (4.4.110) and the main condition (4.4.2), and then the condition (4.4.126) and the estimate (4.4.104), we see that

$$I_N(b_N) - I_N^\partial(b(\overline{b}, \partial)) \leq I_N(b_N) \leq (\alpha|W_N|)^{-1}\int_{b_N}^{E_N(0)} (b - b_N)\, db \\ = (\alpha|W_N|)^{-1}\frac{(E_N(0) - b_N)^2}{2} \leq \frac{\alpha^{-1}c^2}{2}|W_N|^{(\nu-1)/(\nu+1)} \\ \leq \frac{\alpha^{-1}c^2}{2}(c_1)^{-1}\max(|\partial|, |W^-(\partial)|)^{(\nu-1)/\nu} \leq |\partial|. \tag{4.4.127}$$

In the last inequality we assume that for fixed β and c_1 the constant c in the statement of the proposition is so small that

$$(1/2)L_1\alpha^{-1}(c_1)^{-\nu-1/\nu}c^2 \leq 1. \tag{4.4.128}$$

Relations (4.4.93), (4.4.96), (4.4.125), (4.4.127) cover all possible cases. So we obtain that if the condition (4.4.73) is satisfied, then for sufficiently large β, all b_N, and $\overline{b}$ such that $0 \leq \overline{b} \leq [\lambda|W^-(\partial)|]$, we have

$$I_N(b_N) - I_N^\partial(b(\overline{b}, \partial)) \leq \beta|\partial|. \tag{4.4.129}$$

To estimate the variance in (4.4.63), we can again use (4.4.110). So we obtain from (4.4.63) and (4.4.72) that

$$q_N(\overline{b}, \partial) \leq (\alpha|W_N|)^{-1}\exp\{\beta|\partial|\} \quad \text{if } 0 \leq \overline{b} \leq [\lambda|W^-(\partial)|]. \tag{4.4.130}$$

Now we can return to the critical estimate (4.4.48). Since $|W^-(\partial)| \leq |W_N|$, we obtain from (4.4.130) that

$$\sum_{\overline{b}=0}^{[\lambda|W_N^-(\partial)|]} q_N(\overline{b}, \partial) \leq \overline{C} \exp\{\beta|\partial|\}, \tag{4.4.131}$$

where $\overline{C} = \overline{C}(\nu, \beta)$ is a constant. So (4.4.131) and (4.4.48) imply that

$$\begin{aligned} &\frac{P_N^{+,h}\left(\sigma \in \Omega_{V_N} : S_{W_N}(\sigma) = b_N, \max_{\Gamma \in \mathscr{G}(b) : \operatorname{Int}\Gamma \cap W_N \neq \varnothing} q(\Gamma) > \omega_N\right)}{P_N^{+,h}(\sigma \in \Omega_{V_N} : S_{W_N}(\sigma) = b_N)} \\ &\qquad \leq \overline{C} \sum_{\partial \in D_N : \partial \neq \varnothing} \exp\{-(\beta - C)|\partial|\}. \end{aligned} \tag{4.4.132}$$

For any $\partial \in D_N$, $\partial \neq \varnothing$, we have $|\partial| \geq (\omega_N)^{\nu-1}$ (see (3.1.5)). Using the estimate (4.4.15), we obtain the main statement of Proposition 4.4 from the estimate (4.4.132).

§4.5. Absence of large contours

The results of the previous sections are sufficient to obtain the main results formulated in Chapter 1 for the case of classical behavior of large deviations for +-boundary conditions and +-pure Gibbs state. We start with the proof of a statement that is very close to the statement of Proposition 4.4 and that easily implies the statement of Theorem 1.8 about the absence of large contours.

PROPOSITION 4.5. *Let $h \geq 0$, V_N be a sequence of contractible volumes satisfying the Van Hove conditions (see §§1.4, 1.5) and b_N be a sequence of integers satisfying the condition* (1.4.4) *and either the condition* (1.5.2) *or the conditions* (1.5.5), (1.5.6), (1.5.7). *Then statement* (1.8.3) *of Theorem* 1.8 *for the case of +-boundary conditions in volumes V_N is valid provided that β and C are sufficiently large.*

In a similar way, if we set (cf. (1.8.2))

$$A_N^\infty = \left\{\sigma \in \Omega_{\mathbb{Z}^d} : \max_{\Gamma \in \mathscr{G}(\sigma) : \operatorname{Int}\Gamma \cap V_N \neq \varnothing} |\Gamma| \leq C \ln |V_N|\right\}, \tag{4.5.1}$$

where $\mathscr{G}(\sigma)$, $\sigma \in \Omega_{\mathbb{Z}^d}$, is the set of all contours of configuration σ defined (cf. §3.1) *as the set of all connected components of the boundary of this configuration (now $\mathscr{G}(\sigma)$ can contain infinite contours). Suppose the conditions stated above are again satisfied, but with E_N^+ changed to $E_N^{+,\infty}$, where $E_N^{+,\infty}$ is the mean value of S_{V_N} (see* (3.2.8)) *induced by the limit +-pure Gibbs state $P^{+,\beta,h}$ (see* (1.3.8)). *Then (cf.* (1.8.1)) *for the limit of conditional probabilities we have*

$$\lim_{N\to\infty} P^{+,\beta,h}(A_N^\infty \mid S_{V_N} = b_N) = 1. \tag{4.5.2}$$

PROOF. We consider first the case of $+$-boundary condition. In this case we apply Proposition 4.4 in the case $W_N \equiv V_N$. As we noted in §3.2, in the case under consideration, when V_N is contractible and $\omega_N = \infty$, the probability distribution $P_N^{+,h}$ considered in §4.4 coincides with the probability distribution $P_N^{+,h}$ considered in §1.8. Recalling the estimate (3.1.5) we see that the statement (4.4.3) of Proposition 4.4 applied with $\omega_N = (C(2\nu)^\nu)^{1/(\nu-1)}(\ln|V_N|)^{1/(\nu-1)}$ implies the statement (1.8.3) of Theorem 1.8. So we must only verify that the conditions (1.5.2) and (1.5.5)–(1.5.7) together with the condition (1.4.4) imply the condition (4.4.2) of Proposition 4.4. We observe that the mean value E_N^+ used in §1.5 corresponds to the mean value $E(V_N, V_N, +, \beta, 0, \infty)$ in the notation of §3.2 and the mean value $E_N(0)$ used in the condition (4.4.2) corresponds to the mean value $E(V_N, V_N, +, \beta, 0, \omega_N)$, and so it follows from the estimate (3.19.7) that for β sufficiently large

$$\lim_{N\to\infty} |V_N|^{-\nu/(\nu+1)}|E_N^+ - E_N(0)| = 0. \tag{4.5.3}$$

So the estimate $b_N \geq E_N(0) - c|V_N|^{\nu/(\nu+1)}$, which is a part of (4.4.2), follows from the condition (1.5.2) or from the conditions (1.5.5)–(1.5.7) if N is sufficiently large. Finally, the upper bound $\tanh(C\beta)|V_N| \geq b_N$ in (4.4.2) is equivalent to the second condition in (1.4.4).

We prove statement (4.5.2) about the limit pure $+$-state using Proposition 4.4 in the case when $W_N \subset V_N$ and V_N is a sequence of cubes with diameters going to infinity sufficiently fast. For notational consistency between Propositions 4.4 and 4.5 we rename the set V_N in Proposition 4.5 to be W_N, and vice versa. The main additional point needed for the proof of the relation (4.5.2) is the statement that for a fixed sequence W_N, a sequence V_N can be chosen in such a way that

$$\lim_{N\to\infty} \Big(P^{+,\beta,h}(A_N^\infty \mid S_{W_N} = b_N)) - P_N^{+,\beta,h}\left(\sigma \in \Omega_{V_N} : \max_{\substack{\Gamma\in\mathscr{G}(\sigma):\\ \operatorname{Int}\Gamma\cap W_N\neq\varnothing}} |\Gamma| \leq C\ln|V_N| \;\Big|\; S_{W_N}(b) = b_N)\right) = 0. \tag{4.5.4}$$

Then the statement (4.4.3) of Proposition 4.4 evidently implies (4.5.2). To prove (4.5.4), we observe that if

$$W_N' = \{t \in \mathbb{Z}^\nu : \operatorname{dist}(t, W_N) \leq 2C\ln|W_N|\}, \tag{4.5.5}$$

then

$$A_N^\infty = \left\{\sigma \in \Omega_{\mathbb{Z}^\nu} : \max_{\Gamma\in\mathscr{G}(\sigma):\,\operatorname{Int}\Gamma\cap W_N\neq\varnothing} |\Gamma \cap \overline{W}_N'| \leq C\ln|W_N|\right\}, \tag{4.5.6}$$

where $\overline{W}_N'$ is the set of all faces separating points of W_N'. (If there is a contour Γ that intersect W_N and such that $|\Gamma| > C\ln|V_N|$, then at least

$C \ln |V_N|$ of its faces lie in $\overline{W}'_N$.) But in the new formulation the event A_N^∞ is described by the behavior of configurations in a finite volume W'_N; hence the main definition (1.3.9) of the pure +-state implies that for a sufficiently large V_N the probabilities of the events $A_N^\infty \cap \{S_{W_N} = b\}$ and $\{S_{W_N} = b_N\}$ in this pure +-ensemble and in the ensemble in the finite volume V_N are arbitrarily close. This implies (4.5.4). We must also check the condition (4.4.2), but in the case under consideration the situation is completely similar to the situation in the previous case.

§4.6. Classical behavior of deviations for the case of +-boundary conditions

In this section we prove the statements of Theorems 1.5.1, 1.5.2 related to the classical behavior of probabilities of deviations.

PROPOSITION 4.6. *Let V_N be a contractible sequence of volumes satisfying the Van Hove condition* (1.4.1). *Let a sequence b_N satisfy the condition* (1.4.4) *and $\beta \geq \beta_1$, where β_1 is sufficiently large. If $h \geq 0$ and the condition* (1.5.2) *holds, then the asymptotic relations* (1.2.27), (1.2.29), (1.2.31), (1.2.33), (1.2.35), (1.2.37), (1.2.38) *hold both in the case of the +-boundary conditions and the pure +-Gibbs states. If $h = 0$ and the conditions* (1.5.5), (1.5.6), (1.5.7) *are valid, then the asymptotic relations* (1.2.33), (1.2.35) *again hold both in the case of +-boundary conditions and pure +-Gibbs state.*

PROOF. First we consider the case of +-boundary conditions. As we observed in §2.2, in the considered case of contractible volumes V_N the probability distributions $P_N^{+,h}$ used in §1.5 coincide with the probability distribution used in Proposition 4.2 for the case $V_N = W_N$, $h^1 = h$, $\omega_N = \infty$. We start with the case when a condition holds which is slightly stronger than the condition (1.5.2). Namely, we assume that for some $\varepsilon > 0$

$$b_N - E_N^+ \geq \varepsilon |V_N|. \tag{4.6.1}$$

In this case we can directly use Proposition 4.2. Indeed, to apply this proposition it is sufficient to check the conditions of Proposition 4.3. The condition (4.3.2) follows from the condition (1.4.4). Comparing the definitions (1.5.3) and (4.2.2) we see that $E_N^+ = E_N(0)$, hence the conditions (4.3.1) and (4.3.3) are valid. As we have observed in the proof of Proposition 4.2 (see (4.2.13)), the right-hand side of (4.2.7) coincides in the considered case with the right-hand side of the relation (1.2.27). To derive the implications of this relation included in the formulations of Proposition 4.6, some additional properties of the logarithmic generating function were used in §2.1 and they have to be verified in the considered case. By (4.2.13), they reduce to the similar properties of the function $\overline{H}_N(\overline{h})$ defined in (4.2.9). Consider (cf. (1.2.6)) the limit function

$$\overline{H}(\overline{h}) = \lim_{N\to\infty} |V_N|^{-1} \overline{H}_N(\overline{h}). \tag{4.6.2}$$

Because of the phase transition at the point $\overline{h} = 0$ we must restrict the domain of definition of this function to a complex neighborhood of the real half-line $\overline{h} > \varepsilon_0$, where $\varepsilon_0 > 0$. In this domain the limit (4.6.2) exists and defines a holomorphic function, and the sequence of functions $|V_N|^{-1}\overline{H}_N(\overline{h})$ is uniformly bounded. This statement is a reformulation of the statements (3.19.5a), (3.19.9) about the function G^0. The strict convexity of the limit function (4.6.2) follows from the equality

$$\frac{\partial^2\overline{H}(\overline{h})}{\partial\overline{h}^2} = \beta^{-2}\lim_{N\to\infty}|V_N|^{-1}D(V_N, V_N, +, \beta, \overline{h}_N, \infty) \tag{4.6.3}$$

(see (3.19.11)), where $\overline{h}_N = \{h_t \equiv \overline{h}, t \in V_N\}$, and the statement (4.1.6) of Proposition 4.1. For our purposes, it suffices to consider only the case $\overline{h} > \varepsilon_0 > 0$, because it follows from (4.6.1), (4.2.4), and (4.3.6) that $\overline{h}_N(b_N) \geq \varepsilon_0 > 0$ for sufficiently small ε_0 and the same is true for a limit value $\overline{h}(b) = \lim_{N\to\infty}\overline{h}_N(b_N)$, which corresponds to the limit value $b_N/|V_N| \to b$.

In the general case the construction is more complicated, since the conditions of Proposition 4.3 are true for cutoff ensembles only. To reduce the problem to a cutoff case, we observe, using notation of §3.2, that in the case of +-boundary conditions and any cutoff level ω_N we have

$$\begin{aligned}
P_N^{+,h}(b_N) &= P_{V_N}^{+,\beta,h,\infty}(\sigma : S_{V_N}(\sigma) = b_N, q(\Gamma) \leq \omega_N, \Gamma \in \mathscr{G}(\sigma)) \\
&\quad + P_{V_N}^{+,\beta,h,\infty}(\sigma : S_N(\sigma) = b_N, \max_{\Gamma\in\mathscr{G}(\sigma)} q(\Gamma) > \omega_N) \\
&= (Z(V_N, +, \beta, h, \omega_N))^{-1} \sum_{\substack{\sigma : S_{V_N}(\sigma)=b_N, \\ q(\Gamma)\leq\omega_N, \Gamma\in\mathscr{G}(\sigma)}} \exp\{-\beta H_h(\sigma)\}\frac{Z(V_N, +, \beta, h, \omega_N)}{Z(V_N, +, \beta, h, \infty)} \\
&\quad + P_{V_N}^{+,\beta,h,\infty}(\sigma : S_{V_N}(\sigma) = b_N, \max_{\Gamma\in\mathscr{G}(\sigma)} q(\Gamma) > \omega_N) \\
&= P_{V_N}^{+,\beta,h,\infty}(\sigma : q(\Gamma) \leq \omega_N, \Gamma \in \mathscr{G}(\sigma))P_{V_N}^{+,\beta,h,\omega_N}(\sigma : S_{V_N}(\sigma) = b_N) \\
&\quad + P_{V_N}^{+,\beta,h,\infty}(\sigma : S_{V_N}(\sigma) = b_N, \max_{\Gamma\in\mathscr{G}(\sigma)} q(\Gamma) > \omega_N).
\end{aligned} \tag{4.6.4}$$

We set

$$\omega_N = C(\ln|V_N|)^{1/(\nu-1)}, \tag{4.6.5}$$

where C is a sufficiently large constant. If a configuration σ contains a contour Γ such that $q(\Gamma) > \omega_N$, it also contains an external contour Γ' such that $q(\Gamma') > \omega_N$. So applying Proposition 3.20 (in the case $h = 0$ it is possible to use the standard Peierls estimates instead of this proposition) and then using the usual exponential estimate for the number of contours of a fixed length containing a fixed face, we find that, for a sufficiently large β,

$$\begin{aligned} P_{V_N}^{+,\beta,h,\infty}(\sigma : q(\Gamma) \le \omega_N , \Gamma \in \mathscr{G}(\sigma)) &= 1 - P_{V_N}^{+,\beta,h,\infty}(\sigma : \max_{\Gamma\in\mathscr{G}(\sigma)} q(\Gamma) > \omega_N) \\ &\ge 1 - \sum_{\Gamma\in G_{V_N} : q(\Gamma)>\omega_N} P_{V_N}^{+,\beta,h,\infty}(\sigma : \Gamma \in \mathscr{G}(\sigma)) \\ &\ge 1 - \sum_{\Gamma\in G_{V_N} : q(\Gamma)>C\ln|V_N|} \exp\{-2(\beta-\beta_0)\gamma|\Gamma|\} \to 1 \quad (\text{as } N\to\infty). \end{aligned} \tag{4.6.6}$$

We have shown in the proof of Proposition 4.5 that the conditions of Proposition 4.4 are fulfilled in the considered situation. So we can use (4.4.3), which implies that

$$\lim_{N\to\infty} \frac{P_{V_N}^{+,\beta,h,\infty}\left(\sigma : S_{V_N}(\sigma) = b_N , \max_{\Gamma\in\mathscr{G}(\sigma)} q(\Gamma) > \omega_N\right)}{P_N^{+,h}(b_N)} = 0. \tag{4.6.7}$$

(This is the point in the proof where the main restriction (1.5.6) on the order of moderate deviations is essentially used!) Applying relations (4.6.4), (4.6.6), (4.6.7) we find that

$$\lim_{N\to\infty} \frac{P_{V_N}^{+,\beta,h,\omega_N}(\sigma : S_{V_N}(\sigma) = b_N)}{P_N^{+,h}(b_N)} = 1 ; \tag{4.6.8}$$

hence we can consider the case of the cutoff ensembles only. In the proof of Proposition 4.5 we showed that in the case when the conditions of this proposition coincide with the conditions of Proposition 4.6, the inequality (4.4.2) is satisfied, Recalling (4.6.5), we see that the condition (4.4.2) implies the relation (4.3.1). The condition (4.3.4) is also valid. So we can apply Proposition 4.2, which, combined with (4.6.8), shows that

$$P_N^{+,h}(b_N) = \frac{2}{\sqrt{2\pi D_N^{\omega_N}(\overline{h}_N^{\omega_N}(b_N))}} \exp\{-I_N^{\omega_N}(b_N)\}(1+o_N(1)), \tag{4.6.9}$$

where $I_N^{\omega_N}(b_N)$ is given by (4.2.8) and (4.2.9), and $D_N^{\omega_N}(\overline{h}_N^{\omega_N}(b_N))$ is given by (4.2.10). Of course this is not yet the desired statement (1.2.27) of the Principle of Large Deviations, because we defined I_N and D_N for the ω_N-cutoff ensemble instead of the original ensemble. We emphasize this by adding the superscript ω_N to all notation.

We will get a proof of the statement of the proposition for the case of the condition (1.5.2) if we prove that

$$\begin{gathered} \lim_{N\to\infty} (I_N^{\omega_N}(b_N) - I_N^{\infty}(b_N)) = 0 , \\ \lim_{N\to\infty} \frac{D_N^{\omega_N}(\overline{h}_N^{\omega_N}(b_N))}{D_N^{\infty}(\overline{h}_N^{\infty}(b_N))} = 1 , \end{gathered} \tag{4.6.10}$$

where the upper index ∞ means that the corresponding quantities were considered for the ∞-cutoff ensemble. Observe that the condition (1.5.2) implies that

$$\overline{h}_N^{\infty}(b_N) \geq 0. \tag{4.6.11}$$

So we can apply Proposition 3.19 to compare the mean values for the ω_N- and ∞-cutoff ensembles for the case $h^0 = h$, $h^1 = \overline{h}_N^{\infty}(b_N)$ (see (3.19.4)). Comparing the mean values, we obtain from (3.19.7), (4.6.5), and the definition (4.2.2) that

$$|E_N^{\infty}(\overline{h}_N^{\infty}(b_N)) - E_N^{\omega_N}(\overline{h}_N^{\infty}(b_N))| \leq |V_N| \exp\{-\alpha\beta C \ln |V_N|\}\} = o(|V_N|^{-1}) \tag{4.6.12}$$

for $N \to \infty$ if β is sufficiently large. So recalling (see (4.2.4)) that

$$E_N^{\infty}(\overline{h}_N^{\infty}(b_N)) = E_N^{\omega_N}(\overline{h}_N^{\omega_N}(b_N)) = b_N, \tag{4.6.13}$$

we find that

$$|E_N^{\omega_N}(\overline{h}_N^{\omega}(b_N)) - E_N^{\omega_N}(\overline{h}_N^{\infty}(b_N))| = o(|V_N|^{-1}). \tag{4.6.14}$$

The conditions of Proposition 4.2, which we have just checked, imply the conditions of Proposition 4.1. So the estimate (4.1.6) shows that

$$\frac{dE_N^{\omega_N}(h)}{dh} = \beta^{-1} D_N^{\omega_N}(h) \geq \beta^{-1}\alpha |V_N| \tag{4.6.15}$$

for all h between $\overline{h}_N^{\omega_N}(b_N)$ and $\overline{h}_N^{\infty}(b_N)$, and therefore (4.6.14) implies that

$$|\overline{h}_N^{\omega_N}(b_N) - \overline{h}_N^{\infty}(b_N)| = o(|V_N|^{-1}), \qquad N \to \infty. \tag{4.6.16}$$

Comparing (3.19.3) with (4.2.9), we see that

$$\begin{aligned} &\overline{H}_N^{\omega_N}(\overline{h}_N^{\omega_N}(b_N)) - \overline{H}_N^{\infty}(\overline{h}_N^{\infty}(b_N)) \\ &\quad = G^0(V_N, V_N, +, \beta, h_N^{\omega_N}, \omega_N) - G^0(V_N, V_N, +, \beta, h_N^{\infty}, \omega_N), \end{aligned} \tag{4.6.17}$$

where $h_N^{\omega_N}$, h_N^{∞} are defined by (3.19.1) for $V = W = V_N$ and $h^1 = \overline{h}_N^{\omega_N}(b_N)$, resp. $\overline{h}_N^{\infty}(b_N)$. Now applying the estimates (3.19.7), (4.6.16) and the definition (4.2.8) we see that indeed

$$\lim_{N\to\infty} (I_N^{\omega_N}(b_N) - I_N^{\infty}(b_N)) = 0. \tag{4.6.18}$$

Using the definition (4.2.10) and arguments similar to the previous one, we see that we can apply the bound (3.19.6) to estimate the difference of variances. It follows from (4.6.16) and (4.6.5) that for sufficiently large β

$$|D_N^{\omega_N}(\overline{h}_N^{\omega_N}(b_N)) - D_N^{\infty}(\overline{h}_N^{\infty}(b_N))| = o(|V_N|^{-1}). \tag{4.6.19}$$

Using (4.6.15) again, we see that (4.6.19) implies the desired equality

$$\lim_{N\to\infty} \frac{D_N^{\omega_N}(\overline{h}_N^{\omega_N}(b_N))}{D_N^{\infty}(\overline{h}_N^{\infty}(b_N))} = 1. \tag{4.6.20}$$

So we have proved the statement (1.2.27) of the principle of large deviations for the case when the condition (1.5.2) is satisfied. The asymptotic relations (1.2.31) and (1.2.38) are evident implications of (1.2.27). The case of moderate deviations, when the condition (1.2.30) is valid, has to be discussed separately because the approach of §1.2 essentially uses the analyticity of $\overline{H}(\overline{h})$, which of course is not valid near the point $\overline{h} = 0$. If $h > 0$, then the sequence $|V_N|^{-1} E_N^+$ has a positive limit and so the condition (1.2.30) implies the condition (4.6.1). In such a situation we can use the analyticity arguments, but, if $h = 0$, the case of moderate deviations requires a special construction.

Now we assume that $h = 0$ and so $E_N^+ = E_N(0)$. We consider the case when (cf. (1.2.32) and (1.5.7))

$$\lim_{N\to\infty} \frac{|b_N - E_N^+|}{|V_N|^{k/(k+1)}} = 0, \tag{4.6.21}$$

for some integer k with $k \le \nu$ if $b_N - E_N^+ < 0$. We again consider a cutoff ensemble with the cutoff given by (4.6.5). The conditions of Propositions 4.3 and 4.2 are valid in this situation, hence we get the asymptotic relation (4.6.9). Applying estimates similar to those used in §4.3 (see (4.3.5)–(4.3.7)), we see that the condition (4.6.21) implies that

$$\lim_{N\to\infty} \overline{h}_N^{\omega_N}(b_N)|V_N|^{1/(k+1)} = 0. \tag{4.6.22}$$

Using the estimates (3.19.6) and (4.6.15), we find that

$$\lim_{N\to\infty} \frac{D_N^{\omega_N}(\overline{h}_N^{\omega_N}(b_N))}{D_N^{\omega_N}(0)} = 1. \tag{4.6.23}$$

The functions $E_N^{\omega_N}(\overline{h})$ and $\overline{H}_N^{\omega_N}(\overline{h})$ defining $\overline{h}_N^{\omega_N}(b_N)$ and $I_N^{\omega_N}(b_N)$ (see (4.2.4) and (4.2.8)) are holomorphic in the disk of radius $\rho(\omega_N)^{-1}$, where $\rho > 0$ is a constant, with center at the origin (see Proposition 3.3). So the function $I_N^{\omega_N}(b)$ is also holomorphic in b in the disk

$$|b - E_N^+| \le c_0 |V_N| (\omega_N)^{-1}, \tag{4.6.24}$$

where c_0 is a new constant (cf. the derivation of (4.2.5) from (4.3.1)). So we can expand $I_N^{\omega_N}(b_N)$ into a Taylor series around the point $b_N = E_N^+$. We have (cf. (1.2.16))

$$\begin{aligned} I_N^{\omega_N}(b_N) = {} & \frac{(b_N - E_N^+)^2}{2D_N^{\omega_N}(0)} + \sum_{j=3}^{k+1} \frac{K_N^{j,\omega_N}}{j!} \left(\frac{b_N - E_N^+}{D_N^{\omega_N}(0)} \right)^j \\ & + O\left(\frac{|b_N - E_N^+|^{k+2}}{|V_N|^{k+1}} (\ln |V_N|)^{(k+2)/(\nu-1)} \right). \end{aligned} \tag{4.6.25}$$

Here the coefficients K_N^{j,ω_N} are defined by the relation (1.2.18) from semiinvariants of the ω_N-cutoff Gibbs distribution with $h = 0$ (we add the index ω_N to emphasize their dependence of ω_N). The remainder of the Taylor expansion was estimated by the Cauchy formula, where we changed $D_N^{\omega_N}(0)$ to $|V_N|$ using the estimate (4.6.15) and where we have the extra factor $|V_N|^{-(k+1)}(\ln|V_N|)^{(k+2)/(\nu-1)}$ because of the restriction (4.6.24) on the domain of analyticity of the function $I_N^{\omega_N}(b_N)$, the relation (4.6.5) and the *a priori* estimate

$$|I_N^{\omega_N}(b_N)| \le C|V_N|, \tag{4.6.26}$$

which follows from the definition (4.2.8) and the estimate (3.19.5a) for G^0. Using the upper bounds (3.19.5) for semiinvariants and the lower bound (4.1.6) for the variance, we find from the relation (1.2.18) that $K_N^{j,\omega_N} = O(|V_N|)$ as $N \to \infty$. So the main condition (4.6.21) implies that the last two terms in (4.6.25) tend to zero, so that

$$I_N^{\omega_N}(b_N) = \frac{(b_N - E_N^+)^2}{2D_N^{\omega_N}(0)} + \sum_{j=3}^{k} \frac{K_N^{j,\omega_N}}{j!}\left(\frac{b_N - M_N^+}{D_N^{\omega_N}(0)}\right)^j + o_N(1), \tag{4.6.27}$$

where $o_N(1) \to 0$ as $N \to \infty$. Since in the case under consideration $h^1 = 0$ and the main condition (3.19.4) is valid for $\omega_N \equiv \infty$ also, it follows from the estimate (3.19.6) that for any integer m we can choose $C = C(m)$ in (4.6.5) so large that the limit of the difference of semiinvariants

$$\lim_{N\to\infty} |G^j(V_N, V_N, +, \beta, h, \omega_N) - G^j(V_N, V_N, +, \beta, h, \infty)||V_N|^m = 0. \tag{4.6.28}$$

(This fact can be obtained also from the usual Peierls estimate.) Applying this relation for $j = 2\,\mathrm{k}$, we see that

$$\lim_{N\to\infty} |V_N|^m |D_N^{\omega_N}(0) - D_N| = 0, \tag{4.6.29}$$

where D_N is the variance in the original ensemble. So it follows from the main relation (1.2.18) that for any fixed j and m

$$\lim_{N\to\infty} |K_N^{j,\omega_N} - K_N^j||V_N|^m = 0, \tag{4.6.30}$$

where K_N^j are the coefficients computed for the original Gibbs ensemble without a cutoff, and the expansion (4.6.27) can be rewritten as

$$I_N^{\omega_N}(b_N) = \frac{(b_N - E_N^+)^2}{2D_N} + \sum_{j=3}^{k} \frac{K_N^j}{j!}\left(\frac{b_N - E_N^+}{D_N}\right)^j + o_N(1). \tag{4.6.31}$$

Relations (4.6.31), (4.6.23), together with the asymptotic expression (4.6.8), prove the desired equality (1.2.33). The relation (1.2.35) is a special case of the relation (1.2.33). So we have proved Proposition 4.6 for the case of +-boundary conditions.

For the case of pure +-Gibbs state the arguments are completely similar. Indeed, if we let

$$P^{+,\beta,h}(b_N) = P^{+,\beta,h}(\sigma : S_{V_N} = b_N), \tag{4.6.32}$$

we see from comparing the definitions of §§1.3 and 3.2 that

$$P^{+,\beta,h}(b_N) = \lim_{|W_N|\to\infty} P^{+,\beta,h,\infty}_{W_N}(\sigma \in \Omega_{W_N} : S_{V_N} = b_N), \tag{4.6.33}$$

where $h = (h_t \equiv h, t \in V_N)$. So there exists a sequence $W_N \to \infty$ growing so quickly that

$$\lim_{N\to\infty} (P^{+,\beta,h}(b_N))^{-1} P^{+,\beta,h,\infty}_{W_N}(\sigma \in \Omega_{W_N} : S_{V_N} = b_N) = 1, \tag{4.6.34}$$

and we need only to study the probabilities $P^{+,\beta,h,\infty}_{W_N}(\sigma \in \Omega_{W_N} : S_{V_N} = b_N)$. The arguments repeat almost literally the previous one (cf. the second part of §4.5). We note only that we have to interchange the volumes V_N and W_N in applications of the results of the previous sections and, instead of reducing to the cutoff ensembles defined by the set of contours $G^{\omega_N}_{V_N}$ (see (3.2.1)), we must reduce to the case of the cutoff ensemble defined by the set of contours $G^{\omega_N}_{V_N, W_N}$ (see (3.2.2)).

§4.7. Classical behavior of deviations for symmetric boundary conditions

In this section the statements of Theorem 1.6.1 about classical behavior of large deviations for the cases of periodic and free boundary conditions are proved.

PROPOSITION 4.7.1. *Let V_N be a contractible sequence of volumes satisfying the Van Hove condition* (1.4.1). *Let a sequence b_N satisfy the condition* (1.4.4) *and $\beta \geq \beta_1$, where β_1 is sufficiently large. Let $h \geq 0$, and let condition* (1.5.2) *hold. Then the asymptotic relations* (1.2.27), (1.2.29), (1.2.31), (1.2.33), (1.2.35), (1.2.37), (1.2.38) *are true in the case of the periodic boundary conditions, if in all definitions of* §1.2 *the logarithmic generating function* (1.6.4) *is used instead of the logarithmic generating function* (1.2.4).

If $h = 0$ and the conditions (1.5.5), (1.5.6), (1.5.7) *are valid, then the asymptotic relations* (1.2.33), (1.2.35) *are true in the case of periodic boundary conditions, again with the logarithmic generating function* (1.6.4).

A similar result holds for free boundary conditions if we assume that the sets V_N are convex.

PROOF. The proof is a modification of the previous constructions developed for the case of +-boundary conditions, and differs only in some details. So we emphasize these differences only. We note also that a similar construction for the case $\nu = 2$ and periodic boundary conditions was used in the book [DKS]. We begin with the case of periodic boundary conditions.

Take a configuration $\sigma \in \Omega_{\mathbf{T}_N}$. Its boundary $\mathscr{G}(\sigma)$ is defined again (cf. §3.1) as the set of faces Q of the dual torus $\hat{\mathbf{T}}_N$ such $\sigma_s \neq \sigma_t$ for points s, t of the torus adjacent to Q, where the notions of faces and adjacent points are defined with respect to the natural geometry of a torus. The contours are again connected components of the boundary. The main specific feature of the periodic case is that now the boundary $\mathscr{G}(\sigma)$ and the set of contours of a configuration do not determine the configuration σ uniquely: there are two configurations σ and $-\sigma$ with the same set $\mathscr{G}(\sigma)$. Another problem is that there is no natural way to define the interior of a contour. For example, there are evident examples of contours for which any pair of points of the torus can be connected by a path which does not intersect the contour.

To avoid the problems described above and to define a cutoff Gibbs ensemble we restrict ourselves to the set $\overline{G}_{\mathbf{T}_N}$ of contours on $\mathbf{T}_N$ of diameters smaller than $N/2$. It is clear that for each contour $\Gamma \in \overline{G}_{\mathbf{T}_N}$, there is a unique connected component of its complement that contains more than half of the points of the torus $\mathbf{T}_N$. We call the set of points of $\mathbf{T}_N$ lying in this component the exterior of Γ and the set of points belonging to other components the interior of Γ. Let $\overline{D}_{\mathbf{T}_N}$ be the set of all systems of mutually disjoint contours $\mathscr{C} = \{\Gamma_1, \dots, \Gamma_k\} \subset \overline{G}_{\mathbf{T}_N}$. It follows from simple geometric considerations that the intersection $\bigcap_{i=1}^k \operatorname{Ext}\Gamma_i$ is a nonempty connected subset of $\mathbf{T}_N$ and there is a unique configuration $\sigma^+ = \sigma^+(\mathscr{C}) \in \Omega_N$ such that $\sigma_t^+ \equiv 1$ for $t \in \bigcap_{i=1}^k \operatorname{Ext}\Gamma_i$, and the set of this configuration $\mathscr{G}(\sigma^+) = \mathscr{C}$. Fixing cutoff levels ω_N, we denote by $\overline{\Omega}_{\mathbf{T}_N}^{\omega_N}$ the set of all configurations $\sigma \in \Omega_{\mathbf{T}_N}$ such that $\sigma = \sigma^+(\mathscr{C})$ for some set $\mathscr{C} = \{\Gamma_1, \dots, \Gamma_k\} \in \overline{D}_{\mathbf{T}_N}$ of contours for which $q(\Gamma_i) \le \omega_N$, $i = 1, \dots, k$ (cf. §3.2). Now we can define ω_N-cutoff Gibbs distribution

$$P_{\mathbf{T}_N}^{+,\mathrm{per},h,\omega_N}(\sigma) = \begin{cases} \dfrac{\exp\left\{-\beta\left(2\sum\limits_{\Gamma\in\mathscr{G}(\sigma)}|\Gamma| - h\sum\limits_{t\in\mathbf{T}_N}\sigma_t\right)\right\}}{Z^{+,\mathrm{per}}(\mathbf{T}_N, \beta, h, \omega_N)} & \text{if } \sigma \in \overline{\Omega}_{\mathbf{T}_N}^{\omega_N}, \\ 0 & \text{if } \sigma \notin \overline{\Omega}_{\mathbf{T}_N}^{\omega_N}, \end{cases} \tag{4.7.1}$$

where, as usual, the partition function is given by

$$Z_{\mathbf{T}_N}^{+,\mathrm{per}}(\beta, h, \omega_N) = \sum_{\sigma\in\overline{\Omega}_{\mathbf{T}_N}^{\omega_N}} \exp\left\{-\beta\left(2\sum_{\Gamma\in\mathscr{G}(\sigma)}|\Gamma| - h\sum_{t\in\mathbf{T}_N}\sigma_t\right)\right\}. \tag{4.7.2}$$

We set

$$P_{\mathbf{T}_N}^{+,\mathrm{per}}(b) = \sum_{\sigma\,:\,S_{\mathbf{T}_N}(\sigma)=b} P_{\mathbf{T}_N}^{+,\mathrm{per},h,\omega_N}(\sigma). \tag{4.7.3}$$

Using this definition of a cutoff ensemble, we can repeat literally all the constructions and statements of Chapter 3 and §§4.1–4.3.

The generalization of Proposition 4.4 requires additional work. One should replace the sets W_N and V_N with $\mathbf{T}_N$, and the main statement (4.4.3) becomes the relation

$$\frac{P_{\mathbf{T}_N}^{\mathrm{per},h}(\sigma \in \Omega_{\mathbf{T}_N} \setminus \overline{\Omega}_{\mathbf{T}_N}^{\omega_N} : S_{\mathbf{T}_N}(\sigma) = b_N)}{P_{\mathbf{T}_N}^{\mathrm{per},h}(\sigma : S_{\mathbf{T}_N}(\sigma) = b_N)} \leq \exp\{-2(\beta - \beta_0)(\omega_N)^{\nu-1}\}. \quad (4.7.4)$$

In the construction used in the proof we have to do the following changes. The system $\mathscr{G}^L(\sigma)$ (cf. the definition before (4.4.5)) is defined now as the system of contours $\mathscr{G}(\sigma) \cap ((G_{\mathbf{T}_N} \setminus \overline{G}_{\mathbf{T}_N}) \cup \{\Gamma \in \overline{G}_{\mathbf{T}_N} : q(\Gamma) > \omega_N\})$, and as in §4.4 we define D_N^L to be the set of all systems of contours ∂^L such that $\partial^L = \mathscr{G}^L(\sigma)$ for some $\sigma \in \Omega_{\mathbf{T}_N}$. The essential difference from the situation of §4.4 is that now there are two configurations $\sigma \in \Omega_{\mathbf{T}_N}$ such that $\partial^L(\sigma) = \sigma$, which can be obtained from each other by change of signs. There are two disjoint sets $U^1(\partial^L)$, $U^2(\partial^L)$ such that $\mathbf{T}_N = U^1(\partial^L) \cup U^2(\partial^L)$, and these configurations take constant values on these sets. So it is necessary to apply the construction of §4.4 to each of two cases: $U^1(\partial^L) = U^-(\partial^L)$ and $U^2(\partial^L) = U^+(\partial^L)$, and vice versa. Correspondingly, there are two possibilities D_N^1, D_N^2 of the systems of contours D_N defined in §4.4 according to these two choices of the set $U^-(\partial^L)$, and the main representation (4.4.29) must be replaced by the representation

$$\begin{aligned} &P_{\mathbf{T}_N}^{\mathrm{per},0}(\sigma \in \Omega_{\mathbf{T}_N} \setminus \overline{\Omega}_{\mathbf{T}_N}^{\omega_N} : S_{\mathbf{T}_N} = b_N) \\ &\quad = \sum_{i=1}^{2} \sum_{\substack{\partial \in D_N^i : \\ \partial \neq \varnothing}} \sum_{b=0}^{[\lambda |W_N^{-,i}(\partial)|]} P_{\mathbf{T}_N}^{\mathrm{per},0}(\sigma \in \Omega_{\mathbf{T}_N} : S_{\mathbf{T}_N} = b_N, \qquad (4.7.5) \\ &\qquad S_{W^{-,i}}(\partial) = -|W^{-,i}(\partial)| + \overline{b}, \overline{\mathscr{G}}(\sigma) = \partial, \sigma_t \equiv 1 \text{ for } t \in \Delta^{+,i}(\partial)), \end{aligned}$$

where the additional index i indicates that the corresponding sets $W(\partial)$ and $\Delta^+(\partial)$ are defined by the construction of §4.4 with $U^-(\partial^L) = U^i(\partial^L)$. (As in §4.4, we can assume without loss of generality that $h = 0$.) All other constructions used in §4.4 require evident modifications implied by the one above.

In §4.6 we applied the estimates of §3.20 to be able to use the results obtained for the case of the cutoff ensemble in case of the cutoff level $\omega_N = \infty$, and then, using the fact that the ∞-cutoff ensemble coincides with the original Gibbs ensemble for the considered case of contractible volume, we obtained the main result. But in the periodic case the ∞-cutoff ensemble is essentially different from the original one and so we need an additional result, extending Proposition 3.20.

PROPOSITION 4.7.2. *Let* $G_N^{n,+,\mathrm{per},h}$, $n = 1, 2, \ldots$, *be the semiinvariants of order* n *of the random variable* $S_{\mathbf{T}_N}$ *defined by the probability distribution*

(1.6.3), *and* $G_N^{n,+,\mathrm{per},h.\omega_N}$ *be the semiinvariants of order* n *of the same random variable governed by the probability distribution* (4.7.1). *Let* (*cf.* (3.19.3))

$$G_{N.}^{0,+,\mathrm{per},h}(z) = \ln F_N^{+,\mathrm{per},h}(z),$$
$$G^{0,+,\mathrm{per},h,\omega_N}(z) = \ln F_N^{+,\mathrm{per},h,\omega_N}(z), \tag{4.7.6}$$

where the generating functions are given by (*cf.* (1.6.4))

$$F_N^{+,\mathrm{per},h}(z) = \sum_{b\geq 0} P_N^{+,\mathrm{per},h}(b)e^{zb},$$
$$F_N^{+,\mathrm{per},h,\omega_N}(z) = \sum_{b\geq 0} P_N^{+,\mathrm{per},h,\omega_N}(b)e^{zb}, \tag{4.7.7}$$

and where we suppose that the cutoff level

$$\omega_N \geq (\log N)^2. \tag{4.7.8}$$

Then for any $Q' < \infty$, $m > 0$ *there exist* $\beta_0 = \beta_0(\nu, Q', m)$ *such that for* $\beta \geq \beta_0$, *and any* $n = 1, 2, \dots$, *uniformly for* $0 \leq h \leq Q'$

$$\lim_{N\to\infty} |\mathbf{T}_N|^m |G_N^{n,+,\mathrm{per},h} - G_N^{n,+,\mathrm{per},h,\omega_N}| = 0 \tag{4.7.9}$$

and for real λ *such that* $0 \leq h + \beta^{-1}\lambda \leq Q'$ *uniformly with respect to* h *and* λ

$$\lim_{N\to\infty} |\mathbf{T}_N|^m |G_N^{0,+,\mathrm{per},h}(\lambda) - G_N^{0,+,\mathrm{per},h,\omega_N}(\lambda)| = 0. \tag{4.7.10}$$

PROOF. We begin with the proof of the following statement about the partition functions (see the notation of §1.3 and (4.7.2)): for any m and $\beta \geq \beta_0$, where $\beta_0 = \beta_0(\nu, c)$ is sufficiently large

$$\lim_{N\to\infty} |\mathbf{T}_N|^m \left| \frac{Z_{\mathbf{T}_N}^{+,\mathrm{per}}(\beta, 0, \omega_N)}{Z_{\mathbf{T}_N}^{\mathrm{per}}(\beta, 0)} - \frac{1}{2} \right| = 0. \tag{4.7.11}$$

Indeed, comparing definitions of the partition functions and the corresponding probability distributions and recalling that in the case $h = 0$ two configurations with the same set $\mathscr{G}(\sigma)$ are equiprobable, we see that

$$\begin{aligned}\frac{Z^{+,\mathrm{per}}(\mathbf{T}_N, \beta, 0, \omega_N)}{Z_{\mathbf{T}_N}^{\mathrm{per}}(\beta, 0)} &= P_{\mathbf{T}_N}^{\mathrm{per},\beta,0}(\sigma : \sigma \in \overline{\Omega}_{\mathbf{T}_N}^{\omega_N}) \\ &= \frac{1}{2} P_{\mathbf{T}_N}^{\mathrm{per},\beta,0}(\sigma : \mathscr{G}(\sigma) \in \overline{G}_{T_N}, q(\Gamma) \leq \omega_N, \Gamma \in \mathscr{G}(\sigma)).\end{aligned} \tag{4.7.12}$$

Let $\hat{G}^{\omega_N} = (G_{\mathbf{T}_N} \setminus \overline{G}_{\mathbf{T}_N}) \cup \{\Gamma \in \overline{G}_{\mathbf{T}_N} : q(\Gamma) \geq \omega_N\}$ be the set of contours and let $\hat{D}^{\omega_N}$ be the set of all systems of disjoint contours $\mathscr{C} \subset \hat{G}^{\omega_N}$. Observe that for any $\mathscr{C} \in \hat{D}^{\omega_N}$,

$$P_{\mathbf{T}_N}^{\mathrm{per},\beta,0}(\sigma : \mathscr{G}(\sigma) \cap \hat{G}^{\omega_N} = \mathscr{C}) \leq \exp\Big\{ -2\beta \sum_{\Gamma\in\mathscr{C}} |\Gamma| \Big\}. \tag{4.7.13}$$

This estimate can be proved by the usual Peierls transformation, to be used for deleting from the system of contours $\mathcal{G}(\sigma)$ the contours belonging to $\mathcal{C}$. So

$$\begin{aligned} P_{\mathbf{T}_N}^{\mathrm{per},\beta,0}&(\sigma : \mathcal{G}(\sigma) \in \overline{G}_N, q(\Gamma) \le \omega_N, \Gamma \in \mathcal{G}(\sigma)) \\ &\ge 1 - \sum_{\mathcal{C} \in \hat{D}^{\omega_N}} P_{\mathbf{T}_N}^{\mathrm{per},\beta,0}(\sigma : \mathcal{G}(\sigma) \cap \hat{G}^{\omega_N} = \mathcal{C}) \\ &\ge 1 - \sum_{\mathcal{C} \in \hat{D}^{\omega_N}} \exp\Big\{ -2\beta \sum_{\Gamma \in \mathcal{C}} |\Gamma| \Big\}. \end{aligned} \tag{4.7.14}$$

A standard combinatorial estimate shows that the number of different systems of contours $\mathcal{C} = (\Gamma_1, \dots, \Gamma_k)$ with a fixed value of $|\mathcal{C}| = \sum_{\Gamma \in \mathcal{C}} |\Gamma|$ does not exceed $|\mathbf{T}_N|^k K^{|\mathcal{C}|}$, where $K = K(\nu)$ is a constant. These estimates, together with the condition (4.7.8), imply that the last sum in (4.7.14) decreases as $O(N^{-m})$ for any $m > 0$ and so the desired relation (4.7.11) follows from the relations (4.7.12)–(4.7.14).

Comparing the definitions in §1.3 and in this section, we see that

$$P_{\mathbf{T}_N}^{+,\mathrm{per},h,\omega_N}(b_N) = P_{\mathbf{T}_N}^{\mathrm{per},h}(\sigma \in \overline{\Omega}_{\mathbf{T}_N}^{\omega_N} : S_{\mathbf{T}_N}(\sigma) = b_N) \frac{Z_{\mathbf{T}_N}^{\mathrm{per}}(\beta, h)}{Z_{\mathbf{T}_N}^{+,\mathrm{per}}(\beta, h, \omega_N)}. \tag{4.7.15}$$

So it follows from (4.7.4) that for any $m > 0$ and for β sufficiently large,

$$\lim_{N\to\infty} |\mathbf{T}_N|^m \left| \frac{P_{\mathbf{T}_N}^{+,\mathrm{per},h,\omega_N}(b_N)}{P_{\mathbf{T}_N}^{\mathrm{per},h}(b_N)} \frac{Z_{\mathbf{T}_N}^{+,\mathrm{per}}(\beta, h, \omega_N)}{Z_{\mathbf{T}_N}^{\mathrm{per}}(\beta, h)} - 1 \right| = 0 \tag{4.7.16}$$

if (recall (4.4.2))

$$\tanh(C\beta)|\mathbf{T}_N| \ge b_N \ge E_N(0) - c|\mathbf{T}_N|^{\nu/(\nu+1)}. \tag{4.7.17}$$

In the case $h = 0$ we can use (4.7.11) and see that under the same condition (4.7.17)

$$\lim_{N\to\infty} |\mathbf{T}_N|^m \left| \frac{P_{\mathbf{T}_N}^{+,\mathrm{per},0,\omega_N}(b_N)}{2P_{\mathbf{T}_N}^{\mathrm{per},0}(b_N)} - 1 \right| = 0. \tag{4.7.18}$$

If the condition (4.7.17) is valid, we can apply Theorem 4.2 to the ω_N-cutoff ensemble. (Its conditions can be verified in the same way as it was done for the case of +-boundary condition in §4.6.) We observe that

$$\lim_{N\to\infty} |\mathbf{T}_N|^m \left| \sum_{\substack{b : \tanh(C\beta)|T_N| \ge b \\ \ge E_N(0) - c|\mathbf{T}_N|^{\nu/(\nu+1)}}} P_{\mathbf{T}_N}^{+,\mathrm{per},0,\omega_N}(\sigma \in \Omega_{\mathbf{T}_N} : S_{\mathbf{T}_N} = b) - 1 \right| = 0. \tag{4.7.19}$$

One can obtain this estimate by summing up the asymptotic relations (4.2.7) or by using the integral version of the Large Deviations Theorem, which can be proved in the same way as its local version (4.2.7). Using this relation together with (4.7.18) and the symmetry of the periodic Gibbs measure, we see that

$$\lim_{N\to\infty} |\mathbf{T}_N|^m \left| \sum_{\substack{b\,:\,\tanh(C\beta)|\mathbf{T}_N|\ge|b| \\ \ge E_N(0)-c|T_N|^{\nu/(\nu+1)}}} P^{\mathrm{per},0}_{\mathbf{T}_N}(\sigma\in\Omega_{\mathbf{T}_N} : S_{\mathbf{T}_N}(\sigma)=b_N) - 1\right| = 0. \tag{4.7.20}$$

In particular, using the symmetry we see that

$$\begin{aligned}\lim_{N\to\infty} |\mathbf{T}_N|^m |P^{\mathrm{per},0}_{\mathbf{T}_N}(\sigma\in\Omega_{\mathbf{T}_N} : S_{\mathbf{T}_N}(\sigma)\ge 0) - 1/2| \\ = \lim_{N\to\infty} |\mathbf{T}_N|^m P^{\mathrm{per},0}_{\mathbf{T}_N}(0)/2 = 0.\end{aligned} \tag{4.7.21}$$

So it follows from the definition (1.6.3) and the relations (4.7.18)–(4.7.21) that

$$\lim_{N\to\infty} |\mathbf{T}_N|^m \sum_{b\ge 0} |P^{+,\mathrm{per},0}_{\mathbf{T}_N}(b) - P^{+,\mathrm{per},0,\omega_N}_{\mathbf{T}_N}(b)| = 0. \tag{4.7.22}$$

Now we want to extend the estimate (4.7.22) for the variation distance to the case of an arbitrary $h\ge 0$. To do it we observe that

$$P^{\mathrm{per},h}_{\mathbf{T}_N}(\sigma\in\Omega_{\mathbf{T}_N} : S_{\mathbf{T}_N}(\sigma)\ge 0) = \frac{Z^{+,\mathrm{per}}_{\mathbf{T}_N}(\beta, h)}{Z^{\mathrm{per}}_{\mathbf{T}_N}(\beta, h)}, \tag{4.7.23}$$

where we set (cf. (1.3.7))

$$Z^{+,\mathrm{per}}_{\mathbf{T}_N}(\beta, h) = \sum_{\sigma\in\Omega_{\mathbf{T}_N} : S_{\mathbf{T}_N}(\sigma)\ge 0} \exp\{-\beta H^{h,\mathrm{per}}_{\mathbf{T}_N}(\sigma)\}. \tag{4.7.24}$$

Using (4.7.23) and the definition (1.6.3), we can rewrite (4.7.16) as

$$\lim_{N\to\infty} |\mathbf{T}_N|^m \left| \frac{P^{+,\mathrm{per},h,\omega_N}_{\mathbf{T}_N}(b_N)}{P^{+,\mathrm{per},h}_{\mathbf{T}_N}(b_N)} \frac{Z^{+,\mathrm{per}}_{\mathbf{T}_N}(\beta, h, \omega_N)}{Z^{+,\mathrm{per}}_{\mathbf{T}_N}(\beta, h)} - 1\right| = 0. \tag{4.7.25}$$

Now we want to estimate the second factor in (4.7.25) and prove that

$$\lim_{N\to\infty} |\mathbf{T}_N|^m \left| \frac{Z^{+,\mathrm{per}}_{\mathbf{T}_N}(\beta, h, \omega_N)}{Z^{+,\mathrm{per}}_{\mathbf{T}_N}(\beta, h)} - 1\right| = 0. \tag{4.7.26}$$

To do this we show, using (4.7.7), that

$$\begin{aligned} Z^{+,\mathrm{per}}_{\mathbf{T}_N}(\beta, h, \omega_N) &= Z^{+,\mathrm{per}}_{\mathbf{T}_N}(\beta, 0, \omega_N) F^{+,\mathrm{per},0,\omega_N}_N(h\beta), \\ Z^{+,\mathrm{per}}_{\mathbf{T}_N}(\beta, h) &= Z^{+,\mathrm{per}}_{\mathbf{T}_N}(\beta, 0) F^{+,\mathrm{per},0}_N(h\beta). \end{aligned} \tag{4.7.27}$$

It follows from (4.7.18) and (4.7.21) that for any positive λ

$$\lim_{N\to\infty} |\mathbf{T}_N|^m \left| \frac{\sum_{b\,:\,\tanh(C\beta)|\mathbf{T}_N|\ge b\ge E_N(0)-c|\mathbf{T}_N|^{\nu/(\nu+1)}} P^{+,\mathrm{per},0,\omega_N}_{\mathbf{T}_N}(b)e^{\lambda b}}{\sum_{b\,:\,\tanh(C\beta)|\mathbf{T}_N|\ge b\ge E_N(0)-c|\mathbf{T}_N|^{\nu/(\nu+1)}} P^{+,\mathrm{per},0}_{\mathbf{T}_N}(b)e^{\lambda b}} - 1\right| = 0. \tag{4.7.28}$$

Therefore, to estimate the ratio of the generating functions we need to estimate contributions of "tails" only. Observe that

$$(F_N^{+,\mathrm{per},0}(\lambda))^{-1} \sum_{0\le b<E_N(0)-c|\mathbf{T}_N|^{\nu/(\nu+1)}} e^{\lambda b} P_{\mathbf{T}_N}^{+,\mathrm{per},0}(b) \le \frac{\sum_{0\le b<E_N(0)-c|\mathbf{T}_N|^{\nu/(\nu+1)}} e^{\lambda b} P_{\mathbf{T}_N}^{+,\mathrm{per},0}(b)}{\sum_{E_N(0)-c|\mathbf{T}_N|^{\nu/(\nu+1)}\le b} e^{\lambda b} P_{\mathbf{T}_N}^{+,\mathrm{per},0}(b)} \le \frac{\sum_{0\le b<E_N(0)-c|\mathbf{T}_N|^{\nu/(\nu+1)}} P_{\mathbf{T}_N}^{+,\mathrm{per},0}(b)}{\sum_{E_N(0)-c|\mathbf{T}_N|^{\nu/(\nu+1)}\le b} P_{\mathbf{T}_N}^{+,\mathrm{per},0}(b)}, \quad (4.7.29)$$

and so it follows from the estimate (4.7.20) that

$$\lim_{N\to\infty} |\mathbf{T}_N|^m (F_N^{+,\mathrm{per},0}(\lambda))^{-1} \sum_{0\le b<E_N(0)-c|\mathbf{T}_N|^{\nu/(\nu+1)}} e^{\lambda b} P_{\mathbf{T}_N}^{+,\mathrm{per},0}(b) = 0. \quad (4.7.30)$$

In a similar way, using (4.7.19) we find that

$$\lim_{N\to\infty} |\mathbf{T}_N|^m (F_N^{+,\mathrm{per},0,\omega_N}(\lambda))^{-1} \sum_{0\le b<E_N(0)-c|\mathbf{T}_N|^{\nu/(\nu+1)}} e^{\lambda b} P_{\mathbf{T}_N}^{+,\mathrm{per},0,\omega_N}(b) = 0. \quad (4.7.31)$$

We need also similar estimates for $b > \tanh(C\beta)|\mathbf{T}_N|$. The first of them states that for any $\lambda > 0$ there exists $C = C(\nu, \lambda) < \infty$ such that for any $m > 0$ we have

$$\lim_{N\to\infty} |\mathbf{T}_N|^m (F_N^{+,\mathrm{per},0,\omega_N}(\lambda))^{-1} \sum_{\tanh(C\beta)|\mathbf{T}_N|<b\le|\mathbf{T}_N|} e^{\lambda b} P_{\mathbf{T}_N}^{+,\mathrm{per},0,\omega_N}(b) = 0. \quad (4.7.32)$$

This estimate is a consequence of the following two statements. First of all, the estimate (4.7.19) implies that

$$F_N^{+,\mathrm{per},0,\omega_N}(\lambda) \ge e^{E_N(0)\lambda(1+o_N(1))}, \quad (4.7.33)$$

where $o_N(1) \to 0$ as $N \to \infty$. Next, we can use the fact that the estimates of Theorem 4.2 can be applied to the ω_N-cutoff ensemble. The estimate (4.2.11) of this theorem implies that for any $\varkappa < \infty$ there exists $C' = C'(\nu, \varkappa) < \infty$ such that

$$\sum_{\tanh(C'\beta)|\mathbf{T}_N|<b\le|\mathbf{T}_N|} P_{\mathbf{T}_N}^{+,\mathrm{per},0,\omega_N}(b) \le e^{-\varkappa|\mathbf{T}_N|} \quad (4.7.34)$$

for sufficiently large N. We need also a similar estimate for the ensemble without cutoff:

$$\lim_{N\to\infty} |\mathbf{T}_N|^m (F_N^{+,\mathrm{per},0}(\lambda))^{-1} \sum_{\tanh(C\beta)|\mathbf{T}_N|<b\le|\mathbf{T}_N|} e^{\lambda b} P_{\mathbf{T}_N}^{+,\mathrm{per},0}(b) = 0. \quad (4.7.35)$$

The proof consists of two steps similar to those used in the derivation of the estimate (4.7.32). The estimate similar to the estimate (4.7.33) follows now from the estimate (4.7.20). The derivation of the analog of (4.7.34) requires the construction used in the derivation of the estimate (4.7.4). It is necessary to fix a system of contours ∂^L and then to apply (4.2.11) to a resulting cutoff ensemble. It follows from (4.7.28), (4.7.30), (4.7.31), (4.7.32), and (4.7.34) that for any $\lambda > 0$

$$\lim_{N\to\infty} |\mathbf{T}_N|^m \left| \frac{F_N^{+,\mathrm{per},0,\omega_N}(\lambda)}{F_N^{+,\mathrm{per},0}(\lambda)} - 1 \right| = 0. \tag{4.7.36}$$

Since it follows from (4.7.24) that

$$\frac{Z_{\mathbf{T}_N}^{+,\mathrm{per}}(\beta, 0)}{Z_{\mathbf{T}_N}^{\mathrm{per}}(\beta, 0)} = P_{\mathbf{T}_N}(\sigma : S_{\mathbf{T}_N}(\sigma) \ge 0), \tag{4.7.37}$$

the relations (4.7.27), (4.7.11), (4.7.21), and (4.7.36) imply the desired relation (4.7.26).

The relations (4.7.26) together with the estimate (4.7.25) imply that under the condition (4.7.17)

$$\lim_{N\to\infty} |\mathbf{T}_N|^m \left| \frac{P_{\mathbf{T}_N}^{+,\mathrm{per},h,\omega_N}(b_N)}{P_{\mathbf{T}_N}^{+,\mathrm{per},h}(b_N)} - 1 \right| = 0. \tag{4.7.38}$$

Using (4.7.38), the relation

$$\sum_{b\,:\,b\notin[E_N(0)-c|\mathbf{T}_N|^{\nu/(\nu+1)},\,\tanh(C\beta)|\mathbf{T}_N|]} P_{\mathbf{T}_N}^{+,\mathrm{per},h}(b) = \frac{\displaystyle\sum_{b\,:\,b\notin[E_N(0)-c|\mathbf{T}_N|^{\nu/(\nu+1)},\,\tanh(C\beta)|\mathbf{T}_N|]} e^{h\beta} P_{\mathbf{T}_N}^{+,\mathrm{per},0}(b)}{F_N^{+,\mathrm{per},0}(h\beta)}, \tag{4.7.39}$$

which easily follows from the corresponding definitions, a similar relation for the ω_N-cutoff ensemble, and estimates (4.7.28), (4.7.30), (4.7.31), (4.7.32), (4.7.35), we see that

$$\lim_{N\to\infty} |\mathbf{T}_N|^m \sum_{b\ge 0} |P_{\mathbf{T}_N}^{+,\mathrm{per},h}(b) - P_{\mathbf{T}_N}^{+,\mathrm{per},0,\omega_N}(b)| = 0 \tag{4.7.40}$$

for all $h \ge 0$.

Since the semiinvariants are polynomials of the momenta and the moment of order m cannot exceed $|\mathbf{T}_N|^m$, the statement (4.7.9) of the proposition follows immediately from the estimate (4.7.40). It is clear from the corresponding definitions that

$$
\begin{aligned}
F_N^{+,\mathrm{per},h}(\lambda) &= \frac{F_N^{+,\mathrm{per},0}(\lambda+\beta h)}{F_N^{+,\mathrm{per},0}(\beta h)}, \\
F_N^{+,\mathrm{per},h,\omega_N}(\lambda) &= \frac{F_N^{+,\mathrm{per},0,\omega_N}(\lambda+\beta h)}{F_N^{+,\mathrm{per},0,\omega_N}(\beta h)}.
\end{aligned}
\tag{4.7.41}
$$

So the statement (4.7.10) of the proposition follows from relation (4.7.36).

All the arguments in §4.6 can be extended literally to the periodic case if we use Proposition 4.7.2 instead of Proposition 3.20. So the statement of the main Proposition 4.7.1 is valid for the periodic case.

The case of free boundary conditions requires only minor changes compared to the periodic case. The only additional problem is due to the existence of nonclosed contours ending on the boundary of the volume V_N. It is necessary to define the interior of such contours. As was mentioned in §1.6, we restrict ourselves to convex volumes V_N. In the definition of the cutoff ensembles we use the set $\overline{G}_{V_N}$ of only those nonclosed contours Γ that have diameters smaller than $1/4$ of the diameter of the volume V_N. Consider a set of connected components into which such a contour can divide the volume V_N. Among them, there is exactly one containing more than half the points of the volume V_N. We call it the exterior of Γ and its complement the interior of Γ. There are two properties of interiors that are essentially used in our constructions. First, if Γ_1 and Γ_2 are two compatible contours and $\operatorname{Int}\Gamma_1 \cap \operatorname{Int}\Gamma_2 \neq \varnothing$, then either $\operatorname{Int}\Gamma_1 \subset \operatorname{Int}\Gamma_2$ or $\operatorname{Int}\Gamma_2 \subset \operatorname{Int}\Gamma_1$. For the definition given above this property evidently holds. Second, we want to have an estimate of the interior of a contour in terms of its size, similar to the estimate (3.1.5). It is also evidently valid. (The constant $(2\nu)^{-\nu}$ in (3.1.5) has to be changed to another constant depending on ν, but this does not influence applications of this estimate.)

In §1.6 it was stated that the limit functions $I(h)$ coincide for all considered boundary conditions. This follows from the fact that the limit values of the free energies are the same for all these cases, which is easy to prove by comparing expansions (3.3.7) for these cases.

5. Nonclassical Behavior of Large Deviations

In this chapter we prove the statements of Theorems 1.5.1 and 1.5.2 on the nonstandard behavior of large deviations caused by the existence of phase transitions.

§5.1. Lower bounds

PROPOSITION 5.1.1. *Let V_N be a sequence of contractible volumes satisfying the Van Hove condition* (1.4.1) *and the thickness condition* (1.5.1) *and let b_N be a sequence of integers satisfying the condition* (1.4.4). *Consider the case of* *the +-boundary conditions. Suppose that* $h = 0$ *and the conditions* (1.5.8), (1.5.9) *are satisfied. Then for some* $\beta_1 < \infty$, *all* $\beta > \beta_1$, *and some* $\underline{c}$ *that depends on the sequence of volumes* V_N *and* β *only we have*

$$\ln P_N^{+,0}(b_N) \geq -\underline{c}(E_N^+ - b_N)^{(\nu-1)/\nu}. \tag{5.1.1}$$

If the conditions (1.5.11), (1.5.12) *are satisfied, then for the same* β

$$\liminf_{N\to\infty} \frac{-\ln P_N^{+,0}(b_N)}{I_N^0(-b_N)} \geq 1. \tag{5.1.2}$$

If $h > 0$ *and the conditions* (1.5.21), (1.5.22) *are satisfied, then for some* $\beta_1 = \beta_1(h)$ *and all* $\beta > \beta_1$

$$\liminf_{N\to\infty} \frac{\ln P_N^{+,h}(b_N)}{|V_N|} \geq -\tilde{I}^h(\hat{b}). \tag{5.1.3}$$

Here the function $\tilde{I}^h(\hat{b})$ *is defined by the relation* (1.5.24).

Similar statements hold for the cases of limit pure +-state, periodic and free boundary conditions.

PROOF. The arguments for all boundary conditions are completely similar, so we consider in detail only the case of +-boundary condition.

We fix a number δ such that

$$0 < \delta < 1/(\nu+1)^2. \tag{5.1.4}$$

Let integers b_N be such that the condition (1.4.4) is satisfied and

$$-E_N^+ + |V_N|^{(\nu+2\delta-1)/\nu} \leq b_N \leq E_N^+, \tag{5.1.5}$$

where E_N^+ is the mean value of S_N in the case $h = 0$ (see (1.5.3)). It is easy to see that $E_N^+ > 0$ for sufficiently large N and β (see (1.3.14) and (3.19.9)). Consider numbers $\overline{K}_N^{\pm}$, $\overline{b}_N^{\pm}$ such that

$$\overline{K}_N^+ + \overline{K}_N^- = |V_N|, \qquad \overline{b}_N^+ - \overline{b}_N^- = b_N, \tag{5.1.6}$$

and

$$\frac{E_N^+}{|V_N|} = \frac{\overline{b}_N^{\pm}}{\overline{K}_N^{\pm}}. \tag{5.1.7}$$

It is easy to see that these equations have a unique solution

$$\overline{b}_N^{\pm} = \frac{1}{2}(E_N^+ \pm b_N), \qquad \overline{K}_N^{\pm} = \frac{|V_N|}{2}\left(1 \pm \frac{b_N}{E_N^+}\right), \tag{5.1.8}$$

which is nonnegative under the condition (5.1.5). Now we modify the numbers $\overline{K}_N^{\pm}$, $\overline{b}_N^{\pm}$, changing them to nonnegative integers $K_N^{\pm}$, $b_N^{\pm}$ satisfying

$$(b_N^{\pm} - K_N^{\pm}) \equiv 0 \pmod 2, \tag{5.1.9}$$

while again

$$K_N^+ + K_N^- = |V_N| \tag{5.1.10}$$

and

$$b_N^+ - b_N^- = b_N. \tag{5.1.11}$$

In defining the numbers $K_N^\pm$, $b_N^\pm$ we consider three different cases: the first two correspond to the region (5.1.5) and the last one to the complementary region.

CASE 1. Here we assume that

$$E_N^+/2 \le b_N \le E_N^+ \tag{5.1.12}$$

and change $\overline{K}_N^\pm$, $\overline{b}_N^\pm$ only slightly. Namely, we choose $K_N^\pm$, $b_N^\pm$ to be integers such that the conditions (5.1.9), (5.1.10), (5.1.11), and the inequalities

$$|K_N^\pm - \overline{K}_N^\pm| \le 2, \tag{5.1.13}$$

$$|b_N^\pm - \overline{b}_N^\pm| \le 2. \tag{5.1.14}$$

are satisfied. Evidently, this is possible.

CASE 2. Here we consider the case

$$-E_N^+ + |V_N|^{(\nu+2\delta-1)/\nu} \le b_N < E_N^+/2, \tag{5.1.15}$$

and again choose $K_N^\pm$, $b_N^\pm$ to be integers satisfying the conditions (5.1.9), (5.1.10), (5.1.11) and the inequality (5.1.13), but now we increase b_N^+ and decrease b_N^-. So instead of (5.1.14) we impose the condition

$$|b_N^\pm \mp |V_N|^{(\nu+\delta-1)/\nu} - \overline{b}_N^\pm| \le 2. \tag{5.1.16}$$

CASE 3. Here we consider the case which is complementary to (5.1.5), namely

$$b_N < -E_N^+ + |V_N|^{(\nu+2\delta-1)/\nu} \tag{5.1.17}$$

and instead of the previous construction we define

$$K_N^+ = 0, \qquad K_N^- = |V_N|, \qquad b_N^+ = 0, \qquad b_N^- = -b_N. \tag{5.1.18}$$

In this case conditions (5.1.10), (5.1.11) are evidently valid. The condition (5.1.9) for b_N^- follows from the condition (1.4.4), and for b_N^+ it holds evidently. Observe that if $h = 0$, then in all cases considered in the proposition, the values of b_N satisfy the assumptions of one of our Cases 1, 2, or 3 for all sufficiently large N.

Using the thickness condition (1.5.1) we choose a sequence of subvolumes $V_N' \subset V_N$ such that $|V_N'| = K_N^-$ and the estimate (1.5.1) holds. Let $V_N^- = V_N'$ and $V_N^+ = V_N \setminus V_N^-$. Let $\Gamma_N^b \in G_{V_N}$ be the contour consisting of all the faces separating points of the set V_N' from the points of its complement $V_N'^c$ (see §2.1). Consider the set of configurations $\Omega_N^b \subseteq \Omega_{V_N}$ such that $\sigma \in \Omega_N^b$

if and only if the contour $\Gamma_N^b \in \mathcal{G}(\sigma)$ and all other contours of $\mathcal{G}(\sigma)$ are inner contours of V_N^+ or of V_N^-. Using (5.1.11) we find that

$$P_N^{+,0}(b_N) \geq P_N^{+,0}(\sigma \in \Omega_N^b : S_{V_N^+}(\sigma) = b_N^+ , S_{V_N^-}(\sigma) = -b_N^-). \qquad (5.1.19)$$

This inequality is a starting point of the following estimates for the case $h = 0$.

Let us denote

$$\Delta_N^\pm = V_N^\pm \cap \Delta(\Gamma_N^b) \qquad (5.1.20)$$

(see (3.1.2)). Observe that the configurations $\sigma \in \Omega_N^b$ take the values ± 1 at points $t \in \Delta_N^\pm$. Let also $\hat{V}_N^\pm = V_N^\pm \setminus \Delta_N^\pm$. It follows from the definitions of §3.2 and the symmetry of the Gibbs distribution in the case $h = 0$ that

$$\begin{aligned} &P_N^{+,0}(\sigma \in \Omega_N^b : S_{V_N^+}(\sigma) = b_N^+ , S_{V_N^-}(\sigma) = -b_N^-) \\ &\quad = \exp\{-2\beta|\Gamma_N^b|\} \frac{Z(\hat{V}_N^+ , \beta , 0 , + , \infty) Z(\hat{V}_N^- , \beta , 0 , + , \infty)}{Z(V_N , \beta , 0 , + , \infty)} \\ &\quad \times P_{\hat{V}_N^+}^{+,\beta,0,\infty}(\sigma : S_{\hat{V}_N^+}(\sigma) = b_N^+ - |\Delta_N^+|) \\ &\quad \times P_{\hat{V}_N^-}^{+,\beta,0,\infty}(\sigma : S_{\hat{V}_N^-}(\sigma) = b_N^- - |\Delta_N^-|). \end{aligned} \qquad (5.1.21)$$

(If the set $\hat{V}_N^+$ is empty so that the corresponding probability in (5.1.21) is not defined, we put it to be 1.)

From the definition of the contour Γ_N^b and the thickness condition (1.5.1) it follows that

$$|\Gamma_N^b| \leq L|V_N^-|^{(\nu-1)/\nu}. \qquad (5.1.22)$$

Using (5.1.20) and (5.1.22) we find that for a constant C_1

$$|\Delta_N^\pm| \leq C_1|V_N^-|^{(\nu-1)/\nu}. \qquad (5.1.23)$$

(All constants in this proof can depend on the parameters ν, β, h, L used in its formulation.) So using the expansion (3.3.7) for partition functions, Lemma 3.4, and then the estimate (5.1.23) we find that for some constants K_1, K_2 and for all sufficiently large β

$$\begin{aligned} &\exp\{-2\beta|\Gamma_N^b|\} \left| \frac{Z(\hat{V}_N^+ , \beta , 0 , + , \infty) Z(\hat{V}_N^- , \beta , 0 , + , \infty)}{Z(V_N , \beta , 0 , + , \infty)} \right| \\ &\quad \geq \exp\{-K_1|\Gamma_N^b|\} \geq \exp\{-K_2|V_N^-|^{(\nu-1)/\nu}\}. \end{aligned} \qquad (5.1.24)$$

We must estimate the probabilities in (5.1.21). Applying the thickness condition (1.5.1) to the case $|V_N'| = |V_N|$, we see that

$$|\partial V_N| \leq L|V_N|^{(\nu-1)/\nu}. \qquad (5.1.25)$$

It follows from (5.1.22) that

$$|\partial \hat{V}_N^-| \leq L_1|\hat{V}_N^-|^{(\nu-1)/\nu} \qquad (5.1.26)$$

for some constant L_1. Finally, for a new constant L_2 we have

$$|\partial \hat{V}_N^+| \le |\partial V_N| + |\Delta_N^+| \le L_2 |V_N|^{(\nu-1)/\nu}. \tag{5.1.27}$$

Using the mean values and the variances for the ∞-cutoff ensembles, we abbreviate the notation (3.2.10) and (3.2.11) and set

$$\begin{aligned} E^+(V) &= E(V, V, +, \beta, 0, \infty), \\ D^+(V) &= D(V, V, +, \beta, 0, \infty). \end{aligned} \tag{5.1.28}$$

Since the volumes V_N are contractible, the comparison of the definitions of §3.2 and the definition (1.5.3) shows that

$$E^+(V_N) = E_N^+. \tag{5.1.29}$$

We need the following estimates, which follow from (3.19.10), (5.1.25), (5.1.26), (5.1.27), and (3.19.36): for some constant C,

$$\left| \frac{E_N^+}{|V_N|} - m^+(\beta, 0) \right| \le C|V_N|^{-1/\nu}, \tag{5.1.30}$$

$$\left| \frac{E^+(\hat{V}_N^-)}{|\hat{V}_N^-|} - m^+(\beta, 0) \right| \le C|\hat{V}_N^-|^{-1/\nu}, \tag{5.1.31}$$

$$\left| \frac{E^+(\hat{V}_N^+)}{|\hat{V}_N^+|} - m^+(\beta, 0) \right| \le C\frac{|V_N|^{(\nu-1)/\nu}}{|\hat{V}_N^+|}. \tag{5.1.32}$$

We also need the lower bounds for variances that follow from Theorem 4.1. Namely, for a positive constant $\alpha > 0$ we have

$$D^+(V_N) \ge \alpha |V_N|, \tag{5.1.33}$$

and

$$D^+(\hat{V}_N^\pm) \ge \alpha |\hat{V}_N^\pm|. \tag{5.1.34}$$

Consider now Case 1 when the condition (5.1.12) is satisfied. We shall estimate first the probability $P_{\hat{V}_N^-}^{+,\beta,0,\infty}(\sigma : S_{\hat{V}_N^-}(\sigma) = b_N^- - |\Delta_N^-|)$. Using (5.1.26) and the equality $|V_N^-| = K_N^-$, we obtain from (5.1.7), (5.1.13), and (5.1.14) that

$$\left| b_N^- - E_N^+ \frac{|\hat{V}_N^-|}{|V_N|} \right| \le 4. \tag{5.1.35}$$

for a sufficiently large N. Therefore, using the estimates (5.1.30), (5.1.31), and (5.1.23), we find that

$$\begin{aligned} &|b_N^- - |\Delta_N^-| - E^+(\hat{V}_N^-)| \\ &\le \left| b_N^- - E_N^+ \frac{|\hat{V}_N^-|}{|V_N|} \right| + \left| \frac{E_N^+}{|V_N|} - \frac{E^+(\hat{V}_N^-)}{|\hat{V}_N^-|} \right| |\hat{V}_N^-| + |\Delta_N^-| \\ &\le 4 + C|V_N|^{-1/\nu}|\hat{V}_N^-| + C|\hat{V}_N^-|^{(\nu-1)/\nu} + C_1|V_N^-|^{(\nu-1)/\nu} \\ &\le \overline{C}|\hat{V}_N^-|^{(\nu-1)/\nu}, \end{aligned} \tag{5.1.36}$$

where $\overline{C}$ is a new constant. Observe now that it follows from (5.1.8), (5.1.13), and (5.1.30) that

$$|V_N^-| \sim \frac{1}{2m^+(\beta, 0)}(E_N^+ - b_N) \tag{5.1.37}$$

as $N \to \infty$ and $E_N^+ - b_N \to \infty$. (Here and later we use the symbol $\sim$ to connect quantities whose ratio tends to 1 as $N \to \infty$.) It follows from the conditions (1.5.8) or (1.5.11) on b_N that under the conditions of the present proposition

$$E_N^+ - b_N \geq c_1 |V_N|^{\nu/(\nu+1)} \tag{5.1.38}$$

for a sufficiently large N and a positive constant c_1. By (5.1.37) and (5.1.38), $|V_N^-| \to \infty$ as $N \to \infty$. Using (5.1.23) we see that also $|\hat{V}_N^-| \to \infty$ and that the Van Hove condition (1.4.1) for the sequence $\hat{V}_N^-$ holds. Finally, since the set V_N^- is contractible, the set $\hat{V}_N^-$ is also contractible. So we can use the Classical Large Deviations Principle proved in the previous chapter. It follows from the estimate (5.1.36) that the event $\{S_{\hat{V}_N^-} = b_N^- - |\Delta_N^-|\}$ lies in the domain of moderate deviations of the probability distribution in $\hat{V}_N^-$ and that the conditions (1.5.2) or (1.5.5), (1.5.6) of Theorem 1.5.1 hold with $b_N^- - |\Delta_N^-|$, $E^+(\hat{V}_N^-)$ instead of b_N, E_N^+ respectively. So we can use the weak implication (1.2.37) of the Classical Large Deviations Principle to obtain that

$$\ln P_{\hat{V}_N^-}^{+,\beta,0,\infty}(\sigma : S_{\hat{V}_N^-} = b_N^- - |\Delta_N^-|) \sim -\frac{\left|b_N^- - |\Delta_N^-| - E^+(\hat{V}_N^-)\right|^2}{2D^+(\hat{V}_N^-)} \tag{5.1.39}$$

as $N \to \infty$. Hence the estimate (5.1.36) together with the estimate (5.1.34) shows that for some constant $C < \infty$ we have

$$\ln P_{\hat{V}_N^-}^{+,\beta,0.\infty}(\sigma : S_{\hat{V}_N^-} = b_N^- - |\Delta_N^-|) \geq -C|\hat{V}_N^-|^{(\nu-2)/\nu}. \tag{5.1.40}$$

Now let us turn to the factor $P_{\hat{V}_N^+}^{+,\beta,0,\infty}(\sigma : S_{\hat{V}_N^+}(\sigma) = b_N^+ - |\Delta_N^+|)$ in (5.1.21). It follows from the conditions (5.1.12) and (5.1.8), (5.1.13) that in Case 1 we now have $|V_N^+| \geq |V_N|/3$ for sufficiently large N. Using this estimate we can literally repeat all the previous arguments with $\hat{V}_N^+$ instead of $\hat{V}_N^-$ and b_N^+ instead of b_N^-, and obtain similarly to (5.1.40) that

$$\ln P_{\hat{V}_N^+}^{+,\beta,0,\infty}(\sigma : S_{\hat{V}_N^+} = b_N^+ - |\Delta_N^+|) \geq -C|\hat{V}_N^+|^{(\nu-2)/\nu}. \tag{5.1.41}$$

If we substitute the estimates (5.1.24), (5.1.40), (5.1.41) into the relation (5.1.21) and then use the inequality (5.1.19), we obtain that

$$P_N^{+,0}(b_N) \geq \exp\{-K_2|V_N^-|^{(\nu-1)/\nu} - C|\hat{V}_N^-|^{(\nu-2)/\nu} - C|\hat{V}_N^+|^{(\nu-2)/\nu}\}. \tag{5.1.42}$$

It follows from the relations (5.1.37), (5.1.38) that

$$\liminf_{N\to\infty} \frac{|V_N^-|}{|V_N|^{\nu/(\nu+1)}} > 0. \tag{5.1.43}$$

So the evident inequality

$$\frac{\nu-1}{\nu+1} > \frac{\nu-2}{\nu} \tag{5.1.44}$$

shows that the first term in the exponent in (5.1.42) asymptotically dominates the other terms. Therefore, the inequality (5.1.42) together with the asymptotic equality (5.1.37) gives the desired estimate (5.1.1) in Case 1.

In Case 2 the situation becomes slightly more complicated, because if we define the sets V_N^+ in the same way as before they can become relatively small and so the deviations which we must estimate can lie well outside the interval of negative deviations in which the Classical Principle of Large Deviations can be applied. To avoid this difficulty we increased the value b_N^+ in the definition (5.1.16). It follows from (5.1.15), (5.1.8), (5.1.13), and (5.1.30) that in the considered case

$$|V_N^+| \geq c_1 |V_N|^{(\nu+2\delta-1)/\nu} \tag{5.1.45}$$

for a positive constant c_1. Together with the estimate (5.1.27), this relation implies that for some positive constant c_2,

$$|\hat{V}_N^+| \geq c_2 |V_N|^{(\nu+2\delta-1)/\nu}. \tag{5.1.46}$$

So the estimate (5.1.27), together with positiveness of δ, implies the Van Hove condition (1.4.1) for the sequence $\hat{V}_N^+$. Similarly to (5.1.35), we use (5.1.7) and (5.1.16) to find that

$$\left| b_N^+ - |V_N|^{(\nu+\delta-1)/\nu} - (E_N^+ |V_N^+|)/|V_N| \right| \leq 4. \tag{5.1.47}$$

An estimate similar to (5.1.36) shows that

$$\left| b_N^+ - |V_N|^{(\nu+\delta-1)/\nu} - |\Delta_N^+| - E^+(\hat{V}_N^+) \right| \leq \overline{C} |\hat{V}_N^+|^{(\nu-1)/\nu}. \tag{5.1.48}$$

This estimate and the positiveness of δ imply that for N sufficiently large the deviation considered in (5.1.41) is positive:

$$b_N^+ - |\Delta_N^+| - E^+(\hat{V}_N^+) > 0, \tag{5.1.49}$$

for which case Theorem 1.5.1 was already proven. From (5.1.46) and (5.1.48) we find that

$$\frac{b_N^+ - |\Delta_N^+| - E^+(\hat{V}_N^+)}{|\hat{V}_N^+|} \leq \frac{|V_N|^{(\nu+\delta-1)/\nu}}{c_2 |V_N|^{(\nu+2\delta-1)/\nu}} + \frac{\overline{C} |\hat{V}_N^+|^{(\nu-1)/\nu}}{|\hat{V}_N^+|} \to 0 \tag{5.1.50}$$

as $N \to \infty$. So we have again a moderate deviation case, and to estimate its probability we can use the approximation (1.2.37), which shows (cf. (5.1.39)) that

$$\ln P_{\hat{V}_N^+}^{+,\beta,0,\infty}(\sigma : S_{\hat{V}_N^+} = b_N^+ - |\Delta_N^+|) \sim -\frac{|b_N^+ - |\Delta_N^+| - E^+(\hat{V}_N^+)|^2}{2D^+(\hat{V}_N^+)} \tag{5.1.51}$$

as $N \to \infty$. The last ratio can be estimated similarly to (5.1.50). Using also (5.1.33), we find that

$$\frac{\left|b_N^+ - |\Delta_N^+| - E^+(\hat{V}_N^+)\right|^2}{2D^+(\hat{V}_N^+)} \leq 2\alpha^{-1}\left(\frac{(|V_N|^{(\nu+\delta-1)/\nu})^2}{c_2|V_N|^{(\nu+2\delta-1)/\nu}} + \frac{(\overline{C}|\hat{V}_N^+|^{(\nu-1)/\nu})^2}{|\hat{V}_N^+|}\right) \leq C|V_N|^{(\nu-1)/\nu} \tag{5.1.52}$$

for a constant $C < \infty$. So we obtain from (5.1.51) that for sufficiently large N

$$\ln P_{\hat{V}_N^+}^{+,\beta,0,\infty}(\sigma : S_{\hat{V}_N^+} = b_N^+ - |\Delta_N^+|) \geq -C|V_N|^{(\nu-1)/\nu}. \tag{5.1.53}$$

It follows from the condition (5.1.15) and the relations (5.1.8), (5.1.13), (5.1.30) that in Case 2

$$|V_N^-| \geq c|V_N| \tag{5.1.54}$$

for some $c > 0$. Similarly to the estimate (5.1.36), but using the condition (5.1.16) instead of (5.1.14), we find that for a constant $\overline{C} < \infty$

$$|b_N - |\Delta_N^-| - E^+(\hat{V}_N^-)| \leq \overline{C}|V_N|^{(\nu+\delta-1)/\nu}. \tag{5.1.55}$$

It follows now from the restriction (5.1.4) on δ that we are again inside the domain where the conditions (1.5.5), (1.5.6) of Theorem 1.5.1 are fulfilled, so applying the Gaussian approximation (1.2.37) we find (cf. (5.1.40)) that for an appropriate choice of the constant $C < \infty$ we have

$$\ln P_{\hat{V}_N^-}^{+,\beta,0.\infty}(\sigma : S_{\hat{V}_N^-} = b_N^- - |\Delta_N^-|) \geq -C|V_N|^{(\nu+2\delta-2)/\nu}. \tag{5.1.56}$$

Similarly to (5.1.42), we obtain from (5.1.19), (5.1.21), (5.1.24), (5.1.53), (5.1.54), and (5.1.56) that

$$P_N^{+,0}(b_N) \geq \exp\{-cK_2|V_N|^{(\nu-1)/\nu} - C|V_N|^{(\nu+2\delta-2)/\nu} - C|V_N|^{(\nu-1)/\nu}\}. \tag{5.1.57}$$

Due to the condition (5.1.4), the second term in the exponent in (5.1.57) is asymptotically irrelevant, and since in our case the difference $E_N^+ - b_N$ grows linearly with $|V_N|$ (see (5.1.15) and (5.1.30)), we obtain that the desired estimate (5.1.1) is valid in Case 2 also.

Now we consider Case 3. In this case, $V_N^- = V_N$, $b_N^- = -b_N$, and the set V_N^+ is empty, so we need to estimate the probability of the event $\{S_{\hat{V}_N} = -b_N - |\Delta_N|\}$, where we set $\hat{V}_N = \hat{V}_N^-$, $\Delta_N = \Delta_N^-$. We consider two subcases separately. First we assume that (cf. (1.5.9) and (5.1.15)) that

$$-E_N^+ - Q|V_N|^{(2\nu-1)/2\nu} \leq b_N < -E_N^+ + |V_N|^{(\nu+2\delta-1)/\nu}. \tag{5.1.58}$$

Using (5.1.23), (5.1.30), (5.1.31), and (5.1.58) we find that for a constant $q_1 < \infty$

$$\begin{aligned} -b_N - |\Delta_N| - E^+(\hat{V}_N) &\geq -|V_N|^{(\nu+2\delta-1)/\nu} - |E_N^+ - E^+(\hat{V}_N)| - |\Delta_N| \\ &\geq -q_1|V_N|^{(\nu+2\delta-1)/\nu}. \end{aligned} \tag{5.1.59}$$

Using the restriction (1.5.4) on δ we see that if the deviation is negative, then it is sufficiently small, so that we can apply Theorem 1.5.1 and the Gaussian approximation (1.2.37). For positive deviations we obtain in a similar way a rougher estimate

$$-b_N - |\Delta_N| - E^+(\hat{V}_N) \leq q_2 |V_N|^{(2\nu-1)/2\nu}, \tag{5.1.60}$$

where the constant q_2 depends on the constant Q from (5.1.58). But for positive deviations we can apply the Gaussian approximation (1.2.37) in the entire domain of moderate deviations. Substituting this approximation together with the estimate (5.1.24) into the main estimates (5.1.21) and (5.1.19), we find with the help of (5.1.33) that

$$P^{+,0}(b_N) \geq \exp\{-K_2|V_N|^{(\nu-1)/\nu} - q_1\alpha^{-1}|V_N|^{(\nu+4\delta-2)/\nu} - q_2\alpha^{-1}|V_N|^{(\nu-1)/\nu}\}. \tag{5.1.61}$$

Using the condition (1.5.4), we see that the second term in the exponent is irrelevant, and so we again obtain the desired estimate (5.1.1).

Observe that any sequence of values of b_N satisfying the conditions (1.5.8), (1.5.9) of our proposition can be subdivided into three subsequences such that the condition (5.1.12) is satisfied for the first subsequence, the condition (5.1.15) is satisfied for the second subsequence, and the condition (5.1.58) is satisfied for the third subsequence. So we proved the estimate (5.1.1) in the situation stated in the proposition.

Now we shall consider the subcase of Case 3 when the conditions (1.5.11), (1.5.12) are valid and we must obtain the estimate (5.1.2). Again it is sufficient to estimate the probability of the event $\{S_{\hat{V}_N} = -b_N - |\Delta_N|\}$ induced by the probability distribution $P^{+,\beta,0,\infty}_{\hat{V}_N}$. It follows from the condition (1.5.11) and the estimates (5.1.30), (5.1.31) that

$$-b_N - |\Delta_N| - E^+_N(\hat{V}_N) > 0 \tag{5.1.62}$$

for sufficiently large N. The fulfillment of the second of the conditions (1.4.4) for the sequence $\hat{V}_N$ follows from the same condition for the sequence V_N and the estimate (5.1.23). So Theorem 1.5.1 permits us to apply the Classical Large Deviation Principle in the Strong Form (1.2.27), which states that

$$P^{+,\beta,0,\infty}_{\hat{V}_N}(\sigma : S_{\hat{V}_N}(\sigma) = -b_N - |\Delta_N|) \sim \frac{2}{\sqrt{2\pi \hat{D}_N(\hat{h}_N(b_N))}} \exp\{-\hat{I}^0_N(-b_N - |\Delta_N|)\}, \tag{5.1.63}$$

where $\hat{h}_N(b)$ is the solution of the equation

$$\hat{E}_N(\hat{h}_N(b)) = b. \tag{5.1.64}$$

Here we use the notation of §4.2 with superscript 0 to indicate that $h = 0$, add the "hat" to the notation of the action function $I_N(b)$, of the mean value

$E_N(h)$, and of the variance $D_N(h)$, and change the bar in (4.2.4) to the hat in (5.1.64) to emphasize that we take the reduced volume $\hat{V}_N$ instead of the original volume V_N. Now we must compare the values of the action function $\hat{I}_N^0(-b_N - |\Delta_N|)$ used in (5.1.63) and of the function $I_N^0(-b_N)$ used in the desired estimate (5.1.2). Calculating the derivative of the action function $\hat{I}_N(b)$ in b, we can write (cf. (4.4.78), (4.4.82))

$$\hat{I}_N^0(-b_N - |\Delta_N|) = \beta \int_{\hat{E}_N(0)}^{-b_N-|\Delta_N|} \hat{h}_N(b)\, db \tag{5.1.65}$$

and

$$I_N^0(-b_N) = \beta \int_{E_N(0)}^{-b_N} h_N(b)\, db, \tag{5.1.66}$$

where $h_N(b)$ is a solution of the equation

$$E_N(h_N(b)) = b \tag{5.1.67}$$

and where the absence of embellishments means that the corresponding quantities are defined for the original volumes V_N. (Comparing the notation of §4.2 used in these relations and the notation used in the rest of this section, we observe that $E_N(0) = E_N^+$ and $\hat{E}_N(0) = E^+(\hat{V}_N)$.) It follows from (5.1.65) and (5.1.66) that

$$\begin{aligned} &|\hat{I}_N^0(-b_N - |\Delta_N|) - I_N^0(-b_N)| \\ &\le \beta\left(\left|\int_{E_N(0)}^{-b_N}(\hat{h}_N(b)-h_N(b))\,db\right| + \left|\int_{-b_N}^{-b_N-|\Delta_N|}\hat{h}_N(b)\,db\right| + \left|\int_{E_N(0)}^{\hat{E}_N(0)} h_N(b)\,db\right|\right). \end{aligned} \tag{5.1.68}$$

Now we present some facts useful for an estimation of these integrals. It follows from (1.4.4) that for some constant $q < 1$ we have $|b_N| \le q|V_N|$. The estimate (5.1.23) implies that for any constant $q' > q$ we have $|b_N| \le q'|\hat{V}_N|$ for all sufficiently large N. So we can use (5.1.30) and (5.1.31) to verify that the condition (4.3.2) of Proposition 4.3 holds, which implies that for some constant $C = C(q)$ and for all values of b belonging to domains over which the integrals in (5.1.68) are taken,

$$|\hat{h}_N(b)| \le C, \qquad |h_N(b)| \le C. \tag{5.1.69}$$

It follows from the estimates (3.19.7) and (5.1.23) that for a constant $C' = C'(q)$

$$|\hat{E}_N(h) - E_N(h)| \le C'|V_N|^{(\nu-1)/\nu} \quad \text{if } |h| \le C. \tag{5.1.70}$$

Finally, using (5.1.64), (5.1.67), (5.1.70), and (5.1.23), we find that for the same set of values of b as in (5.1.69)

$$|E_N(h_N(b)) - E_N(\hat{h}_N(b))| = |E_N(\hat{h}_N(b)) - \hat{E}_N(\hat{h}_N(b))| \le C'|V_N|^{(\nu-1)/\nu}. \tag{5.1.71}$$

We have

$$(E_N(h))' = \beta D_N(h) \ge c|V_N| \tag{5.1.72}$$

for $|h| \le C$, where $c = c(C) > 0$ (see (4.1.6)). So the estimate (5.1.71) implies that

$$|h_N(b) - \hat{h}_N(b)| \le c^{-1} C' |V_N|^{-1/\nu}. \tag{5.1.73}$$

Now we can substitute the estimates (5.1.23), (5.1.69), (5.1.70) for $h = 0$ and (5.1.73) in the integrals at the right-hand side of the estimate (5.1.68). We see that for a new constant $K = K(q)$

$$|\hat{I}_N^0(-b_N - |\Delta_N|) - I_N^0(-b_N)| \le K |V_N|^{(\nu-1)/\nu}. \tag{5.1.74}$$

On the other hand, using the Gaussian approximation for the action function $I_N^0(-b_N)$ in the domain of moderate deviations (cf. (1.2.37)) and lower bound for the variance (cf. (5.1.72)) we see that for a constant $c_1 = c_1(\varepsilon) > 0$ and sufficiently large N

$$I_N^0(-b_N) \ge c_1 \frac{(b_N + E_N^+)^2}{|V_N|} \quad \text{provided} \quad \frac{|b_N + E_N^+|}{|V_N|^{1-\varepsilon}} \to 0 \text{ with } \varepsilon > 0. \tag{5.1.75}$$

Using the monotonicity of the action function $I_N^0(-b_N)$ for $b_N < -E_N^+$ and recalling the main restriction (1.5.11) and the estimate (5.1.74), we find that

$$\liminf_{N\to\infty} \frac{\hat{I}_N^0(-b_N - |\Delta_N|)}{|V_N|^{(\nu-1)/\nu}} \ge \liminf_{N\to\infty} \left(\frac{I_N^0(-b_N)}{|V_N|^{(\nu-1)/\nu}} - K \right) = \infty. \tag{5.1.76}$$

Using this estimate and the lower estimate (4.1.6) for the variance, we see that the coefficient before the exp in (5.1.63) is asymptotically negligible compared with the exponential itself, and so we find from (5.1.63) and (5.1.74) that

$$-\ln P^+_{\hat{V}_N, \beta, 0, \infty}(\sigma : S_{\hat{V}_N}(\sigma) = -b_N - |\Delta_N|) = I_N^0(-b_N)(1 + o_N(1)) \tag{5.1.77}$$

as $N \to \infty$, where $o_N(1) \to 0$ as $N \to \infty$. Recalling the estimates (5.1.19), (5.1.21), and (5.1.24) we see that

$$P_N^{+,0}(b_N) \ge \exp\{-K_2 |V_N|^{(\nu-1)/\nu} - I_N^0(-b_N)(1 + o_N(1))\}. \tag{5.1.78}$$

The relation (5.1.76) shows that, contrary to previous cases, the first term corresponding to the contribution of the large contour is dominated by the second one describing deviations in the minus-phase, and so we obtain the desired estimate (5.1.2).

Now we turn to the case $h > 0$. First we observe that it follows from the definition (1.3.2)–(1.3.4) of the Gibbs distribution that for any b such that

$$(b - |V_N|) \equiv 0 \pmod 2 \tag{5.1.79}$$

we have

$$\frac{P_N^{+,h}(b_N)}{P_N^{+,0}(b_N)} = \exp\{\beta h(b_N - b)\} \frac{P_N^{+,h}(b)}{P_N^{+,0}(b)}. \tag{5.1.80}$$

Now we choose $b = b_N^0$ in such a way that the condition (5.1.79) is valid and

$$0 \le b_N^0 - E_N^+ \le 2. \tag{5.1.81}$$

Then the condition (1.5.2) is satisfied, and we can apply Theorem 1.5.1 about the Classical Large Deviation Principle (in fact, the local central limit theorem would suffice here) to obtain that

$$\lim_{N\to\infty} \frac{\ln P_N^{+,0}(b_N^0)}{|V_N|} = 0. \tag{5.1.82}$$

In a similar way, applying Theorem 1.5.2 in the case of classical region (which was proved in §4.6) we obtain that

$$\lim_{N\to\infty} \frac{\ln P_N^{+,h}(b_N^0)}{|V_N|} = -I^h(m^+(\beta, 0)). \tag{5.1.83}$$

It follows from (5.1.80), (5.1.81), (5.1.82), and (5.1.83) that

$$\lim_{N\to\infty} \frac{\ln P_N^{+,h}(b_N) - \ln P_N^{+,0}(b_N) - \beta h(b_N - E_N^+)}{|V_N|} = -I^h(m^+(\beta, 0)). \tag{5.1.84}$$

Recalling that $E_N^+/|V_N| \to m^+(\beta, 0)$ as $N \to \infty$ (see (5.1.30)) and the definition (1.5.24) of the function $\tilde{I}^h(\hat{b})$, we find that the desired estimate (5.1.3) follows from relations (5.1.84) and (5.1.1) in the case $m^+(\beta, 0) > \hat{b} > -m^+(\beta, 0)$ and from relations (5.1.84) and (5.1.2) in the case $-m^+(\beta, 0) > \hat{b} > -1$. There are two special intermediate cases. If $\hat{b} = -m^+(\beta, 0)$, we must observe additionally that $I^0(m^+(\beta, 0)) = 0$ and so the desired statement is valid for sequences b_N satisfying either of the conditions (1.5.9) and (1.5.11). Finally, from any sequence b_N such that $b_N/|V_N| \to -m^+(\beta, 0)$ it is possible to choose a subsequence that satisfies either the condition (1.5.9) or the condition (1.5.11). Similarly, for $\hat{b} = m^+(\beta, 0)$ we can choose a subsequence that satisfies either the condition (1.5.2) or the conditions (1.5.5)–(1.5.7) or finally, the condition (1.5.8), and it suffices to verify the desired statement for this subsequence.

So we have proved all statements of the proposition for +-boundary conditions. The considerations for other boundary conditions are completely similar. We mention only that in the cases of periodic and free boundary conditions it is necessary to include in the definition of the set of configuration Ω_N^b the condition $\sigma_t = -1$ for $t \in \Delta_N^-$, and that for $h = 0$ Case 3 does not arise here because of the condition $b_N \geq 0$ in Theorem 1.6.1.

§5.2. The upper bound

PROPOSITION 5.2.1. *All statements of Proposition* 5.1.1 *remain valid after reversing the signs of its estimates; the only change needed is the replacement of the constant* $\underline{c}$ *by* $\overline{c}$ *and limits* lim inf *by* lim sup.

We again consider in detail the case of +-boundary condition only. We shall use a special estimate for large deviations. For the case $\nu = 2$ it was obtained earlier in [DKS].

LEMMA 5.2.2. *Let $b_N \le E_N^+$. For a sufficiently large β we have*

$$P_N^{+,\beta,0,\omega_N}(\sigma : S_N(\sigma) = b_N) \le \exp\{-K(E_N^+ - b_N)h_N\} \tag{5.2.1}$$

with $K = K(\beta) > 0$ if h_N and ω_N satisfy the conditions

$$0 \le h_N \le \min\{(E_N^+ - b_N)|V_N|^{-1}, (E_N^+ - b_N)^{1/2}|V_N|^{-1/2}(\ln|V_N|)^{-3/2}\}, \tag{5.2.2}$$

$$C\ln|V_N| \le \omega_N \le c/h_N, \tag{5.2.3}$$

where C is sufficiently large and $c < 1$.

PROOF OF LEMMA 5.2.2. The starting point of the proof is the following identity, which immediately follows from the definitions (3.2.3)–(3.2.5):

$$P_N^{+,\beta,0,\omega_N}(\sigma : S_N(\sigma) = b_N) = \frac{Z(V_N, +, \beta, h_N, \omega_N)}{Z(V_N, +, \beta, 0, \omega_N)} \times \exp\{-\beta h_N b_N\} P_N^{+,\beta,h_N,\omega_N}(\sigma : S_N(\sigma) = b_N). \tag{5.2.4}$$

(To agree with the notation of §3.2 we assume here that $h_N = \{h_t \equiv h_N, t \in V_N\}$.) This obviously implies that

$$P_N^{+,\beta,0,\omega_N}(\sigma : S_N(\sigma) = b_N) \le \frac{Z(V_N, +, \beta, h_N, \omega_N)}{Z(V_N, +, \beta, 0, \omega_N)} \exp\{-\beta h_N b_N\}. \tag{5.2.5}$$

Let us introduce another cutoff level

$$\omega_N' = C\ln|V_N|. \tag{5.2.6}$$

The condition (5.2.3) allows us to apply Proposition 3.3. So the expansion (3.3.7) together with relations (3.3.8), (3.3.9) shows that

$$\ln\frac{Z(V_N, +, \beta, h_N, \omega_N')}{Z(V_N, +, \beta, h_N, \omega_N)} = \sum_{\Lambda\subset V_N : |\Lambda|>(2\nu\omega_N')^\nu} \Phi(\Lambda) \tag{5.2.7}$$

Applying the main estimate (3.3.11) and Lemma 3.4, we find that

$$\frac{Z(V_N, +, \beta, h_N, \omega_N')}{Z(V_N, +, \beta, h_N, \omega_N)} \to 1 \quad \text{as } N \to \infty. \tag{5.2.8}$$

Since $\omega_N' \le \omega_N$, we have $Z(V_N, +, \beta, 0, \omega_N) \ge Z(V_N, +, \beta, 0, \omega_N')$. Hence from (5.2.5) and (5.2.8) it follows that

$$P_N^{+,\beta,0,\omega_N}(\sigma : S_N(\sigma) = b_N) \le \frac{Z(V_N, +, \beta, h_N, \omega_N')}{Z(V_N, +, \beta, 0, \omega_N')} \exp\{-\beta h_N b_N\}(1 + o_N(1)). \tag{5.2.9}$$

From Proposition 3.3 we know that logarithms of partition functions in (5.2.9) are analytic functions of h_N in the region $|h_N| \le c(\omega_N')^{-1}$ and satisfy the bound (3.3.4). The Taylor expansion up to the third term at the point $h = 0$ and the Cauchy formula for the remainder provide us with the formula

$$\frac{Z(V_N, +, \beta, h_N, \omega_N')}{Z(V_N, +, \beta, 0, \omega_N')} = \exp\{\beta h_N E_N^+ + \beta^2 h_N^2 D_N^+/2 + \beta^3 K_N' |h_N|^3 (\omega_N')^3 |V_N|\}. \tag{5.2.10}$$

Here $D_N^+ = D(V_N, +, \beta, 0, \omega_N')$ is the variance (see (3.2.11)) and $|K_N'| \le K$ with K independent of h_N and N. The estimate (3.19.5) implies that $D_N^+ \le C|V_N|$ for some constant C. Now the condition (5.2.2) implies that the sum of last two terms in (5.2.10) does not exceed $\frac{1}{2}\beta|h_N|(E_N^+ - b_N)$, proving Lemma 5.2.2.

PROOF OF PROPOSITION 5.2.1. We begin with the proof of the converse to (5.1.1): under the conditions (1.5.8), (1.5.9) for a sufficiently large β and a corresponding choice of the constant $\bar{c}$,

$$\ln P_N^{+,0}(b_N) \le -\bar{c}(E_N^+ - b_N)^{(\nu-1)/\nu}. \tag{5.2.11}$$

Actually, we will see that the last estimate also holds outside the region defined by the constrains (1.5.8), (1.5.9), but there it can be improved. To prove it we first set the cutoff level ω_N to be

$$\omega_N = \gamma(E_N^+ - b_N)^{1/\nu}, \tag{5.2.12}$$

with a sufficiently small constant $\gamma = \gamma(\beta)$. It is clear that

$$\begin{aligned} P_N^{+,0}(b_N) &= P_N^{+,\beta,0,\infty}(\sigma : S_N(\sigma) = b_N) \\ &\le P_N^{+,\beta,0,\infty}(\sigma : S_N(\sigma) = b_N, \max_{\Gamma \in \mathcal{G}(\sigma)} q(\Gamma) \le \omega_N) \\ &\quad + P_N^{+,\beta,0,\infty}(\sigma : S_N(\sigma) = b_N, \max_{\Gamma \in \mathcal{G}(\sigma)} q(\Gamma) > \omega_N). \end{aligned} \tag{5.2.13}$$

We start with the first summand. It follows immediately from the definition of cutoff ensembles (see §3.2) that

$$P_N^{+,\beta,0,\infty}(\sigma : S_N(\sigma) = b_N, \max_{\Gamma \in \mathcal{G}(\sigma)} q(\Gamma) \le \omega_N) \le P_N^{+,\beta,0,\omega_N}(\sigma : S_N(\sigma) = b_N). \tag{5.2.14}$$

To estimate the last probability we use Lemma 5.2.2. Our choice of h_N is

$$h_N = \varkappa(E_N^+ - b_N)^{-1/\nu} \tag{5.2.15}$$

with $\varkappa$ sufficiently small. Condition (1.5.8) ensures that the condition (5.2.2) holds for sufficiently large N. The condition (5.2.3) for sufficiently large N follows from (5.2.12) and (5.2.15). The estimate (5.2.5) shows that

$$P_N^{+,\beta,0,\infty}(\sigma : S_N(\sigma) = b_N, \max_{\Gamma \in \mathcal{G}(\sigma)} q(\Gamma) \le \omega_N) \le \exp\{-K\varkappa(E_N^+ - b_N)^{(\nu-1)/\nu}\}. \tag{5.2.16}$$

The second summand in (5.2.13) can be bounded by the probability of the event that a configuration contains a contour Γ of size $q(\Gamma) > \omega_N$. This probability can be estimated by the usual Peierls procedure (or using a more general Proposition 3.20). Recalling that the number of contours Γ containing a fixed face and having length $|\Gamma|$ does not exceed $C^{|\Gamma|}$, $C = C(\nu)$, and using the estimate (3.1.5), we obtain that

$$\begin{aligned} &P_N^{+,\beta,0,\infty}\left(\sigma : S_N(\sigma) = b_N\,,\ \max_{\Gamma\in\mathscr{G}(\sigma)} q(\Gamma) > \omega_N\right) \\ &\quad\le P_N^{+,\beta,0,\infty}\left(\sigma : \max_{\Gamma\in\mathscr{G}(\sigma)} q(\Gamma) > \omega_N\right) \\ &\quad\le |V_N| \exp\{-C_1\beta(E_N^+ - b_N)^{(\nu-1)/\nu}\}. \end{aligned} \tag{5.2.17}$$

Estimates (5.2.12), (5.2.13), (5.2.16), (5.2.17) prove (5.2.11).

To prove the bound

$$\limsup_{N\to\infty} \frac{-\ln P_N^{+,0}(b_N)}{I_N^0(-b_N)} \le 1\,, \tag{5.2.18}$$

(cf. (5.1.2)) we observe first that by flipping all the spins inside V_N we obtain the estimate

$$P_N^{+,0}(b_N) \le \exp\{2\beta|\Gamma_N|\} P_N^{+,0}(-b_N)\,, \tag{5.2.19}$$

where Γ_N is the contour separating the volume V_N and its complement. Indeed, this flip can decrease the energy of any configuration by at most $2|\Gamma_N|$. To estimate the probability $P_N^{+,0}(-b_N)$ we can use the already proved part of Theorem 1.5.1, which tells us that the probability in question obeys the Large Deviation Principle in Strong Form, with the exponent $I_N^0(-b_N)$. An estimate similar to (5.1.76) tells us that because of the condition (1.5.11) the second factor in (5.2.19) is asymptotically dominant. Also the factor before the exponent in the Large Deviation Principle is asymptotically negligible because of the lower bound on the variance from §4.1. This proves (5.2.18).

The proof of the last inequality

$$\limsup_{N\to\infty} \frac{\ln P_N^{+,h}(b_N)}{|V_N|} \le -\tilde{I}^h(\hat{b}) \tag{5.2.20}$$

is obtained by repeating the proof of the opposite estimate (5.1.3) in the previous section. The only change is that we must use (5.2.11) instead of (5.1.1) and (5.2.18) instead of (5.1.2).

Proofs for other boundary conditions are similar to the previous one. We mention only that in the periodic case it is necessary to use the cutoff distribution defined by the relation (4.7.1). Instead of (5.2.17) we must prove the estimate

$$P_{\mathbf{T}^N}^{\mathrm{per},\beta,0}(\sigma \in \Omega_{\mathbf{T}_N} \setminus \overline{\Omega}_{\mathbf{T}_N}^{\omega_N} : S_{\mathbf{T}_N}(\sigma) = b_N) \le |V_N| \exp\{-C_1\beta(E_N^+ - b_N)^{(\nu-1)/\nu}\}. \tag{5.2.21}$$

To this end we observe that in the case $\sigma \in \overline{\Omega}_{\mathbf{T}_N}^{-,\omega_N}$, when the set of configurations $\overline{\Omega}_{\mathbf{T}_N}^{-,\omega_N}$ is defined in the same way as the set $\overline{\Omega}_{\mathbf{T}_N}^{\omega_N}$ in §4.7 but with the condition $\sigma_t \equiv -1$ for $t \in \bigcap_{i=1}^{k} \operatorname{Ext}\Gamma_i$, the condition $b_N \geq 0$ is not satisfied. So it suffices to estimate the probability that a configuration σ contains large contours, and this can be done by the usual Peierls estimate. In the case of free boundary condition we must use the modified definition of cutoff ensembles described at the end of §4.7.

REFERENCES

[B] A. A. Borovkov, *Probability theory*, 2nd ed., "Nauka", Moscow, 1976; German transl., Birkhäuser, Basel, 1976.

[Br] W. Bryc, *On the large deviation principle for stationary weakly dependent random fields*, Ann. Probab. **20** (1992), 1004–1030.

[BD] E. Bolthausen and J.-D. Deuschel, *Critical large deviations for Gaussian fields in the phase transition regime.* I, Ann. Probab. **21** (1993), 1876–1920; II, unpublished manuscript, 1993.

[BI] Ch. Borgs and J. Imbrie, *A unified approach to phase diagrams in field theory and statistical mechanics*, Comm. Math. Phys **123** (1989), 305-328.

[Č] S. Čepulēnas, *Probabilities of large deviations for random fields*, Litovsk. Mat. Sb. **25** (1985), no. 4, 164–176; English trans., Lithuanian Math. J. **25** (1985), 381–390.

[Co] F. Comets, *Large deviation*, Probab. Theory Related Fields **80** (1989), 407–432.

[Cr] H. Cramér, *Sur un nouveau théorème-limite de la théorie des probabilités,* Les Sommes et les Fonctions de Variables Aléatoires (Colloque Théor. Probab., Geneva, 1937), Actualités Sci. Indust., vol. 736, Hermann, Paris, 1938, pp. 5–23.

[CS1] N. R. Chaganty and J. Sethuraman, *Large deviations local limit theorems for arbitrary sequences of random variables*, Ann. Probab. **13** (1985), 97–114.

[CS2] ———, *Limit theorems in the area of large deviations for some dependent random variables.*, Ann. Probab. **15** (1987), 628–645.

[DeSt] J.-D. Deuschel and D. Stroock, *Large deviations*, Pure and Appl. Math., vol. 137, Academic Press, New York, 1989.

[DKS] R. L. Dobrushin, R. Kotecký, and S. Shlosman, *The Wulff construction: a global shape from local interactions*, Transl. Math. Monographs, vol. 104, Amer. Math. Soc., Providence, RI, 1992.

[DN] R. L. Dobrushin and B. S. Nakhapetyan, *Strong convexity of the pressure for lattice systems of classical statistical physics*, Teoret. Mat. Fiz. **20** (1974), 223–234; English transl. in Theoret. and Math. Phys. **20** (1974).

[DS1] R. L. Dobrushin and S. B. Shlosman, *Completely analytic interactions. Constructive description*, J. Statist. Phys. **46** (1987), 983–1014.

[DS2] ———, *Large deviation behavior of statistical mechanics models in the multiphase regime*, Mathematical Physics, X (Proc. Tenth IAMP Congr., Leipzig, 1991; K. Schmüdgen, editor), Springer-Verlag, Berlin, 1992, pp. 328–333.

[DT] R. L. Dobrushin and B. Tirozzi, *The central limit theorem and the problem of the equivalence of ensembles*, Comm. Math. Phys. **54** (1977), 173–192.

[DZ] A. Dembo and O. Zeitouni, *Large deviations techniques and applications*, Jones and Bartlett Publishers, Boston, MA, and London, 1992.

[FO1] H. Föllmer and S. Orey, *Large deviations for the empirical field of a Gibbs measure*, Ann. Probab. **16** (1988), 961–977.

[FO2] H. Föllmer and M. Ort, *Large deviations and surface entropy for Markov fields*, Colloque Paul Lévy sur les Processus Stochastiques (Palaiseau, 1987), Astérisque, vol. 157–158, Soc. Math. France, France, 1988, pp. 173–190.

[G1] H.-O. Georgii, *Large deviations and maximum entropy principle for interacting random fields on* $\mathbb{Z}^d$, Ann. Probab. **21** (1993), 1845–1875.

[G2] ———, *Large deviations and the equivalence of ensembles for Gibbsian particle sytems with superstable interactions*, Ann. Probab. (to appear).

[GZ] H.-O. Georgii and H. Zessin, *Large deviations and the maximum entropy principle for marked point random fields*, Probab. Theory Related Fields **96** (1993), 177–204.

[GKK] K. Gawedski, R. Kotecký, and A. Kupiainen, *Coarse-graining approach to the first-oder phase transitions*, J. Statist. Phys. **47** (1987), 701–724.

[I] S. N. Isakov, *Nonanalytic features of the first order phase transition in the Ising model*, Comm. Math. Phys. **95** (1984), 427–443.

[IL] I. A. Ibragimov and Yu. I. Linnik, *Independent and stationary sequences of random variables*, Walters-Noordhoff, Groningen, 1977.

[Kh] A. I. Khinchin, *Uber einen neuen Grenzwertsatz der Wahrscheinlichkeitsrechnung*, Math. Ann. **101** (1929), 745–752.

[KP] R. Kotecký and D. Preiss, *Cluster expansion for abstract polymer models*, Comm. Math. Phys. **103** (1986), 491–498.

[MM] V. A. Malyshev and R. A. Minlos, *Gibbs states; cluster expansions*, Kluwer, Dordrecht, 1991.

[MS] R. A. Minlos and Ya. G Sinai, *The phenomenon of "phase separation" at low temperatures in some lattice models of a gas*, I, Mat. Sb. **73** (1967), 375–448; English transl., Math. USSR-Sb. **2** (1967), 335–395; II, Trudy Moskov. Mat. Obshch. **19** (1968), 113–178; English transl. in Trans. Moscow Math. Soc. **19** (1968), 121–196.

[O] S. Olla, *Large deviations for Gibbs random fields*, Probab. Theory Related Fields **77** (1988), 343–357.

[P] V. V. Petrov, *Sums of independent random variables*, Springer-Verlag, Berlin, 1975.

[R] D. Ruelle, *Thermodynamic formalism*, Addison-Wesley, Reading, MA, 1978.

[Ri] W. Richter, *Local limit theorem for large deviations*, Theory Probab. Appl. **2** (1957), 206–219.

[S] Ya. G. Sinai, *Theory of phase transitions: Rigorous results*, Pergamon Press, Oxford, 1982.

[Sch] R. Schonmann, *Second order large deviation estimates for ferrromagnetic systems in the phase coexistence region*, Comm. Math. Phys. **112** (1987), 409–422.

[Sh] S. B. Shlosman, *The droplet in the tube: A case of phase transition in the canonical ensemble*, Comm. Math. Phys. **125** (1989), 81–90.

[Sm] N. Smirnoff, *Uber Wahrscheinlichkeiten grosser Abweichungen*, Mat. Sb. **40** (1933), 441–455.

[SS] L. Saulis and V. A. Statulevičius, *Limit theorems for large deviations*, Kluwer, Dordrecht, 1991.

[V] S. R. S. Varadhan, *Large deviations and applications*, SIAM, Philadelphia, PA, 1984.

[VF] A. D. Ventzel and M. I. Freidlin, *Random perturbations of dynamical systems*, Springer-Verlag, Berlin, 1984.

[Za] M. Zahradnik, *An alternative version of Pirogov-Sinai theory*, Comm. Math. Phys. **93** (1984), 559–581.

R. D.: Institute for Problems of Information Transmission, Russian Academy of Science, Moscow, Russia

S. S.: Institute for Problems of Information Transmission, Russian Academy of Science, Moscow, Russia

Hebrew Maimonides University, Moscow, Russia

Department of Mathematics, University of California, Irvine, California, USA

ADVANCES IN SOVIET MATHEMATICS
Volume 20, 1994

Asymptotically Additive Integrals of Motion for Particles with Nonpairwise Interaction in Dimension One

B. M. GUREVICH

§1. Introduction

Asymptotically additive integrals of motion of particles with pairwise translation invariant interaction are defined and studied in the paper [1] for any dimension $d > 1$. For the reader's convenience, here we repeat some of the basic definitions and results of [1].

For $n = 1, 2, \ldots$ define the Hamiltonian H_n of an n-particle system by

$$H_n(x_1, \ldots, x_n) = \begin{cases} |p_1|^2/2, & n = 1, \\ \sum_{i=1}^n |p_i|^2/2 + \sum_{1 \leqslant i < j \leqslant n} U(|q_i - q_j|), & n \geqslant 2, \end{cases} \tag{1.1}$$

where $x_i = (q_i, p_i)$ are the coordinate and the momentum of the ith particle, $q_i, p_i \in \mathbb{R}^d$, and U is the potential of pairwise interaction. The function U is defined on the ray (δ, ∞), $\delta \geqslant 0$, and satisfies the following conditions:

(i) $U \in C^\infty(\delta, \infty)$;
(ii) $\lim_{r\to\infty} U(r) = 0$;
(iii) $\lim_{r\to\delta+0} \sup U(r) = \infty$.

The phase space of this system is of the form

$$M_n = \{(x_1, \ldots, x_n) : x_i = (q_i, p_i) \in \mathbb{R}^d \times \mathbb{R}^d, \ |q_i - q_j| > \delta, \ 1 \leqslant i < j \leqslant n\}. \tag{1.2}$$

1.1. **Definitions.** A sequence $I = \{I_n, n \geqslant 1\}$ of functions $I_n : M_n \to \mathbb{R}$ is called an *asymptotically additive integral of motion* (AAIM) of particles with interaction U (see [1]) if for $n = 1, 2, \ldots$ the following conditions are satisfied:

(i) I_n is a smooth symmetric function of $x_1, \ldots, x_n$;
(ii) for all $(x_1, \ldots, x_n) \in M_n$, $p \in \mathbb{R}^d$, we have

$$\lim_{|q|\to\infty} [I_{n+1}(x_1, \ldots, x_n, (q, p)) - I_n(x_1, \ldots, x_n) - I_1(q, p)] = 0; \tag{1.3}$$

1991 *Mathematics Subject Classification.* Primary 58F05; Secondary 70F10.

(iii) I_n is a first integral of the Hamiltonian system with Hamiltonian H_n, i.e., $\{I_n, H_n\} = 0$, where $\{\cdot, \cdot\}$ is the Poisson bracket.

The set of all AAIM's corresponding to a given U is clearly a linear space.

All the classical integrals of motion, such as the number of particles, the total energy (see condition (iii) on U), and the components of both the total momentum and the kinetic momentum are AAIM's. They are known to play a fundamental role in classical mechanics. Besides, AAIM's are involved in the description of natural invariant measures for the dynamical systems of statistical mechanics [2] and hence they are closely related to the problem of irreversibility and convergence to equilibrium in statistical mechanics.

In dealing with an AAIM, it is convenient to consider along with the sequence $I = \{I_n\}$ another sequence $g = \{g_n, n \geqslant 1\}$, where

$$g_n(x_1, \dots, x_n) = \sum_{k=1}^{n} (-1)^{n-k} \sum_{1 \leqslant i_1 < \cdots < i_k \leqslant n} I_k(x_{i_1}, \dots, x_{i_k}), \quad n \geqslant 1, \quad (x_1, \dots, x_n) \in M_n. \tag{1.4}$$

The function g_n, $n \geqslant 1$, is defined on M_n, and the following inversion formula holds:

$$I_n(x_1, \dots, x_n) = \sum_{k=1}^{n} \sum_{1 \leqslant i_1 < \cdots < i_k \leqslant n} g_k(x_{i_1}, \dots, x_{i_k}), \quad n \geqslant 1, \quad (x_1, \dots, x_n) \in M_n. \tag{1.5}$$

We call g and g_n a *generating sequence* and an *n-particle generating function* respectively. Note that the relationship of I_n with $g_1, \dots, g_n$ is precisely the same as that of the total energy of n interacting particles with the potentials of k-particle interactions, $k = 1, \dots, n$.

An AAIM with the generating function g is said to be *n-particle* if $g_k \equiv 0$ for all $k > n$. All the classical integrals of motion except the total energy are clearly 1-particle, while the latter is 2-particle (when U satisfies (ii)).

In the physical literature there seems to be no definition of AAIM similar to the one given above. However, there one can find the assertion that first integrals should be referred to as additive if they have the following property: for a system consisting of two practically noninteracting subsystems such an integral splits (or asymptotically splits) into two similar integrals corresponding to each subsystem taken separately (see [3]). The above definition of the AAIM can be thought of as a formalization of this description. By the way, we assume the splitting of the AAIM only in the case where one of the two subsystems consists of a single particle. As we will see below, such an *a priori* extension of the class of possible AAIM's does not lead to its actual extension.

The problem of describing all the AAIM's belonging to a sufficiently large and natural class of first integrals was originally considered in [4] in connection with the problem of stationary solutions of the Bogolyubov hierarchy equations (BHGKY hierarchy) (another definition was used there, and although it was not stated explicitly, it is actually equivalent to that in [1]). Under the assumption that U is a function with bounded support and under appropriate smoothness assumptions it was stated in [4] and proved in [5] that in every dimension d any finite-particle (i.e., n-particle for some $n \geqslant 1$) AAIM is a linear combination of the classical integrals. The method used there is not applicable to potentials that are not compactly supported. At the same time, for $d = 1$ there are rapidly decreasing (but not compactly supported) potentials U for which finite-particle AAIM's not reducible to classical ones do exist. The function $U(r) = (\sinh r)^{-2}$ gives an example of such a potential (see [6, 7], where first integrals are found for the n-particle system with the Weierstrass function $\wp$ as the interaction potential).

The new method used in [1] allowed the complete elimination of the condition that U is compactly supported and the substantial extension of the class of AAIM's under consideration.

1.2. **Notation.** Denote by $\mathscr{I}_d(U)$ the set of AAIM $I = \{I_n, n = 1, 2, \dots\}$ of particles in $\mathbb{R}^d$ with interaction potential U such that

(a) $I_n \in C^\infty(M_n)$, $n = 1, 2, \dots$,
(b) I_n is a polynomial in $p_1, \dots, p_n$ (its coefficients depend on $q_1, \dots, q_n$) for infinitely many n.

Clearly, every finite-particle AAIM satisfies the assumption (b).

1.3. THEOREM [1]. *When $d > 1$, the set $\mathscr{I}_d(U)$ with U satisfying conditions* (i)–(iii) *above consists of linear combinations of classical integrals.*

1.4. THEOREM [1]. *When $d = 1$, the assertion of the previous theorem holds for all potentials except $U(r) = a/r^2$ and $U(r) = a/[\sinh(br)]^2$, $a, b =$* const.

Theorems 1.3 and 1.4 together show that only for $d = 1$ do there exist exceptional pair potentials for which the assertion of Theorem 1.3 is not true.

With this in mind, it is natural to study the set of AAIM's for a one-dimensional system with nonpairwise translation invariant interaction.

In this paper we consider the case where the interaction between particles is described by a pair potential $U(q_1, q_2)$ and a 3-particle potential $V(q_1, q_2, q_3)$ satisfying the conditions below.

1.5. **Conditions on the interaction** (U, V).

(a) U and V are functions of class C^∞ defined on the sets

$$\Gamma_2 = \{q_1, q_2 \in \mathbb{R} : q_1 \neq q_2\} \quad \text{and}$$
$$\Gamma_3 = \{q_1, q_2, q_3 \in \mathbb{R} : q_i \neq q_j, \ 1 \leqslant i < j \leqslant 3\}$$

respectively (clearly, Γ_n is an open subset of $\mathbb{R}^n$, $n = 2, 3$);

(b) U and V are symmetric, i.e., invariant with respect to all permutations of their arguments;

(c) U and V are translation invariant, i.e.,

$$\frac{\partial}{\partial q_1} U(q_1, q_2) + \frac{\partial}{\partial q_2} U(q_1, q_2) = 0, \qquad (q_1, q_2) \in \Gamma_2, \tag{1.6}$$

$$\begin{aligned} \frac{\partial}{\partial q_1} V(q_1, q_2, q_3) + \frac{\partial}{\partial q_2} V(q_1, q_2, q_3) \\ + \frac{\partial}{\partial q_3} V(q_1, q_2, q_3) = 0, \qquad (q_1, q_2, q_3) \in \Gamma_3; \end{aligned} \tag{1.7}$$

(d) $$\lim_{|q_1 - q_2| \to \infty} U(q_1, q_2) = 0, \qquad \limsup_{|q_1 - q_2| \to 0} U(q_1, q_2) = \infty, \tag{1.8}$$

(e) $$\lim_{|q_i| \to \infty} V(q_1, q_2, q_3) = 0, \qquad i = 1, 2, 3; \tag{1.9}$$

(f) V is a restriction to Γ_3 of a function $V_1 \in C^\infty(\mathbb{R}^3)$ such that $DV_1 \in L^2(\mathbb{R}^3)$ for every differential operator D of the form

$$D = \frac{\partial^{k+l+m}}{\partial q_1^k \partial q_2^l \partial q_3^m}, \qquad k, l, m \in \{0, 1, 2, \dots\}.$$

1.6. **Remark.** Each of the conditions (a)–(e) seems to be essential for some of the further arguments, while condition (f) can probably be replaced by that of the local boundedness of V (see §6 below).

The Hamiltonian of the n-particle system now has the form (cf. (1.1))

$$H_n(x_1, \dots, x_n) = \begin{cases} p_1^2/2, & n = 1, \\ (p_1^2 + p_2^2)/2 + U(q_1, q_2), & n = 2, \\ \sum_{i=1}^n p_i^2/2 + \sum_{1 \leqslant i < j \leqslant n} U(q_i, q_j) & \\ \qquad + \sum_{1 \leqslant i < j \leqslant n} V(q_i, q_j, q_k), & n \geqslant 3, \end{cases} \tag{1.10}$$

while the phase space M_n is defined just as before (see (1.2) with $\delta = 0$). All the above definitions concerning the notion of AAIM's are also preserved.

1.7. **Notation.** Denote by $\mathcal{I}_1(U, V)$ the set of AAIM's of particles in $\mathbb{R}^1$ with interaction (U, V) satisfying the conditions (a), (b) in 1.2.

Our main result is the following.

1.8. THEOREM. *If U, V satisfy the conditions* 1.5(a)–(d) *and if $V \not\equiv 0$, then $\mathcal{I}_1(U, V)$ consists of linear combinations of classical integrals.*

1.9. **Remark.** Comparing Theorems 1.3, 1.4, and 1.8, one can say that increase of dimension affects the set of AAIM's in just the same way as the complication of the interaction does.

We prove Theorem 1.8 by the method of [1] based on the following assertion (cf. Theorem 1 in [1]).

1.10. THEOREM. *Let $I \in \mathcal{J}_1(U, V)$, and let $g = \{g_n, n \geqslant 1\}$ be the generating sequence associated with I. Assume that U, V satisfy* 1.5(a), (b), (d), (f). *Then for $n = 1, 2, \ldots$ the function g_n is a polynomial in $p_1, \ldots, p_n$.*

The proof of this theorem is the most complicated part of the paper. It will be achieved in §§3–5. Section 2 contains some basic definitions and preliminary results. The main Theorem 1.8 is proved in §6.

§2. Basic definitions and preliminary results

2.1. **Notation.** For $n = 1, 2, \ldots$ we set $x^n = (x_1, \ldots, x_n)$ whenever $(x_1, \ldots, x_n) \in M_n$ and denote by $x^n \setminus x_k$ and $x^n \setminus (x_k, x_l)$ the sequences obtained from x^n by removing x_k (if $n \geqslant 2$, $k \leqslant n$) and x_k, x_l (if $n \geqslant 3$; $k, l \leqslant n$, $k \neq l$) respectively.

Obviously, $x^n \setminus x_k \in M_{n-1}$, $x^n \setminus (x_k, x_l) \in M_{n-2}$.

2.2. THEOREM. *Let $I = \{I_n, n \geqslant 1\}$ be an AAIM of particles with interaction (U, V) and $g = \{g_n, n \geqslant 1\}$ be its generating function. Then*

(i) *for every $n \geqslant 2$ the function g_n is symmetric;*

(ii) *for every $n \geqslant 1$, $x^n = (x_1, \ldots, x_n) \in M_n$, and $p \in \mathbb{R}$, we have*

$$\lim_{|q| \to \infty} g_{n+1}(x_1, \ldots, x_n, (q, p)) = 0; \tag{2.1}$$

(iii) *for every $n \geqslant 1$ and $x^n \in M_n$ the following relation holds:*

$$\begin{aligned} &\{g_n(x^n), H_n(x^n)\} \\ &\quad + \sum_{k=1}^{n} \left\{ g_{n-1}(x^n \setminus x_k), \sum_{i \in [1, n] \setminus \{k\}} U(q_i, q_k) + \sum_{\substack{i, j \in [1, n] \setminus \{k\} \\ i \neq j}} V(q_i, q_j, q_k) \right\} \\ &\quad + \sum_{\substack{k, l \in [1, n] \\ k \neq l}} \left\{ g_{n-2}(x^n \setminus (x_k, x_l)), \sum_{\substack{i \in [1, n] \\ i \neq k, i \neq l}} V(q_i, q_k, q_l) \right\} = 0, \end{aligned} \tag{2.2}$$

where the first and the second external sums are assumed to be zero when $n = 1$ and $n = 1, 2$, respectively.

PROOF. (i) Let us show that $g_n(x_1, \ldots, x_n)$ does not change when we exchange x_i and x_j. For each term $I_k(x_{i_1}, \ldots, x_{i_k})$ on the right-hand side of (1.4) there are three possibilities:

1) neither i nor j occur among $i_1, \ldots, i_k$;
2) both i and j are among $i_1, \ldots, i_k$;
3) exactly one of them occurs there.

In the first case the term under consideration is independent of x_i, x_j; in the second case it is symmetric in x_i, x_j by property (i) of I_k (see §1); in the third case to this term there corresponds another term of the inner sum differing from the former one only in that x_i is replaced by x_j and vice versa. So in all cases the inner sum on the right-hand side of (1.4) is symmetric.

(ii) When $n = 2$, $x_2 = (q, p)$, relation (1.4) takes the form

$$g_2(x_1, (q, p)) = -I_1(x_1) - I_1(q, p) + I_2(x_1, (q, p)),$$

which by (1.3) (with $n = 1$) implies that

$$\lim_{|q|\to\infty} g_2(x_1, (q, p)) = 0.$$

Thus (2.1) holds for $n = 1$. Assume now that (2.1) holds for all $n \leqslant k$, where $k \geqslant 1$. By (1.5),

$$\begin{aligned} &g_{k+2}(x_1, \dots, x_{k+1}, (q, p)) \\ &\quad = [I_{k+2}(x_1, \dots, x_{k+1}, (q, p)) - I_{k+1}(x_1, \dots, x_{k+1}) - I_1(q, p)] \\ &\qquad - \sum_{l=1}^{k} \sum_{1\leqslant i_1<\dots<i_l\leqslant k+1} g_{l+1}(x_{i_1}, \dots, x_{i_l}, (q, p)). \end{aligned}$$

By (1.3) and our assumption, the right-hand side of this equality tends to zero as $|q| \to \infty$. So we obtain (2.1) for $n = k + 1$.

(iii) Due to property (iii) of the functions $I_1, I_2, \dots$, for any $n \geqslant 1$ we have the identity

$$\sum_{k=0}^{n-1}(-1)^k \sum_{1\leqslant i_1<\dots<i_{n-k}\leqslant n} \{I_{n-k}(x_{i_1}, \dots, x_{i_{n-k}}), H_{n-k}(x_{i_1}, \dots, x_{i_{n-k}})\} \equiv 0, \tag{2.3}$$

where H_l for $l = 1, \dots, n$ is defined by (1.10). Substituting the right-hand sides of (1.5) and (1.10) for I_{n-k} and H_{n-k} respectively (with $n - k$ replacing n in both cases) we can represent the above expression as a linear combination of functions of the following three types:

$$\{g_l(x_{i_1}, \dots, x_{i_l}), p_j^2/2\}, \quad j \in \{i_1, \dots, i_l\}, \tag{2.4}$$

$$\{g_l(x_{i_1}, \dots, x_{i_l}), U(q_{j_1}, q_{j_2})\}, \quad \{j_1, j_2\} \cap \{i_1, \dots, i_l\} \neq \varnothing, \tag{2.5}$$

$$\{g_l(x_{i_1}, \dots, x_{i_l}), V(q_{j_1}, q_{j_2}, q_{j_3})\}, \quad \{j_1, j_2, j_3\} \cap \{i_1, \dots, i_l\} \neq \varnothing, \tag{2.6}$$

and find the corresponding coefficients $c(i_1, \dots, i_l; j)$, $c(i_1, \dots, i_l; j_1, j_2)$ and $c(i_1, \dots, i_l; j_1, j_2, j_3)$. Clearly, when $l \leqslant n - k$, there are

$$\binom{n-l}{n-k-l}$$

terms of the inner sum on the left-hand side of (2.3) that contain the expression (2.4) and, when $l > n - k$, there are no such terms. It follows that

$$\begin{aligned} c(i_1, \dots, i_l; j) &= \sum_{k=0}^{n-l}(-1)^k \binom{n-l}{n-k-l} \\ &= \begin{cases} 0, & l < n, \\ 1, & l = n, \end{cases} \qquad j \in \{i_1, \dots, i_l\}. \end{aligned} \tag{2.7}$$

In exactly the same way one can prove that $c(i_1, \dots, i_l; j_1, j_2)$ is also equal to the right-hand side of (2.6) when $j_1, j_2 \in \{i_1, \dots, i_l\}$. If $j_1 \in \{i_1, \dots, i_l\}$, $j_2 \notin \{i_1, \dots, i_l\}$, then there are

$$\binom{n-l-1}{n-k-l-1}$$

terms of the above sum containing expression (2.5) if $l+1 \leqslant n-k$ and no such terms otherwise. It follows that

$$c(i_1, \dots, i_l; j_1, j_2) = \sum_{k=0}^{n-l-1} (-1)^k \binom{n-l-1}{n-k-l-1} = \begin{cases} 0, & l < n-1, \\ 1, & l = n-1, \end{cases}$$
$$j_1 \in \{i_1, \dots, i_l\}, \ j_2 \notin \{i_1, \dots, i_l\}. \tag{2.8}$$

One can similarly prove that $c(i_1, \dots, i_l; j_1, j_2, j_3)$ is equal to the right-hand side of (2.7) or (2.8) when $j_1, j_2, j_3 \in \{i_1, \dots, i_l\}$ and $j_1, j_2 \in \{i_1, \dots, i_l\}$, $j_3 \notin \{i_1, \dots, i_l\}$ respectively, and that

$$c(i_1, \dots, i_l; j_1, j_2, j_3) = \sum_{k=0}^{n-l-2} (-1)^k \binom{n-l-2}{n-k-l-2} = \begin{cases} 0, & l < n-2, \\ 1, & l = n-2, \end{cases}$$

when $j_1 \in \{i_1, \dots, i_l\}$, $j_2, j_3 \notin \{i_1, \dots, i_l\}$. Using the formulas obtained for the coefficients in (2.3), we come to (2.2), completing the proof of the theorem.

2.3. **Remark.** When $V \equiv 0$, equation (2.2) turns into equation (2.1) in [1] and is thus a generalization of the latter.

2.4. **Notation and terminology.** In what follows we shall refer to equation (2.2) for a fixed n as (2.2; n) and call the terms on the left-hand side of (2.2; n) *higher*, *middle*, and *lower* ones if they contain derivatives of the functions g_n, g_{n-1}, and g_{n-2} respectively as factor. We shall differentiate equation (2.2; n) repeatedly with respect to various collections of variables. In order to avoid cumbersome formulas, it is convenient to introduce, following [1], specific notation for some differential operators. We shall write $\partial(q^u)$, $\partial(p^v)$ and $\partial(q_1^{u_1}, p_1^{v_1}, \dots, q_n^{u_n}, p_n^{v_n})$ instead of $\partial^u/\partial q^u$, $\partial^v/\partial p^v$, and

$$\prod_{i=1}^{n} \frac{\partial^{u_i+v_i}}{\partial q_i^{u_i} \partial p_i^{v_i}}$$

respectively. Applying the latter operator to a function of the variables $x_1 = (q_1, p_1), \dots, x_n = (q_n, p_n)$, we shall also denote this operator by

$$\partial_n([u_1 \mid v_1], \dots, [u_n \mid v_n])$$

and replace $[u_i \mid v_i]$ by v_i for short whenever $u_i = 0$. Thus, for instance,

$$\frac{\partial^3}{\partial p_1^2 \partial q_2} f(x_1, x_2) \equiv \partial(p_1^2, q_2) f(x_1, x_2) \equiv \partial_2(2, [1 \mid 0]) f(x_1, x_2).$$

For integers $n \geqslant 1$, $l \in [0, n-1]$ and an operator

$$D = \partial_n([u_1 \mid v_1], \dots, [u_n \mid v_n])$$

we write

$$h_l = h_l(D) = v_1 + \dots + v_{l+1},$$

$$\mathfrak{m} = \begin{cases} (n, l, h_l, [u_{l+1} \mid v_{l+1}], v_{l+2}, v_n) & \text{when } 0 \leqslant l \leqslant n-3, \\ (n, l, h_l, [u_{l+1} \mid v_{l+1}], v_n) & \text{when } l = n-2, \\ (n, l, h_l, [u_{l+1} \mid v_{l+1}]) & \text{when } l = n-1, \end{cases}$$

and call the sequence $\mathfrak{m} = \mathfrak{m}_l(D)$ the *mixed l-mark* of D, while the reduced sequence

$$\mathfrak{m} = \mathfrak{m}_l(D) = \begin{cases} (n, l, h_l, v_{l+2}, v_n) & \text{when } 0 \leqslant l \leqslant n-3, \\ (n, l, h_l, v_n) & \text{when } l = n-2, \\ (n, l, h_l) & \text{when } l = n-1, \end{cases}$$

will be called the *pure l-mark* of D. All the l-marks (mixed and pure) of a given operator D for $l = 0, 1, \dots, n-1$ will be simply called its *marks*.

For a mark $\mathfrak{m} = (n, l, \dots)$ denote by $\mathscr{D}(\mathfrak{m})$ the set of all operators $D = \partial_n([u_1 \mid v_1], \dots, [u_n \mid v_n])$ such that

(i) $\mathfrak{m}_l(D) = \mathfrak{m}$;
(ii) $u_i = 0$ for every i, when $\mathfrak{m}$ is a pure mark;
(iii) $u_i = 0$ for every $i \neq l+1$, when $\mathfrak{m}$ is a mixed mark.
(iv) $v_i = v_{l+2}$ for $i \in [l+2, n-1]$.

We call a mark $\mathfrak{m} = (n, l, \dots)$ an *annihilator* if $Dg_n \equiv 0$ for each $D \in \mathscr{D}(\mathfrak{m})$. Denote by $\mathscr{A}$ the set of all annihilators.

Let us say that an operator $\partial_n([u_1 \mid v_1], \dots, [u_n \mid v_n])$ *dominates* an operator $\partial_n([u_1' \mid v_1'], \dots, [u_n' \mid v_n'])$, if $u_i \geqslant u_i'$, $v_i \geqslant v_i'$ for all $i \in [1, n]$. Similarly, we say that a mixed mark $(n, l, h, [u \mid v], w, z)$ with $l \leqslant n-3$ dominates a mixed mark $(n', l', h', [u' \mid v'], w', z')$, if $n' = n$, $l' = l$, $h \geqslant h'$, $u \geqslant u'$, $v \geqslant v'$, $w \geqslant w'$, $z \geqslant z'$. In exactly the same way this relation can be introduced for mixed marks $(n, l, \dots)$ with $l = n-2$, $l = n-1$ and for all pure marks. Clearly, a mark $\mathfrak{m}$ is an annihilator if there exists an annihilator $\mathfrak{m}'$ dominated by $\mathfrak{m}$.

2.5. **Remark.** Let D be a differential operator denoted by $\partial_n([u_1 \mid v_1], \dots, [u_n \mid v_n])$ when it acts on functions of $x_1, \dots, x_n$. Assume that $u_i = v_i = 0$ for some $i \in [1, n]$. Then, according to the above definition, the same D, when regarded as acting on a function of the variables x_j, $j \in [1, n]\backslash\{i\}$, can be denoted by

$$\partial_{n-1}([u'_1 \mid v'_1], \dots, [u'_{n-1} \mid v'_{n-1}]),$$

where $u'_j = u_j$, $v'_j = v_j$ for $j \in [1, i-1]$ and $u'_j = u_{j+1}$, $v'_j = v_{j+1}$ for $j \in [i+1, n]$. Let, for instance, $D = \partial^3/\partial p_2^2 \partial q_3$, $\varphi = \varphi(x_1, x_2, x_3)$, and $\psi = \psi(x_2, x_3)$. Then

$$D\varphi = \partial_3(0, 2, [1 \mid 0])\varphi, \qquad D\psi = \partial_2(2, [1 \mid 0])\psi,$$

and D has both the marks $(3, 2, 2, [1 \mid 0])$ and $(2, 1, 2, [1 \mid 0])$. This situation will repeatedly occur below.

2.6. **Notation and terminology (continued).** Along with the differential operators mentioned above, in this work we use (in contrast to [1]) operators of another, more complicated, type. Every such operator is determined by a family of nonnegative integers: $n \geqslant 2$, $k \in [0, n-1]$, $v_1, \dots, v_{n+1}$, $\theta \geqslant 0$. It acts on smooth functions defined on M_n (see §1) and takes them into functions defined on the subset of the direct product $M_{n+1} \times \mathbb{R}$ consisting of all pairs (x^{n+1}, q), $x^{n+1} = (x_1, \dots, x_{n+1})$, $x_i = (q_i, p_i)$, $1 \leqslant i \leqslant n+1$, $q \in \mathbb{R}$, such that $q \neq q_i$ for $i = 1, \dots, n+1$. Denote this operator by $\partial_{n,k,\theta}(v_1, \dots, v_{n+1})$ and define it as follows:

$$\begin{aligned}(\partial_{n,k,\theta}(v_1, \dots, v_{n+1})f)(x^{n+1}, q) \\ = \sum_{l=1}^{k+1} \sum_{i=k+2}^{n+1} \partial(p_i)\partial(p_1^{v_1}, \dots, p_{n+1}^{v_{n+1}}) f(x^{n+1} \setminus x_l)\partial(q^\theta, q_i) V(q_l, q_i, q).\end{aligned} \tag{2.9}$$

Define the *mark* of the operator $D = \partial_{n,k,\theta}(v_1, \dots, v_{n+1})$ to be the sequence

$$\mathrm{m}(D) = \begin{cases} (n, k, v_1 + \dots + v_{k+1}, v_{k+2}, v_{n+1})\theta & \text{when } k \leqslant n-2, \\ (n, k, v_1 + \dots + v_{k+1}, v_{n+1})\theta & \text{when } k = n-1. \end{cases}$$

The mark just defined will be called *compound*, in contrast to pure and mixed ones. For every compound mark $\mathrm{m} = (n, k, a, b, c)\theta$ denote by $\mathscr{D}(\mathrm{m})$ the set of operators $D = \partial_{n,k,\theta}(v_1, \dots, v_{n+1})$ such that $v_1 + \dots + v_{k+1} = a$, $v_i = b$ for $i \in [k+2, n]$, and $v_{n+1} = c$. In the same way one defines $\mathscr{D}(\mathrm{m})$ for $\mathrm{m} = (n, k, a, c)\theta$, when $k = n-1$. If $Dg_n \equiv 0$ for every $D \in \mathscr{D}(\mathrm{m})$, we call m (as above) an *annihilator* and write $\mathrm{m} \in \mathscr{A}$. We say that an operator $\partial_{n,k,\theta}(v_1, \dots, v_{n+1})$ is *dominated* by $\partial_{n',k',\theta'}(v'_1, \dots, v'_{n+1})$ whenever $n' = n$, $k' = k$, $v'_i \geqslant v_i$ for $i = 1, \dots, n+1$, and $\theta' \geqslant \theta$. Similarly, a mark $(n, k, a, b, c)\theta$ with $k \leqslant n-2$ is *dominated* by $(n', k', a', b', c')\theta'$, whenever $n' = n$, $k' = k$, $a' \geqslant a$, $b' \geqslant b$, $c' \geqslant c$, $\theta' \geqslant \theta$ (for $k = n-1$

obvious alterations are required). For compound marks, the assertion that an annihilator can be dominated by annihilators only still holds. The following fact will also be used below.

2.7. PROPOSITION. *Assume that for some integers* $n \geqslant 3$, $k \in [1, n-1]$, $a \geqslant 1$, $b, c, \theta \geqslant 0$ *we are given a compound mark* $\mathfrak{m}$ *and an operator* $D = \partial_{n,k,\theta}(r_1, \dots, r_{n+1}) \in \mathscr{D}(\mathfrak{m})$, *where* $\mathfrak{m} = (n, k, a, b, c)\theta$ *if* $k \leqslant n-2$, *and* $\mathfrak{m} = (n, k, a, c)\theta$ *if* $k = n-1$. *For a fixed* $k' \in [0, k-1]$ *set* $D' = \partial_{n,k',\theta}(r_1, \dots, r_{n+1})$,

$$\mathfrak{m}_1 = \begin{cases} (n, k-1, a+1, b, c) & \text{when } k \leqslant n-2, \\ (n, k-1, a+1, c) & \text{when } k = n-1, \end{cases}$$

and assume that (a) $D'g_n \equiv 0$, (b) $\mathfrak{m}_1 \in \mathscr{A}$, *and* (c) $r_l > 0$ *for all* $l \in [k'+2, k+1]$. *Then* $Dg_n \equiv 0$.

PROOF. By definition,

$$(D'g_n)(x^{n+1}, q)$$
$$= \sum_{l=1}^{k'+1} \sum_{i=k'+2}^{n+1} \partial(p_i)\partial(p_1^{r_1}, \dots, p_{n+1}^{r_{n+1}})\, g_n(x^{n+1} \setminus x_l)\partial(q^\theta, q_i)V(q_l, q_i, q)$$

and we must prove that the right-hand side of this relation does not change if the summation is performed over l from 1 to $k+1$ and over i from $k+2$ to $n+1$. The former replacement can be done according to assumption (c). To deal with the latter one, we assume that $i \in [k'+2, k+1]$, $l \in [1, k+1]$ and set $D_i = \partial(p_i)\partial(p_1^{r_1}, \dots, p_{n+1}^{r_{n+1}})$. If either $r_l > 0$ or $i = l$, then clearly $D_i g_n(x^{n+1} \setminus x_l) \equiv 0$. If $r_l = 0$, $i \neq l$, then the operator D acting on $g_n(x^{n+1} \setminus x_l)$ belongs to $\mathscr{D}(\mathfrak{m}_1)$ (see Remark 2.5) and by assumption (b) we again have $D_i g_n(x^{n+1} \setminus x_l) \equiv 0$. Hence $D'g_n \equiv Dg_n$, and it remains to use assumption (a). This completes the proof.

2.8. **Plan of the proof of Theorem 1.10.** In order to prove Theorem 1.10, it is sufficient to show that if for some $n_0 \geqslant 3$ the function g_{n_0+1} is polynomial in the momenta, then the same is true for all g_n with $n \leqslant n_0$, i.e., for $n = 1, \dots, n_0 + 1$ there exists $s(n) \geqslant 1$ such that $\partial(p_i^{s(n)})\, g_n \equiv 0$ for $i = 1, \dots, n$. By the symmetry of g_n (see Theorem 2.2 (i)), it is sufficient to prove this identity for $i = n$ only. Assume that $s(n_0+1)$ exists and denote it by $m_0 + 1$, i.e., $\partial(p_{n_0+1}^{m_0+1})\, g_{n_0+1} \equiv 0$. Using the notation and terminology of 2.4, one can say that the mark $(n_0+1, 0, 0, 0, m_0+1)$ is an annihilator, i.e.,

$$(n_0 + 1, 0, 0, 0, m_0 + 1) \in \mathscr{A}. \tag{2.10}$$

Theorem 1.10 will be proved if for $n = n_0, n_0 - 1, \dots, 1$ we can find an integer $m_n \geqslant 0$ such that $\mathfrak{m} \in \mathscr{A}$, where

$$\mathfrak{m} = (n, n-1, m_n + 1). \tag{2.11}$$

Indeed, $\mathscr{D}(\mathfrak{m})$ contains the operator $\partial_n(0, \ldots, 0, m_n + 1)$, and so one can set $s(n) = m_n + 1$.

Starting with (2.10), we shall successively extend the set of marks $\mathfrak{m}$ such that $\mathfrak{m} \in \mathscr{A}$. For any mark $(n, l, h, [u \mid v], w, z)$ or (n, l, h, w, z) call the numbers $n_0 - n$, l and h its *depth, length,* and *height* respectively. The depth of the mark (2.10) is -1. So in going from this mark to marks of depth $n_0 - n = 0, 1, \ldots, n_0 - 1$ one must deepen it and some other marks. The first deepening will give the pure mark $\mathfrak{m}_1 = (n_0, 0, 1, 1, m_0+3)$ whose depth and length are both zero and whose height is one. The mark $\mathfrak{m}_1$ can be lowered so that its depth and length and certainly the property of being an annihilator are preserved. This, however, yields a mixed mark, namely the mark $\mathfrak{m}_2 = (n_0, 0, 0, [1 \mid 0], 2, m_0+3)$, and so the operation just described can be called a relative lowering. One must purify the mark $\mathfrak{m}_2$, i.e., replace it by a pure mark whose depth, length, and height are all zero. As the result, an absolute lowering of the mark $\mathfrak{m}_1$ is performed, and we get $\mathfrak{m}_3 \in \mathscr{A}$, where $\mathfrak{m}_3$ is a mark of the form $(n_0, 0, 0, w, z)$, $w \geqslant 1$, $z \geqslant m_0 + 3$. Since the set $\mathscr{D}(\mathfrak{m}_3)$ is exhausted by the operator $D = \partial_{n_0}(0, w, \ldots, w, z)$, we have $Dg_{n_0} \equiv 0$. Note that D also belongs to the set $\mathscr{D}(\mathfrak{m}_4)$, where $\mathfrak{m}_4 = (n_0, 1, w, w, z)$. Together, the inclusion $\mathfrak{m}_1 \in \mathscr{A}$ and the symmetry of g_{n_0} imply that $\widetilde{D}g_{n_0} \equiv 0$ for every operator $\widetilde{D} \in \mathscr{D}(\mathfrak{m}_4)$ other than D. Therefore, $\mathfrak{m}_4 \in \mathscr{A}$. Clearly, the length and height of $\mathfrak{m}_4$ are 1 and w respectively, and so $\mathfrak{m}_4$ must be lowered at the next step. It is worth noting that each operation performed above results in the increase of every element belonging to the mark we are dealing with and situated in it after the height. So it appears as if the nonzero elements of the mark are forced out to the right. Our goal will be achieved when we get an annihilator of the form $(n, l, 0, 0, z)$ or (2.11).

The considerations described above yield only the first steps of an inductive procedure used in the proof of Theorem 1.10, but they contain the basic operations of this procedure: lowering, purification, and deepening. These operations are carried out by applying appropriate differential operators to (2.2). The whole outline is the same as in [1], but it now acts in a more complicated situation, where a nontrivial 3-particle interaction is present, and so it contains some new points.

In the next section we turn to the proof of Theorem 1.10.

§3. Proof of Theorem 1.10: basic lemmas

We assume henceforth that the conditions of Theorem 1.10 are fulfilled.

3.1. LEMMA (relative lowering, 1st case). *Let* n, k, a, b, c, b_1, c_1, b_2, *and* c_2 *be nonnegative integers such that* $n \geqslant 2$, $k \in [0, n-2]$, *and* $a \geqslant 1$. *Then*

(a) *if* $n \geqslant 4$, $k \geqslant 2$ *and if the following inclusions hold*:

$$\begin{aligned} (n, k, a, b, c) &\in \mathscr{A} \quad \text{when } k \leqslant n-3 \\ ((n, k, a, c) &\in \mathscr{A} \quad \text{when } k = n-2), \end{aligned} \tag{3.1}$$

$$\begin{aligned} (n-1, k-1, a, b_1, c_1) &\in \mathscr{A} \quad \text{when } k \leqslant n-3 \\ ((n-1, k-1, a, c_1) &\in \mathscr{A} \quad \text{when } k = n-2), \end{aligned} \tag{3.2}$$

$$\begin{aligned} (n-2, k-2, a, b_2, c_2) &\in \mathscr{A} \quad \text{when } k \leqslant n-3 \\ ((n-2, k-2, a, c_2) &\in \mathscr{A} \quad \text{when } k = n-2), \end{aligned} \tag{3.2}$$

then for any integer $\delta \in [1, a]$ *we have*

$$\begin{aligned} (n, k, a-1, [\delta \mid a-\delta], b', c') &\in \mathscr{A} \quad \text{when } k \leqslant n-3, \\ (n, k, a-1, [\delta \mid a-\delta], c') &\in \mathscr{A} \quad \text{when } k \leqslant n-2; \end{aligned} \tag{3.4}$$

here

$$b' = \max\{b+1, b_1, b_2\}, \qquad c' = \max\{c+1, c_1, c_2\}; \tag{3.5}$$

(b) *if* $n \geqslant 3$, $k = 1$ *and* (3.1), (3.2) *hold, then for any integer* $\delta \in [1, a]$ (3.4) *holds with* $b' = \max\{b+1, b_1\}$, $c' = \max\{c+1, c_1\}$;

(c) *if* $n \geqslant 2$, $k = 0$ *and* (3.1) *holds, then, for* $\delta = 1$, (2.6) *holds with* $b' = b+1$, $c' = c+1$.

PROOF. 1°. Consider the case (a) assuming that $k \leqslant n-3$ (for $k = n-2$ only obvious alternations are needed). Apply the operator to both sides of (2.2; n) and start the analysis of the identity obtained with the lower terms. Due to (3.5) and the assumption $a, b', c' \geqslant 1$, the lower terms that do not contain all the variables $x_{k+1}, \dots, x_n$ vanish. The sum of the remaining lower terms is

$$\begin{aligned} &- \sum_{\substack{l, l' \in [1, k] \\ l \neq l'}} \sum_{i \in [1, n] \setminus \{l, l'\}} D\left[\frac{\partial}{\partial p_i} g_{n-2}(x^n \setminus (x_l, x_{l'})) \frac{\partial}{\partial q_i} V(q_l, q_{l'}, q_i)\right] \\ &= - \sum_{\substack{l, l' \in [1, k] \\ l \neq l'}} \sum_{i \in [1, n] \setminus \{l, l'\}} \frac{\partial}{\partial q_i} V(q_l, q_{l'}, q_i) \frac{\partial}{\partial p_i} D g_{n-2}(x^n \setminus (x_l, x_{l'})). \end{aligned}$$

Clearly, $D \in \mathscr{D}(\mathfrak{m})$, where $\mathfrak{m} = (n-2, k-2, a, b', c')$ (see Remark 2.5). By (3.5), $\mathfrak{m}$ dominates an annihilator $(n-2, k-2, a, b_2, c_2)$ (see (3.3)) and hence all the lower terms on the left-hand side of the identity under consideration vanish.

Now let us pass to the middle terms. After the removal of those middle terms that obviously vanish we get the sum

$$-\sum_{l=1}^{k} \sum_{i\in[1,n]\setminus\{l\}} D \frac{\partial}{\partial p_i} g_{n-1}(x^n\setminus x_l) \frac{\partial}{\partial q_i}\left[U(q_l, q_i) + \sum_{j\in[1,n]\setminus\{l,i\}} V(q_l, q_i, q_j)\right]$$

$$= -\sum_{l=1}^{k} \sum_{i\in[1,n]\setminus\{l\}} \frac{\partial}{\partial q_i}\left[U(q_l, q_i) + \sum_{j\in[1,n]\setminus\{l,i\}} V(q_l, q_i, q_j)\right] \times \frac{\partial}{\partial p_i} D_2 g_{n-1}(x^n \setminus x_l),$$

where D_1 is the same operator D but now rewritten in the form

$$\partial_{n-1}(\underbrace{0, \dots, 0}_{k-1}, a, b', \dots, b', c').$$

Clearly, $D_1 \in \mathscr{D}(\mathfrak{m}_1)$, where $\mathfrak{m}_1 = (n-1, k-1, a, b', c')$. By (3.5), $\mathfrak{m}_1$ dominates the mark $(n-1, k-1, a, b_1, c_1)$, which is an annihilator (see (3.2)). Therefore, all middle terms also vanish.

It remains to consider the higher terms. The higher terms of the form

$$-D \frac{\partial}{\partial p_i} g_n(x^n) \frac{\partial}{\partial q_i} U(q_l, q_i) \quad \text{or} \quad -D\left[\frac{\partial}{\partial p_i} g_n(x^n) \frac{\partial}{\partial q_i} V(q_l, q_i, q_j)\right]$$

vanish by (3.1). Indeed, $D \in \mathscr{D}(\mathfrak{m}_2)$, where $\mathfrak{m}_2 = (n, k, a, b', c')$ and, by (3.5), $\mathfrak{m}_2$ dominates the mark (n, k, a, b, c). Now by (3.1), $\mathfrak{m}_2 \in \mathscr{A}$ and hence $Dg_n \equiv 0$. By the same reasoning all the higher terms of the form $D[p_i(\partial/\partial q_i)\, g_n(x^n)]$ with $i \leqslant k$ also vanish. Further, one can easily check that

$$D\left[p_n \frac{\partial}{\partial q_i} g_n(x^n)\right] = p_n \frac{\partial}{\partial q_n} Dg_n(x^n) + \frac{\partial}{\partial q_n} D_3 g_n(x^n), \tag{3.6}$$

where $D_3 = \partial_n(0, \dots, 0, a, b', \dots, b', c' - 1)$. From (3.1), (3.5) it follows that both terms on the right-hand side of (3.6) vanish. In exactly the same way one can prove that

$$D\left[p_i \frac{\partial}{\partial q_i} g_n(x^n)\right] \equiv 0,$$

when $k+1 \leqslant i \leqslant n+1$. Thus only one of the higher terms remains, namely

$$D\left[p_{k+1} \frac{\partial}{\partial q_{k+1}} g_n(x^n)\right] = p_{k+1} \frac{\partial}{\partial q_{k+1}} Dg_n(x^n) + \frac{\partial}{\partial q_{k+1}} D_4 g_n(x^n) = D_5 g_n(x^n),$$

where

$$D_4 = \partial_n(\underbrace{0, \dots, 0}_{k}, a-1, b', \dots, b', c'),$$

$$D_5 = \partial_n(\underbrace{0, \dots, 0}_{k}, [1 \mid a-1], b', \dots, b', c').$$

Clearly, D_5 is the whole set $\mathscr{D}(\mathfrak{m}_5)$, where

$$\mathfrak{m}_5 = (n, k, a-1, [1 \mid a-1], b', c'),$$

and we get $\mathfrak{m}_5 \in \mathscr{A}$, i.e.,

$$(n, k, a-1, [1 \mid a-1], b', c') \in \mathscr{A}, \tag{3.7}$$

which coincides with (3.4) for $\delta = 1$.

2°. Assume that (3.4) holds for some $\delta \in [1, a-1]$; let us show that

$$(n, k, a-1, [\delta+1 \mid a-\delta-1], b', c') \in \mathscr{A}. \tag{3.8}$$

An arbitrary operator from $D(\mathfrak{m})$, where $\mathfrak{m}$ is the mark in (3.8), has the form $\partial_n(v_1, \dots, v_k, [\delta+1 \mid a-\delta-1], b', \dots, b', c')$ with $v_1, \dots, v_k \geqslant 0$, $v_1 + \dots + v_k = \delta$. Apply the operator

$$D' = \partial_n(v_1, \dots, v_k, [\delta \mid a-\delta], b', \dots, b', c')$$

to (2.2; n). In the same way as above, one easily proves that all the lower and middle terms vanish and that among the higher terms only those of the form

$$D'\left[p_{k+1}\frac{\partial}{\partial_{q_{k+1}}} g_n(x^n)\right] \quad \text{and} \quad D'\left[p_i\frac{\partial}{\partial q_i} g_n(x^n)\right]$$

with $i \in [1, k]$, $v_i \geqslant 1$, can remain. Clearly,

$$D'\left[p_i\frac{\partial}{\partial q_i} g_n(x^n)\right] = p_i\frac{\partial}{\partial q_i} D' g_n(x^n) + \frac{\partial}{\partial q_i} D_1' g_n(x^n), \tag{3.9}$$

where $D_1' = \partial_n(v_1', \dots, v_k', [\delta \mid a-\delta], b', \dots, b', c')$ with $v_j' = v_j - \delta_{ij}$, $1 \leqslant j \leqslant k$ (δ_{ij} being the Kronecker symbol). Since $D' \in \mathscr{D}(\mathfrak{m}')$, $D_1' \in \mathscr{D}(\mathfrak{m}_1')$, where $\mathfrak{m}'$ dominates the mark (n, k, a, b, c) (see (3.5)) and where $\mathfrak{m}_1' = (n, k, a-1, [\delta \mid a-\delta], b', c')$, it follows from (3.1) and from our assumption that the right-hand side of (3.9) vanishes. So, after we apply D to the left-hand side of (2.2; n) the latter takes the form

$$D'\left[p_{k+1}\frac{\partial}{\partial q_{k+1}} g_n(x^n)\right] = p_{k+1}\frac{\partial}{\partial q_{k+1}} D' g_n(x^n) + D_2' g_n(x^n) = D_2' g_n(x^n),$$

where

$$D_2' = \partial_n(v_1, \dots, v_k, [\delta+1 \mid a-\delta-1], b', \dots, b', c').$$

It follows that $D_2'g_n \equiv 0$, and we come to (3.8). Therefore, (3.4) holds for any $\delta \in [1, a]$ (see (3.7)), which gives assertion (a).

3°. In cases (b) and (c) the proof is conducted in the same way as above but becomes simpler: in case (b) the lower terms vanish immediately; in case (c) so do both the lower and middle terms. Besides, in case (c) the above induction is impossible and so (3.4) holds for $\delta = 1$ only.

3.2. LEMMA (relative lowering, 2nd case). *Let n, k, a, b, c, b_{-1}, c_{-1}, b_1, c_1, b', c', $\widehat{b}$, $\widehat{c}$, b_1', c_1', α, γ, θ be nonnegative integers such that $n \geqslant 2$, $k \in [0, n-2]$, and $a \geqslant 2$. Then*

(a) *if $n \geqslant 3$, $k \geqslant 1$, and the following inclusions hold:*

$$\begin{aligned}
(n+1, k+1, a-1, b_{-1}, c_{-1}) \in \mathscr{A} \quad & \textit{when } k \leqslant n-3 \\
((n+1, k+1, a-1, c_{-1}) \in \mathscr{A} \quad & \textit{when } k = n-2),
\end{aligned} \tag{3.10}$$

$$\begin{aligned}
(n, k, a, b, c) \in \mathscr{A} \quad & \textit{when } k \leqslant n-3 \\
((n, k, a, c) \in \mathscr{A} \quad & \textit{when } k = n-2),
\end{aligned} \tag{3.11}$$

$$\begin{aligned}
(n, k, a-1, [\gamma \mid a'], b', c') \in \mathscr{A} \quad & \textit{for any } a' \in [0, a-1], \\
& \textit{when } k \leqslant n-3 \\
((n, k, a-1, [\gamma \mid a'], c') \in \mathscr{A} \quad & \textit{when } k = n-2),
\end{aligned} \tag{3.12}$$

$$\begin{aligned}
(n, k+1, a-1, \widehat{b}, \widehat{c})\theta \in \mathscr{A} \quad & \textit{when } k \leqslant n-3 \\
((n, k+1, a-1, \widehat{c})\theta \in \mathscr{A} \quad & \textit{when } k = n-2),
\end{aligned} \tag{3.13}$$

$$\begin{aligned}
(n-1, k-1, a, b_1, c_1) \in \mathscr{A} \quad & \textit{when } k \leqslant n-3 \\
((n-1, k-1, a, c_1) \in \mathscr{A} \quad & \textit{when } k = n-2),
\end{aligned} \tag{3.14}$$

$$\begin{aligned}
(n-1, k-1, a-1, [\alpha \mid a_1'], b_1', c_1') \in \mathscr{A} \quad & \textit{for any } a_1' \in [0, a-1], \\
& \textit{when } k \leqslant n-3 \\
((n-1, k-1, a-1, [\alpha \mid a_1'], c_1') \in \mathscr{A} \quad & \textit{when } k = n-2),
\end{aligned} \tag{3.15}$$

then for every integer $\delta \in [1, a-1]$

$$\begin{aligned}
(n+1, k+1, a-2, [\tau+\delta \mid a-1-\delta], b_{-1}', c_{-1}') \in \mathscr{A} \quad & \textit{when } k \leqslant n-3 \\
((n+1, k+1, a-2, [\tau+\delta \mid a-1-\delta], \widetilde{c}) \in \mathscr{A} \quad & \textit{when } k = n-2),
\end{aligned} \tag{3.16}$$

where

$$\begin{aligned}
b_{-1}' &= \max\{b_{-1}+1, b, b', b_1, b_1', \widehat{b}\}, \\
c_{-1}' &= \max\{c_{-1}+1, c, c', c_1, c_1', \widehat{c}\},
\end{aligned} \tag{3.17}$$

$$\tau = \max\{\alpha, \gamma+\theta\}; \tag{3.18}$$

(b) *if* $n \geqslant 2$, $k=0$ *and if we have* (3.10), (3.11), (3.13), *and if, moreover,* (3.12) *holds for* $a'=a-1$, *then* (3.16) *holds with*

$$b'_{-1}=\max\{b_{-1}+1, b, b', \widehat{b}\},\quad c'_{-1}=\max\{c_{-1}+1, c, c', \widehat{c}\},\quad \tau=\gamma+\theta.$$

PROOF. 1°. We start with case (a), assuming that $k \leqslant n-3$ as in the proof of the previous lemma (only obvious modifications are needed for $k=n-2$). Apply the operator

$$D=\partial_{n+1}(0, \dots, 0, [\tau \mid a-1], b'_{-1}, \dots, b'_{-1}, c'_{-1})$$

to (2.2; $n+1$) and consider first the lower terms of the resulting identity. Since $a-1, b'_{-1}, c'_{-1}>0$, their sum can be written in the form

$$-\sum_{\substack{l, l' \in[1, k+1] \\ l \neq l'}} \sum_{i \in[1, n+1] \backslash\{l, l'\}} D\left[\frac{\partial}{\partial p_i} g_{n-1}(x^{n+1} \backslash(x_l, x_{l'})) \frac{\partial}{\partial q_i} V(q_l, q_{l'}, q_i)\right]$$

$$\begin{aligned} =-\sum_{\substack{l, l' \in[1, k+1] \\ l \neq l'}}\Bigg[& \sum_{i \in[1, k+2] \backslash\{l, l'\}} \partial(q_{k+2}^\tau) D_i g_{n-1}(x^{n+1} \backslash(x_l, x_{l'})) \\ & \qquad\qquad \times \frac{\partial}{\partial q_i} V(q_l, q_{l'}, q_i) \\ & +\sum_{i \in[k+3, n+1]} \frac{\partial}{\partial q_i} V(q_l, q_{l'}, q_i) \frac{\partial}{\partial p_i} D g_{n-1}(x^{n+1} \backslash(x_l, x_{l'}))\Bigg], \end{aligned}$$

where D_i is defined by $D_i \partial(q_{k+2}^\tau)=D \partial(p_i)$, $i \in[1, k+2] \backslash\{l, l'\}$. Clearly, $D_i \in \mathscr{D}(\mathfrak{m}_1)$ and $D \in \mathscr{D}(\mathfrak{m}_2)$, where

$$\begin{aligned} & \mathfrak{m}_1=(n-1, k-1, a, b'_{-1}, c'_{-1}), \\ & \mathfrak{m}_2=(n-1, k-1, a-1, [\tau \mid a-1], b'_{-1}, c'_{-1}). \end{aligned}$$

By (3.17), $\mathfrak{m}_1$ dominates the mark $\mathfrak{m}_3=(n-1, k-1, a, b_1, c_1)$, while by (3.17), (3.18), $\mathfrak{m}_2$ dominates the mark

$$\mathfrak{m}_4=(n-1, k-1, [\alpha \mid a-1], b'_1, c'_1).$$

Since by (3.14), (3.15) $\mathfrak{m}_3, \mathfrak{m}_4 \in \mathscr{A}$, all the lower terms vanish.

As for the middle terms, let us note that those containing $g_n(x^{n+1} \backslash x_l)$

with $l \geqslant k+2$ are evidently zero. The remaining sum is

$$-\sum_{l=1}^{k+1} \sum_{i\in[1,n+1]\setminus\{l\}} D\left[\frac{\partial}{\partial p_i} g_n(x^{n+1}\setminus x_l) \times \frac{\partial}{\partial q_i}\left(U(q_l, q_i) + \sum_{j\in[1,n+1]\setminus\{l,i\}} V(q_l, q_i, q_j)\right)\right]$$

$$\begin{aligned} = -\sum_{l=1}^{k+1}\Bigg[& \sum_{i\in[1,k+2]\setminus\{l\}} \partial(q_{k+2}^{\tau}) D_i g_n(x^{n+1}\setminus x_l) \\ & \times \frac{\partial}{\partial q_i}\left(U(q_l, q_i) + \sum_{j\in[1,n+1]\setminus\{l,i\}} V(q_l, q_i, q_j)\right) \\ & + \sum_{i\in[k+3,n+1]}\Bigg(\frac{\partial}{\partial q_i} U(q_l, q_i) \\ & \qquad + \frac{\partial}{\partial q_i} \sum_{j\in[1,n+1]\setminus\{l,i,k+2\}} V(q_l, q_i, q_j)\Bigg) \\ & \qquad \times \frac{\partial}{\partial p_i} D g_n(x^{n+1}\setminus x_l) \\ & + \sum_{i\in[k+3,n+1]} D_i \partial(q_{k+2}^{\tau})\left(g_n(x^{n+1}\setminus x_l)\frac{\partial}{\partial q_i} V(q_l, q_i, q_{k+2})\right)\Bigg], \end{aligned} \tag{3.19}$$

where D_i is the operator defined above. Since $D_i \in \mathscr{D}(\mathfrak{m}_5)$, $D \in \mathscr{D}(\mathfrak{m}_6)$, where $\mathfrak{m}_5 = (n, k, a, b'_{-1}, c'_{-1})$, $\mathfrak{m}_6 = (n, k, a-1, [\tau \mid a-1], b'_{-1}, c'_{-1})$ (see Remark 2.5) and since by (3.11), (3.17) and by (3.12), (3.17), (3.18) we have $\mathfrak{m}_5, \mathfrak{m}_6 \in \mathscr{A}$ (both $\mathfrak{m}_5, \mathfrak{m}_6$ dominate some annihilators), only the last sum can remain in the square brackets on the right-hand side of (3.19) (note that here we use (3.12) for $a' = a-1$ only). Therefore, denoting the left-hand side of (3.19) by $S(x^{n+1})$, we have

$$\begin{aligned} S(x^{n+1}) &= -\sum_{l=1}^{k+1}\sum_{i=k+3}^{n+1} D_i\partial(q_{k+2}^{\tau})\left[g_n(x^{n+1}\setminus x_l)\frac{\partial}{\partial q_i} V(q_l, q_{k+2}, q_i)\right] \\ &= \sum_{j=0}^{\tau}\binom{\tau}{j} S_j(x^{n+1}), \end{aligned}$$

where

$$\begin{aligned} S_j(x^{n+1}) = -\sum_{l=1}^{k+1}\sum_{i=k+3}^{n+1} & \partial(p_i)\partial(p_{k+2}^{a-1}, p_{k+3}^{b'_{-1}}, \dots, p_n^{b'_{-1}}, p_{n+1}^{c'_{-1}})\partial(q_{k+2}^{j}) \\ & \times g_n(x^{n+1}\setminus x_l)\partial(q_{k+2}^{\tau-j}, q_i) V(q_l, q_{k+2}, q_i). \end{aligned} \tag{3.20}$$

The operator

$$\partial(p_{k+2}^{a-1}, p_{k+3}^{b'_{-1}}, \dots, p_n^{b'_{-1}}, p_{n+1}^{c'_{-1}})\partial(q_{k+2}^{j})$$

when acting on a function of $x^{n+1}\setminus x_l$, $l \leqslant k+1$, belongs to $\mathscr{D}(\mathfrak{m}_7)$, where

$$\mathfrak{m}_7 = (n, k, a-1, [j \mid a-1], b'_{-1}, c'_{-1}).$$

The mark $\mathfrak{m}_7$ with $j \geqslant \gamma$ dominates the annihilator $(n, k, a-1, [\gamma \mid a-1], b, c)$ (see (3.12), (3.17)). Hence $\mathfrak{m}_7 \in \mathscr{A}$ and

$$S(x^{n+1}) = \sum_{j=0}^{\gamma-1} \binom{\tau}{j} S_j(x^{n+1}) \tag{3.21}$$

(for $\gamma = 0$ the right-hand side of this relation is taken to be zero).

Using (3.13), let us show that all the S_j vanish. According to the notation and terminology of 2.6, (3.13) implies

$$\begin{gathered}\sum_{l=1}^{k+2} \sum_{i=k+3}^{n+1} \partial(p_i)\partial(p_{k+2}^{a-1}, p_{k+3}^{\widehat{b}}, \dots, p_n^{\widehat{b}}, p_{n+1}^{\widehat{c}}) \\ \times g_n(x^{n+1} \setminus x_l)\partial(q^\theta, q_i)V(q_l, q_i, q) = 0, \\ x^{n+1} \in M_{n+1}, \ q \in \mathbb{R}, \ q \neq q_i, \ 1 \leqslant i \leqslant n+1;\end{gathered} \tag{3.22}$$

it is sufficient to put $k+1$ in place of k in (2.9) and set

$$v_1 = \dots = v_{k+1} = 0, \quad v_{k+2} = a-1, \quad v_{k+3} = \dots = v_n = \widehat{b}, \quad v_{n+1} = \widehat{c}.$$

Applying the operator

$$\partial(q_{k+2}^j, q^{\tau-\theta-j})\partial(p_{k+3}^{b'_{-1}-\widehat{b}}, \dots, p_n^{b'_{-1}-\widehat{b}}, p_{n+1}^{c'_{-1}-\widehat{c}}), \qquad 0 \leqslant j < \gamma,$$

to (3.22) and observing that by the condition $a \geqslant 2$ the inner sum vanishes when $l = k+2$, we get

$$\begin{gathered}\sum_{l=1}^{k+1} \sum_{i=k+3}^{n+1} \partial(p_i)\partial(p_{k+2}^{a-1}, p_{k+3}^{b'_{-1}}, \dots, p_n^{b'_{-1}}, p_{n+1}^{c'_{-1}}) \\ \times \partial(q_{k+2}^j) g_n(x^{n+1} \setminus x_l)\partial(q^{\tau-j}, q_i)V(q_l, q, q_i) = 0, \\ x^{n+1} \in M_{n+1}, \ q \in \mathbb{R}, \ q = q_i, \ 1 \leqslant i \leqslant n+1, \ 0 \leqslant j < \gamma.\end{gathered} \tag{3.23}$$

When $q \to q_{k+2}$, the left-hand side of this equality tends to $-S_j(x^{n+1})$ (it is essential here that $l < k+2$). It follows that $S_j(x^{n+1}) \equiv 0$ and hence $S(x^{n+1}) \equiv 0$, i.e., the sum of the middle terms vanishes.

Consider the higher terms. Those containing $\partial g_{n+1}/\partial p_i$ vanish because $D = \partial(q_{k+2}^\tau)D_1$, where $D_1 \in \mathscr{D}(\mathfrak{m}_8)$, $\mathfrak{m}_8 = (n+1, k+1, a-1, b'_{-1}, c'_{-1})$, and because by (3.17) $\mathfrak{m}_8$ dominates $(n+1, k+1, a-1, a_{-1}, c_{-1}) \in \mathscr{A}$ (see (3.10)). By similar reasoning all the higher terms containing $p_i \partial g_{n+1}/\partial q_i$ with $i \neq k+2$ also vanish. So there remains only the term

$$\begin{aligned}D\left[p_{k+2}\frac{\partial g_{n+1}(x^{n+1})}{\partial q_{k+2}}\right] &= \partial(q_{k+2})Dg_{n+1}(x^{n+1}) + D_2 g_{n+1}(x^{n+1}) \\ &= D_2 g_{n+1}(x^{n+1}),\end{aligned}$$

where

$$D_2 = \partial_{n+1}(\underbrace{0, \dots, 0}_{k+1}, [\tau+1 \mid a-2], b'_{-1}, \dots, b'_{-1}, c'_{-1}).$$

Returning to (2.2; $n+1$) we conclude that $D_2 g_{n+1}(x^{n+1}) \equiv 0$, i.e.,

$$(n+1, k+1, a-2, [\tau+1 \mid a-2], b'_{-1}, c'_{-1}) \in \mathscr{A},$$

and so (3.16) is true for $\delta = 1$. The subsequent argument is in many respects similar to the one just developed.

2°. We prove (3.16) by induction on δ. Assuming (3.16) to hold for some $\delta \in [1, a-2]$, we set

$$\mathfrak{m}_9 = (n+1, k+1, a-2, [\tau+\delta+1 \mid a-2-\delta], b'_{-1}, c'_{-1})$$

and note that an arbitrary operator in $\mathscr{D}(\mathfrak{m}_9)$ is of the form

$$\partial_{n+1}(v_1, \ldots, v_{k+1}, [\tau+\delta+1 \mid a-2-\delta], b'_{-1}, \ldots, b'_{-1}, c'_{-1}) \tag{3.24}$$

with $v_1 + \cdots + v_{k+1} = \delta$. Apply the operator

$$D = \partial_{n+1}(v_1, \ldots, v_{k+1}, [\tau+\delta \mid a-1-\delta], b'_{-1}, \ldots, b'_{-1}, c'_{-1})$$

to (2.2; $n+1$). Then the lower terms vanish for the same reason as in the first part of the proof, while the sum of the middle terms has the form (3.19) with the exception that $\partial(q^{\tau}_{k+2})$ must be replaced by $\partial(q^{\tau+\delta}_{k+2})$ (one may certainly sum here over those l for which $v_l = 0$). When $\gamma = 0$, this sum is zero, and when $\gamma \geqslant 1$ it is equal to

$$\begin{aligned} S(x^{n+1}) &= -\sum_{l=1}^{k+1}\sum_{k+3}^{n+1} D_i \partial(q^{\tau+\delta}_{k+2})\left[g_n(x^{n+1}\setminus x_l)\frac{\partial}{\partial q_i}V(q_l, q_{k+2}, q_i)\right] \\ &= \sum_{j=0}^{\gamma-1}\binom{\tau}{j} S_j(x^{n+1}), \end{aligned}$$

where this time (cf. (3.20), (3.21)) we have

$$\begin{aligned} S(x^{n+1}) = -\sum_{l=1}^{k+1}\sum_{i=k+3}^{n+1} &\partial(p_i)\partial(p_1^{v_1}, \ldots, p_{k+1}^{v_{k+1}}, p_{k+2}^{a-1-\delta}, p_{k+3}^{b'_{-1}}, \ldots, p_n^{b'_{-1}}, p_{n+1}^{c'_{-1}}) \\ &\times \partial(q^j_{k+2})\, g_n(x^{n+1}\setminus x_l)\partial(q^{\tau+\delta-j}_{k+2}, q_i)V(q_l, q_{k+2}, q_i) \end{aligned} \tag{3.25}$$

(observe that the inner sum vanishes for l satisfying $v_l > 0$). According to 2.6, inclusion (3.13) and the condition $\delta \leqslant a-2$ imply

$$\begin{gathered} \sum_{l=1}^{k+1}\sum_{i=k+3}^{n+1} \partial(p_i)\partial(p_1^{v_1}, \ldots, p_{k+1}^{v_{k+1}}, p_{k+2}^{a-1-\delta}, p_{k+3}^{\widehat{b}}, \ldots, p_n^{\widehat{b}}, p_{n+1}^{\widehat{c}}) \\ \times g_n(x^{n+1}\setminus x_l)\partial(q^{\theta}, q_i)V(q_l, q, q_i) \equiv 0, \\ x^{n+1} \in M_{n+1},\ q \in \mathbb{R},\ q \neq q_i,\ 1 \leqslant i \leqslant n+1. \end{gathered} \tag{3.26}$$

Applying the operator

$$\partial(q^j_{k+2}, q^{\tau+\delta-\theta-j})\partial(p_{k+3}^{b'_{-1}-\widehat{b}}, \ldots, p_n^{b'_{-1}-\widehat{b}}, p_{n+1}^{c'_{-1}-\widehat{c}}), \qquad 0 \leqslant j < \gamma,$$

to (3.26) and using conditions (3.11), (3.17) in the same way as before, we obtain (cf. (3.23))

$$\sum_{l=1}^{k+1}\sum_{i=k+3}^{n+1} \partial(p_i)\partial(p_1^{v_1}, \dots, p_{k+1}^{v_{k+1}}, p_{k+2}^{a-1-\delta}, p_{k+3}^{b'_{-1}}, \dots, p_n^{b'_{-1}}, p_{n+1}^{c'_{-1}})$$
$$\times \partial(q_{k+2}^j)\, g_n(x^{n+1}\setminus x_l)\,\partial(q^{\tau+\delta-j}, q_i)\, V(q_l, q, q_i) \equiv 0,$$
$$x^{n+1} \in M_{n+1},\ q \in \mathbb{R},\ q \neq q_i,\ 1 \leqslant i \leqslant n+1,\ 0 \leqslant j < \gamma.$$

When $q \to q_{k+2}$, the left-hand side of this equality tends to $-S_j(x^{n+1})$ (see (3.25)). It follows that the sum of the middle terms equals zero.

By (3.10), (3.17), all the higher terms containing $\partial g_{n+1}/\partial p_i$ vanish. For the same reason so do the higher terms of the form $D[p_i \partial g_{n+1}(x^{n+1})/\partial q_i]$ for every $i \geqslant k+3$ and for each $i \leqslant k+1$ such that $v_i = 0$. If, otherwise, $i \leqslant k+2$ and $v_i \geqslant 1$, then

$$D\left[p_i \frac{\partial g_{n+1}(x^{n+1})}{\partial q_i}\right] = p_i \partial(q_i, q_{k+2}^{\tau+\delta}) D' g_{n+1}(x^{n+1}) + \partial(p_i) D_i g_{n+1}(x^{n+1}),$$

where D' and D_i are defined by $\partial(q_{k+2}^{\tau+\delta})D' = \partial(p_i)D_i = D$. Clearly, $D' \in \mathscr{D}(\mathrm{m}_{10})$, $D_i \in \mathscr{D}(\mathrm{m}^i)$, where $\mathrm{m}_{10} = (n+1, k+1, a-1, b'_{-1}, c'_{-1})$ and

$$\mathrm{m}^i = (n+1, k+1, a-2, [\tau+\delta \mid a-1-\delta], b'_{-1}, c'_{-1}), \qquad i \neq k+2.$$

By (3.10), (3.17), we have $\mathrm{m}_{10} \in \mathscr{A}$, and, by the inductive assumption, $\mathrm{m}^i \in \mathscr{A}$. So the remaining higher term is only

$$D\left[p_{k+2} \frac{\partial g_{n+1}(x^{n+1})}{\partial q_{k+2}}\right] = \partial(q_{k+2}) D_{k+2} g_{n+1}(x^{n+1})$$
$$= \partial_{n+1}(v_1, \dots, v_{k+1}, [\tau+\delta+1 \mid a-2-\delta], b'_{-1}, \dots, b'_{-1}, c'_{-1}) g_n(x^{n+1}).$$

By what was shown above, the last expression is the result of applying D to the left-hand side of (2.2; n). Hence it vanishes, and so we have passed from δ to $\delta+1$ in (3.16). This completes the proof in case (a).

3°. Case (b) is simpler than case (a) in the following respects: in the first part of the proof (see 1°) the lower terms vanish immediately, while in the second part (see 2°) so do both the lower and middle terms. Due to the last fact, condition (3.12) in case (b) was assumed to hold only for $a' = a-1$.

3.3. LEMMA (purification). *Let* n, k, a, b_{-1}, c_{-1}, b, c, b_1, c_1, b'_{-1}, c'_{-1}, b', c', b'_1, c'_1, β_{-1}, β, β_1 *be nonnegative numbers such that* $n \geqslant 2$, $k \in [0, n-2]$, $a \geqslant 1$. *Then*

(a) *if* $n \geqslant 3$, $k \geqslant 2$, *and the following inclusions hold*:

$$(n+1, k+1, a-1, b_{-1}, c_{-1}) \in \mathscr{A} \quad \text{when } k \leqslant n-3$$
$$((n+1, k+1, a-1, c_{-1}) \in \mathscr{A} \quad \text{when } k = n-2), \tag{3.27}$$
$$(n+1, k+1, a-2, [\beta_{-1} \mid 0], b'_{-1}, c'_{-1}) \in \mathscr{A} \quad \text{when } a \geqslant 2,\ k \leqslant n-3$$
$$((n+1, k+1, a-2, [\beta_{-1} \mid 0], c'_{-1}) \in \mathscr{A} \quad \text{when } a \geqslant 2,\ k = n-2), \tag{3.28}$$
$$(n, k, a, b, c) \in \mathscr{A} \quad \text{when } k \leqslant n-3$$
$$((n, k, a, c) \in \mathscr{A} \quad \text{when } k = n-2), \tag{3.29}$$
$$(n, k, a-1, [\beta \mid 0], b', c') \in \mathscr{A} \quad \text{when } a \geqslant 2,\ k \leqslant n-3$$
$$((n, k, a-1, [\beta \mid 0], c') \in \mathscr{A} \quad \text{when } k = n-2), \tag{3.30}$$
$$(n-1, k-1, a, b_1, c_1) \in \mathscr{A} \quad \text{when } k \leqslant n-3$$
$$((n-1, k-1, a, c_1) \in \mathscr{A} \quad \text{when } k = n-2), \tag{3.31}$$
$$(n-1, k-1, a-1, [\beta_1 \mid 0], b'_1, c'_1) \in \mathscr{A} \quad \text{when } k \leqslant n-3$$
$$((n-1, k-1, a-1, [\beta_1 \mid 0], c'_1) \in \mathscr{A} \quad \text{when } k \leqslant n-2), \tag{3.32}$$

then

$$(n, k, a-1, \tilde{b}, \tilde{c}+1) \in \mathscr{A} \quad \text{when } k \leqslant n-3$$
$$((n, k, a-1, \tilde{c}+1) \in \mathscr{A} \quad \text{when } k = n-2), \tag{3.33}$$
$$(n-1, k, a-1, \tilde{b}, \tilde{c})\tilde{\beta} \in \mathscr{A} \quad \text{when } k \leqslant n-3$$
$$((n-1, k, a-1, \tilde{c})\tilde{\beta} \in \mathscr{A} \quad \text{when } k = n-2), \tag{3.34}$$

where

$$\tilde{b} = \max\{b_{-1}+1, b'_{-1}, b, b', b_1, b'_1\},$$
$$\tilde{c} = \max\{c_{-1}+1, c'_{-1}, c, c', c_1, c'_1\}, \tag{3.35}$$
$$\tilde{\beta} = \max\{\beta_{-1}, \beta, \beta_1\}, \tag{3.36}$$

when $a \geqslant 2$, *and where*

$$\tilde{b} = \max\{b_{-1}+1, b, b', b_1, b'_1\},$$
$$\tilde{c} = \max\{c_{-1}+1, c, c', c_1, c'_1\}, \tag{3.37}$$
$$\tilde{\beta} = \max\{\beta, \beta_1\}, \tag{3.38}$$

when $a = 1$;

(b) *if* $n \geqslant 3$, $k = 1$, *and either* $a \geqslant 2$ *and inclusions* (3.27)–(3.31) *hold, or* $a = 1$ *and inclusions* (3.27), (3.29)–(3.32) *hold, then* (3.33), (3.34) *are satisfied with*

$$\tilde{b} = \max\{b_{-1}+1, b'_{-1}, b, b', b_1\}, \quad \tilde{c} = \max\{c_{-1}+1, c'_{-1}, c, c', c_1\},$$
$$\tilde{\beta} = \max\{\beta_{-1}, \beta\},$$

when $a \geqslant 2$, *and with*

$$\widetilde{b} = \max\{b_{-1}+1, b, b', b_1, b_1'\}, \quad \widetilde{c} = \max\{c_{-1}+1, c, c', c_1, c_1'\},$$
$$\widetilde{\beta} = \max\{\beta, \beta_1\},$$

when $a = 1$;

(c) *if* $n \geqslant 2$, $k = 0$, *and either* $a \geqslant 2$ *and inclusions* (3.27)–(3.30) *hold, or* $a = 1$ *and inclusions* (3.27), (3.29), (3.30) *hold, then* (3.33) *is satisfied with*

$$\widetilde{b} = \max\{b_{-1}+1, b_{-1}', b, b'\}, \quad \widetilde{c} = \max\{c_{-1}+1, c_{-1}', c, c'\},$$

when $a \geqslant 2$ *and with*

$$\widetilde{b} = \max\{b_{-1}+1, b, b'\}, \quad \widetilde{c} = \max\{c_{-1}+1, c, c'\},$$

when $a = 1$.

PROOF. 1°. As above, only case (a) with $k \leqslant n-3$ will be considered in detail. To (2.2; $n+1$) we apply the operator

$$D = \partial_{n+1}(v_1, \dots, v_{k+1}, [\widetilde{\beta} \mid 0], \widetilde{b}, \dots, \widetilde{b}, \widetilde{c}),$$

where $v_1, \dots, v_{k+1}$ are arbitrary nonnegative integers with $v_1 + \dots + v_{k+1} = a-1$. Then by (3.27), (3.35) the higher terms containing $\partial g_{n+1}/\partial p_i$ vanish and so do the terms of the form

$$D\left[p_i \frac{\partial g_{n+1}(x^{n+1})}{\partial q_i}\right],$$

where either $i \geqslant k+2$, or $i \leqslant k+1$, $v_i = 0$. If $a = 1$, there are no other higher terms. If $a \geqslant 2$ and $i \leqslant k+1$, $v_i \geqslant 1$, then

$$D\left[p_i \frac{\partial g_{n+1}(x^{n+1})}{\partial q_i}\right] = p_i \partial(q_i) D g_{n+1}(x^{n+1}) + \partial(q_i) D_i g_{n+1}(x^{n+1})$$
$$= \partial(q_i) D_i g_{n+1}(x^{n+1}),$$

where D_i is defined by $\partial(p_i) D_i = D$. Clearly, $D_i \in \mathscr{D}(\mathfrak{m}_i)$, where $\mathfrak{m}_i = (n+1, k+1, a-2, [\widetilde{\beta} \mid 0], \widetilde{b}, \widetilde{c})$. Since $\widetilde{\beta} \geqslant \beta_{-1}$, $\widetilde{b} \geqslant b_{-1}'$, $\widetilde{c} \geqslant c_{-1}'$, (3.28) implies $\mathfrak{m}_i \in \mathscr{A}$. Thus, in any case, all the higher terms vanish.

Now let us pass to the middle terms. Since $\widetilde{b}, \widetilde{c} \geqslant 1$, the middle terms containing $\partial g_n(x^{n+1} \setminus x_l)/\partial p_i$ with $l \geqslant k+3$ vanish. The sum of other middle terms (denoted by S_1) is equal to

$$\begin{aligned}S_1(x^{n+1}) &= \sum_{l=1}^{k+2} \sum_{i\in[1,n+1]\setminus\{l\}} D\Bigg[\partial(p_i)\, g_n(x^{n+1}\setminus x_l) \\ &\qquad \times \partial(q_i)\Bigg(U(q_l, q_i) + \sum_{j\in[1,n+1]\setminus\{l,i\}} V(q_l, q_i, q_j)\Bigg)\Bigg] \\ &= -\sum_{l=1}^{k+2} \sum_{i\in[1,n+1]\setminus\{l\}} D_i'\partial(q_{k+2}^{\widetilde{\beta}})\Bigg[g_n(x^{n+1}\setminus x_l)\Bigg(\partial(q_i)\, U(q_l, q_i) \\ &\qquad + \partial(q_i) \sum_{j\in[1,n+1]\setminus\{l,i\}} V(q_l, q_i, q_j)\Bigg)\Bigg], \end{aligned} \tag{3.39}$$

where D_i' is defined by $D_i'\partial(q_{k+2}^{\widetilde{\beta}}) = D\partial(p_i)$. Take any $l \in [1, k+2]$ and $i \neq l$. We have

$$\begin{aligned}&D_i'\partial(q_{k+2}^{\widetilde{\beta}})[g_n(x^{n+1}\setminus x_l)\partial(q_i)\, U(q_l, q_i)] \\ &\qquad = \partial(q_{k+2}^{\widetilde{\beta}})[\partial(q_i)\, U(q_l, q_i) D_i' g_n(x^{n+1}\setminus x_l)].\end{aligned} \tag{3.40}$$

If $v_l \neq 0$, then $D_i' g_n(x^{n+1}\setminus x_l) \equiv 0$. If $v_l = 0$, $i \leqslant k+2$, then the operator D_i' as acting on $g_n(x^{n+1}\setminus x_l)$ belongs to $\mathscr{D}(\mathfrak{m}_i')$, where $\mathfrak{m}_i' = (n, k, a, \widetilde{b}, \widetilde{c})$ (see Remark 2.5) By (3.29), (3.35), and (3.37), again $D_i' g_{n+1}(x^{n+1}\setminus x_l) \equiv 0$.

Now suppose in addition to the above that $l \neq k+2$, i.e., $l \leqslant k+1$. If, besides, $i \geqslant k+3$, then

$$\begin{aligned}&\partial(q_{k+2}^{\widetilde{\beta}})[\partial(q_i)\, U(q_l, q_i) D_i' g_n(x^{n+1}\setminus x_l)] \\ &\qquad = \partial(q_i)\, U(q_l, q_i)\partial(p_i) D g_n(x^{n+1}\setminus x_l).\end{aligned} \tag{3.41}$$

When $v_l = 0$, the operator D as acting on $g_n(x^{n+1}\setminus x_l)$ belongs to $\mathscr{D}(\mathfrak{m})$, where $\mathfrak{m} = (n, k, a-1, [\widetilde{\beta} \mid 0], \widetilde{b}, \widetilde{c})$. From (3.30), (3.35)–(3.38) it follows that $\mathfrak{m} \in \mathscr{A}$ and hence $D g_n(x^{n+1}\setminus x_l) \equiv 0$. Similarly, if $l \leqslant k+1$, $i \geqslant k+3$, $j \neq k+2$, then

$$\begin{aligned}&D_i'\partial(q_{k+2}^{\widetilde{\beta}})[g_n(x^{n+1}\setminus x_l)\partial(q_i)\, V(q_l, q_i, q_j)] \\ &\qquad = \partial(q_i)\, V(q_l, q_i, q_j)\partial(p_i) D g_n(x^{n+1}\setminus x_l) = 0.\end{aligned} \tag{3.42}$$

Due to (3.39)–(3.42) and the above considerations, we have

$$S_1(x^{n+1}) = -\sum_{i=k+3}^{n+1} D_i' g_n(x^{n+1} \setminus x_{k+2}) \partial (q_{k+2}^{\widetilde{\beta}}, q_i) \times \Big[U(q_{k+2}, q_i) + \sum_{j\in[1,n+1]\setminus\{k+2,i\}} V(q_{k+2}, q_i, q_j) \Big] - \sum_{l=1}^{k+1} \sum_{i=k+3}^{n+1} \partial (q_{k+2}^{\widetilde{\beta}}) [D_i' g_n(x^{n+1} \setminus x_l) \partial (q_i) V(q_l, q_i, q_{k+2})]. \tag{3.43}$$

Finally, consider the lower terms and denote their sum by S_2,

$$\begin{aligned} S_2(x^{n+1}) &= \sum_{\substack{l,l'\in[1,k+2]\\ l\neq l'}} \sum_{i\in[1,n+1]\setminus\{l,l'\}} \partial (p_i) D[g_{n-1}(x^{n+1} \setminus (x_l, x_{l'})) \times \partial (q_i) V(q_l, q_{l'}, q_i)] \\ &= -\sum_{\substack{l,l'\in[1,k+2]\\ l\neq l'}} \sum_{i\in[1,n+1]\setminus\{l,l'\}} D_i' \partial (q_{k+2}^{\widetilde{\beta}}) [g_{n-1}(x^{n+1} \setminus (x_l, x_{l'})) \times \partial (q_i) V(q_l, q_{l'}, q_i)] \\ &= -\sum_{\substack{l,l'\in[1,k+2]\\ l\neq l'}} \sum_{i\in[1,n+1]\setminus\{l,l'\}} \partial (q_{k+2}^{\widetilde{\beta}}) [\partial (q_i) V(q_l, q_{l'}, q_i) \times D_i' g_{n-1}(x^{n+1} \setminus (x_l, x_{l'}))]. \end{aligned} \tag{3.44}$$

Certainly, $D_i' g_{n-1}(x^{n+1} \setminus (x_l, x_{l'})) = 0$, provided that either $v_l \neq 0$ or $v_{l'} \neq 0$.

If $v_l = v_{l'} = 0$, and if $i \leqslant k+2$, then the operator D_i' as acting on $g_{n-1}(x^{n+1} \setminus (x_l, x_{l'}))$ belongs to $\mathscr{D}(\mathfrak{m}_i'')$, where $\mathfrak{m}_i'' = (n-1, k-1, a, \widetilde{b}, \widetilde{c})$. By (3.31), (3.35), (3.37) we have $\mathfrak{m}_i'' \in \mathscr{A}$. If $l, l' \leqslant k+1$, $v_l = v_{l'} = 0$ (this would be impossible for $k=1$, $a \geqslant 2$) and $i \geqslant k+3$, then

$$\partial (q_{k+2}^{\widetilde{\beta}}) [\partial (q_i) V(q_l, q_{l'}, q_i) D_i' g_{n-1}(x^{n+1} \setminus (x_l, x_{l'}))] = \partial (q_i) V(q_l, q_{l'}, q_i) \partial (p_i) g_{n-1}(x^{n+1} \setminus (x_l, x_{l'})),$$

where the operator D as acting on $g_{n-1}(x^{n+1} \setminus (x_l, x_{l'}))$ belongs to $\mathscr{D}(\mathfrak{m}_1)$, where $\mathfrak{m}_1 = (n-1, k-1, a-1, [\widetilde{\beta} \mid 0], \widetilde{b}, \widetilde{c})$. By (3.32) and (3.35)–(3.38) we have $\mathfrak{m}_1 \in \mathscr{A}$. Thus

$$S_2(x^{n+1}) = -\sum_{l=1}^{k+1} \sum_{i=k+3}^{n+1} D_i' g_{n-1}(x^{n+1} \setminus (x_l, x_{k+2})) \partial (q_{k+2}^{\widetilde{\beta}}, q_i) V(q_l, q_{k+2}, q_i). \tag{3.45}$$

If we fix here an arbitrary $i = i_0 \in [k+3, n+1]$ as well as all x_j with $j \in [1, n+1] \setminus \{k+2\}$ and let q_{k+2} tend to q_{i_0}, then by the smoothness of V (see §1) it follows that $S_2(x^{n+1})$ has a finite limit.

Now we want to apply the same operation to $S_1(x^{n+1})$ (see (3.43)). To do this, let us first observe that by (3.30), (3.35)–(3.38) for any $l \in [1, k+1]$,

$i \in [k+3, n+1]$, we have

$$\partial(q_{k+2}^{\tilde{\beta}})D_i'g_n(x^{n+1} \setminus x_l) = 0$$

(see above). Hence the function $D_i'g_n(x^{n+1} \setminus x_l)$ is a polynomial in q_{n+2} on every connected component of the set

$$\{q \in \mathbb{R} : q \neq q_j,\ 1 \leqslant j \leqslant n+1,\ j \neq l,\ k+2\}.$$

It follows that for any integer $r \geqslant 0$ the expression $\partial(q_{k+2}^r)D_i'g_n(x^{n+1} \setminus x_l)$ has a finite limit as $q_{k+2} \to q_{i_0}$. On the other hand, by (1.8),

$$\overline{\lim_{q_{k+2}\to q_{i_0}}} |\partial(q_{k+2}^{\tilde{\beta}}, q_{i_0})U(q_{k+1}, q_{i_0})| = \infty.$$

Thus, as $q_{k+2} \to q_{i_0}$, the right-hand side of (3.45) and all the terms except

$$D_{i_0}'g_n(x^{n+1} \setminus x_{k+2})\partial(q_{k+2}^{\tilde{\beta}}, q_{i_0})U(q_{k+2}, q_{i_0}) \tag{3.46}$$

on the right-hand side of (3.43) have finite limits, while (3.46) is unbounded whenever $D_{i_0}'g_n(x^{n+1} \setminus x_{k+2}) \neq 0$. Since (2.2; $n+1$) implies that $S_1(x^{n+1}) + S_2(x^{n+2}) \equiv 0$ (and since the higher terms were seen to vanish), we obtain

$$D_i'g_n(x^{n+1} \setminus x_{k+2}) \equiv 0, \qquad k+3 \leqslant i \leqslant n+1,\ x^{n+1} \in M_{n+1}. \tag{3.47}$$

Every point $x^n \in M_n$ can be represented as $x^{n+1} \setminus x_{k+2}$ with some point $x^{n+1} \in M_n$. So (3.47) implies that

$$D_i'g_n(x^n) = 0, \qquad k+3 \leqslant i \leqslant n+1,\ x^n \in M_n. \tag{3.48}$$

For $i = n+1$ this can be rewritten in the form (see the definition of D_i')

$$\partial_n(v_1, \dots, v_{k+1}, \tilde{b}, \dots, \tilde{b}, \tilde{c}+1)g_n(x^n) = 0, \qquad x^n \in M_n,$$

which leads to (3.33) due to the arbitrariness of $v_1, \dots, v_{k+1}$ with $v_1 + \dots + v_{k+1} = a-1$. From (3.48), (3.43) it also follows that $S_1(x^{n+1}) \equiv 0$ and hence $S_2(x^{n+1}) \equiv 0$, i.e.,

$$\sum_{l=1}^{k+1}\sum_{i=k+3}^{n+1} \partial(p_i)\partial(p_1^{v_1}, \dots, p_{k+1}^{v_{k+1}}, p_{k+3}^{\tilde{b}}, \dots, p_n^{\tilde{b}}, p_{n+1}^{\tilde{c}}) \times g_{n-1}(x^{n+1} \setminus (x_l, x_{k+2}))\partial(q_{k+2}^{\tilde{\beta}}, q_i)V(q_l, q_{k+2}, q_i) = 0,$$

which is equivalent to (3.34).

2°. In cases (b) and (c) the proof differs from that for case (a) only for the lower terms. Namely, if $a \geqslant 2$ in case (b), then the lower terms containing $g_{n-1}(x^{n+1} \setminus (x_l, x_{l'}))$ with $l < l' \leqslant k+1$ vanish not because of (3.32), but because in this case $l = 1$, $l' = 2$. In case (c) all the lower terms vanish immediately and so there appears no relation similar to $S_2 \equiv 0$ leading to (3.34).

3.4. LEMMA (deepening). *If for some integers* $n \geqslant 2$ *and* $b, c \geqslant 1$ *we have*

$$(n, 0, 0, b, c) \in \mathscr{A} \quad \text{when } n \geqslant 3 \quad ((n, 0, 0, c) \in \mathscr{A} \quad \text{when } n = 2), \tag{3.49}$$

then

$$\begin{aligned} (n-1, 0, b+1, b+1, c+2) &\in \mathscr{A} \quad \text{when } n \geqslant 4 \\ ((n-1, 0, b+1, c+2) &\in \mathscr{A} \quad \text{when } n = 3 \\ \text{and } (n-1, 0, c+2) &\in \mathscr{A} \quad \text{when } n = 2). \end{aligned} \tag{3.50}$$

PROOF. To be definite, consider the case $n \geqslant 4$ and apply the operator

$$D = \partial_n(0, b+1, \dots, b+1, c+1) \tag{3.51}$$

to (2.2; n). Then by (3.49) all the higher terms vanish. Clearly, all the lower terms vanish as well, and so do the middle terms containing $g_{n-1}(x^n \setminus x_l)$ with $l \geqslant 1$. We thus obtain

$$\sum_{i=2}^{n} \partial(p_i) D g_{n-1}(x^n \setminus x_1) \partial(p_i) \left[U(q_1, q_i) + \sum_{j \in [2, n] \setminus \{i\}} V(q_1, q_i, q_j) \right] = 0. \tag{3.52}$$

From the properties of U and V (see §1) it follows that as $q_1 \to q_n$ all the terms on the left-hand side of (3.52) are bounded, except the term

$$\partial(p_n) D g_{n-1}(x^n \setminus x_1) \partial(p_n) U(q_1, q_n),$$

which is bounded only when

$$\partial(p_n) D g_{n-1}(x^n \setminus x_1) = 0.$$

This relation combined with (3.51), (3.52) yields (3.50).

3.5. **Remark.** Since (3.49) coincides with (3.29) provided that $k = a = 0$, while the other conditions of Lemma 3.3 make no sense in this case, one may call deepening a particular case of purification, the more so since both operations are used to obtain information about the middle terms of equation (2.2).

Below we shall also need the following lemma about the lower terms of (2.2).

3.6. LEMMA. *Let* n, a, b_{-1}, c_{-1}, b, c, b_1, *and* c_1 *be nonnegative integers such that* $n \geqslant 3$, $a \geqslant 1$ *and the following conditions hold*:

$$\begin{aligned} (n+1, 2, a-1, b_{-1}, c_{-1}) &\in \mathscr{A} \quad \text{when } n \geqslant 4 \\ ((n+1, 2, a-1, c_{-1}) &\in \mathscr{A} \quad \text{when } n = 3), \end{aligned} \tag{3.53}$$

$$\begin{aligned} (n, 1, a, b, c) &\in \mathscr{A} \quad \text{when } n \geqslant 4 \\ ((n, 1, a, c) &\in \mathscr{A} \quad \text{when } n = 3), \end{aligned} \tag{3.54}$$

$$\begin{aligned} (n-1, 0, a+1, b_1, c_1) &\in \mathscr{A} \quad \text{when } n \geqslant 4 \\ ((n-1, 0, a+1, c_1) &\in \mathscr{A} \quad \text{when } n = 3). \end{aligned} \tag{3.55}$$

Then

$$(n-1,1,a,\widetilde{b},\widetilde{c})0\in\mathscr{A}\quad \text{when } n\geqslant 4$$
$$((n-1,1,a,\widetilde{c})0\in\mathscr{A}\quad \text{when } n=3), \tag{3.56}$$

where

$$\widetilde{b}=\max\{b_{-1}+1,b,b_1\},\qquad \widetilde{c}=\max\{c_{-1}+1,c,c_1\}. \tag{3.57}$$

PROOF. Assuming that $n\geqslant 4$, apply to $(2.2; n+1)$ the operator

$$D=\partial_{n+1}(v_1,v_2,0,\widetilde{b},\dots,\widetilde{b},\widetilde{c}),$$

where v_1, v_2 are any nonnegative integers with $v_1+v_2=a$. By (3.53), (3.54), (3.57), the higher and middle terms vanish, and by (3.55), (3.57) the sum of the lower terms is

$$\sum_{l=1}^{2}\sum_{i=4}^{n+1}\partial(p_i)\partial(p_1^{v_1},p_2^{v_2},p_4^{\widetilde{b}},\dots,p_n^{\widetilde{b}},p_{n+1}^{\widetilde{c}})$$
$$\times g_{n-1}(x^{n+1}\setminus(x_l,x_3))\partial(q_i)V(q_l,q_3,q_i).$$

Setting this sum equal to zero, we get (3.56).

If $n=3$ one can develop the same argument for $D=\partial_4(v_1,v_2,0,\widetilde{c})$.

In the remaining part of the proof of Theorem 1.10 equation (2.2) is no longer used as such; instead, all the arguments are based on Lemmas 3.1–3.4 and 3.6.

§4. Proof of Theorem 1.10: consequences of the basic lemmas

Most of the assertions of this section will be called lemmas as above. Any auxiliary assertion independent of the assumptions of Theorem 1.10 will be called a proposition.

4.1. LEMMA (absolute lowering). *Let* n, k, a, b_{-1}, c_{-1}, b, c, b_1, c_1, b_2, c_2, b_3, c_3, $\widehat{b}$, $\widehat{c}$, *and* θ *be nonnegative integers such that*

1) $n\in[2,n_0]$, $k\in[0,n-2]$, $a\geqslant 1$ *and*

2) $c_{-1}\geqslant m_0+1$, *if* $n=n_0$.

Then

(a) *if* $n\geqslant 5$, $k\geqslant 3$, *and the following conditions* (4.1)–(4.5) *hold*:

$$(n+1,k+1,a-1,b_{-1},c_{-1})\in\mathscr{A}\quad \text{when } k\leqslant n-3$$
$$((n+1,k+1,a-1,c_{-1})\in\mathscr{A}\quad \text{when } k=n-2), \tag{4.1}$$
$$(n,k,a,b,c)\in\mathscr{A}\quad \text{when } k\leqslant n-3$$
$$((n,k,a,c)\in\mathscr{A}\quad \text{when } k=n-2), \tag{4.2}$$
$$(n-1,k-1,a,b_1,c_1)\in\mathscr{A}\quad \text{when } k\leqslant n-3$$
$$((n-1,k-1,a,c_1)\in\mathscr{A}\quad \text{when } k=n-2), \tag{4.3}$$
$$(n-2,k-2,a,b_2,c_2)\in\mathscr{A}\quad \text{when } k\leqslant n-3$$
$$((n-2,k-2,a,c_2)\in\mathscr{A}\quad \text{when } k=n-2), \tag{4.4}$$

$$(n-3, k-3, a, b_3, c_3) \in \mathscr{A} \quad \textit{when } k \leqslant n-3$$
$$((n-3, k-3, a, c_3) \in \mathscr{A} \quad \textit{when } k = n-2), \tag{4.5}$$

and if in case $n \neq n_0$, $a \geqslant 2$ *the additional condition*

$$(n, k+1, a-1, \widehat{b}, \widehat{c})\theta \in \mathscr{A} \quad \textit{when } k \leqslant n-3$$
$$((n, k+1, a-1, \widehat{c})\theta \in \mathscr{A} \quad \textit{when } k = n-2) \tag{4.6}$$

also holds, then

$$(n, k, a-1, \widetilde{b}, \widetilde{c}+1) \in \mathscr{A} \quad \textit{when } k \leqslant n-3$$
$$((n, k, a-1, \widetilde{c}+1) \in \mathscr{A} \quad \textit{when } k = n-2), \tag{4.7}$$
$$(n-1, k, a-1, \widetilde{b}, \widetilde{c})\widetilde{\beta} \in \mathscr{A} \quad \textit{when } k \leqslant n-3$$
$$((n-1, k, a-1, \widetilde{c})\widetilde{\beta} \in \mathscr{A} \quad \textit{when } k = n-2), \tag{4.8}$$

where

$$\begin{gathered} \widetilde{b} = \max\{b_{-1}+1, b+1, b_1+1, b_2, b_3, \widehat{b}\}, \\ \widetilde{c} = \max\{c_{-1}+1, c+1, c_1+1, c_2, c_3, \widehat{c}\}, \quad \widetilde{\beta} = 2a+\theta-1, \end{gathered} \tag{4.9}$$

provided that $n \neq n_0$, $a \geqslant 2$, *and where*

$$\begin{gathered} \widetilde{b} = \max\{b_{-1}+1, b+1, b_1+1, b_2, b_3\}, \\ \widetilde{c} = \max\{c_{-1}+1, c+1, c_1+1, c_2, c_3\}, \quad \widetilde{\beta} = 2a-1, \end{gathered} \tag{4.10}$$

provided that either $a = 1$, *or* $n = n_0$, $a \geqslant 2$;

(b) *if* $n \geqslant 4$, $k = 2$, *and conditions* (4.1)–(4.4) *hold and if, in the case* $n \neq n_0$, $a \geqslant 2$, *condition* (4.6) *also holds, then* (4.7), (4.8) *are satisfied with*

$$\begin{gathered} \widetilde{b} = \max\{b_{-1}+1, b+1, b_1+1, b_2, \widehat{b}\}, \\ \widetilde{c} = \max\{c_{-1}+1, c+1, c_1+1, c_2, \widehat{c}\}, \quad \widetilde{\beta} = 2a+\theta-1, \end{gathered} \tag{4.11}$$

when $n \neq n_0$, $a \geqslant 2$, *and with*

$$\begin{gathered} \widetilde{b} = \max\{b_{-1}+1, b+1, b_1+1, b_2\}, \\ \widetilde{c} = \max\{c_{-1}+1, c+1, c_1+1, c_2\}, \quad \widetilde{\beta} = 2a-1, \end{gathered} \tag{4.12}$$

when either $a = 1$ *or* $n = n_0$, $a \geqslant 2$;

(c) *if* $n \geqslant 3$, $k = 1$, *and conditions* (4.1)–(4.3) *hold and if, in the case* $n \neq n_0$, $a \geqslant 2$, *condition* (4.6) *also holds, then* (4.7), (4.8) *are satisfied with*

$$\begin{gathered} \widetilde{b} = \max\{b_{-1}+1, b+1, b_1+1, \widehat{b}\}, \\ \widetilde{c} = \max\{c_{-1}+1, c+1, c_1+1, \widehat{c}\}, \quad \widetilde{\beta} = 2a+\theta-1, \end{gathered} \tag{4.13}$$

when $n \neq n_0$, $a \geqslant 2$, *and with*

$$\begin{gathered} \widetilde{b} = \max\{b_{-1}+1, b+1, b_1+1\}, \\ \widetilde{c} = \max\{c_{-1}+1, c+1, c_1+1\}, \quad \widetilde{\beta} = 2a-1, \end{gathered} \tag{4.14}$$

when either $a = 1$, *or* $n = n_0$, $a \geqslant 2$;

(d) *if* $n \geqslant 2$, $k = 0$, *and conditions* (4.1), (4.2) *hold and if in the case* $n \neq n_0$, $a \geqslant 2$, *condition* (4.6) *also holds, then* (4.7) *is satisfied with*

$$\widetilde{b} = \max\{b_{-1} + 1, b + 1, \widehat{b}\}, \quad \widetilde{c} = \max\{c_{-1} + 1, c + 1, \widehat{c}\}, \tag{4.15}$$

when $n \neq n_0$, $a \geqslant 2$, *and with*

$$\widetilde{b} = \max\{b_{-1} + 1, b + 1\}, \quad \widetilde{c} = \max\{c_{-1} + 1, c + 1\}, \tag{4.16}$$

when either $a = 1$, *or* $n = n_0$, $a \geqslant 2$.

PROOF. 1°. We start with statement (a), assuming to be definite that $k \leqslant n - 3$. By Lemma 3.1(a), inclusions (4.2)–(4.4) imply that for any $\delta \in [1, a]$ we have

$$(n, k, a-1, [\delta \mid a-\delta], b', c') \in \mathscr{A}, \tag{4.17}$$

where

$$b' = \max\{b + 1, b_1, b_2\}, \quad c' = \max\{c + 1, c_1, c_2\}. \tag{4.18}$$

Since the mark $(n, k, a-1, [\delta \mid a-\delta], b', c')$ is dominated by $(n, k, a-1, [a \mid a-\delta], b', c')$, where $\delta \in [1, a]$, (4.12) in its turn implies that

$$(n, k, a-1, [a \mid a-\delta], b', c') \in \mathscr{A}, \qquad \delta \in [1, a]. \tag{4.19}$$

By the same Lemma 3.1(a), inclusions (4.3)–(4.5) imply that for any $\delta \in [1, a]$ we have

$$(n-1, k-1, a-1, [\delta \mid a-\delta], b_1, c_1) \in \mathscr{A},$$

where

$$b_1' = \max\{b_1 + 1, b_2, b_3\}, \quad c_1' = \max\{c_1 + 1, c_2, c_3\}. \tag{4.20}$$

From this, as above, we get

$$(n-1, k-1, a-1, [a \mid a-\delta], b_1', c_1') \in \mathscr{A}, \qquad \delta \in [1, a]. \tag{4.21}$$

If $a \geqslant 2$, $n \neq n_0$, apply Lemma 3.2(a), which says that (4.1)–(4.3), (4.6), (4.19), and (4.21) together imply that

$$(n+1, k+1, a-2, [\tau+\delta \mid a-1-\delta], b_{-1}', c_{-1}') \in \mathscr{A}, \qquad \delta \in [1, a-1], \tag{4.22}$$

where

$$\begin{gathered} b_{-1}' = \max\{b_{-1} + 1, b, b', b_1, b_1', \widehat{b}\}, \\ c_{-1}' = \max\{c_{-1} + 1, c, c', c_1, c_1', \widehat{c}\}, \quad \tau = a + \theta. \end{gathered} \tag{4.23}$$

If $a \geqslant 2$, $n = n_0$, then (4.22) is clearly true as well (see (2.10)) and in this case

$$b_{-1}' = \tau = 0, \qquad c_{-1}' = c_{-1}. \tag{4.24}$$

Taking $\delta = a$ in (4.19), (4.21) and $\delta = a-1$ in (4.22), we get

$$(n, k, a-1, [\beta \mid 0], b', c') \in \mathscr{A}, \tag{4.25}$$

$$(n-1, k-1, a-1, [\beta_1 \mid 0], b_1', c_1') \in \mathscr{A}, \tag{4.26}$$

$$(n+1, k+1, a-2, [\beta_{-1} \mid 0], b_{-1}', c_{-1}') \in \mathscr{A}, \tag{4.27}$$

where

$$\beta = \beta_1 = a, \qquad \beta_{-1} = \tau + a - 1. \tag{4.28}$$

Finally, by Lemma 3.3(a), from (4.1)–(4.3), (4.25), (4.26) combined in the case $a \geqslant 2$ with (4.27), it follows that

$$(n, k, a-1, \tilde{b}, \tilde{c}+1) \in \mathscr{A}, \qquad (n-1, k, a-1, \tilde{b}, \tilde{c})\tilde{\beta} \in \mathscr{A},$$

i.e., (4.7), (4.8) hold with

$$\tilde{b} = \max\{b_{-1}+1, b_{-1}', b, b', b_1, b_1'\},$$

$$\tilde{c} = \max\{c_{-1}+1, c_{-1}', c, c', c_1, c_1'\}, \quad \tilde{\beta} = \max\{\beta_{-1}, \beta, \beta_1\}$$

and with

$$\tilde{b} = \max\{b_{-1}+1, b, b', b_1, b_1'\},$$

$$\tilde{c} = \max\{c_{-1}+1, c, c', c_1, c_1'\}, \quad \tilde{\beta} = \max\{\beta, \beta_1\},$$

provided that $a \geqslant 2$ and $a = 1$ respectively. Using (4.28), (4.18), (4.20), (4.23), (4.24), we come to (4.9), when $n \neq n_0$, $a \geqslant 2$, and to (4.10), when either $a = 1$ or $a \geqslant 2$, $n = n_0$.

2°. Assertions (b), (c), and (d) can be treated in the same way. We only remark that one should apply Lemmas 3.1(a,b), 3.2(a), and 3.3(a) in case (b), Lemmas 3.1(b,c), 3.2(a), and 3.3(a) in case (c), and Lemmas 3.1(c), 3.2(b), and 3.3(c) in case (c).

4.2. Lemma. *Assume that* $n_0 \geqslant 7$ *and for some integers* $\kappa \in [2, n_0-2]$, $\alpha \geqslant 1$ *and* $\gamma \geqslant m_0+1$ *the following conditions hold*:

(i) *for any integer* $l \in [0, \kappa]$

$$(n_0-l, \kappa-l, \alpha+l, \alpha+l, \gamma+2l) \in \mathscr{A} \quad \textit{when } \kappa \leqslant n_0-3$$
$$((n_0-l, \kappa-l, \alpha+l, \gamma+2l) \in \mathscr{A} \quad \textit{when } \kappa = n_0-2), \tag{4.29}$$

(ii) *for any integer* $l \in [0, \kappa-1]$ *there exists an integer* $\theta_\kappa(l) \geqslant 0$ *such that*

$$(n_0-l-1, \kappa-l, \alpha+l, \alpha+l+1, \gamma+2l+2)\theta_\kappa(l) \in \mathscr{A}$$
$$\textit{when } \kappa \leqslant n_0-3$$
$$((n_0-l-1, \kappa-l, \alpha+l+1, \gamma+2l+2)\theta_\kappa(l) \in \mathscr{A}$$
$$\textit{when } \kappa = n_0-2). \tag{4.30}$$

Then

(iii) *for any integers* $l \in [0, \kappa]$, $\rho \in [1, \alpha + l]$ *we have*

$$(n_0 - l, \kappa - l, \alpha + l - \rho, \alpha + \kappa + r(l, \rho), \gamma + 2\kappa + t_\kappa(l, \rho)) \in \mathscr{A} \quad \text{when } \kappa \leqslant n_0 - 3$$

$$((n_0 - l, \kappa - l, \alpha + l - \rho, \gamma + 2\kappa + t_\kappa(l, \rho)) \in \mathscr{A} \quad \text{when } \kappa = n_0 - 2), \tag{4.31}$$

(iv) *for any integers* $l \in [0, \kappa - 1]$, $\rho \in [1, \alpha + l]$ *there exists an integer* $\theta_\kappa(l, \rho) \geqslant 0$ *such that*

$$(n_0 - l - 1, \kappa - l, \alpha + l - \rho, \alpha + \kappa + r_\kappa(l, \rho), \gamma + 2\kappa + t_\kappa(l, \rho) - 1)\theta_\kappa(l, \rho) \in \mathscr{A} \quad \text{when } \kappa \leqslant n_0 - 3$$

$$((n_0 - l - 1, \kappa - l, \alpha + l - \rho, \gamma + 2\kappa + t_\kappa(l, \rho) - 1)\theta_\kappa(l, \rho) \in \mathscr{A} \quad \text{when } \kappa = n_0 - 2); \tag{4.32}$$

here

$$r_\kappa(l, \rho) = 2(\kappa - l - 2) + 3\rho,$$
$$t_\kappa(l, \rho) = 5(\kappa - l - 2) + 7\rho,$$
$$\text{for } \kappa \geqslant 3, \ l \in [0, \kappa - 2]; \tag{4.33}$$

$$r_\kappa(\kappa - 1, \rho) = \begin{cases} 2\rho, & \rho = 1, 2, \\ 3\rho - 2, & \rho \geqslant 3, \end{cases}$$
$$t_\kappa(\kappa - 1, \rho) = \begin{cases} 4, & \rho = 1, \\ 7\rho - 5, & \rho \geqslant 2, \end{cases}$$
$$\text{for } \kappa \geqslant 3; \tag{4.34}$$

$$r_\kappa(\kappa, \rho) = \begin{cases} 2\rho - 1, & \rho = 1, 2, \\ 3\rho - 4, & \rho \geqslant 3, \end{cases}$$
$$t_\kappa(\kappa, \rho) = \begin{cases} 4\rho - 2, & \rho = 1, 2, \\ 7\rho - 10, & \rho \geqslant 3, \end{cases}$$
$$\text{for } \kappa \geqslant 3; \tag{4.35}$$

$$r_2(l, \rho) = 2\rho - l + 1,$$
$$t_2(l, \rho) = \begin{cases} 2, & l = 2, \ \rho = 1, \\ 5\rho - 3l + 2, & \text{otherwise}. \end{cases} \tag{4.36}$$

PROOF. 1°. This proof may become easier to understand if one uses the following geometrical image that makes sense for any $\kappa \in [0, n_0 - 3]$ and (with obvious simplifications) for $\kappa = n_0 - 2$ as well. In a system of Cartesian coordinates on the plane, consider the graph of the function that equals $\alpha + l - 1$ on every interval $(l, l + 1)$, $0 \leqslant l \leqslant \kappa$. Decompose the part of the plane between this graph and the horizontal coordinate axis into unit squares by horizontal and vertical lines (see Figure 1). Denote the set of

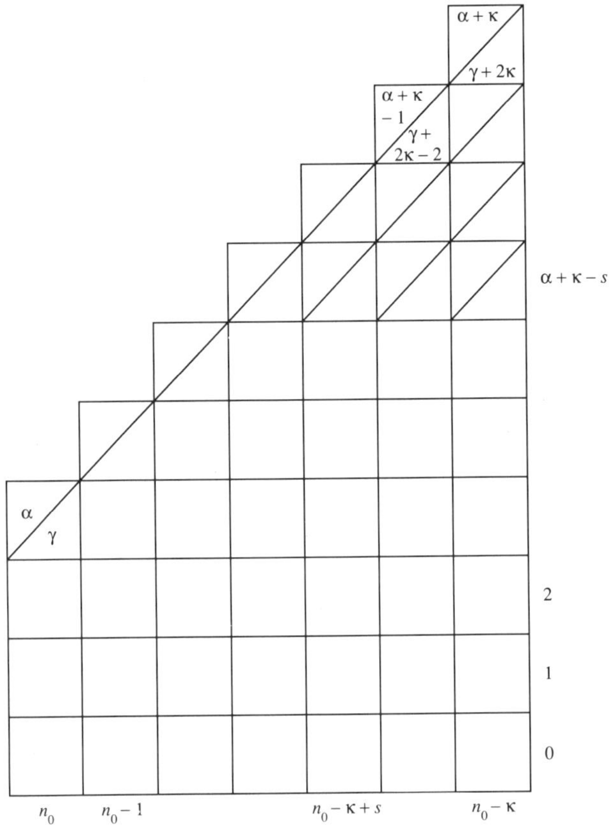

FIGURE 1. $(\kappa = 6, \alpha = 3, s = 3)$

squares just obtained by Σ_κ. Identify the segment $[l, l+1]$ of the horizontal coordinate axis, the segment $[i, i+1]$ of the vertical coordinate axis, and their product $[l, l+1] \times [i, i+1]$ with the numbers $n_0 - l$, i, and the pair $(n_0 - l, i)$ respectively, where $0 \leqslant l \leqslant \kappa$, $i = 0, 1, \dots$. In every square of the form $(n_0 - l, \alpha + l)$ (the upper square of the lth column) inscribe the number $\alpha + l$ (above the diagonal) and $\gamma + 2l$ (below the diagonal), as shown in Figure 1. The square "filled" in this way can be identified with the mark in (4.29) because the numbers $n_0 - l$, $\alpha + l$ and $\gamma + 2l$ are contained in this mark and, moreover, determine it (for κ is fixed). Note that together with (4.29) we also have (4.30) whenever $l \leqslant \kappa - 1$. Having proved (4.31) for some ρ, we fill the square $(n_0 - l, \alpha + l - \rho)$ in the same way as above: write the numbers $\alpha + \kappa + r_\kappa(l, \rho)$ and $\gamma + 2\kappa + t_\kappa(l, \rho)$ above and below its diagonal respectively. Simultaneously with (4.31), we prove (4.32) (if $l \neq \kappa$), so in what follows let us call the square $(n_0 - l, \alpha + l - \rho) \in \Sigma_\kappa$ *filled* if both (4.31) and (4.32) (the latter for $l \neq \kappa$) hold. Thus the proof looks like we are consecutively filling the squares from Σ_κ. We proceed in the following order: first fill the highest of the unfilled rows of squares, namely, the $(\alpha - \kappa - 1)$th row (whose unique unfilled square is $(n_0 - \kappa, \alpha + \kappa - 1)$), then pass to the $(\alpha + \kappa - 2)$th row and fill it moving from left to right, and so on.

Below we shall often mention inclusions (4.29) and (4.30) for various values of l and ρ. For short, in such cases let us write (4.29; l), (4.30; l, ρ), etc., substituting for l and ρ their specific values. To be definite, consider the case $\kappa \leqslant n_0 - 3$ (for $\kappa = n_0 - 2$ the proof becomes somewhat simpler).

To fill the square $(n_0 - \kappa, \alpha + \kappa - 1)$ apply Lemma 4.1(d) with

$$n = n_0 - \kappa, \quad k = 0, \quad a = \alpha + \kappa, \quad b = \alpha + \kappa, \quad c = \gamma + 2\kappa,$$
$$b_{-1} = \alpha + \kappa - 1, \quad c_{-1} = \gamma + 2(\kappa - 1), \quad \widehat{b} = \alpha + \kappa,$$
$$\widehat{c} = \gamma + 2\kappa, \quad \theta = \theta_\kappa(\kappa - 1).$$

By (4.29; κ), (4.29; $\kappa - 1$), and (4.30; $\kappa - 1$), for these values of n, k, a, etc., inclusions (4.1), (4.2), and (4.6) hold. Hence (see (4.7), (4.9))

$$(n_0 - \kappa, 0, \alpha + \kappa - 1, \alpha + \kappa + 1, \gamma + 2\kappa + 2) \in \mathscr{A},$$

which coincides with (4.31; κ, 1) (see (4.35), (4.36)).

2°. Consider now the $(\alpha + \kappa - s)$th row with an arbitrary $s \in [1, \kappa - 1]$ (see Figure 1). By (4.29; $\kappa - s$) we have

$$(n_0 - \kappa + s, s, \alpha + \kappa - s, \alpha + \kappa - s, \gamma + 2(\kappa - s)) \in \mathscr{A}. \tag{4.37}$$

Since we have $s + 1 \leqslant \kappa$ by assumption, it follows that (4.29; $\kappa - s - 1$) holds as well, i.e.,

$$(n_0 - \kappa + s + 1, s + 1, \alpha + \kappa - s - 1, \alpha + \kappa - s - 1, \gamma + 2(\kappa - s - 1)) \in \mathscr{A}. \tag{4.38}$$

By (4.30; $\kappa - s - 1$), we have

$$(n_0 - \kappa + s, s + 1, \alpha + \kappa - s - 1, \alpha + \kappa - s, \gamma + 2(\kappa - s))\, \theta_\kappa(\kappa - s) \in \mathscr{A} \tag{4.39}$$

when $\kappa \leqslant n_0 - 4$, and

$$(n_0 - \kappa + s, s + 1, \alpha + \kappa - s - 1, \gamma + 2(\kappa - s))\, \theta_\kappa(\kappa - s) \in \mathscr{A}$$

when $\kappa = n_0 - 3$. Here $\theta_\kappa(\kappa - s)$ is a nonnegative integer.

Assume that the $(\alpha + \kappa - s)$th row has been filled. Applying (4.31) first with $l = \kappa - s + 1$, $\rho = 1$, then with $l = \kappa - s + 2$, $\rho = 2$ (when $s \geqslant 2$, $\kappa \geqslant 3$), and finally with $l = \kappa - s + 3$, $\rho = 3$ (when $s \geqslant 3$, $\kappa \geqslant 4$), we get respectively:

$$\begin{aligned}(n_0 - \kappa + s - 1, s - 1, \alpha + \kappa - s,\\ \alpha + \kappa + r_\kappa(\kappa - s + 1, 1), \gamma + 2\kappa + t_\kappa(\kappa - s + 1, 1)) \in \mathscr{A},\end{aligned} \tag{4.40}$$

$$\begin{aligned}(n_0 - \kappa + s - 2, s - 2, \alpha + \kappa - s,\\ \alpha + \kappa + r_\kappa(\kappa - s + 2, 2), \gamma + 2\kappa + t_\kappa(\kappa - s + 2, 2)) \in \mathscr{A},\end{aligned} \tag{4.41}$$

$$\begin{aligned}(n_0 - \kappa + s - 3, s - 3, \alpha + \kappa - s,\\ \alpha + \kappa + r_\kappa(\kappa - s + 3, 3), \gamma + 2\kappa + t_\kappa(\kappa - s + 3, 3)) \in \mathscr{A},\end{aligned} \tag{4.42}$$

where $r_\kappa(i, j)$ and $t_\kappa(i, j)$ are defined by (4.33)–(4.36). Let

$$\begin{gathered} n = n_0 - \kappa + s, \quad k = s, \quad a = b = \alpha + \kappa - s, \\ c = \gamma + 2(\kappa - s), \quad b_{-1} = \alpha + \kappa - s - 1, \\ c_{-1} = \gamma + 2(\kappa - s - 1), \quad b_1 = \alpha + \kappa + r_\kappa(\kappa - s + 1, 1), \\ c_1 = \gamma + 2\kappa + t_\kappa(\kappa - s + 1, 1), \quad b_2 = \alpha + \kappa + r_\kappa(\kappa - s + 2, 2), \\ c_2 = \gamma + 2\kappa + t_\kappa(\kappa - s + 2, 2), \quad b_3 = \alpha + \kappa + r_\kappa(\kappa - s + 3, 3), \\ c_3 = \gamma + 2\kappa + t_\kappa(\kappa - s + 3, 3), \quad \widehat{b} = \alpha + \kappa - s, \\ \widehat{c} = \gamma + 2(\kappa - s), \quad \theta = \theta_\kappa(\kappa - s). \end{gathered} \tag{4.43}$$

It is easily checked that for these values of n, k, a, etc., inclusions (4.37), (4.38), (4.40), (4.42), and (4.39) coincide with (4.2), (4.1), (4.3), (4.4), (4.5), and (4.6). So by Lemma 4.1(a)–(c), we have

$$(n_0 - \kappa + s, s, \alpha + \kappa - s - 1, \widetilde{b}, \widetilde{c} + 1) \in \mathscr{A}, \tag{4.44}$$

$$(n_0 - \kappa + s - 1, s, \alpha + \kappa - s - 1, \widetilde{b}, \widetilde{c})\widetilde{\beta} \in \mathscr{A}, \tag{4.45}$$

where $\widetilde{b}, \widetilde{c}$ and $\widetilde{\beta}$ are defined by (4.9), (4.11), and (4.13) (combined with (4.43)), when $s \geqslant 3$, $s = 2$, and $s = 1$ respectively. Using (4.33)–(4.36) it is easy to verify that

$$\widetilde{b} = \begin{cases} \alpha + \kappa + 2, & s = 1, \\ \alpha + \kappa + 2s - 1, & s \geqslant 2, \end{cases} \qquad \widetilde{c} = \begin{cases} \gamma + 2\kappa + 3, & s = 1, \\ \gamma + 2\kappa + 5s - 4, & s \geqslant 2, \end{cases} \tag{4.46}$$

when $\kappa \geqslant 3$, and that

$$\widetilde{b} = \alpha + 4, \qquad \widetilde{c} = \gamma + 7, \tag{4.47}$$

when $\kappa = 2$, $s = 1$ (in the latter case the inequality $s > 1$ contradicts the assumptions on s). By substituting (4.46), (4.47) for $\widetilde{b}$ and $\widetilde{c}$ in (4.44), (4.45) and taking into account (4.33)–(4.36), we come to (4.31), (4.32) for $l = \kappa - s$, $\rho = 1$. This allows to fill the second (from the left) square in the $(\alpha + \kappa - s - 1)$th row (see Figure 1). Thus we have passed from the $(\alpha + \kappa - s)$th row to the $(\alpha + \kappa - s - 1)$th one, provided that $1 \leqslant s \leqslant \kappa - 1$.

3°. Let us do the same for $s \in [\kappa, \alpha + \kappa - 1]$ (see Figure 2). Assume once more that the $(\alpha + \kappa - s)$th row is filled (note that when $s = \kappa$ the first square in this row is filled by (4.29; 0) independently of our assumption). It follows that (4.31) holds for $l = 1$, $\rho = s - \kappa + 1$, for $l = 2$, $\rho = s - \kappa - 2$, and for $l = 3$, $\rho = s - \kappa + 3$ (when $\kappa \geqslant 3$), i.e.,

$$(n_0 - 1, \kappa - 1, \alpha + \kappa - s, \alpha + \kappa + r_\kappa(1, s - \kappa + 1), \gamma + 2\kappa + t_\kappa(1, s - \kappa + 1)) \in \mathscr{A}, \tag{4.48}$$

$$(n_0 - 2, \kappa - 2, \alpha + \kappa - s, \alpha + \kappa + r_\kappa(2, s - \kappa + 2), \gamma + 2\kappa + t_\kappa(2, s - \kappa + 2)) \in \mathscr{A}, \tag{4.49}$$

$$(n_0 - 3, \kappa - 3, \alpha + \kappa - s, \alpha + \kappa + r_\kappa(3, s - \kappa + 3), \gamma + 2\kappa + t_\kappa(3, s - \kappa + 3)) \in \mathscr{A}. \tag{4.50}$$

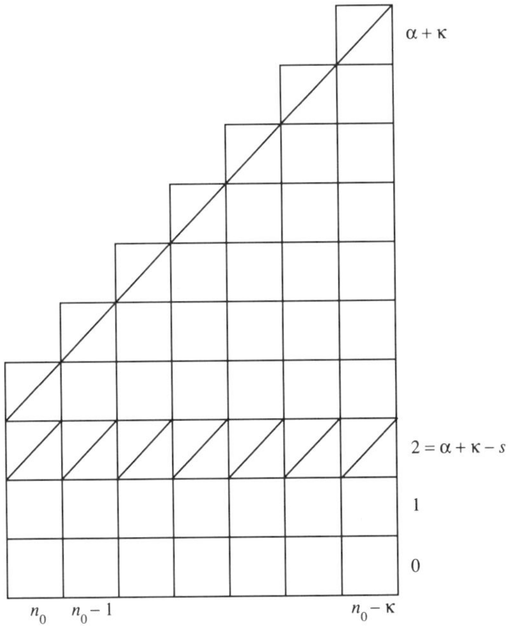

FIGURE 2. $(\kappa = 6, \alpha = 3, s = 7)$

If $s = \kappa$, then (4.31) holds for $l = 0$, $\rho = s - \kappa$ as well, i.e.,

$$(n_0, \kappa, \alpha + \kappa - s, \alpha + \kappa + r_\kappa(0, s - \kappa), \gamma + 2\kappa + t_\kappa(0, s - \kappa)) \in \mathscr{A}. \quad (4.51)$$

If $s = \kappa$, then (4.51) holds too because by (4.29;0) $(n_0, \kappa, \alpha, \alpha, \gamma) \in \mathscr{A}$ and the mark $(n_0, \kappa, \alpha, \alpha, \gamma)$ is dominated by the mark on the left-hand side of (4.51) whenever $s = \kappa$. Note that (2.10) can be rewritten in the form $(n_0 + 1, \kappa + 1, 0, 0, m_0 + 1) \in \mathscr{A}$, which implies that

$$(n_0 + 1, \kappa + 1, \alpha + \kappa - s - 1, 0, m_0 + 1) \in \mathscr{A}. \quad (4.52)$$

Let

$$\begin{gathered} n = n_0, \quad k = \kappa, \quad a = \alpha + \kappa - s, \quad b = \alpha + \kappa + r_\kappa(0, s - \kappa), \\ c = \gamma + 2\kappa + t_\kappa(0, s - \kappa), \quad b_{-1} = 0, \quad c_{-1} = m_0 + 1, \\ b_1 = \alpha + \kappa + r_\kappa(1, s - \kappa + 1), \quad c_1 = \gamma + 2\kappa + t_\kappa(1, s - \kappa + 1), \quad (4.53) \\ b_2 = \alpha + \kappa + r_\kappa(2, s - \kappa + 2), \quad c_2 = \gamma + 2\kappa + t_\kappa(2, s - \kappa + 2), \\ b_3 = \alpha + \kappa + r_\kappa(3, s - \kappa + 3), \quad c_3 = \gamma + 2\kappa + t_\kappa(3, s - \kappa + 3). \end{gathered}$$

With these values of n, k, a, etc., inclusions (4.52), (4.51), (4.48), (4.49), and (4.50) coincide with (4.1)–(4.5) respectively. Hence by Lemma 4.1(a,b)

$$(n_0, \kappa, \alpha + \kappa - s - 1, \tilde{b}, \tilde{c} + 1) \in \mathscr{A}, \quad (4.54)$$

$$(n_0 - 1, \kappa, \alpha + \kappa - s - 1, \tilde{b}, \tilde{c})\tilde{\beta} \in \mathscr{A}, \quad (4.55)$$

where $\widetilde{b}$, $\widetilde{c}$ and $\widetilde{\beta}$ are defined by (4.10), (4.12), and (4.53) combined with (4.33)–(4.36). It follows that

$$\widetilde{b} = \begin{cases} \alpha + 3s - 1, & \kappa \geqslant 3, \\ \alpha + 2s + 1, & \kappa = 2, \end{cases} \qquad \widetilde{c} = \begin{cases} \gamma + 7s - 4, & \kappa \geqslant 3, \\ \gamma + 5s, & \kappa = 2. \end{cases} \tag{4.56}$$

Substituting (4.56) for $\widetilde{b}$, $\widetilde{c}$ in (4.54), (4.55) and using (4.33), (4.36) once more, one can see that (4.54) and (4.55) coincide with $(4.31; 0, s+1-\kappa)$ and $(4.32; 0, s+1-\kappa)$ respectively. Thus we have passed from the $(\alpha+\kappa-s)$th row to the $(\alpha+\kappa-s-1)$th row when $s \geqslant \kappa$ as well. It remains to make a step along a row.

4°. Assume that the entire $(\alpha+\kappa-s)$th row as well as the first i squares of the $(\alpha-\kappa-s-1)$th row have been filled. Our goal is to fill the $(i+1)$th square of the latter row. Here, as above, we distinguish two cases: 1) $s \in [1, \kappa-1]$ and 2) $s \in [\kappa, \alpha+\kappa-1]$. In the first case we have $i \in [2, s+1]$ (because there are precisely $s+2$ squares in the $(\alpha+\kappa-s-1)$th row), while in the second case we have $i \in [1, \kappa]$.

Let us begin with the first case (see Figure 3). By assumption, inclusion (4.31) is true for the following values of l and ρ:

$$1)\quad l = \kappa - s + i - 2, \quad \rho = i - 1;$$
$$2)\quad l = \kappa - s + i - 1, \quad \rho = i - 1;$$

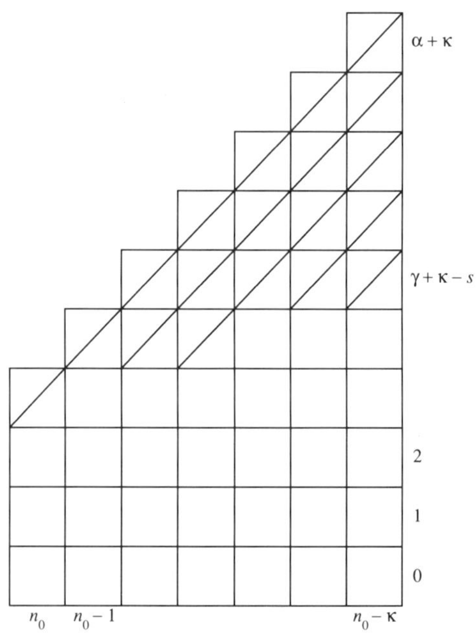

FIGURE 3. $\kappa = 6, \alpha = 3, s = 4, i = 3$)

3) $l = \kappa - s + i, \quad \rho = i \quad$ (when $i \leqslant s$);

4) $l = \kappa - s + i + 1, \quad \rho = i + 1 \quad$ (when $i \leqslant s - 1$);

5) $l = \kappa - s + i + 2, \quad \rho = i + 2 \quad$ (when $i \leqslant s - 2$).

For the same reason, inclusion (4.32) with $l = \kappa - s + i - 2$, $\rho = i - 1$ also holds. Hence we have (under the corresponding conditions on i)

$$(n_0 - \kappa + s - i + 2, s - i + 2, \alpha + \kappa - s - 1, \\ \alpha + \kappa + r_\kappa(\kappa - s + i - 2, i - 1), \gamma + 2\kappa + t_\kappa(\kappa - s + i - 2, i - 1)) \in \mathscr{A}, \tag{4.57}$$

$$(n_0 - \kappa + s - i + 1, s - i + 1, \alpha + \kappa - s, \\ \alpha + \kappa + r_\kappa(\kappa - s + i - 1, i - 1), \gamma + 2\kappa + t_\kappa(\kappa - s + i - 1, i - 1)) \in \mathscr{A}, \tag{4.58}$$

$$(n_0 - \kappa + s - i, s - i, \alpha + \kappa - s, \\ \alpha + \kappa + r_\kappa(\kappa - s + i, i), \gamma + 2\kappa + t_\kappa(\kappa - s + i, i - 1)) \in \mathscr{A}, \tag{4.59}$$

$$(n_0 - \kappa + s - i - 1, s - i - 1, \alpha + \kappa - s, \\ \alpha + \kappa + r_\kappa(\kappa - s + i + 1, i + 1), \gamma + 2\kappa + t_\kappa(\kappa - s + i + 1, i + 1)) \in \mathscr{A}, \tag{4.60}$$

$$(n_0 - \kappa + s - i - 2, s - i - 2, \alpha + \kappa - s, \\ \alpha + \kappa + r_\kappa(\kappa - s + i + 2, i + 2), \gamma + 2\kappa + t_\kappa(\kappa - s + i + 2, i + 2)) \in \mathscr{A}, \tag{4.61}$$

$$(n_0 - \kappa + s - i + 1, s - i + 2, \alpha + \kappa - s - 1, \\ \alpha + \kappa + r_\kappa(\kappa - s + i - 2, i - 1), \\ \gamma + 2\kappa + t_\kappa(\kappa - s + i - 2, i - 1) - 1)\theta_\kappa(\kappa - s + i - 2, i - 1) \in \mathscr{A}. \tag{4.62}$$

Let

$$\begin{aligned}
n &= n_0 - \kappa + s - i + 1, \\
k &= s - i + 1, \\
a &= \alpha + \kappa - s, \\
b &= \alpha + \kappa + r_\kappa(\kappa - s + i - 1, i - 1), \\
c &= \gamma + 2\kappa + t_\kappa(\kappa - s + i - 1, i - 1), \\
b_{-1} &= \alpha + \kappa + r_\kappa(\kappa - s + i - 2, i - 1), \\
c_{-1} &= \gamma + 2\kappa + t_\kappa(\kappa - s + i - 2, i - 1), \\
b_1 &= \alpha + \kappa + r_\kappa(\kappa - s + i, i), \\
c_1 &= \gamma + 2\kappa + t_\kappa(\kappa - s + i, i), \\
b_2 &= \alpha + \kappa + r_\kappa(\kappa - s + i + 1, i + 1), \\
c_2 &= \gamma + 2\kappa + t_\kappa(\kappa - s + i + 1, i + 1), \\
b_3 &= \alpha + \kappa + r_\kappa(\kappa - s + i + 2, i + 2), \\
c_3 &= \gamma + 2\kappa + t_\kappa(\kappa - s + i + 2, i + 2), \\
\widehat{b} &= \alpha + \kappa + r_\kappa(\kappa - s + i - 2, i - 1), \\
\widehat{c} &= \gamma + 2\kappa + t_\kappa(\kappa - s + i - 2, i - 1), \\
\theta &= \theta_\kappa(\kappa - s + i - 2, i - 1).
\end{aligned} \tag{4.63}$$

Then inclusions (4.57)–(4.62) turn into (4.1)–(4.6), and by Lemma 4.1(a)–(d) we obtain

$$(n_0 - \kappa + s - i + 1, s - i + 1, \alpha + \kappa - s - 1, \tilde{b}, \tilde{c} + 1) \in \mathscr{A}, \tag{4.64}$$

$$(n_0 - \kappa + s - i, s - i + 1, \alpha + \kappa - s - 1, \tilde{b}, \tilde{c})\tilde{\beta} \in \mathscr{A}, \tag{4.65}$$

when $i \leqslant s$, and obtain only (4.64) if $i = s + 1$. Here $\tilde{b}, \tilde{c}$ and $\tilde{\beta}$ are defined by (4.9)–(4.16) combined with (4.63) and (4.33)–(4.36). Taking into account the fact that $2 \leqslant i \leqslant s + 2$, it is easy to check that (4.64), (4.65) coincide with (4.31; $\kappa - s + i - 1$, i) and (4.32; $\kappa - s + i - 1$, i) respectively. Thus in the first of the above cases the $(i+1)$th square of the $(\alpha+\kappa-s-1)$th row has been filled.

5°. Let us pass to the second case (see Figure 4). Here we have $\kappa \leqslant s \leqslant \alpha + \kappa - 1$, $1 \leqslant i \leqslant \kappa$. By assumption, (4.31) holds with:

1) $l = i$, $\rho = s - \kappa + i$;
2) $l = i + 1$, $\rho = s - \kappa + i + 1$ (when $i \leqslant \kappa - 1$);
3) $l = i + 2$, $\rho = s - \kappa + i + 2$ (when $i \leqslant \kappa - 2$);
4) $l = i + 3$, $\rho = s - \kappa + i + 3$ (when $i \leqslant \kappa - 3$);
5) $l = i - 1$, $\rho = s - \kappa + i$ (when $i \geqslant 2$);

and (4.32) holds with $l = i - 1$, $\rho = s - \kappa + i$. Thus the following inclusions hold (under the corresponding conditions on i):

$$(n_0 - i, \kappa - i, \alpha + \kappa - s, \alpha + \kappa + r_\kappa(i, s - \kappa + i), \gamma + 2\kappa + t_\kappa(i, s - \kappa + i)) \in \mathscr{A}, \tag{4.66}$$

$$(n_0 - i, \kappa - i + 1, \alpha + \kappa - s - 1, \alpha + \kappa + r_\kappa(i - 1, s - \kappa + 1), \gamma + 2\kappa + t_\kappa(i - 1, s - \kappa + i))\theta_\kappa(i - 1, s - \kappa + i) \in \mathscr{A}, \tag{4.67}$$

$$(n_0 - i - 1, \kappa - i - 1, \alpha + \kappa - s, \alpha + \kappa + r_\kappa(i + 1, s - \kappa + i + 1), \gamma + 2\kappa + t_\kappa(i + 1, s - \kappa + i + 1)) \in \mathscr{A}, \tag{4.68}$$

$$(n_0 - i - 2, \kappa - i - 2, \alpha + \kappa - s, \alpha + \kappa + r_\kappa(i + 2, s - \kappa + i + 2), \gamma + 2\kappa + t_\kappa(i + 2, s - \kappa + i + 2)) \in \mathscr{A}, \tag{4.69}$$

$$(n_0 - i - 3, \kappa - i - 3, \alpha + \kappa - s, \alpha + \kappa + r_\kappa(i + 3, s - \kappa + i + 3), \gamma + 2\kappa + t_\kappa(i + 3, s - \kappa + i + 3)) \in \mathscr{A}, \tag{4.70}$$

$$(n_0 - i + 1, \kappa - i + 1, \alpha + \kappa - s - 1, \alpha + \kappa + r_\kappa(i - 1, s - \kappa + i), \gamma + 2\kappa + t_\kappa(i - 1, s - \kappa + i)) \in \mathscr{A}. \tag{4.71}$$

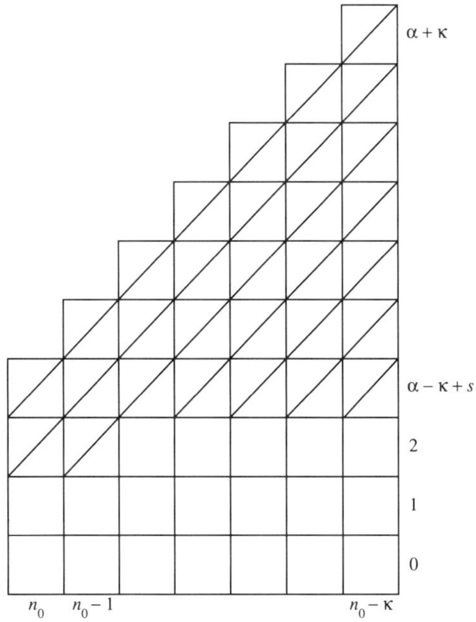

FIGURE 4. $(\kappa = 6, \alpha = 3, s = 6, i = 2)$

By (4.29; 0) we have $(n_0, \kappa, \alpha, \alpha, \gamma) \in \mathscr{A}$, and when $i = 1$ the mark $(n_0, \kappa, \alpha, \alpha, \gamma)$ is dominated by the mark from (4.71); hence (4.71) holds when $i = 1$ as well. Let

$$\begin{aligned}
n &= n_0 - i, \quad k = \kappa - i, \quad a = \alpha + \kappa - s, \\
b &= \alpha + \kappa + r_\kappa(i, s - \kappa + i), \\
c &= \gamma + 2\kappa + t_\kappa(i, s - \kappa + i), \\
b_{-1} &= \alpha + \kappa + r_\kappa(i - 1, s - \kappa + i), \\
c_{-1} &= \gamma + 2\kappa + t_\kappa(i - 1, s - \kappa + i), \\
b_1 &= \alpha + \kappa + r_\kappa(i + 1, s - \kappa + i + 1), \\
c_1 &= \gamma + 2\kappa + t_\kappa(i + 1, s - \kappa + i + 1), \\
b_2 &= \alpha + \kappa + r_\kappa(i + 2, s - \kappa + i + 2), \\
c_2 &= \gamma + 2\kappa + t_\kappa(i + 2, s - \kappa + i + 2), \\
b_3 &= \alpha + \kappa + r_\kappa(i + 3, s - \kappa + i + 3), \\
c_3 &= \gamma + 2\kappa + t_\kappa(i + 3, s - \kappa + i + 3), \\
\widehat{b} &= \alpha + \kappa + r_\kappa(i - 1, s - \kappa + i), \\
\widehat{c} &= \gamma + 2\kappa + t_\kappa(i - 1, s - \kappa + i), \\
\theta &= \theta_\kappa(i - 1, s - \kappa + i).
\end{aligned} \tag{4.72}$$

One can check as above that (4.66)–(4.71) coincide with (4.2), (4.6), (4.3)–

(4.5), and (4.1) respectively. So by Lemma 4.1(a–d) we have

$$(n_0 - i, \kappa - i, \alpha + \kappa - s - 1, \widetilde{b}, \widetilde{c} + 1) \in \mathscr{A}, \tag{4.73}$$

$$(n_0 - i - 1, \kappa - i, \alpha + \kappa - s - 1, \widetilde{b}, \widetilde{c})\widetilde{\beta} \in \mathscr{A} \quad \text{when } i \leqslant \kappa - 1; \tag{4.74}$$

here $\widetilde{b}$, $\widetilde{c}$ and $\widetilde{\beta}$ are defined by (4.9)–(4.16) together with (4.72) and (4.33)–(4.36). Taking into account the fact that $\kappa \leqslant s$ one sees that (4.73), (4.74) coincide with $(4.31; i, s - \kappa + i + 1)$ and $(4.32; i, s - \kappa + i + 1)$ respectively. Thus the $(i + 1)$th square in the $(\alpha + \kappa - s - 1)$th row has been filled in the case under consideration. Since in the first part of the proof we have already filled the $(\alpha + \kappa - 1)$th row, Lemma 4.2 is proved.

4.3. **Remark.** To each square in Σ_κ denoted by (l, i) $(n_0 - \kappa \leqslant l \leqslant n_0,\ 0 \leqslant i \leqslant n_0 - l + \alpha)$, one can assign another pair of integers, $(l, \alpha + \kappa - i)$, and arrange these pairs lexicographically. Observe that the proof just presented uses induction with respect to this lexicographical ordering.

§5. Completion of the proof of Theorem 4.10

We start with the following auxiliary result.

5.1. Proposition. *Let* $\varepsilon_0, \varepsilon_1, \dots$ *be a sequence of real numbers and suppose that for some integers* $l \geqslant 0$, $k \geqslant l$ *and* $i_1 \leqslant i_2 \leqslant \dots \leqslant i_{k-l+2}$ *we have*

$$i_1 + \dots + i_{k-l+2} \geqslant \varepsilon_{k+1} + l. \tag{5.1}$$

Then

(i) *there exists an integer* $k' \in [l, k]$ *such that either*

$$i_1 + \dots + i_{k'-l+1} \leqslant \varepsilon_{k'} + l, \qquad i_{k'-l+2} \geqslant \varepsilon_{k'} + k', \tag{5.2}$$

or

$$i_1 + \dots + i_{k'-l+1} < \varepsilon_{k'} + l, \qquad i_1 + \dots + i_{k'-l+2} \geqslant \varepsilon_{k'+1} + l; \tag{5.3}$$

(ii) *if the following additional conditions hold*:

(a) $\varepsilon_{j+1} \geqslant 2\varepsilon_j + j$ *for any* $j \geqslant 0$,
(b) $k \geqslant l + 1$,
(c) $i_1 = 0$,

then k' *in assertion* (i) *can be chosen so that* $k' > l + 1$.

Proof. (i) This assertion coincides with Proposition 2.2 in [1]. For the reader's convenience we give its proof here.

For any $l \geqslant 0$ apply induction on k. First assume that $k = l$. If we let $k' = l$, then by (5.1) only the first of the inequalities (5.3) can be violated. If it is indeed violated, then $i_1 \geqslant \varepsilon_l + l$ and hence $i_2 \geqslant i_1 \geqslant \varepsilon_l + l$, i.e., both inequalities (5.2) hold. Now let us assume our assertion holds for some $k = \overline{k} \geqslant l$ and prove it for $k = \overline{k} + 1$. Let (5.1) hold for $k = \overline{k} + 1$. Then for $k' = k = \overline{k} + 1$ the second of inequalities (5.3) is true. If the first is violated, i.e., if $i_1 + \cdots + i_{\bar{k}-l+2} \geqslant \varepsilon_{\bar{k}+1} + l$, then (5.1) holds also for $k = \bar{k}$ and by the inductive assumption there exists an appropriate $k' \in [l, k]$. This completes the proof of (i).

(ii) Let us argue as above. Having fixed l, assume first that $k = l + 1$ and let $k' = k = l + 1$. Then by (5.1), the second of inequalities (5.3) is true. If the first is violated, then the first of inequalities (5.2) is true, i.e., $i_1 + i_2 \geqslant \varepsilon_{k'} + k' - 1$, which can be rewritten in the form $i_2 \geqslant \varepsilon_{k'} + k' - 1$ because by our assumption $i_1 = 0$. If, besides, the second of the inequalities (5.2) is violated too, then $i_3 \leqslant \varepsilon_{k'} + k' - 1$ and from the assumption that $i_2 \leqslant i_3$ we get $i_2 = i_3 = \varepsilon_{k'} + k' - 1$. Hence (see condition (a)) $i_1 + i_2 + i_3 = 2(\varepsilon_{k'} + k' - 1) \leqslant \varepsilon_{k'+1} + k' - 2$. This contradicts inequality (5.1), which in our case has the form $i_1 + i_2 + i_3 \geqslant \varepsilon_{k'+1} + k' - 1$. Thus we have proved assertion (ii) for $k = l + 1$. It can be generalized to any $k \geqslant l + 1$ by induction in just the same way as assertion (i) was. This completes the proof of the proposition.

5.2. **Return to the proof of Theorem 1.10.** 1°. One can assume without loss of generality that n_0 is arbitrarily large. We shall proceed from the inclusion $(n_0 + 1, 0, 0, 0, m_0 + 1) \in \mathscr{A}$ (see (2.10)) assuming that $n_0 \geqslant 7$. By Lemma 3.4, it follows from (2.10) that

$$(n_0, 0, 1, 1, m_0 + 3) \in \mathscr{A}. \tag{5.4}$$

Lemma 4.1(d) allows us to lower this mark: it suffices to set $n = n_0$, $k = 0$, $a = 1$ in order to get

$$(n_0, 0, 0, 2, m_0 + 5) \in \mathscr{A} \tag{5.5}$$

(see (4.16)). There are two consequences of (5.5). First, by Lemma 3.4 (with $n = n_0$, $b = 2$, $c = m_0 + 5$) we have

$$(n_0 - 1, 0, 3, 3, m_0 + 7) \in \mathscr{A}. \tag{5.6}$$

Secondly,

$$(n_0, 1, 2, 2, m_0 + 5) \in \mathscr{A}. \tag{5.7}$$

Indeed, let $D \in \mathscr{D}(\mathfrak{m})$, where $\mathfrak{m} = (n_0, 1, 2, 2, m_0 + 5)$. Then

$$D = \partial_{n_0}(v_1, v_2, 2, \ldots, 2, m_0 + 5),$$

where $v_1 + v_2 = 2$. Assume that $v_1 \leqslant v_2$. If, in addition, $v_1 = v_2 = 1$, then D dominates the operator $D_1 = \partial_{n_0}(1, 1, \ldots, 1, m_0 + 3) \in \mathscr{D}(\mathfrak{m}_1)$, where $\mathfrak{m}_1 = (n, 0, 1, 1, m_0 + 3)$. If $v_1 = 0$, then $v_2 = 2$, where $D \in \mathscr{D}(\mathfrak{m}_2)$ with $\mathfrak{m}_2 = (n_0, 0, 0, 2, m_0 + 5)$. By (5.4), (5.5) we have $\mathfrak{m}_1, \mathfrak{m}_2 \in \mathscr{A}$, and so in both cases $Dg_n \equiv 0$. It remains to note that the symmetry of the function g_n allows us to remove the restriction $v_1 \leqslant v_2$. This implies (5.7).

Now observe that (2.10) can be rewritten in the form

$$(n_0 + 1, 2, 0, 0, m_0 + 1) \in \mathscr{A},$$

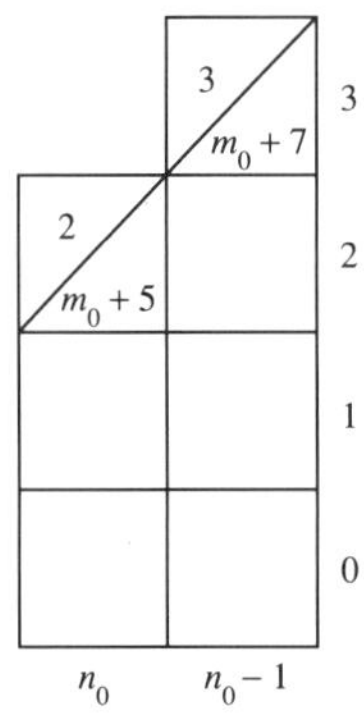

FIGURE 5.

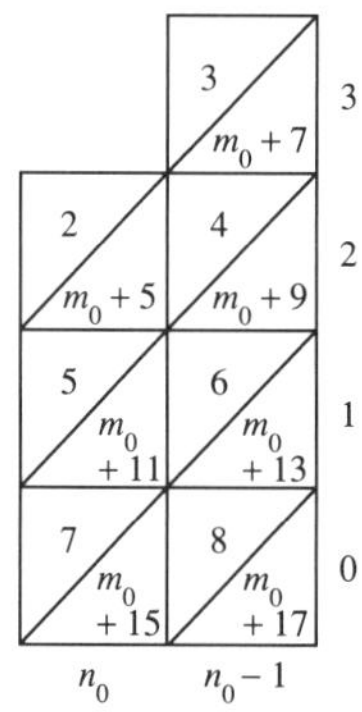

FIGURE 6.

whence it follows that

$$(n_0+1\,,\,2\,,\,1\,,\,0\,,\,m_0+1)\in\mathscr{A}\,. \tag{5.8}$$

Inclusions (5.6)–(5.8) allow us to apply Lemma 3.6 (with $n=n_0$, $a=b=2$, $c=m_0+5$, $b_{-1}=0$, $c_{-1}=m_0+1$, $b_1=3$, $c_1=m_0+7$), which gives

$$(n_0-1\,,\,1\,,\,2\,,\,3\,,\,m_0+7)0\in\mathscr{A}\,. \tag{5.9}$$

According to what was said at the beginning of the proof of Lemma 4.2, we have the picture shown in Figure 5, where $\kappa=1$. It is necessary to fill all the squares in Σ_2.

By Lemma 4.1(d) (with $n=n_0-1$, $k=0$, $a=b=3$, $c=m_0+7$, $b_{-1}=2$, $c_{-1}=m_0+5$, $\widehat{b}=3$, $\widehat{c}=m_0+7$) it follows from (4.7), (5.6), (5.9) that

$$(n_0-1\,,\,0\,,\,2\,,\,4\,,\,m_0+9)\in\mathscr{A}\,. \tag{5.10}$$

Apply Lemma 4.1(c) with $n=n_0$, $k=1$, $a=2$, etc., to deduce

$$(n_0\,,\,1\,,\,1\,,\,5\,,\,m_0+11)\in\mathscr{A}\,, \tag{5.11}$$

$$(n_0-1\,,\,1\,,\,1\,,\,5\,,\,m_0+10)\,1\in\mathscr{A} \tag{5.12}$$

from (5.7), (5.8), (5.10). Applying Lemma 4.1(d) with $n=n_0-1$, $k=0$, $a=2$, etc., we have (see (5.10)–(5.12))

$$(n_0-1\,,\,0\,,\,1\,,\,6\,,\,m_0+13)\in\mathscr{A}\,. \tag{5.13}$$

Now we apply Lemma 4.1(c) with $n=n_0$, $k=1$, $a=1$, etc. To do this, we rewrite (2.10) in the form $(n_0+1\,,\,2\,,\,0\,,\,0\,,\,m_0+1)\in\mathscr{A}$ and use this inclusion together with (5.11), (5.10) to obtain

$$(n_0\,,\,1\,,\,0\,,\,7\,,\,m_0+15)\in\mathscr{A}\,, \tag{5.14}$$

$$(n_0-1\,,\,1\,,\,0\,,\,7\,,\,m_0+14)\,1\in\mathscr{A}\,. \tag{5.15}$$

Finally, we apply Lemma 4.1(d) with $n = n_0 - 1$, $k = 0$, $a = 1$, etc. By this lemma, (5.14) and (5.13) imply

$$(n_0 - 1, 0, 0, 8, m_0 + 17) \in \mathscr{A}. \tag{5.16}$$

Thus all the squares on Figure 5 turn out to be filled (see Figure 6). Our next goal is to pass from Σ_2 to Σ_3.

2°. First by Lemma 4 (with $n = n_0 - 1$) it follows from (5.16) that

$$(n_0 - 2, 0, 9, 9, m_0 + 19) \in \mathscr{A}. \tag{5.17}$$

Let us show that

$$(n_0 - 1, 1, 8, 8, m_0 + 17) \in \mathscr{A}. \tag{5.18}$$

Set $\mathrm{m} = (n_0 - 1, 1, 8, 8, m_0 + 17)$ and suppose that $D(\mathrm{m}) = 0$. Then

$$D = \partial_{n_0-1}(v_1, v_2, 8, \dots, 8, m_0 + 17),$$

where $v_1 + v_2 = 8$. Assuming that $v_1 \leqslant v_2$ and arguing as in the proof of inclusion (5.7) above, it is easy to show that $Dg_{n_0-1} \equiv 0$, which is a consequence of (5.16), (5.13), (5.10) in the following three cases respectively: 1) $v_1 = 0$, $v_2 = 8$; 2) either $v_1 = 1$, $v_2 = 7$, or $v_1 = 2$, $v_2 = 6$; 3) either $v_1 = 3$, $v_2 = 5$, or $v_1 = 4$, $v_2 = 4$. Due to the symmetry of the function g_{n_0-1} the assumption $v_1 \leqslant v_2$ can be lifted and we come to (5.18).

The inclusion

$$(n_0, 2, 7, 7, m_0 + 15) \in \mathscr{A} \tag{5.19}$$

is also true. This can be shown in the same way as (5.18) but with the help of (5.4), (5.7), (5.11), and (5.14).

By Lemma 3.6 (with $n = n_0 - 1$), (5.17)–(5.19) imply

$$(n_0 - 2, 1, 8, 9, m_0 + 19)0 \in \mathscr{A}. \tag{5.20}$$

Let us finally prove that

$$(n_0 - 1, 2, 7, 8, m_0 + 17)1 \in \mathscr{A}. \tag{5.21}$$

An arbitrary operator $D \in \mathscr{D}(\mathrm{m})$ with $\mathrm{m} = (n_0 - 1, 2, 7, 8, m_0 + 17)1$ is of the form

$$\partial_{n_0-1,2,1}(v_1, v_2, v_3, 8, \dots, 8, m_0 + 17),$$

where $v_1 + v_2 + v_3 = 7$. Assume that $v_1 \leqslant v_2 \leqslant v_3$. If $v_1 > 0$, then, clearly, $Dg_{n_0-1} \equiv 0$. If $v_1 = 0$, then one of the following four cases occurs: 1) $v_1 = v_2 = 0$, $v_3 = 7$; 2) $v_1 = 0$, $v_2 = 1$, $v_3 = 6$; 3) $v_1 = 0$, $v_2 = 2$,

$v_3 = 5$; 4) $v_1 = 0$, $v_2 = 3$, $v_3 = 4$. Let

$$\begin{aligned} D^{(1)} &= \partial_{n_0-1,1,1}(0, 0, 7, 8, \dots, 8, m_0+17), \\ \mathfrak{m}^{(1)} &= (n_0-1, 1, 0, 7, m_0+14)1, \\ D^{(2)} &= \partial_{n_0-1,1,1}(0, 1, 6, 8, \dots, 8, m_0+17), \\ \mathfrak{m}^{(2)} &= (n_0-1, 1, 1, 5, m_0+10)1, \\ D^{(3)} &= \partial_{n_0-1,1,1}(0, 2, 5, 8, \dots, 8, m_0+17), \\ \mathfrak{m}^{(3)} &= (n_0-1, 1, 2, 3, m_0+7)0, \\ D^{(4)} &= \partial_{n_0-1,1,1}(0, 3, 4, 8, \dots, 8, m_0+17), \\ \mathfrak{m}^{(4)} &= \mathfrak{m}^{(3)}. \end{aligned}$$

We see that $D^{(i)}$ dominates some operator in $\mathscr{D}(\mathfrak{m}^{(i)})$, $i = 1, 2, 3, 4$. So inclusions (5.15), (5.12), and (5.9) imply that $D^{(i)} g_{n_0-1} \equiv 0$, $1 \leqslant i \leqslant 4$. Now one can apply Proposition 2.7 with $n = n_0 - 1$, $k = 2$, $k' = 1$, $\theta = 1$, $r_s = v_s$ for $s = 1, 2, 3$, $r_s = 8$ for $s \in [4, n_0 - 1]$, and $r_{n_0} = m_0 + 17$ (from (5.18) it follows that $\mathfrak{m}_1 \in \mathscr{A}$). Due to this proposition, $Dg_{n_0-1} \equiv 0$. This has been proved for $v_1 \leqslant v_2 \leqslant v_3$ and is also true for every v_1, v_2, v_3 because of the symmetry of g_{n_0-1}. So we come to (5.21).

Inclusions (5.19), (5.18), (5.17), (5.20), and (5.21) coincide with (4.29; 0), (4.29; 1), (4.29; 2), (4.30; 1), and (4.30; 0) respectively, when one sets $\kappa = 2$, $\alpha = 7$, $\gamma = m_0 + 15$, $\theta_2(0) = 1$, $\theta_2(1) = 0$. Thus the conditions (a) and (b) of Lemma 4.2 are satisfied for $\kappa = 2$ and the indicated values of α and γ.

3°. The following inductive argument is similar to the previous one and deals with the general case. We set

$$\begin{gathered} \alpha_0 = 1, \quad \gamma_0 = m_0 + 3, \quad \alpha_1 = 2, \quad \gamma_1 = m_0 + 5, \\ \theta_1(0, 1) = \theta_1(0, 2) = 1, \quad \alpha_2 = 7, \quad \gamma_2 = m_0 + 15, \end{gathered} \tag{5.22}$$

and for every $\kappa \geqslant 2$ set

$$\begin{aligned} \alpha_{\kappa+1} &= \alpha_\kappa + \kappa + r_\kappa(0, \alpha_\kappa) = \begin{cases} 4\alpha_\kappa + 3\kappa - 4, & \kappa \geqslant 3, \\ 3(\alpha_\kappa + 1), & \kappa = 2, \end{cases} \\ \gamma_{\kappa+1} &= \gamma_\kappa + 2\kappa + t_\kappa(0, \alpha_\kappa) = \begin{cases} \gamma_\kappa + 7\alpha_\kappa + 7\kappa - 10, & \kappa \geqslant 3, \\ \gamma_\kappa + 5\alpha_\kappa + 6, & \kappa = 2, \end{cases} \end{aligned} \tag{5.23}$$

$$\begin{gathered} \theta_\kappa(\kappa - 1) = 0, \quad \theta_\kappa(l) = \max_{\kappa' < \kappa} \left\{ \theta_{\kappa'}(l), \max_{\rho \in [1, \alpha_{\kappa'} + l]} \theta_{\kappa'}(l, \rho) \right\}, \\ \kappa \geqslant 1, \quad 0 \leqslant l \leqslant \kappa - 1. \end{gathered} \tag{5.24}$$

Assume that $n_0 \geqslant 7$ and that for some $\overline{\kappa} \in [2, n_0 - 3]$, for each $\kappa \in [2, \overline{\kappa}]$ and for $\alpha = \alpha_\kappa$, $\gamma = \gamma_\kappa$ (see (5.23)) conditions (a) and (b) of Lemma 4.2 are fulfilled. Let us show that then (a) and (b) are also fulfilled for $\kappa = \overline{\kappa} + 1$, $\alpha = \alpha_{\overline{\kappa}+1}$, $\gamma = \gamma_{\overline{\kappa}+1}$.

Our assumption implies (see Lemma 4.2) that (4.31) holds with $l = \overline{\kappa}$, $\rho = \alpha_{\overline{\kappa}} + \overline{\kappa}$, i.e., we have (see (4.33))

$$(n_0 - \overline{\kappa}, 0, 0, 4(\alpha_{\overline{\kappa}} + \overline{\kappa} - 1), \gamma_{\overline{\kappa}} + 7\alpha_{\overline{\kappa}} + 9\kappa - 10) \in \mathscr{A} \quad \text{when } \overline{\kappa} \geqslant 3,$$
$$(n_0 - \overline{\kappa}, 0, 0, 3\alpha_{\overline{\kappa}} + 2\overline{\kappa} + 1, \gamma_{\overline{\kappa}} + 5\alpha_{\overline{\kappa}} + 4\overline{\kappa} + 2) \in \mathscr{A} \quad \text{when } \overline{\kappa} = 2.$$

By Lemma 3.4 it follows that

$$\begin{aligned}
&(n_0 - \overline{\kappa} - 1, 0, 4(\alpha_{\overline{\kappa}} + \overline{\kappa}) - 3, \\
&\qquad 4(\alpha_{\overline{\kappa}} + \overline{\kappa}) - 3, \gamma_{\overline{\kappa}} + 7\alpha_{\overline{\kappa}} + 9\overline{\kappa} - 8) \in \mathscr{A} \qquad \text{when } 3 \leqslant \overline{\kappa} \leqslant n_0 - 4, \\
&(n_0 - \overline{\kappa} - 1, 0, \\
&\qquad 4(\alpha_{\overline{\kappa}} + \overline{\kappa}) - 3, \gamma_{\overline{\kappa}} + 7\alpha_{\overline{\kappa}} + 9\overline{\kappa} - 8) \in \mathscr{A} \qquad \text{when } \overline{\kappa} = n_0 - 3, \\
&(n_0 - \overline{\kappa} - 1, 0, 3\alpha_{\overline{\kappa}} + 2(\overline{\kappa} + 1), \\
&\qquad 3\alpha_{\overline{\kappa}} + 2(\overline{\kappa} + 1), \gamma_{\overline{\kappa}} + 5\alpha_{\overline{\kappa}} + 4(\overline{\kappa} + 1)) \in \mathscr{A} \quad \text{when } \overline{\kappa} = 2.
\end{aligned}$$

By (5.20) these inclusions can be rewritten in the form

$$(n_0 - \overline{\kappa} - 1, 0, \alpha_{\overline{\kappa}+1} + \overline{\kappa} + 1, \alpha_{\overline{\kappa}+1} + \overline{\kappa} + 1, \gamma_{\overline{\kappa}+1} + 2(\overline{\kappa} + 1)) \in \mathscr{A}$$
$$\text{when } 2 \leqslant \overline{\kappa} \leqslant n_0 - 4,$$
$$(n_0 - \overline{\kappa} - 1, 0, \alpha_{\overline{\kappa}+1} + \overline{\kappa} + 1, \gamma_{\overline{\kappa}+1} + 2(\vec{\kappa} + 1)) \in \mathscr{A} \quad \text{when } \overline{\kappa} = n_0 - 3.$$

Therefore, (4.29) holds with $\kappa = \overline{\kappa} + 1$, $\alpha = \alpha_\kappa$, $\gamma = \gamma_\kappa$, $l = \kappa$.

Let us prove (4.29; l) for the same κ, α, γ and for all the other l. Suppose $l = \overline{l} \in [0, \overline{\kappa}]$ and D is an arbitrary operator from $\mathscr{D}(\mathfrak{m})$, where

$$\mathfrak{m} = \begin{cases} (n_0 - \overline{l}, \overline{\kappa} + 1 - \overline{l}, \alpha_{\overline{\kappa}+1} + \overline{l}, \alpha_{\overline{\kappa}+1} + \overline{l}, \gamma_{\overline{\kappa}+1} + 2\overline{l}), & \overline{\kappa} \leqslant n_0 - 4, \\ (n_0 - \overline{l}, \overline{\kappa} + 1 - \overline{l}, \alpha_{\overline{\kappa}+1} + \overline{l}, \gamma_{\overline{\kappa}+1} + 2\overline{l}), & \overline{\kappa} = n_0 - 3. \end{cases}$$

By definition we have

$$D = \begin{cases} \partial_{n_0 - \overline{l}}(v_1, \dots, v_{\overline{\kappa}+2-\overline{l}}, \alpha_{\overline{\kappa}+1} + \overline{l}, \\ \qquad\qquad \dots, \alpha_{\overline{\kappa}+1} + \overline{l}, \gamma_{\overline{\kappa}+1} + 2\overline{l}), & \overline{\kappa} \leqslant n_0 - 4, \\ \partial_{n_0 - \overline{l}}(v_1, \dots, v_{\overline{\kappa}+2-\overline{l}}, \gamma_{\overline{\kappa}+1} + 2\overline{l}), & \overline{\kappa} = n_0 - 3, \end{cases}$$

where $v_1 + \dots + v_{\overline{\kappa}+2-\overline{l}} = \alpha_{\overline{\kappa}+1} + \overline{l}$. Assume that $v_1 \leqslant \dots \leqslant v_{\overline{\kappa}+2-\overline{l}}$. By Proposition 5.1(i) (with $\varepsilon_j = \alpha_j$, $k = \overline{\kappa}$, $i_s = v_s$, $1 \leqslant s \leqslant \overline{\kappa} + 2 - \overline{l}$) there is a $\kappa' \in [\overline{l}, \overline{\kappa}]$ such that either

$$v_1 + \dots + v_{\kappa' - \overline{l} + 1} \geqslant \alpha_{\kappa'} + \overline{l}, \qquad v_{\kappa' - \overline{l} + 2} \geqslant \alpha_{\kappa'} + \kappa', \tag{5.25}$$

or

$$v_1 + \dots + v_{\kappa' - \overline{l} + 1} = \alpha_{\kappa'} + \overline{l} - \rho, \qquad v_1 + \dots + v_{\kappa' - \overline{l} + 2} \geqslant \alpha_{\kappa'+1} + \overline{l}, \tag{5.26}$$

where $1 \leqslant \rho \leqslant \alpha_{\kappa'} + \overline{l}$. In the former case D dominates some operator from $\mathscr{D}(\mathfrak{m}')$, where

$$\mathfrak{m}' = (n_0 - \overline{l}, \kappa' - \overline{l}, \alpha_{\kappa'} + \overline{l}, \alpha_{\kappa'} + \kappa', \gamma_{\kappa'} + 2\overline{l}).$$

In the latter case D dominates some operator from $\mathscr{D}(\mathfrak{m}'')$, where

$$\mathfrak{m}'' = (n_0 - \bar{l}, \kappa' - \bar{l}, \alpha_{\kappa'} + \bar{l} - \rho, \alpha_{\kappa'+1} - \alpha_{\kappa'} + \rho, \gamma_{\kappa'+1} + 2\bar{l})$$

(here we use the monotonicity of the sequence $v_1, v_2, \dots$ and the same property of the sequence γ_κ, $\kappa = 0, 1, \dots$, which follows from (5.23)). The marks $\mathfrak{m}'$ and $\mathfrak{m}''$ are annihilators. Indeed, the inclusion $\mathfrak{m}' \in \mathscr{A}$ follows from (5.4) when $\kappa' = 0$, $\bar{l} = 0$, from (5.7) when $\kappa' = 1$, $\bar{l} = 0$, from (5.6) when $\kappa' = 1$, $\bar{l} = 1$, and from the inductive assumption when $\kappa' \geqslant 2$ (one has to use (5.22) in the first three cases). Similarly, the inclusion $\mathfrak{m}'' \in \mathscr{A}$ follows from (5.5) when $\kappa' = 0$, $\bar{l} = 0$, $\rho = 1$, from (5.11) and (5.14) when $\kappa' = 1$, $\bar{l} = 0$, $\rho = 1$, and when $\kappa' = 1$, $\bar{l} = 0$, $\rho = 2$, respectively, and from (5.10), (5.13), (5.16) when $\kappa' = 1$, $\bar{l} = 1$, $\rho = 1$, when $\kappa' = 1$, $\bar{l} = 1$, $\rho = 2$, and when $\kappa' = 1$, $\bar{l} = 1$, $\rho = 3$, respectively. Let us show that $\mathfrak{m}'' \in \mathscr{A}$ for any $\kappa' \geqslant 2$.

Using (5.25) and (4.33)–(4.36) it is easy to check that for all $\kappa \geqslant 2$, $l \in [0, \kappa]$, $\rho = [1, \alpha_\kappa + l]$ the following inequalities hold:

$$\alpha_{\kappa+1} - \alpha_\kappa + \rho \geqslant \alpha_\kappa + \kappa + r_\kappa(l, \rho), \qquad \gamma_{\kappa+1} + 2l \geqslant \gamma_\kappa + 2\kappa + t_\kappa(l, \rho). \quad (5.27)$$

It follows that $\mathfrak{m}''$ dominates the mark

$$(n_0 - l, \kappa' - \bar{l}, \alpha_{\kappa'} + \bar{l} - \rho, \alpha_{\kappa'} + \kappa' + r_{\kappa'}(\bar{l}, \rho), \gamma_{\kappa'} + 2\kappa' + t_{\kappa'}(\bar{l}, \rho)).$$

Now let us use inclusion $(4.31; \bar{l}, \rho)$ with $\kappa = \kappa'$, $\alpha = \alpha_\kappa$, $\gamma = \gamma_\kappa$, which follows from the inductive assumption and Lemma 4.2. So we get $\mathfrak{m}'' \in \mathscr{A}$. Together with what has been proved above, this leads to the identity $Dg_{n_0 - \bar{l}} \equiv 0$. Due to the symmetry of the function $g_{n_0 - \bar{l}}$, one can forget the assumption $v_1 \leqslant v_2 \leqslant \cdots$ and so obtain $(4.29; l)$, $0 \leqslant l \leqslant \kappa$. Thus condition (a) of Lemma 4.2 holds for $\kappa = \bar{\kappa} + 1$.

Passing to condition (b), we first remark that, by Lemma 3.6 (with $n = n_0 - \bar{\kappa}$), from $(4.29; \bar{\kappa} - 1)$, $(4.29; \bar{\kappa})$, and $(4.29; \bar{\kappa} + 1)$ (with $\kappa = \bar{\kappa} + 1$, $\alpha = \alpha_{\bar{\kappa}+1}$, $\gamma = \gamma_{\bar{\kappa}+1}$) it follows that $(4.30; \bar{\kappa})$ holds with the same κ, α, γ and with $\theta_{\bar{\kappa}+1}(\bar{\kappa}) = 0$ ($(4.29; \bar{\kappa} - 1)$ and $(4.29; \bar{\kappa})$ hold by the inductive assumption, and $(4.29; \bar{\kappa} + 1)$ was just proved).

Let us prove that $(4.30; l)$ holds for any $l \in [0, \bar{\kappa} - 1]$. We set $l = \bar{l}$ and take an arbitrary operator

$$D = \partial_{n_0 - \bar{l} - 1, \bar{\kappa} + 1 - \bar{l}, \theta}(v_1, \dots, v_{n_0 - \bar{l}}) \in \mathscr{D}(\mathfrak{m}),$$

where

$$\mathfrak{m} = \begin{cases} (n_0 - \bar{l} - 1, \bar{\kappa} + 1 - \bar{l}, \alpha_{\bar{\kappa}+1} + \bar{l}, \\ \qquad \alpha_{\bar{\kappa}+1} + \bar{l} + 1, \gamma_{\bar{\kappa}+1} + 2\bar{l} + 2)\,\theta_{\bar{\kappa}+1}(\bar{l}), & \bar{\kappa} \leqslant n_0 - 4, \\ (n_0 - \bar{l} - 1, \bar{\kappa} + 1 - \bar{l}, \\ \qquad \alpha_{\bar{\kappa}+1} + \bar{l}, \gamma_{\bar{\kappa}+1} + 2\bar{l} + 2)\,\theta_{\bar{\kappa}+1}(\bar{l}), & \bar{\kappa} = n_0 - 3, \end{cases}$$

and

$$\begin{aligned} &v_1 + \cdots + v_{\overline{\kappa}+2-\overline{l}} = \alpha_{\overline{\kappa}+1} + \overline{l}\,, \qquad v_i = \alpha_{\overline{\kappa}+1} + \overline{l} + 1\,, \\ &\overline{\kappa} + 3 - \overline{l} \leqslant i \leqslant n_0 - \overline{l} - 1\,, \qquad v_{n_0 - \overline{l}} = \gamma_{\overline{\kappa}+1} + 2\overline{l} + 2\,. \end{aligned} \tag{5.28}$$

While proving the identity $Dg_{n_0-\overline{l}-1} \equiv 0$, we can assume as above that $0 = v_1 \leqslant v_2 \leqslant \cdots \leqslant v_{\overline{\kappa}+2-\overline{l}}$. Applying Proposition 5.1(ii) with $\varepsilon_j = \alpha_j$, $k = \overline{\kappa}$, $i_s = v_s$, $1 \leqslant s \leqslant \overline{\kappa} + 2 - \overline{l}$ (the condition $\varepsilon_{j+1} \geqslant 2\varepsilon_j + j$ holds here by (5.22) and (5.23), while the conditions $k \geqslant \overline{l} + 1$ and $i_1 = 0$ are ensured by our choice of $\overline{l}$ and v_1) we find $\kappa' \in [\overline{l}+1, \overline{\kappa}]$ such that either (5.25) or (5.26) (with $l = \overline{l}$) is true. We set

$$D' = \partial_{n_0-\overline{l}-1,\kappa'-\overline{l},\theta}(v_1, \ldots, v_{n_0-\overline{l}}) \tag{5.29}$$

and show that in both cases $D' g_{n_0-\overline{l}-1} \equiv 0$.

In the former case it follows from (5.25), (5.24), (5.28) that D' dominates some operator belonging to $\mathscr{D}(\mathfrak{m}')$, where

$$\mathfrak{m}' = (n_0 - \overline{l} - 1, \kappa' - \overline{l}, \alpha_{\kappa'} + \overline{l}, \alpha_{\kappa'} + \kappa', \gamma_{\kappa'} + 2\overline{l} + 2)\theta_{\kappa'}(\overline{l}).$$

When $\kappa' = 1$, $\overline{l} = 0$, it follows from (5.9), (5.22) that $\mathfrak{m}' \in \mathscr{A}$. By the inductive assumption, when $\kappa = \kappa' \geqslant 2$, $\alpha = \alpha_{\kappa'}$, $\gamma = \gamma_{\kappa'}$, $l = \overline{l}$ (note that $\overline{l} \leqslant \kappa' - 1$) the inclusion (4.3; l) holds. It follows that $\mathfrak{m}' \in \mathscr{A}$, and hence $D' g_{n_0-\overline{l}-1} \equiv 0$ in the case where (5.25) holds.

If we have (5.26), then by (5.28), (5.24) D' dominates some operator belonging to $\mathscr{D}(\mathfrak{m}'')$, where

$$\mathfrak{m}'' = (n_0 - \overline{l} - 1, \kappa' - \overline{l}, \alpha_{\kappa'} + \overline{l} - \rho, \alpha_{\kappa'+1} - \alpha_{\kappa'} + \rho, \gamma_{\kappa'+1} + 2\overline{l})\theta_{\overline{\kappa}+1}(\overline{l}).$$

The mark $\mathfrak{m}''$ is an annihilator. When $\kappa' = 1$, $\overline{l} = 0$, $\rho = 1$ and when $\kappa' = 1$, $\overline{l} = 0$, $\rho = 2$, this follows from (5.12) and (5.14) respectively (combined with (5.22) and (5.23)). When $\kappa' \geqslant 2$, by (5.27), (5.24) the mark $\mathfrak{m}''$ dominates the mark

$$\begin{aligned} (n_0 - \overline{l} - 1, \kappa' - \overline{l}, \alpha_{\kappa'} + \overline{l} - \rho, \alpha_{\kappa'} + \kappa' + r_{\kappa'}(\overline{l}, \rho), \\ \gamma_{\kappa'} + 2\kappa' + t_{\kappa'}(\overline{l}, \rho) - 1)\theta_{\kappa'}(\overline{l}, \rho). \end{aligned}$$

The latter is an annihilator, because by the inductive assumption, for $\kappa = \kappa'$, $\alpha = \alpha_\kappa$, $\gamma = \gamma_\kappa$, the assumptions of Lemma 4.2 are satisfied and hence inclusion (4.32) holds. Therefore, $\mathfrak{m}'' \in \mathscr{A}$ for every κ', and so $D' g_{n_0-\overline{l}-1} \equiv 0$ in both cases mentioned above.

Let

$$\mathfrak{m}_1 = \begin{cases} (n_0 - \overline{l} - 1, \overline{\kappa} - \overline{l}, \alpha_{\overline{\kappa}+1} + \overline{l} + 1, \\ \qquad \alpha_{\overline{\kappa}+1} + 2\overline{l} + 1, \gamma_{\overline{\kappa}+1} + 2\overline{l} + 2), & \overline{\kappa} \leqslant n_0 - 4, \\ (n_0 - \overline{l} - 1, \overline{\kappa} - \overline{l}, \\ \qquad \alpha_{\overline{\kappa}+1} + \overline{l} + 1, \gamma_{\overline{\kappa}+1} + 2\overline{l} + 2), & \overline{\kappa} = n_0 - 3. \end{cases}$$

Due to inclusion $(4.29; \bar{l}+1)$ proved above for $\kappa = \bar{\kappa}+1$, the mark is an annihilator. Observe that $v_j > 0$ when $\kappa'+2-\bar{l} \leqslant j \leqslant \bar{\kappa}+2-\bar{l}$. This follows from (5.25) and from (5.26), (5.27) in each of the above cases respectively (in both cases the condition $v_1 \leqslant v_2 \leqslant \cdots$ must be taken into account). Hence when letting

$$n = n_0 - \bar{l} - 1, \quad k = \bar{\kappa} + 1 - \bar{l}, \quad a = \alpha_{\bar{\kappa}+1} + \bar{l},$$
$$b = \alpha_{\bar{\kappa}+1} + \bar{l} + 1, \quad c = \gamma_{\bar{\kappa}+1} + 2\bar{l} + 2, \quad k' = \bar{\kappa} - \bar{l},$$

we see all the assumptions of Proposition 2.2 are fulfilled and, by this proposition, $Dg_{n_0-\bar{l}-1} \equiv 0$.

Summing up, one can now conclude that if conditions (a) and (b) of Lemma 4.2 are satisfied for some $\kappa = \bar{\kappa} \in [2, n_0 - 3]$ and for $\alpha = \alpha_\kappa$, $\gamma = \gamma_\kappa$, then they are also satisfied for $\kappa = \bar{\kappa} + 1$. In part 2° of this proof these conditions were checked for $\kappa = 2$, $\alpha = \alpha_2$, $\gamma = \gamma_2$ (see also (5.22)). So they are satisfied for any $\kappa \in [2, n_0 - 2]$. Apply Lemma 4.2 for $\kappa = n_0 - 2$, taking $\rho = \alpha_\kappa + l$, $l \in [0, \kappa]$ in (4.31), to obtain

$$\begin{aligned}&(n_0 - l, n_0 - 2 - l, 0,\\ &\qquad \gamma_{n_0-2} + 2(n_0-2) + t_{n_0-2}(l, \alpha_{n_0-2} + l)) \in \mathscr{A}, \quad l \in [0, n_0 - 2].\end{aligned} \tag{5.30}$$

When $l = n_0 - 2$ this takes the form

$$(2, 0, 0, \gamma_{n_0-2} + 2(n_0-2) + t_{n_0-2}(n_0 - 2, \alpha_{n_0-2} + n_0 - 2)) \in \mathscr{A}.$$

By Lemma 3.4 it follows that

$$(1, 0, \gamma_{n_0-2} + 2(n_0-2) + t_{n_0-2}(n_0 - 2, \alpha_{n_0-2} + n_0 - 2) + 2) \in \mathscr{A}. \tag{5.31}$$

Inclusions (5.30) and (5.31) together imply that for each $n \in [1, n_0]$ the function $g_n = g_n((q_1, p_1), \dots, (q_n, p_n))$ is a polynomial in $p_1, \dots, p_n$. This completes the proof of Theorem 1.10.

§6. The structure of the set $\mathscr{I}_1(U, V)$. Proof of Theorem 1.8

6.1. **Notation.** In this section we deal with functions $f\colon M_n \to \mathbb{R}$, $n \geqslant 1$, that are polynomials of the momenta with coefficients depending on the coordinates. Every such function has the form

$$f(x^n) = \sum_{i_1, \dots, i_n} c_{i_1, \dots, i_n}(q_1, \dots, q_n) p_1^{i_1} \cdots p_n^{i_n}, \qquad x^n \in M_n, \tag{6.1}$$

where $x^n = (x_1, \dots, x_n)$, $x_i = (q_i, p_i)$, $1 \leqslant i \leqslant n$, $n \geqslant 1$, and the sum is over all sequences of nonnegative integers $i_1, \dots, i_n$ whose sum does not exceed some $d \geqslant 0$. When $f \not\equiv 0$, we assume d to be chosen in such a way that at least one of the higher coefficients, i.e., $c_{i_1, \dots, i_n}$ with $i_1 + \cdots + i_n = d$, does not vanish identically. Call d the *degree of* f and denote it by $\deg f$. When $f \equiv 0$, we let $\deg f = -\infty$.

By Theorem 1.10 each function g_n, $n \geqslant 1$, is of the form (6.1). Fix the notation d_n and $\gamma_{i_1, \dots, i_n}$ for its degree and its coefficients respectively. The function $\gamma_{i_1, \dots, i_n}$ is defined on the open set

$$\Gamma_n = \{(q_1, \dots, q_n) \in \mathbb{R}^n : q_i \neq q_j, \ 1 \leqslant i < j \leqslant n\}, \qquad n = 1, 2, \dots,$$

and belongs to C^∞ there, because of a similar property of g_n (see §1). From the symmetry of g_n and from (2.1) one can easily deduce the following two properties of every $\gamma_{i_1, \dots, i_n}$, $n \geqslant 2$:

$$\gamma_{i_1, \dots, i_n}(q_1, \dots, q_n) = \gamma_{i_{\sigma(1)}, \dots, i_{\sigma(n)}}(q_{\sigma(1)}, \dots, q_{\sigma(n)}), \tag{6.2}$$

$$\lim_{|q_j| \to \infty} \gamma_{i_1, \dots, i_n}(q_1, \dots, q_n) = 0, \qquad j = 1, \dots, n. \tag{6.3}$$

Here σ is an arbitrary permutation of $i_1, \dots, i_n$ and the limit in (6.3) is taken when $j \in [1, n]$ and all q_l, $l \neq j$, are fixed, so that the left-hand side of (6.3) makes sense.

For the rest of the paper we assume the assumptions of Theorem 1.8 to be satisfied.

6.2. PROPOSITION (see [1, Proposition 4.1]). *Let $n \geqslant 1$ and*

$$f_n(x^n) = \sum_{i=1}^n p_i \frac{\partial}{\partial q_i} g_n(x^n).$$

Then $\deg f_n = d_n + 1$.

6.3. LEMMA (see [1, Lemma 4.1]). *Every coefficient of g_1 is a constant, i.e., $\partial g_1(q, p)/\partial q \equiv 0$.*

This follows immediately from equation (2.1; 1), which has the form

$$p \frac{\partial g_1(q, p)}{\partial q} = 0.$$

6.4. LEMMA. (i) *If $d_2 \geqslant 0$, then $d_1 = d_2 + 2$;*
(ii) *if $d_2 = -\infty$, then $d_1 \leqslant 0$.*

PROOF. Equation (2.2; 2) has the form

$$\{g_2(x_1, x_2), H_2(x_1, x_2)\} + \{g_1(x_1), U(q_1, q_2)\} + \{g_1(x_2), U(q_1, q_2)\} = 0,$$

i.e., it can be written as

$$f_2(x_1, x_2) - \varphi_2(x_1, x_2) - \psi_2(x_1, x_2) = 0, \tag{6.4}$$

where

$$
\begin{aligned}
f_2(x_1, x_2) &= p_1 \frac{\partial}{\partial q_1} g_2(x_1, x_2) + p_2 \frac{\partial}{\partial q_2} g_2(x_1, x_2), \\
\varphi_2(x_1, x_2) &= \frac{\partial}{\partial p_1} g_2(x_1, x_2) \frac{\partial}{\partial q_1} U(q_1, q_2) \\
&\quad + \frac{\partial}{\partial p_2} g_2(x_1, x_2) \frac{\partial}{\partial q_2} U(q_1, q_2), \\
\psi_2(x_1, x_2) &= \frac{\partial}{\partial p_1} g_1(x_1) \frac{\partial}{\partial q_1} U(q_1, q_2) + \frac{\partial}{\partial p_2} g_1(x_2) \frac{\partial}{\partial q_2} U(q_1, q_2).
\end{aligned}
\tag{6.5}
$$

By Proposition 6.2 $\deg f_2 = d_2 + 1$, so that (see (6.4)) $\deg(\varphi_2 + \psi_2) = d_2 + 1$. Since, clearly, $\deg \varphi_2 \leqslant d_2 - 1$ it follows that $\deg(\varphi_2 + \psi_2) = \deg \psi_2$, i.e.,

$$\deg \psi_2 = d_2 + 1. \tag{6.6}$$

If either (a) $\deg \psi_2 \geqslant 0$, $d_1 - 1 > \deg \psi_2$ or (b) $\deg \psi_2 = -\infty$, $d_1 > 0$, then the coefficient of $p_1^{d_1 - 1}$ in ψ_2 is equal to zero, i.e., $d_1 \gamma_{d_1} \partial U(q_1, q_2)/\partial q_1 = 0$ (see (6.5) and Lemma 6.3). Since by (1.8) $\partial U(q_1, q_2)/\partial q_1 \not\equiv 0$, it follows that $d_1 \gamma_1 = 0$. In both cases (a) and (b) $d_1 \neq 0$, so $\gamma_{d_1} = 0$, which contradicts the definition of d_1. We now see that $d_1 \leqslant 1 + \deg \psi_2$ when $\deg \psi_2 \geqslant 0$ and that $d_1 \leqslant 0$ when $\deg \psi_2 = -\infty$. Due to (6.6) and to the obvious inequality $d_1 \geqslant 1 + \deg \psi_2$ (see (6.5)), this implies both assertions of the lemma.

6.5. LEMMA. *If $d_3 \geqslant 0$, then $d_3 = d_2$.*

PROOF. Rewrite (2.2; 2) in the form

$$f_3(x_1, x_2, x_3) - \varphi_3(x_1, x_2, x_3) - \psi_3(x_1, x_2, x_3) = 0, \tag{6.7}$$

where

$$
\begin{aligned}
f_3(x_1, x_2, x_3) &= p_1 \frac{\partial}{\partial q_1} g_3(x_1, x_2, x_3) + p_2 \frac{\partial}{\partial q_2} g_3(x_1, x_2, x_3) \\
&\quad + p_3 \frac{\partial}{\partial q_3} g_3(x_1, x_2, x_3), \\
\psi_3(x_1, x_2, x_3) &= \frac{\partial}{\partial p_1} g_1(x_1) \frac{\partial}{\partial q_1} V(q_1, q_2, q_3) \\
&\quad + \frac{\partial}{\partial p_2} g_1(x_2) \frac{\partial}{\partial q_2} V(q_1, q_2, q_3) + \frac{\partial}{\partial p_3} g_1(x_3) \frac{\partial}{\partial q_3} V(q_1, q_2, q_3),
\end{aligned}
$$

and where φ_3 is defined by these equalities combined with (2.2; 3) and (6.7).

By Proposition 6.2,

$$\deg f_3 = d_3 + 1. \tag{6.8}$$

If $d_2 = -\infty$, then $\deg \varphi_3 \leqslant d_3 - 1$ (this follows from the definition of φ_3) and by (6.7),

$$\deg \psi_3 = d_3 + 1. \tag{6.9}$$

From the definition of ψ_3 it is seen that $\deg\psi_3 \leqslant d_1 - 1$. Hence (6.9) together with the assumption $d_3 \geqslant 0$ leads to the inequality $d_1 \geqslant d_3+2$. If, however, $d_2 \geqslant 0$, then by Lemma 6.4(i) we have $d_1 = d_2+2$. So in any case $d_2 \geqslant 2$, and by Lemma 6.4(ii), (i),

$$d_2 \geqslant 0, \qquad d_1 = d_2 + 2. \tag{6.10}$$

Let us show that

$$\deg\psi_3 = d_1 - 1. \tag{6.11}$$

For this it is sufficient to verify the inequality $\deg\psi_3 \geqslant d_1 - 1$. Assume that the opposite inequality is true. Then the coefficient of $p_1^{d_1-1}$ in ψ_3 vanishes, i.e., we have

$$d_1\gamma_{d_1}\frac{\partial}{\partial q_1}V(q_1, q_2, q_3) = 0 \quad \text{for all } (q_1, q_2, q_3) \in \Gamma_3.$$

From (1.9) and the assumption $V \not\equiv 0$ it follows that

$$\frac{\partial}{\partial q_1}V(q_1, q_2, q_3) \not\equiv 0.$$

Hence $d_1\gamma_{d_1} = 0$, and, since $d_1 \geqslant 2$ (see (6.10)), we have $\gamma_{d_1} = 0$. The latter contradicts the definition of d_1. So (6.11) is true.

From (6.7), (6.8), (6.10), (6.11), and the obvious inequality

$$\deg\varphi_3 \leqslant \max(d_3 - 1, d_2 - 1) \tag{6.12}$$

we get

$$\begin{aligned} d_3 + 1 = \deg(\varphi_3 + \psi_3) &\leqslant \max(\deg\varphi_3, \deg\psi_3) \\ &\leqslant \max(d_1 - 1, d_2 - 1, d_3 - 1) = \max(d_1 - 1, d_3 - 1), \end{aligned}$$

which shows that $d_1 - 1 > d_3 - 1$. Then by (6.10)–(6.12) $\deg\varphi_3 < \deg\psi_3$ and hence

$$d_3 + 1 = \deg(\varphi_3 + \psi_3) = \deg\psi_3 = d_1 - 1,$$

i.e., $d_1 = d_3 + 2$. Comparing this equation with (6.10) we come to the required equality.

6.6. LEMMA. *For any $n \geqslant 3$ we have*

$$\max(d_{n-1}, d_{n-2}) \geqslant d_n + 2. \tag{6.13}$$

PROOF. Represent (2.2; n) in the form $f_n - h_n = 0$, where f_n is defined in Proposition 6.2. By this equation and Proposition 6.2,

$$\deg h_n = \deg f_n = d_n + 1.$$

On the other hand, the explicit form of h_n (see (2.2; n)) shows that

$$\deg h_n \leqslant \max(d_n - 1, d_{n-1} - 1, d_{n-2} - 1).$$

Hence $\max(d_n - 1, d_{n-1} - 1, d_{n-2} - 1) \geqslant d_n + 1$, and so

$$\max(d_{n-1} - 1, d_{n-2} - 1) \geqslant d_n + 1.$$

From this inequality we get (6.13).

6.7. LEMMA. *If* $d_3 = -\infty$, *then* $d_1 \leqslant 0$ *and* $d_n = -\infty$ *for all* $n \geqslant 2$.

PROOF. Let us first show that $d_3 = -\infty$ implies $d_2 = -\infty$. If the latter is not true, then by Lemma 6.4 we have $d_1 = d_2 + 2 \geqslant 2$. Hence the coefficient of $p_1^{d_1 - 1}$ on the left-hand side of (2.2; 3) vanishes, i.e.,

$$d_1 \gamma_{d_1} (\partial / \partial q_1) V(q_1, q_2, q_3) \equiv 0.$$

Since $d_1 \geqslant 2$ and $\gamma_{d_1} \neq 0$ (by the definition of d_1), $(\partial / \partial q_1) V(q_1, q_2, q_3) \equiv 0$. This contradicts (1.9) combined with the assumption $V \not\equiv 0$. So $d_2 = -\infty$ and, since $d_3 = -\infty$, Lemma 6.6 implies that $d_n = -\infty$ for all n.

Suppose that $d_1 > 0$. Then the coefficient of $p_1^{d_1 - 1}$ on the left-hand side of (2.2; 2) is $d_1 \gamma_{d_1} (\partial / \partial q_1) U(q_1, q_2)$, and since $\gamma_{d_1} \neq 0$ we have

$$(\partial / \partial q_1) U(q_1, q_2) = 0 \quad \text{for all} \quad (q_1, q_2) \in \Gamma_2.$$

This contradicts (1.8) and hence $d_1 \leqslant 0$ as required.

We now can specify all higher coefficients in the expansion of g_2.

6.8. LEMMA. *If* $d_2 \geqslant 0$, *then*

$$\gamma_{ij}(q_1, q_2) = d_1 \gamma_{d_1} U(q_1, q_2), \qquad (q_1, q_2) \in \Gamma_2, \tag{6.14}$$

for any i, j *such that* $i + j = d_2$.

PROOF. By Lemma 6.4 and the inequality $d_2 > 0$, we have $d_1 = d_2 + 2$. With this in mind, set the coefficients of $p_1^k p_2^l$ on the left-hand side of (2.2; 2) for $k + l = d_2 + 1$, $k \geqslant 1$, equal to zero. This gives

$$\frac{\partial}{\partial q_1} \gamma_{d_2, 0}(q_1, q_2) - d_1 \gamma_{d_1} \frac{\partial}{\partial q_1} U(q_1, q_2) = 0, \tag{6.15}$$

$$\frac{\partial}{\partial q_1} \gamma_{k-1, l}(q_1, q_2) + \frac{\partial}{\partial q_2} \gamma_{k, l-1}(q_1, q_2) = 0, \tag{6.16}$$

$$1 \leqslant k \leqslant d_2, \; l = d_2 + 1 - k.$$

From (6.14) it follows that, when q_2 is fixed, the function $\gamma(q) = \gamma_{d_2, 0}(q, q_2) - d_1 \gamma_{d_1} U(q_1, q_2)$ is constant on each of the two connected components of the set $\{q \in \mathbb{R} : q \neq q_2\}$. Since (6.3) and the assumptions on U together imply that $\gamma(q) \to 0$ as $q \to \pm\infty$, we come to (6.14) with $i = d_2$, $j = 0$. This completes the proof when $d_2 = 0$. If $d_2 > 0$, suppose (6.14) is true for some pair i, j, where $i \in [1, d_2]$, $j = d_2 - i$. Then (6.16) for $k = i$, $l = d_2 + 1 - i$ takes the form

$$\frac{\partial}{\partial q_1} \gamma_{i-1, j+1}(q_1, q_2) - d_1 \gamma_{d_1} \frac{\partial}{\partial q_1} U(q_1, q_2) = 0,$$

which as above implies (6.14) for the pair $(i - 1, j + 1)$. Thus we have proved the lemma by induction.

6.9. LEMMA. $d_3 \leqslant 0$.

PROOF. Let d stand for d_3 and suppose that $d > 0$. Then by Lemmas 6.4, 6.5 we have $d = d_2 = d_1 - 2$. With this in mind look at the coefficients of $p_3^{d+1}, p_3^d p_2, \ldots, p_3 p_2^d, p_2^{d+1}$ on the left-hand side of (2.2; 3). Setting them equal to zero we consecutively obtain

$$\frac{\partial}{\partial q_3} \gamma_{0,0,d}(q_1, q_2, q_3) - (d+2)\gamma_{d+2} \frac{\partial}{\partial q_3} V(q_1, q_2, q_3) = 0, \qquad (6.17;0)$$

$$\frac{\partial}{\partial q_3} \gamma_{0,k,d-k}(q_1, q_2, q_3) + \frac{\partial}{\partial q_2} \gamma_{0,k-1,d-k+1}(q_1, q_2, q_3) = 0, \qquad (6.17;k)$$

$$0 < k < d+1$$

$$\frac{\partial}{\partial q_2} \gamma_{0,d,0}(q_1, q_2, q_3) - (d+2)\gamma_{d+2} \frac{\partial}{\partial q_2} V(q_1, q_2, q_3) = 0. \quad (6.17;d+1)$$

For each $k \in [0, d+1]$ apply the operator

$$(-1)^k \frac{\partial^{d+1}}{\partial q_2^{d+1-k} \partial q_3^k}$$

to (6.17; k) and sum the results over k. After dividing by $-(d+2)\gamma_{d+2}$ we obtain

$$\frac{\partial^{d+2}}{\partial q_2^{d+1} \partial q_3} V(q_1, q_2, q_3) + (-1)^{d+1} \frac{\partial^{d+2}}{\partial q_2 \partial q_3^{d+1}} V(q_1, q_2, q_3) = 0.$$

This equality holds on Γ_3, and since V is the restriction of the smooth function V_1 to Γ_3 (see condition 1.5(f)) it follows that

$$\frac{\partial^{d+2}}{\partial q_2^{d+1} \partial q_3} V(q_1, q_2, q_3) + (-1)^{d+1} \frac{\partial^{d+2}}{\partial q_2 \partial q_3^{d+1}} V_1(q_1, q_2, q_3) = 0,$$
$$(q_1, q_2, q_3) \in \mathbb{R}^3. \quad (6.18)$$

Condition 1.5(f) allows us to integrate (6.18) with respect to q_2 and q_3 successively, obtaining

$$\frac{\partial^d}{\partial q_2^d} V_1(q_1, q_2, q_3) + (-1)^{d+1} \frac{\partial}{\partial q_3^d} V_1(q_1, q_2, q_3) = 0, \qquad (q_1, q_2, q_3) \in \mathbb{R}^3. \quad (6.19)$$

Using condition 1.5(f) once more and the Fubini Theorem, one can find a set $R_0 \subset \mathbb{R}^1$ such that $\operatorname{meas}(\mathbb{R}^1 \setminus R_0) = 0$ and for any $q_1^0 \in R_0$ all the functions

$$\frac{\partial^{n+m}}{\partial q_2^n \partial q_3^m} V_1(q_1^0, q_2, q_3), \qquad n, m = 0, 1, \ldots,$$

belong to $L^2(\mathbb{R}^2)$ as functions in q_2, q_3. Then by fixing a $q_1^0 \in R_0$, we can pass from the function $W(x, y) = V_1(q_1^0, x, y)$, $(x, y) \in \mathbb{R}^2$, to its Fourier image $\widetilde{W}(s, t)$ and get from (6.19) the following equation:

$$[s^d + (-1)^{d+1} t^d] \widetilde{W}(s, t) = 0, \qquad (s, t) \in \mathbb{R}^2,$$

Since $d \geqslant 1$ and $\widetilde{W}$ is a continuous function, we see that $\widetilde{W} \equiv 0$ and hence $W \equiv 0$, i.e., $V_1(q_1^0, q_1, q_2) = 0$ for almost all $q_1^0 \in \mathbb{R}^1$ and all $(q_1, q_2) \in \mathbb{R}^2$. But this is impossible because V_1 is continuous and $V_1 \not\equiv 0$. Thus the inequality $d > 0$ is impossible. □

6.10. **Completion of the proof of Theorem 1.8.** By Lemma 6.9, $d_3 \leqslant 0$. If $d_3 = -\infty$, then by Lemma 6.7 we have $d_1 \leqslant 0$, $d_n = -\infty$ for all $n \geqslant 2$, and hence we deal with an integral proportional to the classical integral of the particle number. So assume that $d_3 = 0$, i.e., $g_3 \equiv \gamma_{0,0,0}$, and set the coefficient of p_3 on the left-hand side of (2.2; 3) equal to zero, obtaining

$$\frac{\partial}{\partial q_3} \gamma_{0,0,0}(q_1, q_2, q_3) = 2\gamma_2 \frac{\partial}{\partial q_3} V(q_1, q_2, q_3), \qquad (q_1, q_2, q_3) \in \Gamma_3.$$

By (1.9), (6.3) it follows that $\gamma_{0,0,0} \equiv 2\gamma_2 V$ on the set Γ_3^3, where

$$\Gamma_3^3 = \{(q_1, q_2, q_3) \in \Gamma_3 : q_3 = \max(q_1, q_2, q_3) \text{ or } q_3 = \min(q_1, q_2, q_3)\}.$$

Since $\gamma_{0,0,0} \equiv g_3$ and V, g_3 are symmetric functions, we have $g_3 = 2\gamma_2 V$ on the whole set Γ_3.

Now turn to the functions g_2 and g_1. By Lemma 6.5, $d_2 = 0$. Then $g_2 \equiv \gamma_{0,0}$ and by Lemma 6.8, $g_2 \equiv 2\gamma_2 U$. By Lemma 6.4(i), $d_1 = 2$ and hence $g_1(q, p) = \gamma_2 p^2 + \gamma_1 p + \gamma_0$, where γ_2, γ_1, γ_0 are constants (see Lemma 6.3). Finally, by Lemma 6.6, $d_n = -\infty$ for all $n \geqslant 4$.

The form of the functions g_n, $n \geqslant 1$, that we have just found means that the integral under consideration is a linear combination of the integrals of energy, momentum, and particle number with coefficients $2\gamma_2$, γ_1, and γ_0 respectively. This concludes the proof of Theorem 1.8.

References

1. B. M. Gurevich, *Asymptotically additive integrals of motion of particles with pairwise interaction*, Trudy Moskov. Mat. Obshch. **52** (1989), 175–228; English transl., Trans. Moscow Math. Soc. **1990**, 177–232.
2. ———, *Invariant measures of dynamical systems of statistical mechanics and first integrals of Hamiltonian systems with finitely many degrees of freedom*, Uspekhi Mat. Nauk **41** (1986), no. 2, 193–194; English transl., Russian Math. Surveys **41** (1986), no. 2, 201–202.
3. L. D. Landau and E. M. Lifshitz, *Mechanics*, Fizmatigiz, Moscow, 1958; English transl., *Course of theoretical physics*. Vol. 1, Pergamon Press, Oxford, and Addison-Wesley, Reading, MA, 1960.
4. B. M. Gurevich and Yu. M. Sukhov, *Stationary solutions of the Bogolyubov hierarchy equations in classical statistical mechanics*, Dokl. Akad. Nauk SSSR **223** (1975), no. 2, 276–279; English transl., Soviet Math. Dokl. **16** (1975), 857–861.
5. ———, *Stationary solutions of the Bogolyubov hierarchy equations in classical statistical mechanics*, 1–4, Comm. Math. Phys. **49** (1976), 63–96; **54** (1977), 81–96; **56** (1977), 225–236; **84** (1982), 333–376.

6. J. Moser, *Three integrable Hamiltonian systems connected with isospectral deformations*, Adv. Math. **16** (1975), 197–220.

7. F. Calogero, *Exactly solvable one-dimensional many-body problems*, Lett. Nuovo Cimento (2) **13** (1975), 411–416.

Moscow State University, Mechanics and Mathematics Department, Moscow 119899, Russia

Translated by the author

ADVANCES IN SOVIET MATHEMATICS
Volume 20, 1994

On a Ground State in the Frenkel–Kontorova Model and Metric Properties of Mappings of Standard Type

L. D. PUSTYL′NIKOV

Let us consider an infinite number of points (particles) P_n on the line $-\infty < q < \infty$ interacting with the potential

$$U = \sum_n g(q_n) + \sum_n (q_{n+1} - q_n - \alpha)^2,$$

where n is an integer, q_n is the coordinate of the point P_n on the line, $g(q) = \gamma G(q)$ is an infinitely differentiable function of q of period 1 with minima at the points $q = n$ and maxima at the points $q = n + 1/2$, γ and α are parameters, $\gamma > 0$, $\alpha > 0$. From the physical point of view this means that the system is a chain of springs in the outer field with potential $g(q)$. If we consider this system having in mind the problem of obtaining the ground state (i.e., the state with the minimal energy) the following questions arise:

(1) how to obtain a limit Gibbs state with temperature tending to zero;
(2) to what extent the distribution depends on the parameter α when the potential $g(u)$ is fixed (see [1–3]).

In his paper [1], Sinai reduced the problem to the consideration of metric properties of a standard transformation of the cylinder. With the help of such a reduction, he showed the existence of phase transitions, i.e., the existence of different limit Gibbs states for $\alpha = 1$ and a certain $\alpha \neq 1$ (see [2, 3]). In the present article we describe the new ground state for $\alpha \neq 1$, but in contrast to the above-mentioned works, here the values of $\alpha \neq 1$ for which the ground state exists are given. We also obtain the ground state in terms of invariant measures of a standard mapping of the cylinder. Besides, the ground state possesses the following properties of stability: it remains the same if α changes in the neighborhood of the value $\alpha = 1/2$ and the function $g(q)$ belongs to an open set in the space $\widehat{K}$ with the properties mentioned above and with the natural norm (see Theorem 3). From the point of view of dynamical systems, the problem is reduced to the following.

1991 *Mathematics Subject Classification*. Primary 82B20; Secondary 46N10, 58F11.

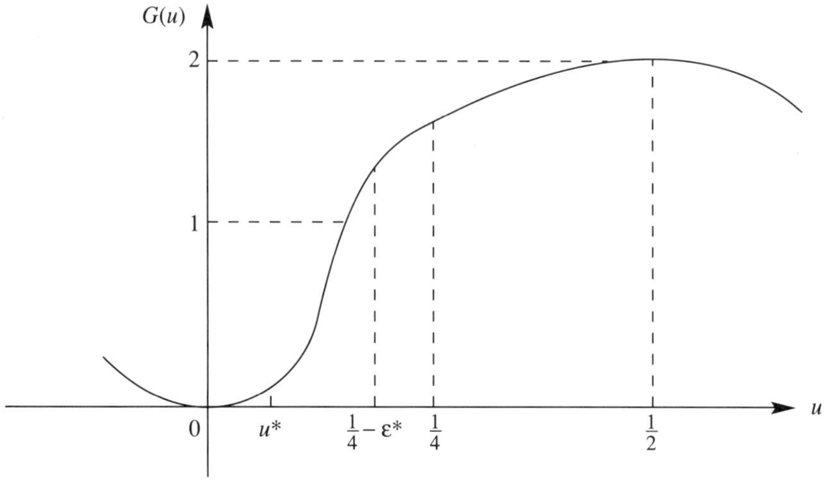

FIGURE 1

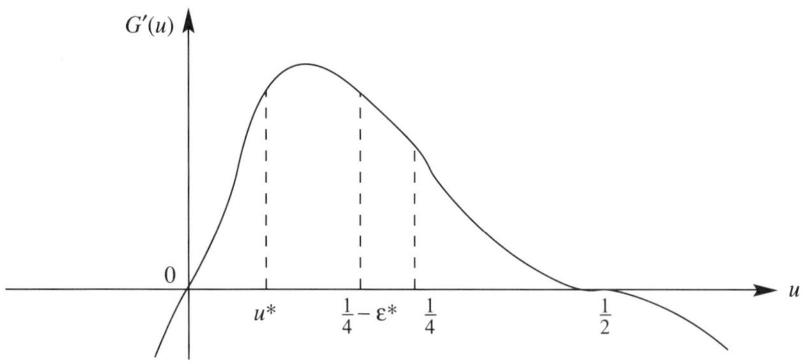

FIGURE 2

Consider the transformation A of the cylinder $C = \{u, z : 0 \leqslant u \leqslant 1, -\infty < z < \infty\}$, of the type

$$A = \begin{cases} u' = u + z' \bmod 1, \\ z' = z + g'(u)/2, \end{cases} \qquad \left(g'(u) = \frac{dg}{du}(u)\right) \tag{1}$$

and the function

$$H = H(u, z) = g(u) + (z - \alpha)^2. \tag{2}$$

It is necessary to find a measure μ in the class of all Borel measures $\widetilde{\mu}$ that are normalized and invariant with respect to the transformation A on C and satisfy the condition that the integral

$$\int_C H \, d\widetilde{\mu} \tag{3}$$

takes the minimal value at $\widetilde{\mu} = \mu$.

Theorem 3 of this article shows that if α belongs to a neighborhood of the value $\alpha = 1/2$ and the function $g(u)$ belongs to an open set in the space $\widehat{K}$, such an invariant measure μ is concentrated at the points $x_1 = (0, 1/2)$, $x_2 = (1/2, 1/2)$ and $\mu(x_1) = \mu(x_2) = 1/2$, whereas if $\alpha = 1$ we have the measure concentrated at the point $(0, 1)$. Let us come back to the problem of the distribution of the particles on the line. From our results it follows that the particles are on the line at integer and semi-integer points when the conditions of Theorems 1, 2, 3 are satisfied. On the other hand, if $\alpha = 1$, particles are at integer points only.

The set of potentials $g(u)$ is described in Theorem 3 (there they are denoted by $\widetilde{g}(u)$) as an open neighborhood of a concrete function $g(u) = \gamma G(u)$ specified in Theorem 2. The graph of $G(u)$ is shown in Figure 1 and the graph of the derivative $G'(u)$ is shown in Figure 2. The function $G(u)$ is even and periodic with period 1, the absolute value of the derivative $|G'(u)|$ in the neighborhood of $u = 0$ decreases slowly as $u \to 0$, as compared with the neighborhood of the point $u = 1/2$ as $u \to 1/2$. Due to this last circumstance, the two periodic points $x_1 = (0, 1/2)$, $x_2 = (1/2, 1/2)$ with respect to the transformation A on C are hyperbolic, which is decisive for proving the stability of the ground state. Theorem 1 states general conditions on the function $G(u)$ which, when $\alpha = 1/2$, provide a ground state such that the invariant measure (with respect to A) minimizing the integral (3) are concentrated at points x_1 and x_2.

The function $G = G(q)$ can be obtained from the function $G^*(q) = 1 - \cos(2\pi q)$ by a deformation such that in the domain $1/4 - \varepsilon < q < 3/4 + \varepsilon$ $(0 < \varepsilon < 1/4)$ we have $G(q) \geqslant G^*(q)$, $G(1/2) = G^*(1/2)$ and in the domain $-1/4 + \varepsilon \leqslant q \leqslant 1/4 - \varepsilon$ we have $G(q) = G^*(q)$.

Note that for the function $g(u) = \gamma G^*(u)$ with a small γ and $\alpha = 1/2$, the Borel normalized measure on C which is invariant with respect to A and minimizes the integral (3) cannot occupy the points $(0, 1/2)$, $(1/2, 1/2)$, because otherwise the second order periodic trajectory of the transformation A which consists of these points would be elliptic. Moreover, the proof of Theorem 1 implies that for the function $g(u) = \gamma G^*(u)$ with a small γ and $\alpha = 1/2$, the ground state cannot possibly occupy the second order trajectory of the transformation A.

Thus the present article reveals a new effect which shows that the ground state changes in general as the result of the deformation of the function $G(q)$, when γ and α are as described above.

The main result of this article was announced in [4].

§1. General definitions

DEFINITION 1. We say that the real-valued function $G = G(u)$ defined on the real line *belongs to the class* $K(\gamma)$ if it possesses the following properties.

(1) The function G is an infinitely differentiable even function of u with period 1.
(2) The function G has a unique minimum in the domain $0 \leqslant u \leqslant 1$ at $u = 0$, and a maximum for $u = 1/2$; $G(0) = 0$, $G(1/2) = 2$.
(3) The function G' vanishes in the domain $0 \leqslant u < 1$ only in the points $u = 0, 1/2$.
(4) $G(1/4 - \varepsilon^*) - 1 > 0$, where $0 < \varepsilon^* < 1/4$.
(5) The system of equation

$$G'(u) = -G'(u'), \qquad u' = u + 1/2 + \gamma G'(u)/2$$

has a solution only if at least one of the following conditions holds:
a) $u' \equiv -u \bmod 1$,
b) $u' \equiv u + 1/2 \bmod 1$,
c) $G(u) + G(u') > 2$.

REMARK. An example of a function G belonging to the class $K(\gamma)$ when $\gamma > 0$ is sufficiently small is given in Figure 1.

DEFINITION 2. Let $g = g(u)$ be a smooth real function of period 1 and α be a real number. The normalized Borel measure μ on the cylinder

$$C = \{u, z : 0 \leqslant u \leqslant 1, -\infty < z < \infty\} \qquad (\mu(C) = 1)$$

will be called a (g, α)-*measure* if μ is invariant with respect to the transformation A defined by (1) and if among all normalized Borel measures $\widetilde{\mu}$ invariant with respect to A the integral (3) of the function H from (3) takes the minimal value at $\widetilde{\mu} = \mu$.

§2. Statements of the theorems

THEOREM 1. *Let $G \in K(\gamma)$, and let c_1 and c_2 be constants such that*

$$c_1 \geqslant \max_{0 \leqslant u \leqslant 1} |G'(u)|, \quad c_2 \geqslant \max_{0 \leqslant u \leqslant 1} |G''(u)|, \qquad \alpha = 1/2, \quad \gamma_0 = \min(1/c_2, 2\varepsilon^*/c_1).$$

If $0 < \gamma < \gamma_0$ and $g = g(u) = \gamma G$, then the (g, α)-measure μ is concentrated at the points $x_1 = (0, 1/2)$ and $x_2 = (1/2, 1/2)$, and $\mu(x_1) = \mu(x_2) = 1/2$.

THEOREM 2. *Let us assume that the function G satisfies the following conditions (Figures 1, 2).*

(1) *Conditions (1)–(4) from Definition 1 in §1 hold.*
(2) *In the domain $0 < u < 1/2$, the function G' has a unique extremum which is a maximum and is contained in the domain $0 < u < 1/4 - \varepsilon^*$.*
(3) *There exists a point u^* in the domain $0 < u < 1/2$ such that $u^* \neq 1/4 - \varepsilon^*$, $G'(u^*) = G'(1/4 - \varepsilon^*)$ and when $0 < u \leqslant u^*$ we have $|G'(u)| > |\widetilde{G}'(2u)|$, where $\widetilde{G}(u) = 2 - G(u + 1/2)$.*
(4) $G''(0) \neq 0$, $G''(0) > \widetilde{G}''(0)$.

Then there exists $\widetilde{\gamma}_0 > 0$ such that for $0 < \gamma < \widetilde{\gamma}_0$, $g = \gamma G$ there exists $\delta = \delta(\gamma)$ such that if $|\alpha - 1/2| < \delta$, then the (g, α)-measure is concentrated at the points $x_1 = (0, 1/2)$, $x_2 = (1/2, 1/2)$.

THEOREM 3. *Consider the space* $\widehat{K}$ *of infinitely differentiable real functions* $g = g(u)$ *of* u *with period* 1, *such that in the domain* $0 \leqslant u \leqslant 1$ *the value at* $u = 0$ *is the minimum point, the value at* $u = 1/2$ *is the maximum point and the value*

$$\|g\| = \max_{0 \leqslant u \leqslant 1} |g(u)| + \sum_{i=1}^{3} \max_{0 \leqslant u \leqslant 1} \left| \frac{d^i g}{du^i}(u) \right|$$

is taken to be the norm $\|g\|$ *of the function* $g \in \widehat{K}$. *Then there is a constant* $\widetilde{\gamma}_0 > 0$ *such that if* $0 < \gamma < \widetilde{\gamma}_0$ *and* $g = \gamma G$, *then there exist constants* $\delta = \delta(\gamma) > 0$, $\varepsilon_0 = \varepsilon_0(\gamma) > 0$ *such that the* $(\widetilde{g}, \alpha)$*-measure is concentrated at the points* $x_1 = (0, 1/2)$, $x_2 = (1/2, 1/2)$ *if* $|\alpha - 1/2| < \delta$, $\widetilde{g} \in \Omega_{\varepsilon_0}$, Ω_{ε_0} *being an* ε_0*-neighborhood of the function* g *in the space* $\widehat{K}$.

§3. Proof of the theorems

PROOF OF THEOREM 1.

LEMMA 1. *Let* $(u', z') = A(u, z)$. *If* $0 < \gamma < \gamma_0$, $g = \gamma G$, $\alpha = 1/2$, *then the function* $f(u, z) = H(u, z) + H(u'(u, z), z'(u, z))$ *has its minima only at the points* $x_1 = (0, 1/2)$, $x_2 = (1/2, 1/2)$.

PROOF OF LEMMA 1. The problem of obtaining the minimum of $f(u, z)$ is equivalent to the problem of finding the conditional minimum of the function

$$F = F(u, z, u', z') = H(u, z) + H(u', z')$$

in four variables u, z, u', z' satisfying the following conditions:

$$\psi = \psi(u, z, u', z') = u' - u - z' = 0, \tag{4}$$

$$\widetilde{\psi} = \widetilde{\psi}(u, z, u', z') = z' - z - \varphi(u) = 0, \tag{5}$$

where $\varphi(u) = g'(u)/2$.

From the definitions of the functions ψ, $\widetilde{\psi}$ it follows that the vectors

$$\psi' = (\psi'_u, \psi'_z, \psi'_{u'}, \psi'_{z'}) \quad \text{and} \quad \widetilde{\psi}' = (\widetilde{\psi}'_u, \widetilde{\psi}'_z, \widetilde{\psi}'_{u'}, \widetilde{\psi}'_{z'})$$

are linearly independent for $\psi'_{u'} = -\psi'_{z'} = 1$, $\widetilde{\psi}'_{z'} = 1$. $\widetilde{\psi}'_{u'} = 0$. Hence the necessary condition for the minimum of the function F under the restrictions (4), (5) is that the Lagrange function

$$\begin{aligned} \mathcal{L} &= \mathcal{L}(u, z, u', z', \lambda_1, \lambda_2) \\ &= F(u, z, u', z') + \lambda_1(u' - u - z') + \lambda_2(z' - z - \varphi(u)), \end{aligned}$$

where λ_1, λ_2 are independent parameters ([5]), have an extremum; in its turn, necessary conditions of the extremum of the function $\mathcal{L}$ have the following form:

$$\frac{\partial \mathcal{L}}{\partial u} = g'(u) - \lambda_1 - \lambda_2\varphi'(u) = 0 \qquad \left(\varphi'(u) = \frac{d\varphi}{du}(u)\right), \tag{6}$$

$$\frac{\partial \mathcal{L}}{\partial z} = 2(z-\alpha) - \lambda_2 = 0, \tag{7}$$

$$\frac{\partial \mathcal{L}}{\partial u'} = g'(u') + \lambda_1 = 0, \tag{8}$$

$$\frac{\partial \mathcal{L}}{\partial z'} = 2(z'-\alpha) - \lambda_1 + \lambda_2 = 0. \tag{9}$$

From (6)–(9) we have

$$\lambda_1 = g'(u) - \lambda_2\varphi'(u),$$

$$z = \lambda_2/2 + \alpha, \tag{10}$$

$$z' = (\lambda_1 - \lambda_2)/2 + \alpha = (g'(u) - \lambda_2\varphi'(u) - \lambda_2)/2 + \alpha, \tag{11}$$

$$g'(u') = -\lambda_1 = \lambda_2\varphi'(u) - g'(u). \tag{12}$$

Substituting (10) and (11) into (5), we obtain

$$\lambda_2(1 + \varphi'(u)/2) = 0. \tag{13}$$

Since for $\gamma < \gamma_0$ we have $|\varphi'(u)| < 1$, it follows from (13) that

$$\lambda_2 = 0. \tag{14}$$

Now substituting (14) into (10), (11), and (12), we get

$$z = \alpha, \tag{15}$$

$$z' = g'(u)/2 + \alpha, \tag{16}$$

$$g'(u') = -g'(u), \tag{17}$$

and from (4) and (16) we obtain

$$u' = u + g'(u)/2 + \alpha. \tag{18}$$

By condition (5) of Definition 1, if $\alpha = 1/2$ the solutions of the system of equations (17), (18) can only be values u, u' that satisfy at least one of the following relations:

$$u' = -u + n, \tag{19}$$

$$u' = u + 1/2 + n, \tag{20}$$

where n is an integer,

$$G(u) + G(u') > 2. \tag{21}$$

Let us consider two cases, I and II.

CASE I. Relation (19) holds.

Substituting (18) into (19) when $\alpha = 1/2$, we obtain

$$-2u - \varphi(u) = 1/2 - n. \tag{22}$$

All the solutions of equation (22) can be split into two classes.

The first class consists of solutions u satisfying (22) when n is even, and the second class consists of solutions u satisfying (22) with n odd. It is obvious that any two solutions in the same class differ by an integer. Thus to obtain the complete solution of equation (22) it is sufficient to find one solution from each class.

For this purpose let us solve (22) for $n = 0$ and for $n = 1$.

(1) Let $n = 0$. If $\gamma < \gamma_0$, then Definition 1 implies that equation (22) has the solution

$$u = -1/4 + k, \tag{23}$$

where $0 < k = -\varphi(u)/2$.

(2) Let $n = 1$. Similarly, when $\gamma < \gamma_0$ equation (22) has solution

$$u = 1/4 - k. \tag{24}$$

CASE II. Relation (20) holds.

Substituting (18) into (20) for $\alpha = 1/2$, we obtain $\varphi = n$, which becomes $\varphi = 0$ when $\gamma < \gamma_0$.

By the condition (3) from Definition 1, the two values of u

$$u = 0, \tag{25}$$

$$u = 1/2 \tag{26}$$

are the solutions of the last equality modulo integers.

Thus we obtain four possible solutions (23)–(26) for u. Using (18), (15), (16) to get the values u', z, z' for each of these values of u, we obtain that the solution (u, u', z, z') of the problem of the minimum of the function F under the conditions (4), (5) either coincides with one of the four vectors

a) $u = -1/4 + k, \quad u' = -u, \quad z = 1/2, \quad z' = \varphi(u) + 1/2;$

b) $u = 1/4 - k, \quad u' = -u + 1, \quad z = 1/2, \quad z' = \varphi(u) + 1/2;$

c) $u = 0, \quad u' = 1/2, \quad z = 1/2, \quad z' = 1/2;$

d) $u = 1/2, \quad u' = 1, \quad z = 1/2, \quad z' = 1/2,$

or inequality (21) holds.

If (21) holds, we have $F(u, z, u', z') > 2\gamma$. Let us show that if $\gamma < \gamma_0$ in the cases a), b), then we have $F(u, z, u', z') > 2\gamma$. Indeed, if $\gamma < \gamma_0$, we have $0 < k < \varepsilon^*$, and by conditions (2), (3) and (4) of Definition 1 in the cases a), b) we get

$$F = g(u) + g(-u) + \varphi^2(u) > 2\gamma.$$

Provided c) and d) hold, we have $F(u, z, u', z') = 2\gamma$.

Thus c) and d) are points of global minimum of the function F. Lemma 1 is proved.

Let $\widetilde{\mu}$ be an arbitrary normalized measure (on the cylinder C) which is invariant with respect to the transformation A. This implies

$$2\int_C H\,d\widetilde{\mu} = \int_C f\,d\widetilde{\mu},$$

where f is the function defined by Lemma 1.

Therefore, to prove Theorem 1 it is sufficient to show that if $0 < \gamma < \gamma_0$ and $\alpha = 1/2$, then for all such measures $\widetilde{\mu}$ the integral $\int_C f\,d\widetilde{\mu}$ takes its least value on the Borel normalized measure μ invariant with respect to A. But due to Lemma 1, this fact obviously follows from the inequality

$$\int_C f\,d\widetilde{\mu} \geqslant f(0, 1/2)\widetilde{\mu}(C) = f(0, 1/2)\mu(C)$$

and from the assertion that $x_1 = (0, 1/2)$ and $x_2 = (1/2, 1/2)$ are periodic points of the second order for the transformation A and $Ax_1 = x_2$.

Theorem 1 is proved.

PROOF OF THEOREM 2. Let c_1 and c_2 be constants, such that

$$c_1 \geqslant \max_{0\leqslant u\leqslant 1} |G'(u)|, \qquad c_2 \geqslant \max_{0\leqslant u\leqslant 1} |G''(u)|,$$

and γ_0 be the number introduced in the statement of Theorem 1. Let us prove that if $0 < \gamma < \gamma_0$, we have $G(u) \in K(\gamma)$. Properties (1)–(4) from Definition 1 obviously follow from the first condition of Theorem 2. Let us prove the validity of condition (5) from Definition 1.

Assume that when $\alpha = 1/2$, the values u, u' satisfy the system (17), (18). We consider three cases: 1) $0 < u \leqslant u^*$, 2) $u^* < u \leqslant 1/4 - \varepsilon^*$, 3) $1/4 - \varepsilon^* < u < 1/2$, and prove that there exists no solution for the above-mentioned system in the cases 1), 2). In the case 3) there can be a solution only if relation a) in item (5) of Definition 1 holds.

CASE 1). From (18) it follows that u' satisfies the inequality $u' > 1/2$, and by the condition $\gamma < \gamma_0 \leqslant 1/c_2$ we have

$$G'(u)/2 = \gamma(G'(u) - G'(0))/2 \leqslant \gamma c_2 u/2 < u. \tag{27}$$

From the definition of $\widetilde{G}$ in condition (3) of Theorem 2 it follows that

$$\widetilde{G}'(u' - 1/2) = -G'(u').$$

Applying (17) to the previous formula we get

$$G'(u) = \widetilde{G}'(u' - 1/2). \tag{28}$$

Then by (27) and (18) the following inequalities hold:

$$u' - 1/2 < 2u, \qquad 0 < (u' - 1/2)/2 < u \leqslant u^*. \tag{29}$$

Thus, applying condition (3) of Theorem 2, we obtain

$$G'((u' - 1/2)/2) > \widetilde{G}'(u' - 1/2).$$

But according to conditions (2), (3) of Theorem 2 the function $G'(u)$ increases monotonically in the domain $0 < u \leqslant u^*$, and thus due to the previous inequality and (29) we have

$$G'(u) > \widetilde{G}'(u' - 1/2),$$

which contradicts (28). Therefore the system (17), (18) has no solution in case 1) for $\alpha = 1/2$.

CASE 2). For $\gamma < \gamma_0$ we have $\gamma G'(u)/2 < \gamma c_1/2 < \varepsilon^*$. Thus it follows from (18) that u' satisfies the inequality

$$1/2 \leqslant u' < 3/4. \tag{30}$$

By conditions (3) and (2) of Theorem 2, for $u^* < u \leqslant 1/4 - \varepsilon^*$, $1/4 < \widetilde{u} \leqslant 1/2$, we have $G'(u) \neq G'(\widetilde{u})$, and thus we obtain $G'(u) \neq -G'(u')$ from (30); since $G(u)$ is even, this means there is no solution of the system (17), (18) in case 2).

CASE 3). For $\gamma < \gamma_0$ we have $|g''(u)/2| < 1$, and hence the function $u' = u'(u) = u + 1/2 + \gamma G'(u)/2$ increases monotonically. Thus if $1/4 - \varepsilon^* < u < 1/2$, we have $1/2 < u' < 1$.

Then condition (2) of Theorem 2 and the definition of the number u^* in condition (3) of Theorem 2 imply that (17) holds in one of the following two cases:

a) $u' = -u + 1$,

b) $1 - u^* < u' < 1$.

Let us prove that the system (17), (18) has no solution for version b). By the condition $\gamma < \gamma_0 \leqslant 1/c_2$, we have

$$|\gamma G'(u)/2| = |\gamma(G'(u) - G'(1/2))/2| \leqslant \gamma c_2(1/2 - u)/2 < (1/2 - u)/2.$$

Thus according to (18),

$$0 < u' - (u + 1/2) < (1/2 - u)/2.$$

But $u' - (u + 1/2) = (1 - (u + 1/2)) - (1 - u') = (1/2 - u) - (1 - u')$ and hence

$$(1/2 - u) - (1 - u') < (1/2 - u)/2. \tag{31}$$

From the definition of the function $\widetilde{G}$ in condition (3) of Theorem 2, it follows that $\widetilde{G}'(1/2 - u) = G'(u)$; but since the function G is even and periodic, we have

$$G'(1 - u') = -G'(u').$$

Thus from (17) we obtain the equation

$$\widetilde{G}'(1/2 - u) = G'(1 - u'). \tag{32}$$

Since $1 - u' < u^*$ holds for case b), according to (31) we have $(1/2 - u)/2 < u^*$, and applying condition (3) of Theorem 2 to inequality (31) we obtain

$$0 < \widetilde{G}'(1/2 - u) < G'((1/2 - u)/2). \tag{33}$$

From the definition of u^* and condition (2) of Theorem 2 it follows that the function $G'(u)$ increases monotonically in the domain $0 \leqslant u < u^*$, so that (31) and the inequality $1 - u' < u^*$ imply

$$G'((1/2 - u)/2) < G'(1 - u').$$

From the previous inequality and (33) we obtain

$$\widetilde{G}'(1/2 - u) < G'(1 - u'),$$

which contradicts (32). Thus there is no solution of the system (17), (18) if $\alpha = 1/2$ for version b), whereas version a) is equivalent to condition a) from part (5) of Definition 1. Applying cases 1)–3) we can show that the system (17), (18) for $\alpha = 1/2$ has solution in the domain $0 < u < 1/2$ only if condition a) from part (5) of Definition 1 holds.

Assume that $-1/2 < u < 0$.

LEMMA 2. *Let*

$$u' = u + 1/2 + \gamma G'(u)/2, \quad \widetilde{u} = -u, \quad \widetilde{u}' = \widetilde{u} + 1/2 + \gamma G'(\widetilde{u})/2.$$

Then $\widetilde{u}' = -u' + 1$.

PROOF OF LEMMA 2. We just add the following two relations:

$$\widetilde{u} + 1/2 = 1 - (u + 1/2), \qquad -\gamma G'(u)/2 = \gamma G'(\widetilde{u})/2.$$

Lemma 2 is proved.

Lemma 2 implies that if u, u' is the solution of the system (17), (18) for $\alpha = 1/2$ in the domain $-1/2 < u < 0$, then $\widetilde{u}$, $\widetilde{u}'$ is the solution of the same system in the domain $0 < \widetilde{u} < 1/2$.

Hence it is possible to solve the system (17), (18) for $\alpha = 1/2$ in the domain $-1/2 < u < 0$ only in the case when condition a) in (5) of Definition 1 holds. Thus we have shown that if $-1/2 \leqslant u < 1/2$, $u \neq -1/2, 0$, then condition (5) of Definition 1 holds. When $u = -1/2, 0$, we obtain the values $u' = 0, 1/2$ respectively by (18) and condition (1) of Theorem 2. These values satisfy condition b) from item (5) of Definition 1 and hence the function G satisfies item (5) of Definition 1 and $G(u) \in K(\gamma)$. Now the statement of Theorem 2 follows from Theorem 1 for $\alpha = 1/2$.

Assume that α belongs to a δ-neighborhood of $1/2$, i.e., $|\alpha - 1/2| < \delta$. From Lemma 1 it is easy to see that if δ is small, $|\alpha - 1/2| < \delta$, and $0 < \gamma < \gamma_0$, then there exist open circular neighborhoods U_1 and U_2 of the points $x_1 = (0, 1/2)$ and $x_2 = (1/2, 1/2)$ respectively such that

$$x = (u, z) \in C \setminus (U_1 \cup U_2), \qquad (u', z') = A(u, z),$$
$$f(x) = f(u, z) = H(u, z) + H(u'(u, z), z'(u, z))$$

implies

$$f(x) > f(x_1) = f(x_2). \tag{34}$$

Let $\widetilde{\mu}$ be a (g, α)-measure, $\widetilde{U} = U_1 \cup U_2 \setminus (x_1 \cup x_2)$. Let us show that if $\widetilde{\gamma}_0$, δ are sufficiently small positive numbers, $0 < \gamma < \widetilde{\gamma}_0$ and $\widetilde{\mu}(\widetilde{U}) \neq 0$, then the following inequality holds:

$$\int_{\widetilde{U}} H d\widetilde{\mu} > \frac{1}{2}\widetilde{\mu}(\widetilde{U}) f(x_1). \tag{35}$$

Let us define the sets $\widetilde{U}_k$ $(k = \pm 1, \pm 2, \dots)$ inductively in the following way. Suppose $\widetilde{U}_1$ is the subset of $\widetilde{U}$ consisting of points $y = (u, z) \in \widetilde{U}$ for which we have $A^2 y \notin \widetilde{U}$. Assuming that the sets $\widetilde{U}_1, \dots, \widetilde{U}_{k-1}$ $(k \geqslant 2)$ are defined, we define $\widetilde{U}_k$ as follows: $\widetilde{U}_k$ is the subset of $\widetilde{U}$ consisting of all points $y = (u, z)$ for which $A^{2k} y \notin \widetilde{U}$, $y \notin \widetilde{U}_s$ $(s = 1, \dots, k-1)$. Thus the sets $\widetilde{U}_k$ are defined for all $k = 1, 2, \dots$. Denote by $\widetilde{U}_{-1}$ the subset $\widetilde{U}$ of points $y = (u, z)$ for which $A^{-2} y \notin \widetilde{U}$ and $y \notin \bigcup_{s=1}^{\infty} \widetilde{U}_s$. Assuming that the sets $\widetilde{U}_{-1}, \dots, \widetilde{U}_{-(k-1)}$ $(k \geqslant 2)$ are defined, we define $\widetilde{U}_{-k}$ as follows: $\widetilde{U}_{-k}$ is the subset of $\widetilde{U}$ of all points y for which $A^{-2k} y \notin \widetilde{U}$, $y \notin \bigcup_{s=-(k-1)}^{\infty} \widetilde{U}_s$, where $\widetilde{U}_0$ is the empty set. Thus all the sets are defined.

Lemma 3. *If neighborhoods U_1, U_2 are sufficiently small, we have $\widetilde{U} = \bigcup_{s=-\infty}^{\infty} \widetilde{U}_s$, where $\widetilde{U}_0$ is the empty set.*

Proof of Lemma 3. The transformation $B = A^2$ preserves area, has two fixed points, and its Jacobian dB has the following form:

$$dB = \begin{pmatrix} (1 + \varphi'(u'))(1 + \varphi'(u)) + \varphi'(u) & \varphi'(u') + 2 \\ \varphi'(u')(1 + \varphi'(u)) + \varphi'(u) & \varphi'(u') + 1 \end{pmatrix},$$

where $\varphi = g'/2$.

Therefore, by condition (4) of Theorem 2, for the trace $\operatorname{tr} dB$ of the matrix dB at the points x_1 and x_2, we obtain the inequality:

$$\operatorname{tr} dB = 2(1 + \varphi'(0) + \varphi'(1/2)) + \varphi'(0)\varphi'(1/2) > 2,$$

for $\widetilde{\gamma}_0$ small and $0 < \gamma < \widetilde{\gamma}_0$.

This means that the fixed points x_1 and x_2 are of hyperbolic type. By the Grobman–Hartman theorem (see [6]), we can show that if the neighborhoods U_1 and U_2 are sufficiently small, then any point from $U_1 \cup U_2$ except the points x_1 and x_2 leaves the domain $U_1 \cup U_2$ as the result of iterations of B or B^{-1}. Thus the equality $\widetilde{U} = \bigcup_{s=-\infty}^{\infty} \widetilde{U}_s$ holds.

Lemma 3 is proved.

Neighborhoods U_1 and U_2 will be small and Lemma 3 will hold if we choose δ to be sufficiently small. Since any two sets $\widetilde{U}_s$ are disjoint, Lemma 3 and invariance of the measure $\widetilde{\mu}$ with respect to A show that

$$\int_{\widetilde{U}} H\,d\widetilde{\mu} = \sum_{s=-\infty}^{\infty} \int_{\widetilde{U}_s} H\,d\widetilde{\mu}$$
$$= \sum_{s=-\infty}^{\infty} \int_{A^{2s}\widetilde{U}_s} H\,d\widetilde{\mu} = \frac{1}{2} \sum_{s=-\infty}^{\infty} \int_{A^{2s}\widetilde{U}_s} f(u, z)\,d\widetilde{\mu}, \tag{36}$$

and by (34) and the definition of $\widetilde{U}_s$, inequality (34) holds on the set $A^{2s}\widetilde{U}_s$ for $s \neq 0$.

Therefore, for $\widetilde{\mu}(\widetilde{U}) \neq 0$ inequality (35) follows from (36) and (34):

$$\int_{\widetilde{U}} H\,d\widetilde{\mu} > \frac{1}{2} f(x_1) \sum_{s=-\infty}^{\infty} \widetilde{\mu}(A^{2s}\widetilde{U}_s) = \frac{1}{2} f(x_1) \sum_{s=-\infty}^{\infty} \widetilde{\mu}(\widetilde{U}_s) = \frac{1}{2} f(x_1)\widetilde{\mu}(\widetilde{U}).$$

Applying inequalities (34), (35), we can show that if μ is a measure concentrated at points x_1 and x_2, $\mu(x_1) = \mu(x_2) = 1/2$, and $\mu \neq \widetilde{\mu}$, it follows that

$$\int_C H\,d\widetilde{\mu} > \int_C H\,d\mu.$$

Indeed, if $\widetilde{\mu}(\widetilde{U}) + \widetilde{\mu}(C \setminus (U_1 \cup U_2)) \neq 0$, we have

$$\begin{aligned}
\int_C H\,d\widetilde{\mu} &= \int_{\widetilde{U}} H\,d\widetilde{\mu} + \int_{x_1 \cup x_2} H\,d\widetilde{\mu} + \int_{C\setminus(U_1\cup U_2)} H\,d\widetilde{\mu} \\
&= \int_{\widetilde{U}} H\,d\widetilde{\mu} + \frac{1}{2} \int_{x_1 \cup x_2} f\,d\widetilde{\mu} + \frac{1}{2} \int_{C\setminus(U_1\cup U_2)} f\,d\widetilde{\mu} \\
&> \frac{1}{2}\widetilde{\mu}(\widetilde{U}) f(x_1) + \frac{1}{2}\widetilde{\mu}(x_1) f(x_1) \\
&\quad + \frac{1}{2}\widetilde{\mu}(x_2) f(x_1) + \frac{1}{2} f(x_1)\widetilde{\mu}(C \setminus (U_1 \cup U_2)) \\
&= \frac{1}{2} f(x_1)\,\mu(C) = \int_C H\,d\mu.
\end{aligned}$$

Theorem 2 is proved.

Proof of Theorem 3. From the definitions of the norm in the space $\widehat{K}$, of the function H in (2), and of the transformation A in (1), it follows that

$$f = f(g) = f(u, z) = H(u, z) + H(A(u, z)),$$

which is a function of g and is continuous in g in this norm, i.e., for any $\varepsilon_1 > 0$ there exists an $\varepsilon_2 > 0$ such that $|f(g_1) - f(\widetilde{g})| < \varepsilon_1$ if $\|g - \widetilde{g}\| < \varepsilon_2$.

Assuming that δ_0 and ε_0 are sufficiently small positive numbers, $\widetilde{g}(u) \in \Omega_{\varepsilon_0}$, $|\alpha - 1/2| < \delta_0$ and using Lemma 1 of the proof of Theorem 1, we see that the following assertion holds: there are open disk neighborhoods U_1 and U_2 of points x_1 and x_2 respectively such that $f(x)$ satisfies inequality (34) if $x \in C \setminus (U_1 \cup U_2)$.

Then the rest of the proof of Theorem 3 is similar to the proof of Theorem 2 for the case $|\alpha - 1/2| < \delta$.

Theorem 3 is proved.

References

1. Ya. G. Sinai, *Commensurate-incommensurate phase transitions in one-dimensional chains*, J. Statist. Phys. **29** (1982), no. 3, 401–425.
2. V. L. Pokrowsky, *Splitting of commensurate-incommensurate phase transitions*, J. Physique **42** (1981), no. 6, 761–766.
3. S. Aubry and P. Y. Le Daeron, *The discrete Frenkel-Kontorova model and its extensions*, Phys. D **8** (1983), no. 3, 381–422.
4. L. D. Pustyl′nikov, *On a ground state in the Frenkel-Kontorova model*, Funktsional. Anal. i Prilozhen. **22** (1988), no. 1, 85–86; English transl. in Functional Anal. Appl. **22** (1988).
5. V. M. Alekseev, V. M. Tikhomirov, and S. V. Fomin, *Optimal control*, "Nauka", Moscow, 1979; English transl., Plenum Press, New York, 1987.
6. Philip Hartman, *Ordinary differential equations*, Wiley, New York, 1964.

Moscow

Translated by the author